Models of
the
Visual Cortex

Models of the Visual Cortex

Edited by

DAVID ROSE
Department of Psychology
University of Surrey

and

VERNON G. DOBSON
Department of Psychology
Brunel University

A Wiley-Interscience Publication

JOHN WILEY & SONS
Chichester · New York · Brisbane · Toronto · Singapore

Library of Congress Cataloging in Publication Data:
Main entry under title:

Models of the visual cortex.
'A Wiley–Interscience publication.'
Includes index.
1. Visual cortex. 2. Biological models. I. Rose, David II. Dobson, Vernon G. [DNLM: 1. Models, Biological. 2. Visual Cortex—physiology. 3. Visual Perception—physiology. WL 307 M689]
QP382.022M63 1985 612'.8255 84–29143

ISBN 0 471 90697 2

British Library Cataloguing in Publication Data:
Models of the visual cortex.
1. Visual cortex
I.Rose, David, *1946–* II.Dobson, Vernon G.
599.01'88 QL739.2

ISBN 0 471 90697 2

Phototypeset in 10/12pt Linotron Times
by Input Typesetting Ltd, London
Printed and bound in Great Britain

For Selina and
Rosemarie and
Lizzie and Dianne

Contents

List of Contributors

K. ALBUS	*Abt fur Neurobiologie, Max-Planck Institut fur Biophysikalische Chemie, Postfach 968, D-3400 Gottingen, Federal Republic of Germany*
S. AMARI	*Department of Mathematical Engineering and Instrumentation Physics, University of Tokyo, Bunkyo-ku, Tokyo 113, Japan*
H. B. BARLOW	*Kenneth Craik Laboratory, Physiological Laboratory, Downing Street, Cambridge, CB2 3EG, UK*
R. BLAKE	*Departments of Psychology and Neurobiology/Physiology, Northwestern University, Evanston, IL 60201, USA*
A. B. BONDS	*Department of Electrical and Biomedical Engineering, Vanderbilt University, Nashville, TN 37235, USA*
V. BRAITENBERG	*Max-Planck Institut fur Biologische Kybernetik, Spemannstrasse 38, D-7400 Tubingen, Federal Republic of Germany*
P. CAVANAGH	*Department de Psychologie, Universite de Montreal, Montreal, Quebec, H3C 3J7, Canada*
M. C. CITRON	*Children's Hospital of Los Angeles, Neurology Research, PO Box 54700, Los Angeles, CA 90054, USA*
L. N. COOPER	*Center for Neural Sciences and Physics Department, Brown University, Providence, RI 02912, USA*
J. D. COWAN	*Mathematics Department, The University of Chicago, Chicago, IL 60637, USA*
O. CREUTZFELDT	*Abt fur Neurobiologie, Max-Planck Institut fur Biophysikalische Chemie, Postfach 968, D-3400 Gottingen-Nikolausberg, Federal Republic of Germany*
J. G. DAUGMAN	*Division of Applied Sciences, 321 Pierce, Harvard University, Cambridge, MA 02138, USA*

T. L. DAVIS	*Department of Anatomy, School of Medicine, University of Pennsylvania, Philadelphia, PA 19104, USA*
A. F. DEAN	*The Physiological Laboratory, Downing Street, Cambridge, CB2 3EG, UK*
E. J. DEBRUYN	*Department of Electrical and Biomedical Engineering, Vanderbilt University, Nashville, TN 37235, USA*
V. G. DOBSON	*Department of Psychology, Brunel University, Uxbridge, Middlesex, UB8 3PH, UK*
R. C. EMERSON	*Center for Visual Science, University of Rochester, Rochester, New York 14627, USA*
D. J. FELLEMAN	*Department of Biology, 216-76, California Institute of Technology, Pasadena, CA 91125, USA*
K. H. FOSTER	*Department of Neurology, University of Massachusetts Medical School, Worcester, MA 01605, USA*
Y. FRÉGNAC	*Laboratoire de Neurobiologie du Developpment, Bat. 440, Universite Paris XI, 91405 Orsay, Cedex, France*
J. P. GASKA	*Department of Neurology, University of Massachusetts Medical School, Worcester, MA 01605, USA*
M. A. GEORGESON	*Department of Psychology, University of Bristol, 8–10 Berkeley Square, Bristol, BS8 1HH, UK*
V. D. GLEZER	*Laboratory of Vision Physiology, IP Pavlov Institute of Physiology, Academy of Sciences of USSR, N Makarowa 6, 199164 Leningrad, USSR*
P. GOURAS	*Department of Ophthalmology, Columbia University, 630 West 168 Street, NY 10032, USA*
P. HAMMOND	*Department of Communication and Neuroscience, University of Keele, Keele, Staffordshire, ST5 5BG, UK*
G. HARTMANN	*Universität –GH– Paderborn, Pohlweg 47–49, 4790 Paderborn, Federal Republic of Germany*
P. HEGGELUND	*Neurobiology Laboratory, University of Trondheim, N-7055 Dragvoll, Norway*
G. H. HENRY	*Department of Physiology, JCSMR, Australian National University, PO Box 334, Canberra City, ACT 2601, Australia*
H. V. B. HIRSCH	*Neurobiology Research Center, State University of New York, Albany, NY 12222, USA*
S. HOCHSTEIN	*Hebrew University of Jerusalem, Institute of Life Sciences, Jerusalem, Israel*
A. JOHNSTON	*Department of Psychology, Brunel University, Kingston Lane, Uxbridge, Middlesex, UB8 3PH, UK*
J. P. JONES	*Department of Anatomy, School of Medicine, University of Pennsylvania, Philadelphia, PA 19104, USA*
J. H. KAAS	*Department of Psychology, Vanderbilt University, 134 Wesley Hall, Nashville, TN 37240, USA*

C. KOCH	*Centre for Biological Information Processing and Artificial Intelligence Laboratory, MIT, 545 Technology Square, Cambridge, MA 02139, USA*
J. J. KULIKOWSKI	*Visual Sciences Laboratory (Jackson), Ophthalmic Optics Department, UMIST, PO Box 88, Manchester, M60 1QD, UK*
C. R. LEGÉNDY	*ITT Avionics Division, 390 Washington Ave., Nutley, NJ 07110, USA*
A. G. LEVENTHAL	*Department of Anatomy, University of Utah School of Medicine, Salt Lake City, Utah 84132, USA*
D.M. LEVI	*University of Houston, College of Optometry, University Park, 4901 Calhoun Blvd., Houston, TX 77004, USA*
D. M. MACKAY	*Department of Communication and Neuroscience, University of Keele, Keele, Staffs, ST5 5BG, UK*
L. MAFFEI	*Instituto di Neurofisiologia del CNR, Via S Zeno 51, 56100 Pisa, Italy*
K. A. C. MARTIN	*Department of Experimental Psychology, South Parks Road, Oxford, OX1 3UD, UK*
N. MAXWELL	*History and Philosophy of Science Department, University College, Gower Street, London, WC1E 6BT, UK*
C. R. MICHAEL	*Department of Physiology, Yale Medical School, 333 Cedar Street, New Haven, CT 06510, USA*
G. J. MITCHISON	*Kenneth Craik Laboratory, Department of Physiology, Downing Street, Cambridge, CB2 3EG, UK*
W. H. MULLIKIN	*Department of Anatomy, School of Medicine, University of Pennsylvania, Philadelphia, PA 19104, USA*
I. J. MURRAY	*Visual Sciences Laboratory (Jackson), Ophthalmic Optics Department, UMIST, PO Box 88, Manchester, M60 1QD, UK*
J. I. NELSON	*Department of Ophthalmology, New York University Medical Center, 550 First Avenue, NY 10016, USA*
D. E. NIELSEN	*Institute of Neurophysiology, The Panum Institute, Blegdamsvej 3C, DK-2200 Copenhagen N, Denmark*
G. A. ORBAN	*KUL Laboratorium voor Neuro en Psychofysiologie, Katholieke Universiteit Leuven, Campus Gasthuisberg, Herestraat, B-3000 Leuven, Belgium*
L. A. PALMER	*Department of Anatomy, School of Medicine, University of Pennsylvania, Philadelphia, PA 19104, USA*

A. Peters — *Department of Anatomy, Boston University, School of Medicine, 80 East Concord Street, Boston, MA 02118, USA*

T. Poggio — *Centre for Biological Information Processing and Artificial Intelligence Laboratory, MIT, 545 Technology Square, Cambridge, MA 02139, USA*

D. A. Pollen — *Department of Neurology, University of Massachusetts Medical School, Worcester, MA 01605, USA*

D. Rose — *Department of Psychology, University of Surrey, Guildford, Surrey, GU2 5XH, UK*

P. H. Schiller — *Department of Psychology, MIT, Cambridge, MA 02139, USA*

E. L. Schwartz — *Brain Research Laboratories, Department of Psychiatry, New York University Medical Center, 550 First Avenue, NY 10016, USA*

S. M. Sherman — *Department of Neurobiology and Behavior, State University of New York at Stony Brook, NY 11794, USA*

A. M. Sillito — *Department of Physiology, University College, PO Box 78, Cardiff, CF1 1XL, UK*

W. Singer — *Max-Planck Institut fur Hirnforschung, Postfach 710409, Deutschordenstrasse 46, D-6000 Frankfurt, Federal Republic of Germany*

M. E. Sloane — *Department of Psychology, University of Alabama in Birmingham, Birmingham, AL 35294, USA*

P. Somogyi — *Department of Experimental Psychology, South Parks Road, Oxford, OX1 3UD, UK*

H. Spitzer — *Hebrew University of Jerusalem, Institute of Life Sciences, Jerusalem, Israel*

P. Sterling — *Department of Anatomy, School of Medicine, University of Pennsylvania, Philadelphia, PA 19104, USA*

N. V. Swindale — *Dept. of Physiology, Dalhousie University, Halifax, Nova Scotia B3H 4J1, Canada*

S. B. Tieman — *Neurobiology Research Center, State University of New York, Albany, NY 12222, USA*

D. J. Tolhurst — *The Physiological Laboratory, Downing Street, Cambridge, CB2 3EG, UK*

K. Toyama — *Department of Physiology, Kyoto Prefactural School of Medicine, Kawaramachi-Hirokoji, Kamigyoko, Kyoto 602, Japan*

T. R. Vidyasagar — *Abt Neurobiologie, Max-Planck Institut fur Biophysikalische Chemie, Postfach 968, D-3400 Göttingen, Federal Republic of Germany*

C. von der Malsburg — *Max-Planck Institut fur Biophysikalische Chemie, Postfach 968, D-3400 Göttingen, Federal Republic of Germany*

J. Wolfe	*Department of Psychology, MIT, Cambridge, MA 02139, USA*
M. J. Wright	*Department of Psychology, Brunel University, Kingston Lane, Uxbridge, Middlesex, UB8 3PH, UK*

Preface

A quarter of a century on from the pioneering work of Hubel and Wiesel, their discoveries have rightly become recognized as of the greatest importance and are seen as central to fields of study as diverse as cognitive psychology, artificial intelligence, cortical physiology and anatomy, clinical ophthalmology and the philosophy of perception and mind. Yet doubts have been expressed about the exact nature of the underlying mechanisms in the visual cortex. In fact, it would be true to say that a profusion of blooming and buzzing ideas permeates the minds of those workers in the field who are unfortunate enough to start to worry about such things. In recognition of this, we felt a need to clarify and facilitate the communication of the relevant concepts, many of which at the moment exist only in the heads of researchers or which, if published at all, are available only in obscure or esoteric journals. We therefore wrote to 75 of the most prominent theorists in the field asking them if they would be so kind as to present their ideas and conclusions in clear, concise articles for this book. The vast majority proved to be more than willing to do so, and we think that the chapters that they have produced for this volume will be a great revelation to many readers and a refreshing stimulus to further research in this field.

Our brief to our authors was to write in a simple manner which would be understood by workers in all the disciplines involved in this field (e.g. anatomy, physiology, psychology and cybernetics). One of the problems that arises with communicability comes from the use of mathematics, which it seems people either understand clearly or not at all. We have chosen to eliminate as much mathematics as possible, assuming only that our readers have a knowledge of calculus but not of more advanced procedures (however we have included and explained a few of these in the Glossary). The only restriction we asked our authors for as to subject matter was to concentrate on area 17, because the book would not be large enough to include detailed discussion of other areas of visual cortex; nevertheless, it is impossible to describe area 17 without discussing its relationship with the other areas of cortex, so several articles in the book do compare area 17 with extrastriate areas. Even so, and with such a large number of contributors, there are still several issues within the study of the striate cortex which have not been covered to the extent we would have liked (such as stereopsis, lesion studies, evolution and intracellular recording) and further material and ideas remain which could fill at least another volume of this size! We can only apologize to the large number of workers whose ideas we have not been able to include in the present volume.

The articles in this volume do nevertheless cover a multitude of problems, and the cross-connections between the themes of the various chapters are many and complex. Therefore, it has not proved possible to order the chapters consistently according to their subject matter or to divide them into sections, putting together, for example, all chapters on orientation selectivity or development or binocularity or anatomy. Instead we have placed the chapters in approximate (but by no means monotonic) order of level of description, beginning with articles on overall function and running through to the detailed microanatomy. At the end we have added a discussion in which we propose some new possibilities for the course of future research.

Our aim originally was to elicit all the available theories about the visual cortex, and to compare and classify these systematically, with a view to facilitating the emergence of the 'correct' model. However, we realized that there was such a diversity of models available that this could not be accounted for solely by the variety of experimental evidence and techniques. It must also reflect differences in people's ideas and preconceptions about the overall function of the visual system, and deeper issues about scientific goals and methods. This has led us into the realms of the philosophy and sociology of science, and despite the stigma attached to such disciplines by most experimental scientists, we have no qualms about presenting several chapters in which these issues are discussed. The philosophy of science is simply concerned with the examination of what scientists do and what perhaps they should do in future to improve their productivity and effectiveness.

Our first realization was that the 'correct' model might not even be included among those that we had elicited. The only way to ensure that one of the models under examination is the 'correct' one is to develop a technique for systematically and exhaustively generating all possible models (or, at least, all types of model). We propose that this can be achieved by listing all the alternative possible metaphysical backgrounds and functional roles of brain systems, and using this as the basis for classifying the specific models of the visual cortex. This process is, however, not merely one of classification: as pointed out by the philosopher of science Nick Maxwell in this volume, the process of 'interconnecting' problem solving at different levels of analysis (from the metaphysical nature of the world to the specific physiological and anatomical mechanisms in the brain) is as important as the construction of the individual models themselves. This leads to questions about how well models from the same or different levels of analysis fit together to form functionally consistent systems, and which of these systems of models are the most progressive, successful or able to subserve the overall aims of the brain. This is in addition to and alongside questions about which individual models are consistent with empirical evidence, simple, efficient, effective, and so on.

At the same time, our available methodological strategies for model building and assessment must similarly be considered systematically and comprehensively and improved upon where possible. We have been bequeathed from the past a philosophy of science which, while it has been most successful in dealing with relatively simple physical systems, may need modification and improvement to cope with the more complex, self-organizing, goal-directed systems which are studied in the behavioural and brain sciences. It is essential that all concerned in these fields should participate in this improvement process (certainly the philosophy of science is too important to be left to philosophers) and we hope therefore that our arguments will spur further debate on this fundamental and vital question.

We owe a large debt of gratitude to many people for the success of this venture. First, to our authors, for their great enthusiasm as well as their hard work and their tolerance of our editorial demands (and especially to the twelve who submitted their manuscripts before the deadline). We received encouragement and help from many sources, including Julia Maidment, Mick Wright, Alan Johnston, T. R. Vidyasagar, Jonathon Stone, Bob Phelps, Nigel Gilbert, and of course Michael Dixon and Carolyn White in the offices of John Wiley. VGD was supported by the MRC.Our ability to make overall sense of the contents was triggered and nurtured in many enlivening and stimulating conversations and correspondence with Nick Maxwell. The administrative load would have been impossible without the secretarial help of Jane Roberts, Cynthia Joel, Sarah Fenton, Elizabeth Bruce and Amanda Lambert-Gorwin, whose heroic efforts to cope with both WordStar Mail-Merge and the NEC Spinwriter were a marvel to behold. Finally, we owe much to David Hubel and Torsten Wiesel for starting the whole thing off and erecting such a far-reaching beacon to guide the rest of us.

David Rose
Vernon G. Dobson

'I'll make my report as if I told a story, for I was taught as a child . . . that Truth is a matter of the imagination. The soundest fact may fail or prevail in the style of its telling: like that singular organic jewel of our seas, which grows brighter as one woman wears it and, worn by another, dulls and goes to dust. Facts are no more solid, coherent, round, and real than pearls are. But both are sensitive.

The story is not all mine, nor told by me alone. Indeed I am not sure whose story it is; you can judge better. But it is all one, and if at moments the facts seem to alter with an altered voice, why then you can choose the fact you like best; yet none of them are false, and it is all one story.'

Ursula K. LeGuin

Models of the Visual Cortex
Edited by D. Rose and V. G. Dobson

CHAPTER 1

Introduction: the recent history and current state of research on the visual cortex

DAVID ROSE
Department of Psychology, University of Surrey, Guildford, Surrey GU2 5XH, UK

At the end of the 1950s the time was right for a breakthrough in the study of the visual cortex. Jung and Baumgartner (1955), Hubel (1959) and others had demonstrated that responses to light stimuli could be recorded from single cells in the visual cortex, and Hubel (1960) had investigated the receptive fields of cells in the lateral geniculate nucleus (LGN), the immediate source of visual input to the cortex, finding them to be concentric, much like the receptive fields of retinal ganglion cells. In related fields of study, Mountcastle (1957) found a columnar organization of functional activity in somatosensory cortex, while new conceptions of perceptual processes were being developed in simpler systems (Lettvin *et al.*, 1959; Young, 1960) and in artificial intelligence (Selfridge, 1959).

Against this background, the discovery of orientation-specific responses and columnar organization in area 17 (Hubel and Wiesel, 1959, 1962) fitted perfectly. The realization that cortical cells respond only to straight lines or edges presented a major clue to the way in which visual information is encoded, while the anatomical arrangement of cells with identical preferred orientations of the stimulus edge—vertically above one another in 'columns'—was further evidence that certain mechanisms may exist universally in all parts of the cortex. Hubel and Wiesel went on to suggest a model of the connections within the visual cortex, prompted by two further aspects of their work. First was the division of receptive fields into two types: 'simple' having parallel adjacent areas in each of which the cell responded either to a bar of light turned on or off, with spatial summation within each area; and 'complex' having somewhat larger, uniform receptive fields wherein responses could in some cells be evoked both at light on and at off. Second was the observation that mainly simple cells were found near the region where the LGN inputs terminated in the middle layer of the cortex, while complex cells were rare in this layer. This could all be explained if simple cells are each driven directly by a number of LGN cells with receptive fields lined up in a row and with, for example, all on-centre receptive fields, and perhaps flanked on one or both sides by a parallel row of off-centre receptive fields (or similarly with on- and off-centre reversed). Complex cells would then be driven through vertically running connections by a number of simple cells with their receptive fields offset slightly from one another but with identical preferred orientations of stimulus. This model thus postulated a serial or hierarchical flow of information from LGN cells to simple cells to complex cells. Later, a third cortical stage came to be added: that of 'hypercomplex' cells, which responded only to short bars or edges and which, it was suggested, were both driven and inhibited by complex cells with different receptive field locations (Hubel and Wiesel, 1965, 1968).

The propensity of many cells to respond only to one direction of movement of the stimulus could be accounted for at the first stage of the hierarchy, in that it depended on whether the stimulus moved from an area of inhibition to one of excitation within the simple cell's receptive field, or vice versa. At the same level, the responsiveness of many neurones to stimuli presented to either eye could be related to convergent input from two sets of LGN cells, each set conveying information from one eye. These binocular cells and their more monocular counterparts also formed a system of columns. Yet another major and far-reaching principle arose from the demonstration of plasticity in the binocularity system, in that closure of one eye during infancy reduced the ability of stimuli presented to that eye to activate cells in the cortex after that eye had been reopened (Wiesel and Hubel, 1963).

In retrospect, Hubel and Wiesel's discoveries may appear inevitable to some, given the background against which they were made. However, at the time they were far from expected, and it was the high quality, as well as the quantity, of Hubel and Wiesel's work which immediately won over the scientific community to their conclusions. The elegance of their results, and of the model they proposed to explain them, left a lasting impression on all who read their papers. This is shown by the number of textbooks being issued even today which describe Hubel and Wiesel's findings almost verbatim as the definitive account of the visual cortex.

However, the last 20 years have not seen a condition of stasis in this field. As with all seminal works, a proliferation of investigations has followed both within the framework of Hubel and Wiesel's original studies and also extending far beyond it. By the late 1960s and on into the 1970s new pictures were beginning to emerge, with novel ideas about binocular integration (Barlow, Blakemore and Pettigrew, 1967; Pettigrew, Nikara and Bishop, 1968), cortical plasticity (Blakemore and Cooper, 1970; Hirsch and Spinelli,

1970), anatomical structure (Colonnier, 1964; Szentagothai, 1969; Garey and Powell, 1971; Hubel and Weisel, 1972, 1974; Albus, 1975); perceptual theory (Campbell and Robson, 1968; Barlow, 1972; Marr, 1976), psychophysical mechanisms (Sutherland, 1961; Julesz, 1964; McCulloch, 1965; Campbell and Kulikowski, 1966; Andrews, 1967; Pantle and Sekuler, 1968; Gilinsky and Doherty, 1969) and nerve network interactions (von der Malsberg, 1973; Dobson, 1975; Nass and Cooper, 1975; Perez, Glass and Shlaer, 1975). Most central to the debate, perhaps, were physiological studies of receptive fields and the origins of stimulus selectivity (e.g. Bishop, Coombs and Henry, 1971a, 1971b; Hoffmann and Stone, 1971; Benevento, Creutzfeldt and Kuhnt, 1972; Blakemore and Tobin, 1972; Toyama *et al.*, 1974; Hammond and MacKay, 1975; Hess, Negishi and Creutzfeldt, 1975; Sillito, 1975; Schiller, Finlay and Volman, 1976—to name but a few).

A full history of this development is not possible here (further references can be found in the list of review articles given at the end of this Introduction). It becomes even more difficult to summarize events as the field expanded through the late 1970s and up to the present day, when the use of techniques such as histochemistry (e.g. Humphrey and Hendrickson, 1983; Somogyi *et al.*, 1983) and the application of powerful conceptual tools such as efficiency and entropy (Barlow, 1980; Marcelja, 1980; Daugman, 1983) bring about radical advances in our knowledge from one month to the next.

Today there are thousands of scientists working on the visual cortex (Hubel, 1982a) and tens of thousands of published articles on the subject. So what conclusions (if any) can now be drawn about the visual cortex? In one view it is the job of scientists to test hypotheses and observations; how well have Hubel and Wiesel's results and their model stood up to those tests? Although Hubel (1982b) still sees the question as an open one, it is true that the views of most of the workers cited above are different from those of Hubel and Wiesel, and not just on points of detail, but often standing in diametrical opposition. This range of alternative possibilities, many with much supporting evidence, is only rarely (and then incompletely) considered in textbooks and reviews of the field. We must therefore ask about the status of *all* the various ideas which have been put forward, and how they relate to one another.

The sheer number of published articles and models causes severe practical problems for any *objective* attempt to review the field, or for any newcomer or outsider to understand what is going on. We must consider our methodology or tactics very carefully; i.e. *how* are we best to go about drawing conclusions? We believe that it is important to consider *all* the possible models or ideas before making an evaluative decision, for reasons which will be elaborated below. Two main areas of difficulty exist here: finding out what possibilities there are for a model of the visual cortex and then resolving the conflicts that exist between them.

We have already mentioned the vast and detailed literature which must be searched to find the multiplicity of ideas which exist, but additional factors limit the success of any such project. One is that not all the ideas are available in published articles. Many journals regard themselves as repositories for reports of experiments, and deny their authors the space to describe their model, on the grounds that articles must be kept brief and 'speculation' should be eliminated. Another problem is that the literature on the visual cortex is so diverse: it is found in a wide range of journals from *Perception and Psychophysics* to *Developmental Brain Research*, and from the *Journal of Neurocytology* to the *Transactions of the Institute of Electrical and Electronics Engineers*. This not only makes it difficult to find new models when they appear, but outsiders to the particular discipline of the journal are unable to tell whether the specialized methods described in the article are highly sophisticated or woefully inadequate, assuming of course that they can understand the technical jargon in which the paper is written. This last point can be particularly frustrating, for while one can place some trust in the journal's referees to guarantee the quality of the work, it is a waste of time reading the article if you do not understand it at the end. One of our objectives in preparing this book has therefore been to aid the search for models by eliciting unpublished as well as published ideas, and then presenting them in a manner which will be comprehensible to everyone who is interested in the field. During the course of the book's preparation, we have also considered possible ways of ensuring the comprehensiveness of such a search for models, and we discuss this issue as well in the final two chapters.

The second area of difficulty for reviewers arises from the contradictions that occur between the models which have been proposed. Discussion of such contradictions is conspicuous by its absence in much of the literature. The models are developed in isolation by their protagonists, and although mention may be made of whether the model is consistent with that of Hubel and Wiesel, the other possible models are not given similar consideration and the sources of any inconsistency with the established model are not explained. This is partly because of restrictions on publishing space, partly through lack of awareness of the other models (see above) and partly because of an inability to explain the discrepancies. On the last of these points, excuses are sometimes put forward about differences in technique between laboratories causing the differences in the models generated. There is likely some truth in this, because another problem in the field is that almost every experimental finding has been disputed. Perhaps this can indeed be attributed to differences in technique (e.g. in the anaesthetic used in a physiological experiment) or in the expectations of the experimenter (e.g. in the way cells are classified). It is not known how large an effect that changes in any one of a large number of such variables might have on the outcome of the very complicated experiments which are done in this field. However, the underlying conceptual models and conclusions are often based on experiments which are not similar; they differ deliberately in their technique, for example by the introduction of some innovation to the field (conditioning stimulus, microiontophoresis, immunocytochemistry, antidromic activation, etc.), or the models may even arise from entirely different disciplines such as anatomy, physiology, psychology or cybernetics. This difficulty in comparing ideas is known to philosophers of science as incommensurability (e.g. Feyerbend, 1970; Kuhn, 1970). The underlying assumptions of each model's proponents about which experimental results they consider reliable, or consistent with their metaphysical point of view, are often not apparent to other people. The problem of how to compare the models may then appear intractable.

There is a general hope that in time one model will emerge simply by having an overwhelming weight of evidence in its favour, or by the application of some convincing new technique which attacks the problem more directly than previous attempts have done; the other models will then (hopefully) either be explained away or else be absorbed into a more comprehensive scheme, or even simply forgotten. There are many precedents for this in the history of science. However, there are also precedents for a whole field's becoming so snarled and entangled that no clear picture ever emerges, until eventually people start to work on different, apparently easier problems. After 20 years of research on the visual cortex, which of those outcomes looks more likely? Many would say that the answer is the latter. The discovery of a simple unifying theory has long since become a pipe dream in this field. How then can we change things to ensure that a clear picture does emerge? It is one of our aims in preparing this book to propose some guidelines as to how this might be done.

To achieve this it is necessary to go back and examine the methods by which models are generated, compared and assessed. Most models are suggested retrospectively to explain a particular set of experimental observations. Models are assessed according to such criteria as how many other observations they can explain as well, whether they predict new observations and whether they appear simple or elegant. Comparison with other models is usually not attempted until the model has been developed to a stage where it can fulfil many of the above criteria. By this time a lot of work has been done on the model, which may have become quite complicated and inflexible. Also, sociological factors (such as the need to win patronage or maintain reputations or friendships) can bias the opinions of the people who are making the comparisons (see Dobson and Rose, chapters 3, 56 and 57 in this volume; and McBurney, 1983, Chaps. 1, 2 and 12, for further descriptions of ulterior factors in science).

The final outcome of all this is that a multitude of models exist, each of which explains some but not all of the empirical phenomena, so that in a comparison process it is difficult to reject any particular model clearly. The new procedure which we will present in more detail in the final chapters of this volume seeks to use a more *a priori* approach to model building, wherein *all the possible classes of model which could underlie the system are considered before most of the experiments are done.* This apparently radical departure from normal practice develops both from the trial-and-error experience of the editors in their investigations of cortical mechanisms (Dobson, 1981; Rose, 1978) and from recent thinking on the philosophy and methodology of science (Maxwell, 1977, 1980, and Chapter 2 in this volume: rule 2). The procedures to be followed are: firstly to articulate and define the problem which is to be solved (e.g. what does the visual cortex do? or how does orientation selectivity arise?), secondly to generate *all* the possible classes of solution or model by which the problem could (in theory) be solved, thirdly to decide by what criteria these models will be assessed (e.g. by experimental tests of the consequences of each model) and fourthly to apply these criteria to the models until all the unworkable models have been eliminated. Note that the criteria to be applied are not limited to empirical success. Considerations about a model's simplicity, efficiency, reliability, ability to have developed from previous stages, internal logic coherence and relatability to both wider and narrower problems (rule 4 of Maxwell, Chapter 2 in this volume) are also a part of rational science (Toulmin, 1972, Maxwell, 1980, 1984; Gregory, 1981; Zahar, 1983; Rose and Dobson, Chapters 56 and 57 in this volume). As MacKay (1981) says:

> . . . our primary theoretical question has not been 'How can we predict the behaviour of this system?' but again rather 'How ought we to *conceive* of its functioning?' As with a car or a computer of unknown design, we would be very happy at this stage to settle for the kind of theoretical understanding that is offered by a maker's handbook.
>
> If I am right, then mere success in quantitative prediction may be the least relevant criterion of merit to apply to our present theorizing. What matters much more is the clarity with which it helps us to *know what to make* of each experimental result as it comes along, whether as predicted or not.

To this end we have included an article by Nicholas Maxwell in which he discusses the general problem of finding solutions in science. In the final chapters of this volume we will describe the new principles in detail and present examples of how they can be applied in practice to the modelling of the visual cortex. Because we consider Maxwell's article to be of fundamental relevance to the rest of the book, we have placed his chapter next.

REFERENCES

Albus, K. (1975). A quantitative study of the projection area of the central and paracentral visual field in area 17 of the cat. II. The spatial organization of the orientation domain. *Exp. Brain Res.*, **24**, 181–202.

Andrews, D. P. (1967). Perception of contour orientations in the central fovea. Part 1: short lines. *Vision Res.*, **7**, 975–997.
Barlow, H. B. (1972). Single units and sensation: a neuron doctrine for perceptual psychology? *Perception*, **1**, 371–394.
Barlow, H. B. (1980). The absolute efficiency of perceptual decisions. *Phil. Trans. Roy. Soc. Lond. B*, **290**, 71–82.
Barlow, H. B., Blakemore, C., and Pettigrew, J. D. (1967). The neural mechanism of binocular depth discrimination. *J. Physiol*, **193**, 327–342.
Benevento, L. A., Creutzfeldt, O. D., and Kuhnt, U. (1972). Significance of intracortical inhibition in the visual cortex. *Nature New Biology*, **238**, 124–126.
Bishop, P. O., Coombs, J. S., and Henry, G. H. (1971a). Responses to visual contours: spatio-temporal aspects of excitation in the receptive fields of simple striate neurones. *J. Physiol*, **219**, 625–657.
Bishop, P. O., Coombs, J. S., and Henry, G. H. (1971b). Interaction effects of visual contours on the discharge frequency of simple striate neurones. *J. Physiol.*, **219**, 659–687.
Blakemore, C., and Cooper, G. F. (1970). Development of the brain depends on the visual environment. *Nature*, **228**, 477–478.
Blakemore, C. and Tobin, E. A. (1972). Lateral inhibition between orientation detectors in the cat's visual cortex. *Exp. Brain Res.*, **15**, 439–440.
Campbell, F. W., and Kulikowski, J. J. (1966). Orientational selectivity of the human visual system. *J. Physiol.*, **187**, 437–445.
Campbell, F. W., and Robson, J. G. (1968). Application of Fourier analysis to the visibility of gratings. *J. Physiol*, **197**, 551–566.
Colonnier, M. (1964). The tangential organization of the visual cortex. *J. Anat.*, **98**, 327–344.
Daugman, J. G. (1983). Six formal properties of two-dimensional anisotropic visual filters: structural principles and frequency/orientation selectivity. *IEEE Trans. SMC*, **13**, 882–887.
Dobson, V. (1975). Pattern learning and the control of behaviour by all-inhibitory network hierarchies. *Perception*, **4**, 35–50.
Dobson, V. G. (1981). A model of the development of functional organization underlying psychophysical performance in the visual system. Ph.D. thesis, University of Brunel.
Feyerabend, P. (1970). Consolations for the specialist. *In Criticism and the Growth of Knowledge* (Eds. I. Lakatos and A. Musgrave), Cambridge University Press, Cambridge, pp. 197–230.
Garey, L. J., and Powell, T. P. S. (1971). An experimental study of the termination of the lateral geniculo-cortical pathway in the cat and monkey. *Proc. Roy. Soc. Lond. B*, **179**, 41–63.
Gilinsky, A. S., and Doherty, R. S. (1969). Interocular transfer of orientational effects. *Science*, **164**, 454–455.
Gregory, R. L. (1981). *Mind in Science*, Weidenfeld and Nicolson, London.
Hammond, P., and MacKay, D. M. (1975). Differential responses of cat visual cortical cells to textured stimuli. *Exp. Brain Res.*, **22**, 427–430.
Hess, R., Negishi, K., and Creutzfeldt, O. (1975). The horizontal spread of intracortical inhibition in the visual cortex. *Exp. Brain Res.*, **22**, 415–419.
Hirsch, H. V. B., and Spinelli, D. N. (1970). Visual experience modifies distribution of horizontally and vertically oriented receptive fields in cats. *Science*, **168**, 869–871.
Hoffmann, K-P., and Stone, J. (1971). Conduction velocity of afferents to cat visual cortex: a correlation with cortical receptive field properties. *Brain Res.*, **32**, 460–466.

Hubel, D. H. (1959). Single unit activity in striate cortex of unrestrained cats. *J. Physiol.*, **147**, 226–238.
Hubel, D. H. (1960). Single unit activity in lateral geniculate body and optic tract of unrestrained cats. *J. Physiol.*, **150**, 91–104.
Hubel, D. H. (1982a). Cortical neurobiology: a slanted historical perspective. *Ann. Rev. Neurosci.*, **5**, 363–370.
Hubel, D. H. (1982b). Exploration of the primary visual cortex 1955–1978. *Nature*, **299**, 515–524.
Hubel, D. H. and Wiesel, T. N. (1959). Receptive fields of single neurones in the cat's striate cortex. *J. Physiol*, **148**, 574–591.
Hubel, D. H., and Wiesel, T. N. (1962). Receptive fields, binocular interaction and functional architecture in the cat's visual cortex. *J. Physiol*, **160**, 106–154.
Hubel, D. H., and Wiesel, T. N. (1965). Receptive fields and functional architecture in two nonstriate visual areas (18 and 19) of the cat. *J. Neurophysiol.*, **28**, 229–289.
Hubel, D. H., and Wiesel, T. N. (1968). Receptive fields and functional architecture of monkey striate cortex. *J. Physiol.*, **195**, 215–242.
Hubel, D. H. and Wiesel, T. N. (1972). Laminar and columnar distribution of geniculo-cortical fibers in the macaque monkey. *J. comp. Neurol.*, **146**, 421–450.
Hubel, D. H., and Wiesel, T. N. (1974). Sequence regularity and geometry of orientation columns in the monkey striate cortex. *J. comp. Neurol.*, **158**, 267–294.
Humphrey, A. L., and Hendrickson, A. E. (1983). Background and stimulus-induced patterns of high metabolic activity in the visual cortex (area 17) of the squirrel and macaque monkey. *J. Neurosci*, **3**, 345–358.
Julesz, B. (1964). Binocular depth perception without familiarity cues. *Science*, **145**, 356–362.
Jung, R., and Baumgartner, G. (1955). Hemmungsmechanismen und bremsende Stabilisierung an einzelnen Neuronen des optischen Cortex: ein Beitrag zur Koordination corticaler Erregungsvorgange. *Pflügers Arch. ges. Physiol.*, **261**, 434–456.
Kuhn, T. S. (1970). Reflections on my critics. In *Criticism and the Growth of Knowledge* (Eds. I. Lakatos and A. Musgrave), Cambridge University Press, Cambridge, pp. 231–278.
Lettvin, J. Y., Maturana, H. R., McCulloch, W. S., and Pitts, W. H. (1959). What the frog's eye tells the frog's brain. *Proc. Inst. Radio Engrs.*, **47**, 1940–1951.
McBurney, D. H. (1983). *Experimental Psychology*, Wadsworth, Belmont.
McCulloch, C. (1965). Color adaptation of edge-detectors in the human visual system. *Science*, **149**, 1115–1116.
MacKay, D. M. (1981). Where are the growing points? In *Neural Communication and Control* (Eds. G. Szekely, E. Labos and S. Damjanovich), Pergamon, Oxford, pp. 333–334.
Marcelja, S. (1980). Mathematical description of the responses of simple cortical cells. *J. opt. Soc. Amer.*, **70**, 1297–1300.
Marr, D. (1976). Early processing of visual information. *Phil. Trans. Roy. Soc. Lond. B*, **275**, 483–519.
Maxwell, N. (1977). Articulating the aims of science. *Nature*, **265**, 2.
Maxwell, N. (1980). Science, reason, knowledge, and wisdom: a critique of specialism. *Inquiry*, **23**, 19–81.
Maxwell, N. (1984). *From Knowledge to Wisdom.* Blackwell, Oxford
Mountcastle, V. B. (1957). Modalities and topographical properties of single neurons of cat's somatic sensory cortex. *J. Neurophysiol.*, **20**, 408–434.
Nass, M. M., and Cooper, L. N. (1975). A theory for the development of feature detecting cells in visual cortex. *Biol. Cybern.*, **19**, 1–18.

Pantle, A., and Sekuler, R. (1968). Size-detecting mechanisms in human vision. *Science*, **162**, 1146–1148.
Perez, R., Glass, L., and Shlaer, R. (1975). Development of specificity in the cat visual cortex. *J. math. Biol.*, **1**, 275–288.
Pettigrew, J. D., Nikara, T., and Bishop, P. O. (1968). Binocular interactions on single units in cat striate cortex: simultaneous stimulation by single moving slit with receptive fields in correspondence. *Exp. Brain Res.*, **6**, 391–410.
Rose, D. (1978). Functional interactions in the visual cortex. Ph.D. thesis, University of Cambridge.
Schiller, P. H., Finlay, B. L., and Volman, S. F. (1976). Quantitative studies of single-cell properties in monkey striate cortex. *J. Neurophysiol.*, **39**, 1288–1374.
Selfridge, O. (1959). Pandemonium: a paradigm for learning. In *Symposium on the Mechanization of Thought Processes*, HMSO, London, pp. 511–531.
Sillito, A. M. (1975). The contribution of inhibitory mechanisms to the receptive field properties of neurones in the striate cortex of the cat. *J. Physiol.*, **250**, 305–329.
Somogyi, P., Cowey, A., Kisvarday, Z. F., Freund, T. F., and Szentagothai, J. (1983). Retrograde transport of γ-amino[^{3}H]butyric acid reveals specific interlaminar connections in the striate cortex of the monkey. *Proc. Nat. Acad. Sci. USA*, **80**, 2385–2389.
Sutherland, N. S. (1961). Figural after-effects and apparent size. *Quart. J. exp. Psychol.*, **13**, 222–228.
Szentagothai, J. (1969). Architecture of the cerebral cortex. In *Basic Mechanisms of the Epilepsies* (Eds. H. H. Jasper, A. A. Ward and A. Pope), Little, Brown & Co., Boston, pp. 13–28.
Toulmin, S. (1972). *Human Understanding*, Vol. 1, Clarendon, Oxford.
Toyama, K., Matsunami, K., Ohno, T., and Tokashiki, S. (1974). An intracellular study of neuronal organization in the visual cortex. *Exp. Brain Res.*, **21**, 45–66.
von der Malsberg, C. (1973). Self-organization of orientation-sensitive cells in the striate cortex. *Kybernetik*, **14**, 85–100.
Wiesel, T. N., and Hubel, D. H. (1963). Single-cell responses in striate cortex of kittens deprived of vision in one eye. *J. Neurophysiol.*, **26**, 1003–1017.
Young, J. Z. (1960). The visual system of *Octopus*. (1) Regularities in the retina and optic lobes of *Octopus* in relation to form discrimination. *Nature*, **186**, 836–839.
Zahar, E. (1983). Logic of discovery or psychology of invention? *Brit. J. Phil. Sci.*, **34**, 243–261.

REVIEW ARTICLES

For the early work on the physiology of the visual cortex, see:

Brooks, B., and Jung, R. (1973). Neuronal physiology of the visual cortex. In *Handbook of Sensory Physiology* (Ed. R. Jung), Vol. VII/3/B, Springer, Berlin, pp. 325–440.

A selection of monographs published in the last few years is as follows:

DeValois, R. L., and DeValois, K. K. (1980). Spatial vision. *Ann. Rev. Psychol*, **31**, 309–341.
Frégnac, Y., and Imbert, M. (1984). Development of neuronal selectivity in the primary visual cortex of cat. *Physiol. Rev.*, **64**, 325–434.
Gilbert, C. D. (1983). Microcircuitry of the visual cortex. *Ann. Rev. Neurosci.*, **6**, 217–247.
Howard, I. P. (1982). *Human Visual Orientation*, Wiley, Chichester.
Hubel, D. H. (1982). Cortical neurobiology: a slanted historical perspective. *Ann. Rev. Neurosci.*, **5**, 363–370.
Hubel, D. H. (1982). Exploration of the primary visual cortex 1955–1978. *Nature*, **299**, 515–524.
Julesz, B., and Schumer, R. A. (1981). Early visual perception. *Ann. Rev. Psychol.*, **32**, 575–627.
Kelly, D. H., and Burbeck, C. A. (1984). Critical problems in spatial vision. *CRC Crit. Rev. Biomed. Engng.*, **10**, 125–177.
Lennie, P. (1980). Parallel visual pathways: a review. *Vision Res.*, **20**, 561–594.
Marr, D. (1982). *Vision*, Freeman, San Francisco.
Movshon, J. A., and Van Sluyters, R. C. (1981). Visual neural development. *Ann. Rev. Psychol.*, **32**, 477–522.
Orban, G. A. (1984). *Neuronal Operations in the Visual Cortex*, Springer, Berlin.
Palm, G. (1982). *Neural Assemblies*, Springer, Berlin.
Poggio, G. F., and Poggio, T. (1984). The analysis of stereopsis. *Ann. Rev. Neurosci.*, **7**, 379–412.
Sherman, S. M., and Spear, P. D. (1982). Organization of visual pathways in normal and visually deprived cats. *Physiol. Rev.*, **62**, 738–855.
Spoehr, K. T., and Lehmkuhle, S. W. (1982). *Visual Information Processing*, Freeman, San Francisco.
Stone, J. (1983). *Parallel Processing in the Visual System*, Plenum, New York.
Westheimer, G. (1984). Spatial vision. *Ann. Rev. Psychol.*, **35**, 201–226.
Wiesel, T. N. (1982). Postnatal development of the visual cortex and the influence of environment. *Nature,* **299**, 583–591.

Finally, the following edited books or symposia each contain many articles of direct relevance:

Braddick, O. J., and Sleigh, A. C. (1983). *Physical and Biological Processing of Images*, Springer, Berlin.
Harris, C. S. (1980). *Visual Coding and Adaptability*, Lawrence Erlbaum, Hillsdale.
Longuett-Higgins, H. C., and Sutherland, N. S. (1980). The psychology of vision. *Phil. Trans. Roy. Soc. Lond., B* **290**, 1–218.
Orban, G. A. (1984). Symposium on movement perception. *Vision Res.*, **24**, 1–62.
Peters, A., and Jones, E. G. (1984). *Cerebral Cortex*, Plenum, New York.
Pettigrew, J. D., Levick, W. R., and Sanderson, K. J. (1985). *Visual Neuroscience. Festschrift for P.O. Bishop*, Cambridge University Press, Cambridge.
Sanderson, A. C., and Zeevi, Y. Y. (1983). Special issue on neural and sensory information processing. *IEEE Trans. SMC*, **13**, 666–1047.
Spillman, L., and Wooten, B. R. (1984). *Sensory Experience, Adaptation, and Perception. Festschrift for Ivo Kohler*, Lawrence Erlbaum, Hillsdale.

Models of the Visual Cortex
Edited by D. Rose and V. G. Dobson

Chapter 2

Methodological problems of neuroscience

NICHOLAS MAXWELL
History and Philosophy of Science Department, University College, Gower Street, London WC1E 6BT, UK

In this paper I argue that neuroscience has been harmed by the widespread adoption of seriously inadequate methodologies or philosophies of science—most notably inductivism and falsificationism.

Any branch of inquiry, in order to be rational, must at the very least obey the following rules:

1. Articulate and seek to improve the articulation of the basic problem(s) to be solved.
2. Propose and critically examine alternative possible solutions.

Many basic intellectual problems are, however, too intractable to be solved by means of this direct approach alone. It proves necessary to crcatc a host of preliminary, subordinate, specialized problems, whose resolution leads gradually and progressively towards a resolution of the basic problem to be solved. Especially important is the strategy of tackling problems that are analogous to but simpler and more solvable than the basic problem to be solved—in this way progressively developing problem-solving methods and capacities which lead eventually to the solution of the basic problem. Indeed, all problem solving may be said to exploit this principle in one way or another: inevitably in solving a new problem we discover how to relate it to analogous already solved problems in such a way that the solutions may be adapted to provide a solution to the new problem. We thus have an important third rule of rational problem solving:

3. Where necessary, break the basic problem up into a number of preliminary, simpler, analogous, subordinate, specialized problems (to be tackled in accordance with rules 1 and 2), in an attempt to work gradually towards a solution to the basic problem to be solved.

The danger in putting this third rule into practice is that the activity of solving preliminary, specialized problems may obliterate all concern for the original, basic problem(s). We need therefore a fourth rule to counteract this danger:

4. Interconnect attempts to solve basic and specialized problems, so that basic problem solving may guide, and be guided by, specialized problem solving.

All science, and indeed all inquiry, needs to put these four elementary methodological rules into practice (Maxwell, 1980, 1984).

Two historically important but seriously defective methodological views—*inductivism* and *falsificationism*—have tended, as a result of being widely accepted by scientists and non-scientists alike, to prevent the above four rules from being put into practice in science, to some extent at least.

Inductivism holds that science begins with observation and experimentation, and only gradually and cautiously moves from observational and experimental knowledge to theoretical knowledge. Inductivism is in effect an exaggerated version of rule 3. It demands of scientists that they restrict themselves, in the first instance at least, to solving preliminary, subordinate problems of observational and experimental knowledge, solutions to such problems only subsequently and gradually leading to the solutions of more general, theoretical problems of knowledge. Francis Bacon, an important proponent of inductivism, was quite explicit on this point. He argued that if we are to acquire genuine knowledge of Nature of real value then we must abandon the sterile theoretical speculations of traditional philosophy about ultimate problems and seek instead to acquire much more limited, but genuine, knowledge soundly based on observation and experiment.

Inductivism is of value to the extent that it does endorse rule 3. Otherwise it is damagingly irrational, in that it violates rules 1, 2 and 4. The rational procedure is to *interconnect* philosophical speculation concerning fundamental problems and much more restricted observational and experimental problem solving, as rule 4 stipulates. Pursuing science in accordance with inductivism is profoundly damaging in that it leads to the acquisition of vast amounts of observational and experimental data devoid of any theoretical interest or importance. This is a direct consequence of the irrationality of inductivism—its failure to interconnect theoretical and empirical problem solving.

Falsificationism (or hypothetico-deductivism), as expounded especially by Karl Popper (1959, 1963), holds that science begins not with observation and experiment but rather with *problems* generated by *theories*. Science proceeds by proposing solutions to these problems, namely new theories, which are then critically assessed, especially by experimental testing. Scientific method thus amounts to a process of theoretical conjecture and empirical refutation.

Scientific theories cannot be verified empirically, but they can be empirically falsified. There is thus the possibility of detecting and eliminating error and of making progress towards greater (conjectural) knowledge. In order to exploit this possibility for making progress, however, science must restrict itself to considering theories that are capable of being falsified empirically. Untestable philosophical, metaphysical and methodological ideas must be excluded from science (in accordance with Popper's criterion demarcating science from non-science).

Falsificationism thus demands of scientists that they propose and criticize empirically falsifiable possible solutions to theoretical problems: to this extent it endorses rules 2 and 3 and is a great improvement over inductivism. Falsificationism stipulates, however, that only empirically testable ideas can enter into the intellectual domain of science; this ensures that untestable ideas designed to help clarify and solve basic scientific problems are excluded from the intellectual domain of science. To this extent falsificationism violates rules 1, 2 and 4, and is thus damagingly irrational (Maxwell, 1972, 1974, 1979). The long process of articulating basic (philosophical or metaphysical) problems of knowledge and understanding, and of proposing and criticizing possible solutions to such problems—which is such a vital part of science and which can lead eventually to important new empirically testable theories—is banished from the explicit intellectual domain of science altogether. To this extent fundamental and specialized theorizing and problem solving are dissociated from one another, in violation of rule 4—philosophy and empirical science becoming dissociated to the detriment of both. One bad consequence of this is that the process of discovery in science becomes a mystery, an irrational affair, as Popper himself acknowledges (Popper, 1959, pp. 31–32).

Inductivism and falsificationism both uphold what Popper has called '*the principle of empiricism*, which asserts that in science, only observation and experiment may decide upon the *acceptance or rejection* of scientific statements, including laws and theories' (Popper, 1963, p. 54). Even if the principle of empiricism were tenable, inductivism and falsificationism would still both be unacceptable for the reason just given, namely for their violation of elementary methodological rules of rational problem solving 1 to 4. What makes the matter much more serious is that the principle of empiricism is *untenable*. Given any scientific theory, however extensively corroborated empirically, not all its predictions will have been tested. By arbitrarily modifying these untested predictions, we can create as many rival theories as we please, all just as successful empirically as the given theory (Maxwell, 1974, pp. 127–136; 1984, pp. 206–214). Thus any honest attempt to pursue science in accordance with the principle of empiricism would overwhelm science with infinitely many different, horribly complex, *ad hoc* theories—all, at any given stage, equally acceptable because of equal empirical success.

This would instantly bring science to a standstill. In practice science usually avoids this disaster by considering only simple, coherent, unified, explanatory or comprehensible theories that meet with empirical success—thus at a stroke violating the principle of empiricism, which states that empirical considerations *alone* determine choice of theory in science. The scientific enterprise is obliged, in other words, to presuppose that the universe is comprehensible in some way or other; in order to be acceptable a theory must at least be compatible with the best current version of this presupposition (or more compatible than any rival theory). Thus two kinds of consideration govern the choice of theory in science: (a) considerations of empirical success and failure; (b) non-empirical considerations concerning the inherent unity, explanatory character or comprehensibility of the theory in question.

The best current version of the metaphysical assumption (or conjecture) that the world is comprehensible, implicit in the current basic concepts and methodology of science, inevitably exercises a profound influence over the whole of science. This assumption must also, however, for obvious reasons, be profoundly problematic. Even if the world is comprehensible in some way or other, almost certainly it is not comprehensible in just the way it is assumed to be by science at any given stage of its development. An elementary requirement for intellectual rigour, for rationality, is that problematic and influential assumptions be made explicit so that they can be criticized and, we may hope, improved. Therefore, if science is to comply with even the most elementary of requirements for intellectual rigour, it is essential that profoundly influential and problematic metaphysical assumptions about how the world is comprehensible be made explicit in science. If such metaphysical assumptions are articulated, criticized and improved within science, we may hope to improve *the aims and methods* of science as we proceed. As we improve our scientific knowledge of the world, we may hope to improve our knowledge about how to improve knowledge—scientific progress accelerating as a result (Maxwell, 1974, 1979, 1984, Chaps. 5 and 9).

All this once again illustrates rules 1 to 4, and especially rule 4, the importance of interconnecting basic and specialized problem solving.

Inductivism and falsificationism, however, both seek to exclude influential and problematic metaphysical assumptions about comprehensibility from science. They seek to do this in a misguided attempt to preserve the intellectual rigour, the scientific character, of science. Actually they do the exact opposite: they *undermine* the intellectual rigour, the scientific character, of science. They render influential and problematic assumptions undiscussable within science.

The rules 1 to 4 are, of course, put into practice in science to a very great extent: without this, science would not have achieved the success that it has achieved. Widespread attempts to pursue science in accordance with inductivism or falsificationism have, nevertheless, had damaging consequences for science (Maxwell, 1976, 1984). This holds for the physical sciences to some extent at least; it holds to a greater extent for the biological sciences, and is especially pronounced, I wish to argue, for the neurosciences. In the physical sciences falsificationism nowadays predominates; inductivism at least has been almost universally repudiated. Biological science, by contrast, has not yet reached even this degree of methodological sophistication. Here inductivism still predominates. Biological scientists are still reluctant to propose empirically unsupported, falsifiable speculations. In neuroscience this inductivist reluctance has resulted in the accumulation of a vast amount of empirical knowledge in the almost complete absence of any testable, empirically progressive theory as to how the brain works overall—any theory, that is, that is comparable in stature to the great unifying, explanatory theories of physics. If physics was like neuroscience in this respect, then we would have in physics a vast amount of empirical knowledge, but we would be without Newton's theory of gravitation, Maxwell's theory of the electromagnetic field, Einstein's special and general theories of relativity, and the quantum theories of Bohr, Heisenberg, Schrödinger, Dirac, Schwinger, Feynman, Weinberg and Salem.

Not only is there in neuroscience an *inductivist* reluctance to publish falsifiable speculations, in additon there is a *falsificationist* reluctance to publish unfalsifiable speculations as to how the basic problems of neuroscience are to be conceived and solved, in violation of rules 1 and 2. Exceptions to this do of course exist, e.g. Eccles (1970) or Young (1978). On the whole, however, in order to find such unfalsifiable speculations one has to look elsewhere, to the extensive philosophical literature on the mind–body problem: see, for example, Broad (1925), Ryle (1949), Smart (1963), Vesey (1964), Feigl (1967), Armstrong (1968), Campbell (1970), Popper (1977) and Dennett (1979). In this way, philosophical discussion of the basic mind–body problem tends to be harmfully dissociated from scientific discussion of more specialized problems of neuroscience, in violation of rule 4. This division persists even where deliberate attempts are made to overcome it: see, for example, Popper and Eccles (1977).

How then ought neuroscience to proceed, granted that it puts rules 1 to 4 into practice? And how, in more detail, does the current failure to put these rules into practice, as a result of the adoption of inductivism and falsificationism, serve to impede progress in neuroscience?

The first step is to formulate the basic problem of neuroscience, in accordance with rule 1. This ought not to be difficult, as long as we do not attempt to formulate the problem too precisely. It might be put like this. How do our brains enable us to do all the different sorts of things that we can do in life—see, hear, smell, feel, experience, understand, walk, speak, write, love, hate, be conscious of, choose, plan, reason, communicate?

The next step is to put forward diverse possible solutions to the problem, in accordance with rule 2. This has been attempted in the philosophical literature, indicated above.

At once the problem arises as to how a preferred possible solution is to be selected from all these candidates, to guide more detailed neuroscientific research. We need to choose that conjecture which seems to be the most strikingly implicit in, and borne out by, specialized neuroscientific research and which, at the same time, seems to hold out the greatest hope for progress in neuroscience, if true. Of all the proposed solutions to the basic mind–body problem so far put forward, there is one, I suggest, which best satisfies these methodological requirements. It might be called *the control theory of mind and brain*. It asserts that it is our brain, operating in accordance with physical law, which guides or controls us to perform and experience all that we do perform and experience in life. The mind is the brain: it is the control aspect of the brain. Our inner experiences, thoughts, feelings, states of awareness are complex neurological processes construed from the standpoint of their role in guiding or controlling our actions. There is more to us than can ever even in principle be described and explained in purely physical terms, but this is because physics seeks only to describe a selected aspect of all that there is; it deliberately omits experiential, purposive or control aspects of reality (Maxwell, 1966, 1984, Chap. 10).

This metaphysical conjecture about the nature of mind ought, I suggest, to influence and be influenced by neuroscientific research in much the same way as metaphysical conjectures about how the universe is comprehensible influence and are influenced by research in physics, in the way indicated briefly above (in accordance with rules 3 and 4).

Once this control theory of mind is conjecturally adopted, the basic problem of neuroscience becomes to specify in detail how neurological processes occurring in our brains both correspond to our inner experiences, thoughts, feelings and states of awareness, and guide us to act in the ways that we do in response to our inner experiences, thoughts, feelings and states of awareness.

This reformulated version of the basic problem of neuroscience is, however, profoundly intractable—if for no other reason than that there are an immense number of neurons in the human brain interconnected in incredibly numerous and complex ways. We need, then, to tackle the problem by attempting in the first instance to solve easier, analogous problems, thus putting rule 3 into practice. An important and by no means obvious matter is to choose the best possible, easier, analogous problems to try to solve, and the best possible route to take to the resolution of the basic problem we wish to solve.

In order to discover the best possible way to put rule 3 into practice, the vital point that needs to be remembered is that human brains have been designed by the twin evolutionary mechanisms of random variation and natural selection—the earliest, simplest kind of nervous system in existence being subjected to a vast number of small modifications over millions of years until eventually the human brain resulted. My suggestion is that it is something like this evolutionary path that we should seek to retrace in attempting to solve progressively the problem of how the human brain works. We need to begin by attempting to understand how the simpler nervous systems work—those of jellyfish, for example, or sea anemones—progressively moving on to more and more sophisticated and complex nervous systems until we come to those of humans.

In putting this evolutionary research programme into practice we do not need to retrace precisely the path taken by evolution in developing the human brain. Rather, the basic idea is to develop progressively problem-solving capacities (in accordance with rule 3) by moving from simpler to progressively more complex brains, in a way that is roughly in accordance with evolutionary development. Thus the fact that species from which we have evolved have long become extinct does not constitute a major obstacle for the *methodologically* evolutionary research programme proposed here, even though it does constitute a serious obstacle for those who seek to retrace precisely the path taken by evolution in developing the human brain.

This latter problem of how nervous systems have actually evolved has received considerable attention: see, for example, Sarnat and Netsky (1981). The evolutionary research programme proposed here has not, however, received the sustained and coordinated attention that it deserves—essentially because not enough explicit attention has been given to the problem of how rule 3 ought to be put into practice, due to the prevalence of inductivism or falsificationism amongst neuroscientists, as opposed to the methodology of rational problem solving outlined above.

Five striking indications of the failure even to attempt to put this evolutionary research programme into practice are the following.

First, in order to put the programme into practice, it is essential that there is close collaboration between the specialized disciplines of evolutionary biology, ethology, neuroanatomy, neurophysiology, artificial intelligence, psychology and the philosophy of mind. This absolutely essential collaboration has not always been very apparent—philosophy of mind and artificial intelligence, especially, and psychology, to a lesser extent, being pursued somewhat independently of the biological sciences.

Second, artificial intelligence has quite strikingly failed to adopt the evolutionary path to understanding the human brain. Instead, almost without discussion, rule 3 has been put into practice in a quite different way. Artificial intelligence has sought to design artefacts which imitate more or less elementary *fragments* of intelligent human activity—recognizing patterns and objects, manipulating objects manually, reasoning, chess playing, speaking, translating—the hope being, presumably, that these fragments of human activity can be put together to form eventually an artefact that can imitate *all* that we do. This way of putting rule 3 into practice fails lamentably, however, to tackle the problem of how the brain achieves *overall* control, in easy stages, from elementary to highly complex, sophisticated versions of the problem. This problem of how the brain achieves overall control is the problem of understanding the primary control system of the brain—that which activates subordinate control systems to guide the animal, from moment to moment, to act as it does in its given environment. Only the evolutionary application of rule 3 can enable us to tackle this basic problem of overall control in a progressive fashion, from simple to complex versions of the problem by easy stages. Thus, in a quite elementary way, the entire research programme of artificial intelligence has been misconceived, due to a failure to consider intelligently how rule 3 is to be put into practice. The very title of the discipline is indicative of this mistake: 'artificial intelligence' ought to be called 'artificial life', or perhaps 'artificial control' or 'artificial goal pursuing'.

Third, artificial intelligence has failed in an elementary way to tackle the particular kind of control problems that arise in connection with the nervous system of animals and people. It is quite obvious that even the simplest action performed by an animal or person involves what may be called 'hierarchical-parallel' control. There is the psychologically elementary decision to run, hunt or whatever, at the highest level of control. This initiates a large number of low-level control systems controlling contractions and relaxations of individual muscles. These low-level control systems must, however, work together harmoniously if individual muscle contractions are to add up to the overall intended action—running, hunting or whatever. Low-level control systems operating in parallel must presumably communicate with each other, and with higher level control systems, if the animal or person is to act in a way that is intelligently responsive to the particularities of the environment. All this, it deserves to be noted, beautifully exemplifies rules 1 to 4, problem solving being a special case of goal pursuing. Indeed, rules 1 to 4 might almost be said to encapsulate the notion of hierarchical-parallel control, goal pursuing or problem solving. Methodology is doubly relevant to neuroscience. It is relevant to the conduct of neuroscience, and it is relevant to the actual subject matter itself of neuroscience. For the brain is itself a problem solver, designed by evolution, we may presume, to solve problems of living in a highly efficient manner—in a manner, that is, that puts into practice a rational methodology of problem solving. However, if living systems operate by means of hierarchical-parallel control, computers and robots built by artificial intelligence experts seem to work according to quite different principles. Such artefacts proceed *sequentially*, one step being performed at a time, rather than by means of *hierarchical-parallelism*. This means that the control problems tackled by artificial intelligence have been largely irrelevant from the standpoint of understanding how animal and human brains work. This, as before, is a result of artificial intelligence being pursued in a way that is dissociated from biology.

Fourth, psychology and the philosophy of mind have not seriously attempted to implement the evolutionary research programme indicated above. If our minds are the control aspect of our brains, and if this control aspect of our brains has evolved over millions of years by means of very many successive modifications produced by random variation and selected by natural selection, then all apparently distinctively human capacities—such as our capacity to experience, to be conscious, to choose freely, to communicate, to use and understand language, to produce art and science, to imagine and reason, and to love—must have evolved gradually, step by step, from early beginnings deep in our animal past. This means that no understanding of these human capacities can be adequate which does not portray them as *capable* of evolving gradually, in response to evolutionary pressures. A basic task for psychology and the philosophy of mind is to develop theories of consciousness, free will, etc., which render these things open to such evolutionary understanding—in close collaboration, of course, with the other branches of neuroscience. This task has not been given the priority it deserves.

Fifth, neuropsychology—somewhat like artificial intelligence—has failed to give priority to the problem of how the brain achieves *overall* control. Much work has been devoted to improving knowledge and understanding of subsystems of the brain—the visual cortex, the motor cortex, the cerebellum, and so on—but, as the editors of this book in effect point out in their discussion, no model of the visual cortex can ultimately be satisfactory which fails to show how the visual cortex is functionally related to the rest of the brain. Ultimately, the job of the visual cortex is to enable the animal to act successfully in its environment: visual information is processed to this end. It is this that models of the visual cortex need to describe and explain. In short, in order to understand how the visual cortex works, we need to understand how the brain achieves *overall* control. In order to solve this problem of overall control we will need to adopt the evolutionary approach advocated above.

Finally, in an attempt to put rule 4 into practice, I conclude with a crude neuropsychological speculation as to how the fundamental problem of overall control is to be solved. In mammals, including humans, overall control is to be associated with the reticular formation. Furthermore, since *consciousness* is what, for us, achieves overall control (Shallice, 1978), consciousness is to be associated with the functioning of the reticular formation in our brain. Diverse neurological processes occurring in the cerebral cortex constitute *subordinate control systems*, which become differentially activated and so more conscious and deactivated and so less conscious—as the primary control system of the reticular formation dictates—as our attention moves from one thing to another.

This reticular formation theory of overall control and consciousness has not, I suspect, been given the attention it deserves because it conflicts with the traditional view that consciousness is to be associated with the cerebral cortex. We differ from other animals in having both enhanced consciousness and an enlarged cerebral cortex. From this the conclusion is reached that consciousness is to be associated with the cerebral cortex. However this argument is invalid (MacKay, 1966). Enhancement of consciousness may well be associated with the development of *subordinate* control systems of the brain, facilitating imagination, planning, speech, and so on. The reasonable conjecture, in line with the evolutionary approach, is to associate consciousness in us with that neurological feature of our brain which most closely corresponds to that which achieves overall control in the simplest mammalian brain.

REFERENCES

Armstrong, D. M. (1968). *A Materialist Theory of Mind*, Routledge and Kegan Paul, London.

Broad, C. D. (1925). *The Mind and Its Place in Nature*, Kegan Paul, London.

Campbell, K. (1979). *Body and Mind*, Macmillan, London.

Dennett, D. C. (1979). *Brainstorms*, Harvester Press, Sussex.

Eccles, J. C. (1970). *Facing Reality*, Springer-Verlag, Berlin.

Feigl, H. (1967). *The 'Mental' and the 'Physical'*, University of Minnesota Press, Minneapolis.

MacKay, D. M. (1966). Cerebral organization and the conscious control of action. In *Brain and Conscious Experience*, J. C. Eccles, Springer-Verlag, Berlin, pp. 422–445.

Maxwell, N. (1966). Physics and common sense. *Brit. J. Phil. Sci.*, **16**, 295–311.

Maxwell, N. (1972). A critique of Popper's views on scientific method. *Phil. Sci.*, **39**, 131–152.

Maxwell, N. (1974). The rationality of scientific discovery. *Phil. Sci.*, **41**, 123–153 and 247–295.

Maxwell, N. (1976). *What's Wrong with Science?* Bran's Head Books, Middlesex.

Maxwell, N. (1979). Induction, simplicity and scientific progress. *Scientia*, **114**, 629–653.

Maxwell, N. (1980). Science, reason, knowledge and wisdom: a critique of specialism. *Inquiry*, **23**, 19–81.

Maxwell, N. (1984). *From Knowledge to Wisdom: A Revolution in the Aims and Methods of Science*, Blackwell, Oxford.

Popper, K. R. (1959). *The Logic of Scientific Discovery*, Hutchinson, London.

Popper, K. R. (1963). *Conjectures and Refutations*, Routledge and Kegan Paul, London.

Popper, K. R. (1977). The self and its brain, Part I. In *The Self and Its Brain* (K. R. Popper and J. C. Eccles), Springer-Verlag, London, pp. 1–223.

Popper, K. R., and Eccles J. C. (1977). *The Self and Its Brain*, Springer-Verlag, London.

Ryle, G. (1949). *The Concept of Mind*, Hutchinson, London.
Sarnat, H. B., and Netsky, M. G. (1981). *Evolution of the Nervous System*, Oxford University Press, Oxford.
Shallice, T. (1978). The dominant action system: an information-processing approach to consciousness. In *The Stream of Consciousness* (Eds. K. S. Pope and J. L. Singer), Wiley, Chichester, pp. 117–157.
Smart, J. J. C. (1963). *Philosophy and Scientific Realism*, Routledge and Kegan Paul, London.
Vesey, G. N. A. (Ed.) (1964). *Body and Mind*, Allen and Unwin, London.
Young, J. Z. (1978). *Programs of the Brain*, Oxford University Press, Oxford.

Models of the Visual Cortex
Edited by D. Rose and V. G. Dobson

Chapter 3

Models and metaphysics: the nature of explanation revisited

VERNON G. DOBSON
Department of Psychology, University of Brunel, Uxbridge, Middlesex UB8 3PH, UK
and DAVID ROSE
Department of Psychology, University of Surrey, Guildford, Surrey GU2 5XH, UK

> Logos is the thought that steers all things through all things
>
> Heraclitus (*c.* 500 BC)

Maxwell (Chapter 2 in this volume) is not alone in calling for an explicit metaphysics capable of standing up to the fundamental functional and developmental questions posed by brain science. For example, Longuet-Higgins has argued that 'only the most simple-minded scientist can believe that it is possible to do science without any metaphysical assumptions, and his science will inevitably be coloured by his unconscious metaphysical presuppositions' (Longuet-Higgins *et al*, 1972, p. 5).

The emergence of metaphysics, from the mythologies of ancient Egypt and Babylon, can be seen as a consequence of the ancient Greeks' search for the simplest *a priori* assumptions which can explain how the world is, and came to be, and why it has to be that way (Frankfort *et al.*, 1946). Although there are no perfectly complete and consistent metaphysical theories (Russell, 1961), metaphysical assumptions will be used in this chapter as alternative model-building strategies, or as heuristics for analysing and modelling systems in the brain and behavioural sciences. It will be argued that Craik's (1943) insights into neural symbolism can be applied productively to both main classes of metaphysical system (Heraclitean–Darwinian and Parmenidean–Cartesian) and that they generate a different kind of brain model in each case.

HERACLITEAN CHANCE AND NECESSITY, AND CONSTRAINT ON VARIETY

Heraclitus (*c*. 500 BC) assumed *a priori* that all things are eternally in conflict with one another and with themselves, changing spontaneously and cyclicly into their opposites. This assumption of fundamental dynamic variety makes the development and survival of stable complex systems under such hostile conditions puzzling. However, the spontaneous development of order from chaos can be accounted for by assuming that, by chance, conflicting change processes occasionally cancel out and mutually constrain one another to produce ordered invariants. Empedocles (*c*. 430 BC) developed these ideas further by proposing that, in the past, all possible combinations of forms were generated, but most of these could not survive because they were mutually unstable. Thus the existence of stable systems, in the face of universal flux, can be explained in terms of the control of variety by patterns of constraint, which are produced by chance and necessity. These ideas, of course, were reformulated by Darwin and now constitute the metaphysics of modern biology.

Over the course of the last century, Heraclitean models have taken precedence throughout the physical and biological sciences (Ashby, 1956; Bohm, 1957; Russell, 1961; Waddington, 1977; Weiss, 1969). Thus it is now accepted that material objects have 'wavelengths', that matter is a form of trapped energy, that quantum-mechanical levels are dominated by an irreducible probabilistic indeterminancy, that closed systems move inexorably towards states of maximum disorder or entropy, and that energy must be expended simply to prevent the decay of orderly systems. Chance and necessity, or trial and error, explanations dominate in theories of evolution and of learning.

However, Heraclitean and Darwinian ideas have not been automatically accepted as 'common-sense' for several reasons. One is that they depend on negative concepts which, according to Piaget (Hill, 1972), are the most difficult to follow. They develop last of all, both during individual intellectual growth and also during the development of fields of enquiry, such as arithmetic and logic. At the same time, negative concepts are often more fundamental. For example, -1 and i in arithmetic and the NAND connective in Boolean algebra and logical network theory can be used to generate all positive or negative numbers or logical functions (Von Neumann, 1952) but have no positive equivalent. Another reason for the reaction against Heraclitean ideas is that they emphasize the transience of both individual identities and political systems. Russell (1961) argued that, since Heraclitus proposed his doctrine of eternal flux, philosophy has been engaged in an unsuccessful search for more comforting, 'wish-fulfilling' views which are equally as consistent with experience.

CARTESIAN CREATIONISM AND CAUSE-AND-EFFECT

The advantages of the Heraclitean approach to modelling brain systems are best understood in the light of the problems generated by the alternative anti-Heraclitean metaphysics which stemmed from Parmenides' (*c*. 450 BC) ripost that 'all change is illusion'. Parmenides saw the Universe as a perfected unalterable indivisible unity in which all parts are present everywhere. Democritus (*c*. 420 BC) then proposed that the Universe consists of myriads of unchangeable atoms interacting deterministically in unchanging space, according to invariant rules which conserve the amount of movement in each direction. Each event was caused by preceding circumstances and nothing happened by chance, so that once the world had been constructed, further development was unalterably fixed by mechanical principles (Russell, 1961). As both Ashby (1956) and Waddington (1977) have pointed out, this atomist cause-and-effect approach is highly effective when applied to simple physical systems, but cannot account for the development of complex intelligent systems. Complex, goal-directed 'wholes' always seem to be more than the sums of their material parts, and 'ghosts in the machine' have to be invoked to make atomist–mechanistic explanations plausible.

Aristotle (384–322 BC) was able to supply more comprehensive explanations in thorough-going spiritualistic terms. He assumed a hidden Artificer whose purpose was realized in the course of Nature and who conferred on all things their essence or goal. The Universe and its contents had been designed to be in perfect harmony with each other and with the superordinate goal of their Creator (which was to contemplate his own perfection). Objects naturally returned to a state of rest unless caused to move teleologically by the will or future goal of the Artificer or his agents.

These creationist assumptions became so entrenched that the founders of scientific method, like Francis Bacon (1561–1626), took it for granted that the world had been perfectly designed to operate, like a piece of clockwork, according to a few simple principles. Bacon expected these principles to be so simple that they would all be revealed within a decade, if his empiricist scientific method were to be properly applied (Butterworth, 1949). Despite their constant border disputes, the atomistic and spiritualistic metaphysics are really complementary and interdependent, and a combination of both is required to account for the development of complex, intelligent, goal-directed systems. This combination was put together by Descartes (1596–1650), who proposed two parallel worlds of mind and matter which could be studied independently. Both worlds were governed by deterministic positive feedforward processes and it was assumed that every change must have a cause. In the material plane, mechanical cause-and-effect rules governed bodily reflexes, and the nerves were assumed to 'excite' movement by transmitting fluid pressures to the muscles. The mental plane was governed by teleology; individual human minds were assumed to be 'closed' or cut off from one another, but to be responsive to the will of a benign undeceiving Creator. Knowledge of the external world depended on innate mental processes which construct clear and distinct concepts of creation from unreliable sense data for a suitably passive and admiring Cartesian 'thinker'. Both spiritual and material worlds were assumed to be harmoniously designed mechanisms within which inhibitory processes were unnecessary.

According to Diamond, Balvin and Diamond (1963), Descartes' assumption that neural excitation involved transmission of material fluids and forces led to a misinterpretation of the laws of conservation of matter and energy. When Helmholtz (1847) stated that 'nothing cannot come from something' this was taken to mean that neural inhibition was both a logical and physical impossibility, as it implied the annihilation of excitation. Subsequently, in neurophysiology, explanations in terms of excitation were regarded as more 'parsimonious', and inhibitory effects were required to be accounted for in terms of fatigue, leakage or occlusion of excitatory processes. For these reasons the crucial work on inhibitory processes by Hughlings Jackson and Sherrington was regarded as peripheral by the 'purest' neurophysiologists (Diamond, Balvin and Diamond, 1963).

Nativist spiritual philosophies, implying that all knowledge and authority stems from a Creator, have been widely used for purposes of political control, and materialist and empiricist philosophies can be seen as serving to undermine this source of scholastic power (Morgan, 1977). Materialist approaches eliminated the spiritual plane from Cartesian philosophy, while retaining the mechanistic rules and the passive 'matter' of the material plane. This left the development of complex, intelligent, goal-directed systems unexplainable, and implied that all systems were, in fact, very simple.

The empiricist modifications of Cartesian metaphysics assumed that all knowledge stems from sensation rather than from the Creator. The great empirical success of atomistic metaphysics in the physical sciences was followed up by an atomistic approach to psychology and learning (Rachlin, 1976). Instead of being constructed by the Creator, the mind was assumed to be assembled from sets of elementary sensations into compound percepts forming a veridical model of the world, simply through experience. This led to the problem of specifying simple innate associative mechanisms which could generate realistic models of complex, intelligent, goal-directed systems.

CRAIK'S THEORY OF NEURAL SYMBOLISM

Craik (1943) saw himself as a mechanist, and he took 'causation' and the existence of an objective 'external world' for granted. Unlike the behaviourists, he did not deny the existence of 'mind'. Instead, he assumed that the primary function of the brain is to generate and use 'working models of reality'. These models parallel or imitate the causal structure of the external world, enabling prediction of future events and the investigation and evaluation of alternative possibilities, so that the best options can be selected. The modelling process involves (a) translation of external events into symbols in the model (e.g. by stimulation of the sense organs), (b) calculation, inference or operation of the model (e.g. by 'association'), and (c) retranslation from the results of the calculation back into terms of external events (e.g. by excitation of effectors).

Craik proposed that the firing of specific groups of brain cells represents the occurrence of specific events in the world. The main principle is that of 'unique determination' so that there is an unambiguous one-to-one coding between the firing of cell groups and the corresponding events. Causal relationships between events in the world are represented by excitatory links between corresponding neurons, so that the model can be run in fast time to predict causal sequences.

Craik was also an empiricist, suggesting that the brain develops from a random circuitry, perhaps by selective facilitation of excitatory synapses, and that information was stored by a holograph-like associative process and reconstructed by interpolation. Craik also anticipated modern theories of pattern recognition by suggesting that cells in visual cortex might receive from lines of retinal points and fire only on receipt of the sum of their inputs (Craik, 1943, pp. 67–68). He realized that by simply summing their inputs cells could only represent things in absolute terms, whereas differentiation offered the possibility of adaptation and the representation of relative characteristics and rates of change.

However, Craik (1943) did not even mention neural inhibition and its capacity to perform differentiation; nor did he discuss brain evolution or epigenesis, which are of central interest today. Indeed, Craik's proposal that a causal model of a system is equivalent to an 'explanation' of that system obscures problems of ultimate causation and developmental rationale. The central difficulties in Craik's theory relate to how the brain models could develop to mimic the structure of reality, and how they could serve to guide adaptive behaviour.

Although Craik took the Cartesian framework for granted, he was moving towards a Heraclitean position. He was interested in error-correcting machines and codes. He speculated about the development of physical systems by processes of combination followed by the elimination of structures which were 'mutually unstable' or inconsistent with themselves. He appeared to be searching for the link between physical and logical necessity, assuming, like the early Greeks, that the same principle, or 'logos', governs both existence and knowledge. After Craik's tragic death in 1948, new evidence and theory led several of his contemporaries (e.g. Hebb, 1949,

1977) to reformulate their theories in Heraclitean terms, with inhibitory processes operating on spontaneously active cells. Others (e.g. Pribram, 1971) saw memory and control primarily in inhibitory terms. It is probable that, had Craik survived, he might also have recast his theory of neural symbolism in Heraclitean terms. We will argue below that, had he done so, he would have been able to resolve the problem of how brain models might match the structure of reality and guide adaptive behaviour.

HERACLITEAN 'WORKING MODELS'

In this section it will be argued that, in a Heraclitean world, everything that can develop and survive must embody a selective functional model of the environment upon which it depends for its existence. This rule applies from subatomic particles to brains and societies, and suggests that the differences between matter, body and mind are determined by the level of the representations that they embody.

The perpetual strife and flux which makes the Heraclitean world so hostile to ordered systems is also the source of its creativity through 'chance and necessity'. In any environment consisting of large numbers of natural systems interacting spontaneously, all possible combinations of interactions that can occur will occur, given time, and all possible stable compounds of interactants that can form, will form. Each of these is then subjected to all possible destructive interactions that can occur, and only systems capable of dealing with all these destructive events survive indefinitely. The remainder are broken down into their components, which then interact and reassemble in other ways. The environment as a whole reaches dynamic equilibrium when the rates at which the various compounds are forming are balanced by the rates at which they decay.

This type of explanation is supposed to apply to development in all 'natural' systems, from the formation of atomic nuclei within stars to the random mutation and recombination of genetic alleles during sexual reproduction and evolution. During system formation the amount of variety in the behaviour of system components is markedly reduced or constrained, so that, in this sense, 'wholes' are less than 'the sums of the parts' (Weiss, 1969). For example, if a deep-sea oil-rig is completely dismantled *in situ*, its components would be rapidly dispersed by storm and tide, but when assembled into the rig, the structure persists unchanged. Not only do components lose behavioural variety, because of mutual constraints, but also the resultant subsystems constrain the influence of specific environmental processes threatening the survival of the system as a whole. In a Heraclitean world all viable structures must comply with 'Sod's law', which assumes that, for each system, everything that can go wrong will go wrong, unless specific steps are taken to prevent it. Clearly, for every potentially destructive environmental process, there should be a corresponding constraint in the system structure. Similarly, each component of the system must embody a set of constraints on destructive interactions between itself and other system components.

It should thus be possible to construct a model of any stable system in terms of a balance sheet showing how each process or variable threatening system stability is counteracted, matched or regulated by an opposing process, variable or parameter so that no unconstrained change-inducing processes remain to threaten stability. In such models, system-threatening variables would be 'represented' by the corresponding system components on the unambiguous one-to-one basis required by the 'principle of unique determination' (Craik, 1943). It should also be possible to couch this account in developmental terms, giving the sequence in which each constraint was developed, and the corresponding destructive factors which were controlled.

These arguments apply to all systems mediating functional goal-directed processes or programmes, which must also systematically prevent all events inconsistent with the goal event. This means that, in Heraclitean worlds, strategies for systems analysis and modelling commence with alternative hypotheses concerning the possible goal or function of the system under investigation. These goals can then be translated into alternative programmes, in which each step requires specific operations upon a substrate, and demands specific properties in the components of the system executing the programme. Different functional hypotheses make different predictions about which parameters and parameter values are essential for optimal function, and which are accidental or irrelevant to function. The empirical tests and observations required to distinguish the correct hypothesis can be derived from these differing predictions (Gregory, 1959).

This Heraclitean–Darwinian type of model differs from those that Craik (1943) developed within a Cartesian context in the following ways: the model is already an abstraction, as only environmental parameters relevant to system stability are selectively represented; the model is parsimonious, only representing destabilizing events that normally occur in that environment; the model is dynamic, actively matching and equalling the processes it constrains; the model is negative in that, instead of passively imitating events, it actively prevents or counteracts them through negative feedforward or feedback; and the model is self-sustaining and locked into an interdependent relationship with its environment. The model is also developmental and historical: the process of stepwise trial-and-error differentiation from pre-existing structures imposes severe constraints on the structures that can develop and the order in which they can form.

BODIES, MODELS AND TOOL-KITS

There are several other advantages in the view that all systems embody models of their environments. One is that it is universal, applying to non-living and living systems and to brains and societies. The difference between living and non-living systems is in the complexity of their representations and in the fact that living things significantly influence the rates of their own production by constraining the production of alternative systems from the same precursors. In a Heraclitean world, genes cannot directly 'cause' a trait, nor can they directly 'self-replicate'. They can only operate indirectly by selectively constraining all possibilities, other than the goal events, which are then free to occur spontaneously.

Replication leads both to competition for scarce resources and to all possible mutations. Mutation leads to the possibility of division of labour and cooperation between different replicators to compete more effectively. Orgel and Crick (1980) defined evolution in terms of competitions between replicating entities for limited resources, in which the least efficient replicators are eventually eliminated. If more complex and orderly systems are expensive to maintain against entropy, then selection would be for the simplest systems which can operate spontaneously and reliably over the widest range of conditions.

This leads to the idea of living systems as replicating, autonomously operating tool-kits. At microscopic levels these tool-kits are sets of enzymes with stereospecific sites, at macro-levels they are sets of organs such as horns, claws, fins, genitalia and scales, and at social levels they are roles within a group. The efficiency of an organ or tool depends upon how well it is adapted to fit and deal with its target substrate. The better the design, the less energy and information processing is required for its guidance and operation. Each organ or tool embodies models of both its target and its user. For example, a single tooth or stone tool can indicate a great deal about the diet, habitat and life-style of an extinct animal or hominid.

HERACLITEAN BRAINS AND INHIBITORY CONTROL HIERARCHIES

As the efficiency and complexity of bodily tool-kits increase, so does the probability of conflict between components. Interference between components can be minimized if they are organized into inhibitory control hierarchies in which higher levels monitor system needs and constrain activities at lower levels that conflict with these needs, or with each other. There is no need for higher levels to specify the activity of lower systems completely; they need only prevent the operation of inappropriate or conflicting subsystems or sequences, leaving the remaining subsystems under the control of 'if-then' rules and local information.

Within efficient inhibitory control hierarchies, limiting factors or targets are ranked in terms of their likelihood, and the loss that could result if they are not anticipated. The effort made to search for these targets selectively varies with their priority. The detection of one target leads to the suppression of responses to targets of lower priority, so that resources are always directed towards the current most significant threat or opportunity. Where targets are of equal importance the first target to be detected is 'locked onto' and takes priority.

Monod (1970) has described the operation of inhibitory hierarchies at microbiological levels. For each target substrate there is an 'operon' of enzymes. These are held in check, but 'on the ready' by a suppressor signal, which is itself suppressed by the presence of the target for that operon, thus releasing the operon's activity. Enzymes operate in sequence, with each enzyme in the chain being inhibited by the products of its own operation, which then serve as targets or releasers for the production of the next enzyme in sequence.

In larger animals and cells the distances between receptor and effector organs means that control of behaviour by diffusion of repressor proteins is unreliable and an alternative system of transmitting electrical signals over cell membranes is used. Nerve cells, of course, are specialized for this function. Since Descartes, these signals have been referred to as 'excitatory', but this is really a confusing misnomer. Polarized nerve cell membranes are relatively impermeable to sodium ions and so inhibit the sodium influxes which lead to action potentials; excitatory neurotransmitters open sodium channels and so release the membrane from the constraint on its firing which sodium impermeability imposes. The term 'excitation' thus really refers to a process of disinhibiting action potentials.

TARGET TRACKING AND BRAIN MAPS

Inhibitory hierarchies at different levels have to solve different problems in tool guidance. At microbiological levels enzymes are simply released and reach their targets by diffusion. The stereospecific site of an enzyme matches the shape of its target substrate molecule well enough to engage it selectively and operate upon its structure. At higher levels, single cells manoeuvre within their environments by moving in fixed directions until they receive error signals, which induce random changes in direction until the error signals cease.

Efficient tool operation in larger animals requires multidimensional control systems like those used by modern guided missiles. The first levels of tool guidance are the mosaics of tactile sensors within and over the surface of the tool which can distinguish the substrate target, localize it and guide local effector systems centring the tool at the optimal point and angle of attack for the next stage process to commence. This requires two-dimensional mappings between tactile sensory mosaics and the motor systems which translate target positions into corrective movements, making finer distinctions and adjustments as the tool approaches its target. The conformal logarithmic mapping meets these requirements for a two-dimensional guidance system, which may explain why approximately logarithmic maps of sensory surfaces are common in the brain and visual system (Schwartz, 1977).

In Heraclitean models, military analogies are entirely appropriate. Just as radar systems are used to pick up incoming targets and to guide warplanes or missiles to position in which they can 'lock onto' their targets, so the visual system can be seen as having the high-level general purpose functions of distinguishing and tracking targets, and guiding tools (such as limbs) to a point where their own local sensory systems can 'lock on'.

To perform this function the visual system must first be able to distinguish all targets relevant to survival. This requires the capacity to characterize targets, in terms of a comprehensive and economical range of stimulus domains or parameters. However, in a goal-directed inhibitory control hierarchy, 'higher' systems must selectively inhibit responses to low-priority targets, while optimizing sensitivity to key distinguishing features, or feature-patterns of high-priority targets.

Second, the visual system must be able to 'lock onto' and track targets as they are detected, perhaps by producing enhanced topographically coded responses to target positions which can drive foveation eye movements. Once a target has been locked onto it would be adaptive if its particular individual characteristics were also 'registered' in some short-term store and also locked onto. This would, for example, enable tracking of prey individuals within a herd without distraction and would also facilitate the association of the distinguishing visual features of prey varieties with their rewarding or aversive properties. The visual guidance of tools onto evasive targets requires systematic mappings and codings between the high-level and local control systems. For example, in hand–eye coordination the position of the eyes fixating a target may be read off to guide arm movements towards it, while details of the image shape and texture guide preparatory hand and finger movements.

This suggests several functional hypotheses about the visual cortical regions. One is the generation of a range of feature preferences sufficient to distinguish alternative targets. Another is the facilitation of target detection and recognition by providing detailed analyses of particular stimulus domains. Third, they may mediate the selective inhibition of responses to irrelevant target features, so that the system attends selectively to targets of high priority or to which it is already 'locked on'. Fourth, the regions may be specialized to pick up stimulus features relevant to the active guidance of particular tools and behaviours.

CONSTRAINTS ON BRAIN DEVELOPMENT IN HERACLITEAN WORLDS

Development, in Heraclitean systems, can only occur by a process of stepwise differentiation from more primitive forms in which each intermediate stage of development is viable. Each developmental step involves the imposition of specific constraints on the system structure, which, in turn, improve its capacity to regulate the environment. This means, for example, that during phylogeny and epigenesis, brain structures and functions must develop by initial prolific random growth, followed by the selective elimination of 'wrong' links and the selective inhibition of the 'wrong' responses. This, in turn, requires the development of synaptic control systems which anticipate and detect 'wrong' links and prevent them from forming or surviving. 'Wrong' links may be identified by mismatches in cytospecificities or activity patterns, in conjunction with signals indicating sensitive periods or reward.

Similarly, the development of single-cell stimulus preferences can only proceed from diffuse to precise specificity. To some extent this must involve reducing scatter in the excitatory inputs from the receptors. However, the simplest way to increase fine selectivity is to use inhibitory processes to subtract the response of one input from a slightly different one. In this way broadly tuned multipurpose inputs from the receptors can be used to produce highly specific stimulus preferences.

The inhibitory circuits modulating stimulus preferences must also develop progressively by simple transformations during evolution and early ontogeny. For example, the isotropic lateral inhibitory circuitry underlying the concentric centre-surround receptive field properties in the retina can be modified to produce the surround inhibition required for experientially generated topographic mappings (Amari, Chapter 15 and Dobson, Chapter 18 in this volume). A further simple anisotropic transformation of this circuitry would generate selectivity for orientation, direction and edge contrasts (Dobson, Chapter 18 in this volume). Modifiable inhibitory circuits, in which cells with the same stimulus preferences cut mutual links and adjust the remaining links so as to balance average pre- and postsynaptic firing rates, would produce 'cooperativity', gain control and sensitivity to slight stimulus changes. The same circuitry, with more powerful links, can function as an efficient associative net (Dobson, Chapter 18 in this volume).

INFORMATION PROCESSING IN HERACLITEAN NERVOUS SYSTEMS

Descartes assumed that the Creator had designed the world, and the systems in it, to operate with perfect mathematical precision in accordance with the smallest number of principles. In Heraclitean worlds, systems must also be as simple as possible, but, instead of operating with precision, should operate with the maximum amount of imprecision, or variety, which is consistent with competitive function. This is because the more parameters of a system which have to be defined and the more precisely they have to be specified, the more energy that has to be expended to maintain the system in operational condition in the face of entropy (Gregory, 1981). More orderly systems are thus less cost-effective and less likely to survive evolutionary competitions than less orderly systems with the same performance (Orgel and Crick, 1980). Less orderly functional systems are also more likely to be discovered by chance evolutionary processes than more orderly systems, which are intrinsically more improbable.

Examples of neural network models tending to require precisely specified and standardized parameters are correlative matrix nets (Anderson, 1970) and incrementing associative nets (Marr, 1969, 1970; Willshaw, 1981; Dobson, Chapter 18 in this volume). These nets represent things in terms of arithmetical relationships between their parameter values, so that, for optimal performance, link gains, thresholds, firing rates, connectivity ratios and input pattern sizes must be standardized in strict ratios to one another. Any random variation in these values causes errors which can be reduced by threshold modulating systems which, in turn, require modulation. Information is stored in these nets by quantitatively varying synaptic gains according to synaptic modulation rules. The resultant synaptic gain must then be maintained indefinitely at precisely the same value by some sort of experientially adjustable internal referent or homeostat, so that all synapses of the same type, with the same history, have the same pre-specified gain.

In decrementing associative nets (Dobson, Chapter 18 in this volume) information is stored qualitatively, rather than quantitatively, by selectively cutting links where there is conflicting pre- and postsynaptic activity. By storing information in these 'missing links' the problem of specifying and maintaining their gains is avoided.

The remaining intact links are assumed to be innately programmed to vary their gains, according to the consequences of their own activity, to produce the postsynaptic states required for efficient performance of the net. This only requires error signals from postsynaptic cell activity, indicating that the gain should be incremented or decremented. Thus the gains of intact links in decrementing nets can be determined simply by the physics of their existing structure and the patterns of recent activity at the synapse, as required by Uttley (1979). The actual weight of a link would depend on the threshold of the postsynaptic cell and on the weights of the other links operating in parallel with it. There is no need for 'internal referents' to 'remember' the values of the gains of each link, nor for complex standardized systems of components to encode and decode information in arithmetical form, so that decrementing nets can operate reliably in randomly grown circuitry. In this sense decrementing associative nets operate 'synthetically', while the incrementing and correlative associative nets, which depend on arithmetical coding, operate 'analytically', in terms of the distinction made by Gregory (1968).

LEARNING IN HERACLITEAN ENVIRONMENTS

The empiricist assumption that, within the brain, sense data can assemble itself associatively into models of the world is only plausible in benignly simplified 'user-friendly' environments such as 'blocksworlds' or Skinner boxes. Heraclitean worlds are complex, deceptive and hostile, so that, for learning to be rapid and efficient, educational situations have to be simplified and prestructured to the point where the key variables, targets and accidence-essence distinctions can be determined with minimal opportunity for error. This requires the evolution of genetic, sociocultural and cognitive systems for gathering and transmitting information and for providing heuristics, signals, instructions or examples which simplify the learning task (Plotkin and Odling-Smee, 1981).

Learning, in the context of an inhibitory control hierarchy, would involve associating those cells firing while 'locked onto' a target (and guiding a 'body-tool' towards it) with those cells firing in response to the need which the target satisfies. Subsequent arousal of the same need would then 'set' the system to search for the same target, and repeat the action. Skill learning would involve selectively cutting preexisting inhibitory links between the patterns of cells representing different targets, so that they are organized into an inhibitory control hierarchy which always addresses the target of highest priority (Dobson, 1981).

HERACLITEAN AND DARWINIAN MINDS

In a Heraclitean world, selection pressures favour the evolution of modelling capacities enabling effective group cooperation and competition. Higher cognitive processes, such as those involved in consciously constructing and testing alternative hypotheses, may derive from the capacity to model the models of others (including their views of the modeller), so as to anticipate and influence their goals and actions. The tendencies of preliterate peoples to attribute consciousness and wills to inanimate things, and our own predisposition to see events as 'caused' or 'made' to happen, may also reflect the social origins of human intelligence (Young, 1978).

SUMMARY

The capacity of human brains to develop detailed and deeply 'nested' representations of themselves and other brains suggests that they utilize optimal modelling strategies implemented in the most efficient neural networks. The identification of these most efficient strategies and nets should thus be of central interest to brain theory.

In this article we have argued that there are two basic kinds of metaphysics—Heraclitean and Cartesian—which offer alternative strategies for constructing brain models in accordance with Craik's (1943) theory of neural symbolism. The Cartesian approach has difficulty in accounting for the development of brain models and their capacity to guide action.

However, the Heraclitean approach, in its modern Darwinian form, does not encounter these problems and does appear to offer an alternative explanation for the capacity of brains and bodies to match the structure of their environments, as Craik (1943, p. 99) himself required:

> . . . there is something wonderful in the idea that man's brain is the greatest machine of all, imitating within its tiny network events happening in the most distant stars, predicting their appearances with accuracy, and finding in this power of successful prediction and communication the ultimate feature of consciousness. Further, I see no great difficulty in understanding how anything so 'different' from physical objects as concepts and reasoning can tell us more about those physical objects; for I see no reason to suppose that the processes of reason are fundamentally different from the mechanism of physical nature. On our model theory, neural or other mechanisms can imitate or parallel the behaviour and interaction of physical objects and so supply us with information on physical processes which are not directly available to us. Our thought, then, has objective validity because it is not fundamentally different from objective reality but is specially suited for imitating it—that is our suggested answer.

ACKNOWLEDGEMENTS

VGD was supported by the MRC.

REFERENCES

Anderson, J. A. (1970). Two models for memory organisation using interactive traces. *Mathematical Biosciences*, **8**, 137–160.

Ashby, W. R. (1956). *An Introduction to Cybernetics*, Methuen, London.

Bohm, D. (1957). *Causality and Chance in Modern Physics*, Routledge and Kegan Paul, London.

Butterworth, H. (1949). *The Origins of Modern Science*, Bell, London.

Craik, K. J. W. (1943). *The Nature of Explanation*, University Press, Cambridge.

Diamond, S., Balvin, R. S., and Diamond, F. R. (1963). *Inhibition and Choice*, Harper and Row, New York.

Dobson, V. G. (1981). A model of the development of functional neural organization underlying psychophysical performance in the visual system, Ph.D. thesis, Brunel University.

Frankfort, H. Frankfort, H. A., Wilson, J. A., and Jacobsen, T. (1946). *Before Philosophy*, Pelican, Harmondsworth.

Gregory, R. L. (1959). Models and the localisation of function in the central nervous system. In *The Mechanization of Thought Processes*, Vol. 2, National Physical Laboratory Symposium 50, HMSO, London, pp. 669–689.

Gregory, R. L. (1968). Perceptual illusions and brain models. *Proc. Roy. Soc. Lond. B*, **171**, 278–296.

Gregory, R. L. (1981). *Mind in Science*, Weidenfeld and Nicholson, London.

Hebb, D. O. (1949). *The Organisation of Behaviour*, Wiley, New York.

Hebb, D. O. (1977). Elaborations of Hebb's cell assembly theory. In *Neuropsychology after Lashley* (Ed. J. Orbach), Lawrence Erlbaum, New Jersey, pp. 483–496.
Helmholtz, H. L. F. von (1847). *Uber die Erhaltung der Kraft*, reprinted 1915, Berlin.
Hill, B. (1972). Piaget. *Times Ed. Supt.* (Lond.) Feb. 18, pp. 18–19.
Longuet-Higgins, H. C., Kenny, A. J. P., Lucas, J. R., and Waddington, C. H. (1972). *The Nature of Mind: The Gifford Lectures*, Edinburgh University Press, Edinburgh.
Marr, D. (1969). A theory of cerebellar cortex. *J. Physiol. (Lond.)*, **202**, 437–470.
Marr, D. (1970). A theory for cerebral neocortex. *Proc. Roy. Soc. Lond. B.*, **176**, 161–234.
Monod, J. (1970). *Chances and Necessity*, Collins, London.
Morgan, M. J. (1977). *Molyneux's Question*, Cambridge University Press, Cambridge.
Orgel, L. E. and Crick, F. H. C. (1980). Selfish DNA; the ultimate parasite. *Nature*, **284**, 604–607.
Plotkin, H. C., and Odling-Smee, F. J. (1981). A multiple-level model for evolution and its implications for sociobiology. *The Behavioural and Brain Sciences*, **4**, 225–268.
Pribram, K. H. (1971). *Languages of the Brain*, Prentice-Hall, Englewood Cliffs, New Jersey.
Rachlin, H. (1976). *Introduction to Modern Behaviourism*, W. H. Freeman, San Francisco.
Russell, B. (1961). *History of Western Philosophy*, Allen and Unwin, London.
Schwartz, E. L. (1977). Afferent geometry in the primate visual cortex and the generation of neuronal trigger features. *Biological Cybernetics*, **28**, 1–14.
Uttley, A. (1979). *Information Transmission in the Nervous System*, Academic Press, London.
Von Neumann, J. (1952). *Probabilistic Logics*, California Institute of Technology.
Waddington, C. H. (1977). *Tools of Thought*, Cambridge University Press, Cambridge.
Weiss, P. A. (1969). The living system: determinism stratified. In *Beyond Reductionism* (Eds. A. Koestler and J. R. Smythies), Hutchinson, London.
Willshaw, D. J. (1981). *Holography, Associative Memory and Inductive Generalisation.* In *Parallel Models of Associative Memory* (Eds. G. E. Hinton and J. A. Anderson), Lawrence Erlbaum, Hillsdale, New Jersey.
Young, J. Z. (1978). *Programs of the Brain*, Oxford University Press, Oxford.

Models of the Visual Cortex
Edited by D. Rose and V. G. Dobson

CHAPTER 4

Cerebral cortex as model builder

H. B. Barlow
Kenneth Craik Laboratory, Physiological Laboratory, Downing Street,
Cambridge CB2 3EG, UK

The cerebral cortex is supposed to be responsible for humanity's dominance of the natural world, and in particular for the intellectual preeminence that underlies this position. With the hope that the function of the whole cortex may illuminate the role of the parts devoted to vision, and vice versa, I have asked the following five questions:

1. Why does the cortex everywhere possess a similar structure?
2. What sort of new behaviour does the cortex make possible?
3. What types of operation are required for such new behaviour?
4. Do neurons in visual cortex perform such operations?
5. Can we suggest how the cortex does it?

In this article I attempt to give the simplest and most straightforward answer to each question in turn, and a unified view emerges which makes sense of a wide assortment of facts about the cortex.

WHY DOES THE CORTEX EVERYWHERE POSSESS A SIMILAR STRUCTURE?

The simplest answer is that *it performs the same operation everywhere*, but the input upon which this operation is performed varies in different cortical areas according to the origin of the input fibres, and the effects of the operation also vary according to the destinations of the output fibres. The principal regional differences that one has to account for are those between the primordial areas, which include the sensory projection areas and the motor area, and the other areas. This distinction was originally made by Flechsig (1901) on the basis of the earlier myelinization of the tracts associated with the primordial areas, and it is possible that their most important characteristic is simply that they become functional at an earlier age. Other local variations could result from differences in the number, size, origin and destination of the fibres entering and leaving; regional differences of structure are interesting and important but they do not violate the hypothesis that the cortex performs a uniform operation throughout.

WHAT SORT OF NEW BEHAVIOUR DOES THE CORTEX MAKE POSSIBLE?

Early comparative anatomists such as Elliot-Smith (1924) answered this by saying that animals with well-developed cortices had greater flexibility of response and were able to modify genetically determined behaviour patterns in different circumstances. This required the ability to detect types of association that an animal with a less developed cortex would be unable to use.

Unfortunately this qualitative insight has not been confirmed and extended by modern behavioural psychologists; for instance Macphail (1982) could not find any evidence for improved learning ability correlated with greater cortical development. This was based on a review of laboratory studies; an idea that may explain his unexpected conclusion comes from Craik (1943): he postulated that the brain builds up a working model of the normal environment it lives in, and learning in the laboratory is not usually designed to test this model. How the cortex might form and hold the model will be sketched below, but the suggested answer to our second question is now clear: *the cortex mediates behaviour that requires extensive knowledge and understanding of an animal's environment.*

WHAT TYPES OF OPERATION ARE REQUIRED TO MODEL THE ENVIRONMENT?

In this section it will be argued that the cortex behaves like a gifted detective, noting suspicious coincidences in its afferent input and thereby gaining knowledge of the non-random, causally related, features of its environment. The problem of detecting such coincidences is an immense one since there are 2^N possible patterns of association among the N inputs to the cortex. The hugeness of this number is intimidating, but one conclusion can be drawn from it with confidence: only a small fraction of the possible associations can be used to construct the model, and it follows that there is extensive scope for genetic preselection of those associations that can be detected and used; furthermore, this would still leave plenty of room for modification by experience.

Thus the overall scheme proposed is that preselection is done by the genetically determined pattern of connections to, between, and within cortical regions; these may not be precisely laid down, but loosely guided in the manner that Innocenti (1980) found for pathways crossing in the corpus callosum. There is much overlap among the cortical afferents, so one small cortical region receives the information required to detect associations in the firing of many afferents. The important knowledge about the environment is contained in these associations and their structure, and it is proposed that the elementary operation performed everywhere in the cortex is the extraction of this knowledge.

To understand the nature of this elementary operation it must be realized that there is only one way that it can be done. To quote Fisher (1935): 'Inductive inference is the only way known to us by which essentially new knowledge comes into the world.' Thus we are led to postulate that a process of inductive reasoning is applied to the information provided by the senses; this was long ago advocated by Helmholtz (1897, 1924) and its importance has recently been recognized by perceptual psychologists such as Rock (1983). However, here we want to understand the role of single cortical neurons, or small groups of them, in building the model; we must therefore think of the essential operations of inductive reason being performed on events consisting of the activation of cortical afferents and groups of cortical afferents, rather than on the complex sensory events of which one is consciously aware.

Now inductive inference is done by formulating hypotheses, confronting them by facts, and finding whether the facts violate the hypotheses. I want to suggest that a nerve cell, with its excitatory and inhibitory pattern of connections, represents a hypothesis about the sense organs it ultimately connects with, and its firing rate indicates how strongly this hypothesis is violated by the current state of activation of the sense organs. Since the state of activation of the sense organs depends upon the physical environment, this amounts to the suggestion that the multitude of nerve cells in sensory pathways is constantly testing a multitude of hypotheses about the environment. I have argued elsewhere (Barlow, 1969, 1980) that the properties of the on- and off-centre ganglion cells of the retina conform quite closely to the expectations of this proposal, the hypotheses they embody being of the type 'there has not been a local increase (for the on-type) or decrease (for the off-type) of illumination in my part of the retinal image'. The hypotheses required to test for and establish a model of the environment are more complex than this: instead of postulating that a single variable has a certain value or lies within a certain range, the hypotheses of the model must postulate associations and correlations between variables.

The cortex, in setting about the task of making inductive inferences about the environment, is in rather the same position as a detective attempting to make inferences that go beyond the direct evidence and the statements of his informants. What he knows he must do is to look out for 'suspicious coincidences' in the collected evidence, such as footprints in the sand of a lonely beach and the report of an agonized cry from the same vicinity. Such coincidences may establish unsuspected relationships, and they therefore constitute new knowledge about his special world. In the same way each small region of the cortex should pay special attention to suspicious coincidences in the firing of its afferents; let us look at the definition of a suspicious coincidence.

The coincident occurrence of two events, A and B, is 'suspicious' if they occur jointly more often than would be expected from the probabilities of their individual occurrence, together with the assumption that they are unrelated; i.e. the coincidence A&B is suspicious if $P(\mathrm{A\&B}) << P(\mathrm{A}) \times P(\mathrm{B})$. Any detective knows that, for a coincidence to be suspicious, the events themselves must be rare ones and that, if they are rare enough, even a single occurrence is significant. Adaptation and habituation are commonly found in sensory systems, and this is where they prove to be especially valuable. They ensure that any afferent that is strongly active must signal an event of low prior probability, and therefore any strongly active pair constitutes a significant coincidence.

The importance of the genetically determined afferents to a particular cortical region now becomes apparent, for their pattern is what determines the types of coincidence that can be detected; natural selection will have moulded these connections so that suspicious coincidences are *a priori* likely to occur amongst them.

Now the coincidences that occur regularly and frequently do so because they result from some causal connection in the environment. This is the knowledge that needs to be incorporated in the cortical model, and the way to do this is to establish neurons whose job it is to report the occurrence of these frequent coincidences. Suppose that a cell that responds to coincidental firing of two or more inputs is thereby sensitized to those particular inputs (see the last section below), then after a number of repetitions it will come to be selectively sensitive to that particular suspicious coincidence and will signal whenever it occurs. Since this signal will in many cases go to other cortical regions it should use the same language as the other cortical afferents; the strength of the signal should indicate the degree of confidence that the coincidence occurred and was not due to chance, i.e. it should signal how strongly the 'no coincidence' hypothesis has been violated.

This discussion has led to the conclusion that the sensory system must perform inductive inference on the afferent messages, and has suggested how this may be done. Individual inputs to the system can be, and probably are, handled at precortical levels, whereas the operation characteristic of the cortex itself is thought to be *the detection and signalling of suspicious coincidences*; the cortical neurones in a region must share amongst themselves the task of detecting the coincidences, that occur in the input fibres to that region and must then signal to higher levels when each one occurs. In this way each region of the cortex would discover valid hypotheses about the associative structure in its inputs, and it is proposed that this is the elementary operation performed in every cortical area. Note that, since one cortical region feeds another, this discovery of structure is a sequential, repeated operation; this is important, because even a weak selective force can have powerful consequences when repeated many times.

Finally, it should be realized that this section stems from the ancient idea that the cortex is above all the organ of association. It also builds on more recent developments that have occurred, under the stimulus of information theory, from Attneave (1954), MacKay (1956, 1978) and Uttley (1979).

DO NEURONS IN VISUAL CORTEX DETECT SUSPICIOUS COINCIDENCES?

There are three key questions to confront these proposals with:

(a) Does the known organization of the cortex fit in with the proposed nature of the basic operation?
(b) Is there evidence for the modifiability of a cortical neuron's responsiveness by experience?
(c) Does this modification occur in response to commonly occurring coincidences?

Is the organization of visual cortex appropriate for the hypothesis?

Topographical mapping of the visual field is the most prominent feature of visual cortex; the first question must be whether it makes sense on the current hypothesis. We want to regard this as an example of a more general principle, namely that cortical afferents are grouped together by genetic mechanisms in such a way that suspicious coincidences are *a priori* likely to occur among them. The effect of topographical mapping is to cause neighbouring parts of the visual field to project to neighbouring parts of the cortex, so the question becomes: 'Are suspicious coincidences likely to occur in the excitations of neighbouring parts of the visual field?' The answer must be 'Yes', because objects characteristically have extension and tend to have uniform surface properties; to put it another way, neighbouring points in the visual field are not free to vary independently because they often belong to parts of a single object. Thus it makes sense to look first for coincidences amongst events occurring at neighbouring points in the visual field.

It has been suggested (Barlow, 1981; Zeki, 1978) that other visual areas specialize in colour, motion, disparity or texture, and it would make computational sense to organize these areas as 'parameter spaces' (Ballard, 1984) in which different selections of afferents are brought together by genetic mechanisms because they are similar in colour, motion, disparity or texture.

Like proximity in the visual field, sharing one of these properties obviously increases the prior probability of suspicious coincidences occurring among them.

Are cortical neurons modifiable?

There is no doubt that the pattern-selective properties of neurons in primary visual cortex are modifiable by the patterns of stimulation they receive, provided that this occurs within a certain critical period early in life (for a recent review, see Movshon and Van Sluyters, 1981). Perhaps the best experiments remain those of Hubel and Wiesel (1965, 1970) in which they showed that cortical neurons receiving input from both eyes were common in normal animals, but rare in animals that had not received synchronous input from both eyes. In other words, if coincident excitation of the two eyes was prevented from occurring, then after a period without such excitation very few cells were found that would be capable of signalling binocular excitation. The current interpretation is that normal cells in primary visual cortex are capable of detecting, and thereafter signalling, the 'suspicious coincidence' corresponding to the excitation of the same part of the visual field in both eyes; if this does not occur within the appropriate developmental period, they cease to be able to detect and signal it. Similarly, it has been found that selectivity for orientation (Blakemore and Cooper, 1970; Hirsch and Spinelli, 1970, 1971; Rauschecker and Singer, 1981), motion (Cynader, Berman and Hein, 1973; Olson and Pettigrew 1974) and disparity (Pettigrew, 1974) are influenced by experience and deprivation of the appropriate types of stimulation.

Is it coincidences that modify?

This seems to be the case for binocularity, but can the selectivity of neurons for orientation and motion be explained in the same way? The statistical properties of visual images have not been examined with this question in mind, but it seems plausible. Most people will have seen pictures of natural scenes that have been high-pass filtered to remove low spatial frequencies. Characteristically the edges of objects are represented by lines darker or lighter that the rest of the picture, and the images that reach the cortex, which will have been high-pass filtered by lateral inhibition in the retina and lateral geniculate nucleus (LGN), must also have such lines prominently represented. The frequent occurrence of these lines and edges can reasonably be regarded as a 'suspicious coincidence' to which the orientation selectivity of cortical neurons is an adaptation. The fact that the genetically determined pattern of connections to cortical neurons contributes to their selective properties does not invalidate this interpretation.

It is believed that similar arguments can be made for the motion and disparity selectivity of cortical neurons, and that their modifiability by deprivation and experience supports such arguments.

These forms of plasticity are restricted to critical periods within a few months of birth and it is possible that the acquisition of knowledge about the environment later in life may involve quite different types of learning. It would, however, be plausible for the main work of building up Craik's model to be done early in maturation and development, because much of an animal's environment remains constant throughout life.

HOW MIGHT THE CORTEX BEGIN TO BUILD MODELS?

An outline will be presented of how this might be done as a three-stage developmental process, though it is not necessary that the stages should be strictly sequential, nor that all cortical cells should go through the stages synchronously (see also Phillips, Zeki and Barlow, 1984).

First imagine that a cortical cell receives excitatory connections more or less indiscriminately from many of the afferent fibres projecting to its vicinity. It is easy to see that this will have a blurring or smudging effect on patterns of activity among the input fibres, but on the other hand the arrangement brings with it the possibility of summation: many input fibres can contribute to the activation of one output neuron, which would then respond to an extensive combination of inputs.

Now add to the overlapping inputs a powerful system of mutual inhibition, perhaps mediated through an inhibitory interneuron and possibly confined to one cortical layer. If this inhibition is strong enough, as soon as one cell becomes active it will inhibit all other neighbours in its layer. Thus the effect will be that only one cell can be active in any one small area in one layer, no matter what the state of activity in the input. Now we have seen that extensively overlapping inputs cause some, at least, of the output cells to respond to combinations of their excitatory inputs; hence the addition of strong mutual inhibition will tend to cause all input states to be classified into a finite number of mutually exclusive outputs. This is likely to introduce some ambiguity because, even though there are many more cortical cells in an area than there are afferent inputs, there are not nearly as many as there are possible input states. Also there is nothing special about the classification that has so far been achieved, for it would result simply from the chance overlapping of inputs and the fact that strong mutual inhibition leads to a 'winner-take-all' situation (Feldman, 1982).

The third stage is to add a mechanism for Hebb-like reinforcement of the particular synaptic inputs that are effective in exciting a cortical neuron. Let us suppose that whenever a cortical neuron fires it releases at its recipient synapses a 'synaptic rewarding factor' that has properties akin to those of nerve growth factor. This will be picked up selectively by those presynaptic terminals that have contributed to the activation; it will act locally to stabilize and strengthen the synapses and will also be transported back to the cell bodies where it will trigger changes that cause further growth of that cell and of the terminals picking up the factor released by the postsynaptic cell (Changeux and Danchin, 1976; Hendry and Iversen, 1973; Hendry *et al.*, 1974). This mechanism can give *meaning* to the pattern selectivity of cortical neurons by making them more sensitive to those particular compound events or suspicious coincidences that occur frequently.

Swindale (1982) has made the interesting suggestion that the superimposed, but independent, system of cortical 'columns' leads to the representation at their intersections of many different combinations of features; the shrinkage or expansion of these regions that is observed would, of course, correspond to the development of the appropriate coincidence detectors.

Simulation studies (e.g. Malsburg, 1973) confirm that some of the mechanisms sketched above will perform as specified, but it is likely that the scheme will require much innate knowledge in the form of predetermined connections before it can model the environment satisfactorily. Nonetheless, it is a plausible initial scheme for a model-building cortex.

SUMMARY

The suggested answers to the five questions about the cerebral cortex that we started with are as follows:

1. The cortex has similar structure everywhere because the operation it performs is similar everywhere.
2. It is responsible for building up a working model of the environment that an animal has explored and observed.
3. The operation needed to acquire the new knowledge for building the model is inductive inference; cortical cells are thought to detect, and subsequently signal, 'suspicious coincidences' amongst the sensory and other messages they receive.
4. The hypothesis fits many of the experimental facts on the organization of cortical interconnections, on the modifiability of cortical neurons by experience and on the nature of the patterns to which they respond.
5. A three-stage scheme is suggested for forming sets of cortical neurons adapted to signal the main structural features of the environment an animal has experienced.

Much is uncertain and unproved, but this hypothesis represents a unified view of what the whole cortex does, and it is subjct to experimental testing in almost all branches of science concerned with the function of the cortex.

REFERENCES

Attneave, F. (1954). Informational aspects of visual perception. *Psychol. Rev.*, **61**, 183–193.

Ballard, D. H. (1984). Parameter networks. *Artificial Intelligence*, **22,** 235–267.

Barlow, H. B. (1969). Pattern recognition and the responses of sensory neurons. *Annals of the New York Academy of Sciences*, **156**, 872–881.

Barlow, H. B. (1980). Cortical function: a tentative theory and preliminary tests. In *Neural Mechanisms in Behaviour* (Ed. D. McFadden), Springer, New York, Chap. 4, pp. 143–167.

Barlow, H. B. (1981). The Ferrier Lectures: critical limiting factors in the design of the eye and visual cortex. *Proc. Roy. Soc. B*, **212**, 1–34.

Blakemore, C., and Cooper, G. F. (1970). Development of the brain depends on the visual environment. *Nature*, **228**, 477–478.

Changeux, J-P, and Danchin, A. (1976). Selective stabilization of developing synapses as a mechanism for the specification of neuronal networks. *Nature*, **264**, 705–712.

Craik, K. J. W. (1943). *The Nature of Explanation*, University Press, Cambridge.

Cynader, M., Berman, N., and Hein, A. (1973). Cats reared in stroboscopic illumination: effects on receptive fields in cat visual cortex, *Proc. Nat. Acad. Sci. (Wash.)*, **70**, 1353–1354.

Elliot-Smith, C. (1924). *The Evolution of Man: Essays*, Oxford University Press, Oxford.

Feldman, J. A. (1982). Four frames suffice; a provisionary model of vision and space. Report TR99, Computer Science Department, University of Rochester, New York.

Fisher, R. A. (1935). *The Design of Experiments*, Oliver and Boyd, Edinburgh.

Flechsig, P. (1901). Developmental (myelogenetic) localisation of the cerebral cortex in the human subject. *Lancet*, **II**, 1027–1029.

Helmholtz, H. von (1897, 1924). *Treatise on Physiological Optics*, Vol. **III**. Translated from the 3rd German edition edited by J. P. C. Southall. Optical Society of America.

Hendry, I. A., and Iversen, L. L. (1973) Reduction in the concentration of nerve growth factor in mice after sialectomy and castration. *Nature (Lond.)*, **243**, 500–504.

Hendry, I. A., Stockel, K., Thoenen, H., and Iversen, L. L. (1974). The retrograde axonal transport of nerve growth factor. *Brain Res.*, **68**, 103–121.

Hirsch, H. V. B., and Spinelli, D. N. (1979). Visual experience modifies distribution of horizontally and vertically oriented receptive fields in cats. *Science*, **168**, 869–871.

Hirsch, H. V. B., and Spinelli, D. N. (1971). Modification of the distribution of receptive field orientation in cats by selective visual exposure during development. *Expl. Brain Res.*, **13**, 509–527.

Hubel, D. H., and Wiesel, T. N. (1965). Binocular interaction in striate cortex of kittens reared with artificial squint. *J. Neurophysiol.*, **28**, 1041–1059.

Hubel, D. H., and Wiesel, T. N. (1970). The period of susceptibility to the physicological effects of unilateral eye closure in kittens. *J. Physiol.*, **206**, 419–436.

Innocenti, G. M. (1980). The primary visual pathway through the corpus call-osum: morphological and functional aspects in the cat. *Arch. Ital. Biol.*, **118**, 124–188.

MacKay, D. M. (1956). Towards an information-flow model of human behaviour. *Brit. J. Psychol*, **47**, 30–43.

MacKay, D. M. (1978). The dynamics of perception. In *Cerebral Correlates of Conscious Experience*. Eds. P. A. Buser and A. Rougeul-Buser), Elsevier, pp. 53–68.

Macphail, E. (1982). *Brain and Intelligence in Vertebrates*, Oxford University Press, Oxford.

Malsburg, Ch. von der (1973). Self-organisation of orientation sensitive cells in the striate cortex. *Kybernetik*, **14**, 85–100.
Movshon, J. A., and Van Sluyters, R. C. (1981). Visual neural development. *Ann. Rev. Psychol.*, **32**, 477–522.
Olson, C. R., and Pettigrew, J. D. (1974). Single units in visual cortex of kittens reared in stroboscopic illumination. *Brain Res.*, **70**, 189–204.
Pettigrew, J. D. (1974). The effect of visual experience on the development of stimulus specificity by kitten cortical neurons. *J. Physiol. (Lond.),* **237**, 49–74.
Phillips, C. G., Zeki, S., and Barlow, H. B. (1984). Localisation of function in the cerebral cortex: past, present and future. *Brain*, **107**, 327–361.
Rauschecker, J. P., and Singer, W. (1981). The effects of early visual experience on the cat's visual cortex and their possible explanation by Hebb synapses. *J. Physiol*, **310**, 215–239.
Rock, I. (1983). *The Logic of Perception*, MIT Press, Cambridge, Mass.
Swindale, N. V. (1982). The development of columnar systems in the mammalian visual cortex. *Trends in Neuroscience*, **5**, 235–240.
Uttley, A. M. (1979). *Information Transmission in the Nervous System*, Academic Press.
Zeki, S. M. (1978). Functional specialization in the visual cortex of the rhesus monkey. *Nature*, **274**, 423–428.

Models of the Visual Cortex
Edited by D. Rose and V. G. Dobson

CHAPTER 5

The significance of 'feature sensitivity'

D. M. MacKay
Department of Communication and Neuroscience, University of Keele, Keele, Staffordshire ST5 5BG, UK.

The physiological notion of 'feature detection' goes back to a classic paper by Barlow (1953) on 'Summation and inhibition in the frog's retina', and was brought to prominence by Lettvin *et al.* (1959). Barlow's suggestion was that a visual cell showing centre-surround antagonism, for example, could be thought of as signalling the presence of a small object against a contrasting background and so could serve as a 'fly detector'. When Hubel and Wiesel (1959) reported the sensitivity of their 'simple' and 'complex' cells to the orientation of edges or bars of luminance contrast it seemed natural to dub these cells 'edge or bar detectors'. Theories of global pattern recognition were advanced on the hypothesis that simple cells served to identify the location of specifically oriented component edges or bars, complex cells facilitated recognition of orientation regardless of location, and so on in various hierarchic schemes. At the apex of the hierarchy there would be a population of 'grandmother cells' whose firing would signify the presence of specific gestalten, such as grandmother's face (or whatever), before the eye.

Implicit in such thinking was not only the idea that complex (and higher order) cells depended on simple cells for their input, but also the suggestion that the geometrical form of objects was directly classified and so 'recognized' by the filtering operations performed by such cells.

Geometrical pattern recognition, however, is possible when outlines are depicted by other features than luminance contrast, as Gibson (1950, 1966) was at pains to emphasize. Accordingly, in 1975, Peter Hammond and I decided to test the generality of the geometrical 'feature-detecting' role of simple and complex cells in area 17 of the cat by using as stimuli edges and bars of texture contrast rather than luminance contrast (Hammond and MacKay, 1975, 1976, 1977). We found that (a) moving bars of finely textured 'static noise', which were clearly visible against a similarly textured stationary background, failed to excite simple cells that were responsive to luminance bars of the same orientation; yet (b) most complex cells in the same cortical region were responsive to such stimuli, though not selectively to the orientation of their edges. These first experiments showed conclusively that (a) although information necessary for pattern recognition might well be conveyed by the firing of complex cells (especially if one thought in terms of ensembles of their responses), the firing of a single complex cell in isolation could not signify uniquely to the CNS the presence of one specifically oriented edge or bar either of luminance or of texture; (b) although complex cells doubtless received major inputs from simple cells, they were not exclusively dependent for their excitation on such inputs, and must have parallel access to retinal information via other pathways. Further examination of this line of investigation is given by Hammond (Chapter 33 in this volume).

A long-standing alternative to the 'edge-or-bar detector' interpretation of simple/complex cell function, perceived by many of its advocates (e.g. DeValois, DeValois and Yund, 1979) as an exclusive rival, postulates that these cells work as Fourier analysers (Campbell and Robson, 1968). There is an obvious sense in which any cell with a specific receptive field profile must (for purely mathematical reasons) be more sensitive to some spatially periodic luminance patterns than to others. In particular, receptive fields with the 'Mexican hat' profile typical of simple cells have a spatial frequency response limited to an octave or two, which gives them the property of a bandpass frequency filter (Gabor, 1946; Marcelja, 1980, MacKay, 1981a). Theorists have thus been tempted to advance models of geometrical pattern recognition based on the idea that the firing rate of simple and complex cells signifies for the central nervous system (CNS) the strengths of specific Fourier components and that the CNS identifies the global pattern by means of Fourier synthesis. Because Fourier analysis is ideally independent of the geometrical origin of the coordinate framework chosen, a model on these lines, it was suggested, would ease the problem of recognizing shape regardless of location. (Unfortunately for such arguments, coordinate invariance would be lost if single cells with limited receptive fields were used as analytical elements. Responses of simple cells are spatial-phase sensitive.)

When visual cortical cells are stimulated with sufficiently small-scale checkerboard luminance patterns, those whose preferred orientations (for a single edge) lie parallel to the checkerboard edges do not respond (DeValois, DeValois and Yund, 1979). The same cells, however, respond vigorously when the checkerboard is rotated so that its diagonals lie parallel to the receptive field axis. DeValois, DeValois and Yund pointed out that under Fourier analysis a checkerboard has no spatial frequency components oriented parallel to its edges, but has its strongest components parallel to its diagonals. They argued that this experiment therefore supported a 'Fourier' model of visual cortical function as against an 'edge-detector' model.

MacKay (1981a), arguing on information-theoretic grounds that the apparent antithesis between the putative 'rivals' was in any case a false one, showed that if the cells in question were sensitive to the *polarity* of luminance contrast in their field, the results of DeValois, DeValois and Yund could have been expected, even on 'edge-detector' theories, on the assumption that edges were treated vectorially; i.e. that luminance-gradient vectors with opposite directions in the same receptive field would tend to cancel. The latter hypothesis has now been verified experimentally by Hammond and MacKay (1981, 1983a, 1983b) for both simple and complex cells, though with some interesting qualifications. The stimuli used were conventional black or white bars on a background of 50 per cent black/white static noise, to which a small bar-segment of opposite polarity of luminance (i.e. white or black respectively) and of variable extent was added. In both simple and complex cells the addition of the small opposing segment (or of two segments of the same total length at each end of the stimulating bar) turned out to have a more-than-linear suppressive effect on the response. In many simple cells the response to a short bar could be gated off completely by adding opponent segments totalling only 30 to 40 per cent. of its length. In most complex cells, and in simple cells with large receptive fields, increasing the length of the opponent segment(s) eventually led to a pickup of response, the added segment(s) becoming the main driving stimulus. These findings confirm that the non-responsiveness of such cells to checkerboards with edges parallel to their field axes, demonstrated by DeValois, DeValois and Yund, can be amply accounted for without recourse to Fourier models of their function, and lends no special support to such models.

Both 'edge-detector' and 'frequency-filter' theorists have to admit that, regarded as feature classifiers, simple and complex cells offer rather blunt instruments (e.g. they often have frequency bandwidths of 1 octave or more). How can they produce such a sharp picture of the visual world as we perceive and how is it kept stable as the eyes move? To ask the question in this form, however, is to import some unwarranted assumptions as to the requirements that the visual system has to meet. An organism comes up against the world in two ways. On the one hand, it suffers stimulation of receptors; on the other, it finds its range of effector action limited by the stable structure of the world. It is correlations between these that invite the special kind of adaptation we call perception and cognition (MacKay, 1956, 1978, 1984). The CNS can be thought of as a means of developing and storing the 'organizing subroutines' to enable the organism to adapt and match its actions or plans to the contents of its world, in response to the detected patterns of correlation. Each subroutine may be thought of as embodying a 'conditional readiness to reckon' with an object or other regularity of the type that first called it into being. As such it amounts to an internal *abstract representation* of that object or regularity. On this principle, the world finds itself categorized and depicted in terms of the demands it makes on the planning and execution of relevant action. As I have argued elsewhere (MacKay, 1956, 1959, 1961, 1966, 1967, 1970, 1978, 1984), not only overt anticipatory behaviour but the preparation of indefinitely complex conditional strategies based on abstract conceptualization are within the scope of a self-organizing information system on these lines—a system equipped to spot spontaneously the coherent, and hence cognizable, aspects of the pattern of demand imposed on it by its world and to build up a hierarchical (or perhaps heterarchical) repertoire of subroutines to match them.

In this scheme, the identity of a perceived object is neurally represented not by the firing of some particular descriptor cell (or group of cells) but rather by the particular combination of conditional readinesses set up in response to it, however widely distributed their neural substrate. *Perception* is represented by the activity of updating the conditional state of internal organization to match current sensory data. The focus of *attention* has its correlate in the region of the field of incoming data against which the state of internal organization is currently being evaluated and updated. Perception of *change* in the world is the alteration of this state of internal organization, being normally brought about only when signals are received which indicate that the existing state is inadequate or mismatched, as judged by the current criteria of the evaluator; and so on.

On this basis, then, we can think of visual pattern recognition not as a passive filtering process but as an active internal 'matching response' to demands elaborated by the visual system, whereby the organism's 'conditional readiness' for action is updated to reckon with the presence of the pattern recognized. On this view, patterns (like any object) are classified by the (internal) updating responses they evoke. The function of visual information-processing cells is not to 'name' the stimuli, but to extract feedforward to guide the active matching process that identifies them.

Much of the information needed for this updating purpose is represented by the way in which parallel sensory signals *co-vary* with one another and with ongoing motor activity. For vision the most relevant activity is what may be called 'ocular navigation'—the continual movement of the optical projection of the retina over and through the visual world during oculomotion and locomotion. Updating the conditional readiness for such ocular navigation, it is suggested, amounts to reckoning with, and so internally representing, features of the world as they specifically affect the use of the eyes. As such it offers an appropriate correlate of visual, as distinct from other, modalities of perception (MacKay, 1981a, 1981b).

With a system on these lines the visual world would be perceptually stable during voluntary eye movement, not because the retinal signals of change produced by the movement are suppressed but because they are evaluated as confirming, rather than demanding a change in, the internal representation of the world (state of conditional readiness) in terms of which the movement was planned and executed (MacKay, 1957, 1966, 1973).

Covariational analysis, even by simple coincidence detection, is computationally demanding and potentially inefficient. For coincidences to be informative they should not occur too frequently, so analysers of covariation must be shielded from signals whose coincidences are functionally meaningless. With too narrowly tuned filters on their inputs, on the other hand, coincidence detectors would be underemployed. For maximum informational efficiency, they should receive inputs extracted by filters sufficiently broadly tuned to optimize their information rate.

This suggests an alternative interpretation for the significance of 'feature sensitivity' in visual cortical cells. Instead of seeing it as leading simply to a primitive description of the visual scene in terms of the firing rate profile of 'feature detectors' (and then worrying about the crudity and ambiguity of the resulting analysis), it seems attractive to see it as helping to segregate sensory signals whose main information content has to be extracted by discovering *what co-varies with what* (and in what ways). For this purpose, relatively broad tuning in the filters is positively desirable. The categories to which they are sensitive (edge orientation, spatial periodicity, texture, motion or whatever) must, of course, be those that are likely to co-vary in a functionally meaningful way as the projected retinae rove over the visual world during oculomotion or locomotion. Taken in conjunction with such feedforward as can be extracted directly from the profile of firing rates of feature-sensitive cells, this co-variational analysis could yield just the input needed for an associative net to keep the updating process matched to the content of the visual world.

In summary, what I would emphasize in the present context is our need as theorists to see the visual cortex as subservient to the function of the cerebral information system as a whole. The way we think of and label the responses of striate cortical cells depends on our underlying assumptions as to how the whole system functions; on both experimental and theoretical grounds it now seems timely to reconsider the possible roles of these cells in visual pattern recognition. As individual descriptors of the stimuli falling on their fields, they are imprecise or ambiguous in all the domains of space, frequency and velocity. As members of local cooperative assemblies they could collectively offer more exact descriptions. The third possibility, however, advocated here, is that their job is not to provide a symbolic description of the visual field at all. The task of symbolizing the perceived world could well be a more central process, which requires from the sensory system not pictures or descriptions but an array of selective clues to help it 'home in' on the appropriate conditional readiness to reckon with that world. If so, we must be prepared to look for quite different kinds of link between striate cortical activity and pattern recognition.

REFERENCES

Barlow, H. B. (1953). Summation and inhibition in the frog's retina. *J. Physiol.*, **119**, 69–88.

Campbell, F. W., and Robson J. G. (1968). Application of Fourier analysis to the visibility of gratings. *J. Physiol.*, **197**, 551–566.

DeValois, K. K. DeValois, R. L., and Yund, E. W. (1979). Responses of striate cortex cells to grating and checkerboard patterns. *J. Physiol.*, **291**, 483–505.

Gabor, D. (1946). Theory of communication. *J. Inst. Elec. Engrs.*, **93**, III, 429–457.

Gibson, J. J. (1950). *The Perception of the Visual World*, Houghton Mifflin, Boston.

Gibson, J. J. (1966). *The Senses Considered as Perceptual Systems*, Houghton Mifflin, Boston.

Hammond, P., and MacKay, D. M. (1975). Differential responses of cat visual cortical cells to textured stimuli. *Exp. Brain Res.*, **22**, 427–430.

Hammond, P., and MacKay, D. M. (1976). Functional differences between cat visual cortical cells revealed by use of textured stimuli. In *Afferent and Intrinsic Organization of Laminated Structures in the Brain* (Ed. O. D. Creutzfeldt, Springer-Verlag, Berlin and Heidelberg, pp. 397–402.

Hammond, P., and MacKay, D. M. (1977). Differential responsiveness of simple and complex cells in cat striate cortex to visual texture. *Exp. Brain Res.*, **30**, 275–296.

Hammond, P., and MacKay, D. M. (1981). Suppressive effects of luminance gradient reversal on simple cells in cat striate cortex. *J. Physiol.*, 315, 30P.

Hammond, P., and MacKay, D. M. (1983a). Influence of luminance gradient reversal on simple cells in feline striate cortex. *J. Physiol.*, **337**, 69–87.

Hammond, P. and MacKay, D. M. (1983b). Effects of luminance gradient reversal on complex cells in cat striate cortex. *Exp. Brain Res.*, **49**, 453–456.

Hubel, D. H., and Wiesel, T. N. (1959). Receptive fields of single neurons in the cat's striate cortex. *J. Physiol.*, **149**, 574–591.

Koenderink, J. J. (1984a). Simultaneous order in nervous nets from a functional standpoint. *Biolog. Cybernetics*, **50**, 35–41.

Koenderink, J. J. (1984b). Geometrical structures determined by the functional order in nervous nets. *Biolog. Cybernetics*, **50**, 43–50.

Lettvin, J. Y., Maturana, H. R., McCulloch, W. S., and Pitts, W. H. (1959). What the frog's eye tells the frog's brain. *Proc. IRE*, **47**, 1940–1951.

MacKay, D. M. (1956). Towards an information-flow model of human behaviour. *Brit. J. Psychol.*, **47**, 30–43.

MacKay, D. M. (1957). The stabilization of perception during voluntary activity. *Proc. 15th Int. Congress of Psychol.*, **1957**, 284–285.

MacKay, D. M. (1959). Operational aspects of intellect. In *Mechanization of Thought Processes*. HMSO, London, pp. 37–52.

MacKay, D. M. (1961). Information and learning. In *Learning Automata* (Ed. H. Billing), Oldenbourg, Munich, pp. 40–49.

MacKay, D. M. (1966). Cerebral organization and the conscious control of action. *Brain and Conscious Experience* (Ed. John C. Eccles), Springer, Heidelberg and New York, pp. 422–445.

MacKay, D. M. (1967). Ways of looking at perception. In *Models for the Perception of Speech and Visual Form* (Ed. Weiant Wathen-Dunn), MIT Press, Boston, pp. 25–43.

MacKay, D. M. (1970). Perception and brain function. In *The Neurosciences: Second Study Program* (Ed.-in-chief, F. O. Schmitt), Rockefeller University Press, pp. 303–316.

MacKay, D. M. (1973). Visual stability and voluntary eye movement. In *Handbook of Sensory Physiology* (Ed. R. Jung), Vol. VII/3A, Springer, Heidelberg and New York, pp. 307–331.
MacKay, D. M. (1978). The dynamics of perception. In *Cerebral Correlates of Conscious Experience* (Eds. P. A. Buser and A. Rougeul-Buser), Elsevier, pp. 53–68.
MacKay, D. M. (1981a). Strife over visual cortical function. *Nature*, **289**, 117–118.
MacKay, D. M. (1981b). What kind of neural image? *Freiburger Universitaetsblatter*, **74**, 67–82.
MacKay, D. M. (1984). Evaluation—the missing link between cognition and action. In *Cognition and Motor Processes* (Eds. W. Prinz and A. Sanders), Springer, New York, Heidelberg and Berlin, pp. 175–184.
Marcelja, S. (1980). Mathematical description of the responses of simple cortical cells, *J. Opt. Soc. Amer.*, **70**, 1297–1300.
Phillips, C. G., Zeki, S., and Barlow, H. B. (1984). Localization of function in the cerebral cortex: past, present and future. *Brain*, **107**, 327–361.

NOTE

The idea of analysis of co-variation as a basic function of neural networks was published in a paper (MacKay, 1978) which was precirculated at the Oxford Symposium in 1981 on 'Cerebral localization and cerebral function'. At that Symposium I advanced the further suggestion that the need to segregate associative networks seeking correlations in different categories could explain the multiplicity of sensory areas in each modality (MacKay, 1981b). The Symposium contributions were not published, but an extended report of it (Phillips, Zeki and Barlow, 1984) has just appeared in which both ideas are developed in greater detail, with the help of other participants. Two theoretical papers with a somewhat similar emphasis, though from a different starting point, have just been published by Professor J. Koenderink (1984a, 1984b). Readers interested in pursuing the theme of covariance analysis are recommended to consult them.

Models of the Visual Cortex
Edited by D. Rose and V. G. Dobson

CHAPTER 6

Multiple visual areas: Multiple sensori-motor links

O. Creutzfeldt
Department of Neurobiology, Max-Planck-Institute for Biophysical Chemistry, 3400 Göttingen, Federal Republic of Germany

The concept of serial processing of information from sensory to motor systems in the cerebral cortex is an old concept in neurology. Meynert (1867) coined the term 'association fibres' for intercortical connections, Wernicke (1874) proposed the conceptual serial scheme for language processing, Flechsig (1896) delineated the association areas as the final places of synthesis of the sensory inputs and as the 'organs of mind' and Campbell (1905) saw, in the peristriate visual cortex, which he considered as one large area, a visuopsychic field (the term derived from Munk, 1881), a region responsible for higher cognitive tasks. Complementary to these ideas about the brain mechanisms of perception, which were—and still are—strongly philosophically biased in the sense of Hume's association philosophy, were ideas about the hierarchical organization of the motor and action system mainly formulated by Hughling Jackson at the end of the last century and by H. Liepmann in the beginning of this century. But the 'hierarchical' scheme of motor organization was the first to be challenged when it was realized that electrical stimulation of large parts of the cerebral cortex including sensory cortices produced motor effects in primates (Vogt and Vogt, 1919) and Man (Foerster, 1936). Under the impression of these observations, Foerster imagined the functional organization of these various motor fields as a 'working cooperative' (*Arbeitsgemeinschaft*) rather than a hierarchical command structure. Thus, parallel and serial representation of function are both familiar aspects of the functional organization of the cerebral cortex since the beginning of its scientific exploration.

It is now a well-established fact that sensory organs are not represented only in one or two cortical fields, but several times (for reviews see, for example, Merzenich and Kaas, 1980, Creutzfeldt 1983). It is assumed that in some of these areas specific invariants of a sensory stimulus may be represented, such as colour, movement or certain Gestalt aspects of a visual stimulus (Zeki 1978; Van Essen, 1979), modality of a somatic stimulus (Merzenich *et al.*, 1978) or location in space of an auditory stimulus (in bats) (Suga *et al.*, 1983). This has led to the model of parallel representation of various functional aspects of sensory inputs in contrast to the serial or hierarchical model which received new support from the interpretations of the work of Hubel and Wiesel (1962, 1965). However, considering the fact that there are 8 to 15 representations of each sensory organ found in the cortex, it is still a matter of debate as to what extent these various representations are specifically tuned to certain stimulus aspects and which is their specific function.

It has not yet been settled whether the various visual representations in areas 17, 18, 19 and 21/22 represent a parallel representation of various outputs from area 17 or true parallel representations of retinal inputs transmitted to the cortex through the thalamus after relay in other retinal-recipient zones (colliculus superior, pretectum, area 17). Since the times of Wernicke, Flechsig and Liepmann, it is assumed that the input to association areas comes essentially through association fibres from the sensory fields. If we define—with Meynert—as association areas those cortical fields which do not receive a direct input from a sensory projection nucleus, then the visual fields in area 19, the occipitotemporal transition and the inferotemporal cortex are all 'visual association areas'. They receive, however, also a thalamic input from thalamic 'association' nuclei in the nucleus lateralis posterior/pulvinar complex (LP/pulvinar complex).

The LP/pulvinar complex receives its visual input from the tectum and pretectum, from areas 17 and 18 and, in the cat in its lateral and possibly also in its superior part, directly from the retina. In cats as well as in primates these inputs are topographically restricted to different zones of the LP/pulvinar complex so that retinal, cortical, tectal and pretectal recipient zones may be distinguished (Graybiel and Berson, 1980; Lin and Kaas, 1978; Benevento and Davis, 1975, 1977; Guillery *et al.*, 1980; Itoh, Mizuno and Kudo, 1983). However, at least the cortical and the tectal inputs overlap so that, in the overlap zone, thalamic neurons can be excited from area 17/18 as well as from the superior colliculus (Benedek, Norita and Creutzfeldt, 1983).

The more we learn about the functional properties of thalamic neurons, the more it becomes clear that many of the response and stimulus features in the various cortical areas are already present in the respective thalamic projection nuclei. These properties are, to some extent, enhanced, and a selectivity for some stimulus features, as for example, directional sensitivity of visual cortical neurons, may be added by intracortical mechanisms such as intracortical convergence and inhibition (Hubel and Wiesel, 1962; Creutzfeldt and Ito, 1968; Creutzfeldt, Kuhnt and Benevento, 1974; Sillito, 1977). But the principal properties of the functional organization of the cortical fields are already found in their thalamic projection neurons. In the striate cortex, this applies to the separation of inputs from the two eyes, to orientation sensitivity and to colour selectivity. However, due to excitatory and inhibitory convergence of thalamic and intracortical afferents, some of the geniculate response features are enhanced and new ones are elaborated. Neurons are binocularly innervated and specific mechanisms for the determination of binocular disparity are developed (see, for example, Poggio and Fischer, 1977). The orientation bias of geniculate relay cells (Vidyasagar and Urbas, 1982) is sharpened to orientation sensitivity, and direction sensitivity for moving stimuli is added; in the case of colour sensitivity, only a minority of cortical cells retains colour selectivity comparable to parvocellular cells in the lateral geniculate nucleus, while the majority shows a convergent input from various types of parvo- and magnocellular geniculate afferents and these cells become sensitive to luminance as well as to colour-contrast contours (Gouras and Krüger, 1979; Tanaka, Lee and Creutzfeldt, 1983).

In extrastriate visual areas, the large receptive fields of cortical neurons, their preference for moving stimuli (including direction preferences) and the inaccuracy of the visual map are already found in the respective thalamic projection zones (Mason 1978, 1981; Bender, 1982; Benedek, Lorita and Creutzfeldt, 1983). These data are so compelling that there is, at least in the cat, no need to assume that the extrastriate visual fields receive their major visual excitation and their response bias through association fibres from area 17 (Guedes, Watanabe and Creutzfeldt, 1983; Mucke *et al.*, 1982).

The claim that extrastriate visual association areas receive their visual drive predominantly or even exclusively through association fibres from area 17 is in fact only based on a few experimental observations and was challenged by further experimental evidence. It is necessary to redefine the function of association fibres and to compile more rigorous experimental evidence for their function after learning so much about their anatomical organization in recent years (for a further discussion see Creutzfeldt, 1983).

Much has been said, in the literature, about the possible functional significance of the multiple representation of sensory surfaces in the brain and especially the visual systems (for reviews see Merzenich and Kaas, 1980; Creutzfeldt, 1979; Van Essen, 1979). Some observations of Hubel and Wiesel (1965) suggested that features of increasing abstraction may be represented from areas 17 to 19, but these authors were also disappointed not to find any hint of a 'higher order' representation in the Clare-Bishop area (Hubel and Wiesel, 1969). Zeki (1978) used the metaphor of a division of labour and suggested that different aspects of a visual stimulus are represented in the different extrastriate visual areas such as colour (V4) or movement (STS).

However, he points out that the functional differences between these areas were not as marked as one might expect to see if different sensory features were in fact specifically represented in these various fields. These types of hierarchical models of feature abstraction of increasingly higher order certainly need reconsideration, although they still dominate the thinking and search of many experimentalists as well as of theoreticians (e.g. Marr, 1982). Also the functional significance of weak retinotopy, of various topological transformations of visual space and the lack of exact retinotopy in extrastriate visual areas still waits for a reasonable functional interpretation.

I like to emphasize another important feature of the various representations of the visual field by turning attention to the fact that the various visual areas do not only differ with respect to the origin of their thalamic and cortical afferents but also with respect to the destination of their corticofugal efferents, so that *each area becomes a different link between the sensory organ and the motor output* (Creutzfeldt, 1981, 1983). The outputs from the striate and the extrastriate visual cortex into the motor control system differ insofar as corticocollicular fibres predominate from the striate cortex and corticopontine efferents from the extrastriate cortex (Albus and Donate-Oliver, 1977; Brodal, 1978). Also in other aspects, the anatomical patterns of outputs from the various visual areas into subcortical structures responsible for behavioural or motor control differ significantly (Spatz, Tigges and Tigges, 1970; Kawamura, Sprague and Niimi, 1974; Updyke, 1977; Diamond, 1979; Graham, Lin and Kaas, 1979). The inferotemporal visually excitable cortex is characterized by its significant output into limbic structures (Prelevic, Burnham and Gloor, 1976; Ungerleider and Mishkin, 1980), in addition to its collicular and caudate output (Whitlock and Nauta, 1956).

Neurophysiologically, a general property of neurons in extrastriate visual areas is their responsiveness to moving stimuli with some preference for a certain range of velocities. This is usually interpreted as a neuronal correlate of motion perception, but it also suggests some function in visuomotor behaviour. Indeed, many neurons may not only be characterized by their sensory trigger features but also by their responses in relation to visuomotor responses. Thus, in the awake monkey, neurons are found in the prelunate gyrus which are not only visually driven but discharge also or even exclusively in relation to eye movements (Fischer and Boch, 1981). These eye movement related discharges appear before an eye movement is initiated and they begin when a visual stimulus becomes a target of a saccadic eye movement and usually last through the saccadic movement. The responses appear only if the target is located in a circumscribed area of the visual field (the so-called 'goal field'). The authors of this report furthermore conclude that the visual responses in this part of V4 in the prelunate gyrus must be derived from other than geniculostriate inputs, probably from the retinocollicular-pulvinar projection. Even in areas 17 (Wurtz and Mohler, 1976) and 18 (Robinson, Baizer and Dow, 1980) at least some neurons are found, whose responses to visual stimuli may be stronger before eye movements, but without directional preferences such as those found in the prelunate cortex. In the inferotemporal cortex, some neurons are only activated when a stimulus has behavioural significance, i.e. depending on whether it leads to a behavioural response or not (Ridley, Hester and Ettlinger, 1977; Rolls, Judge and Sanghera, 1977).

These examples and my own experience in recordings from prestriate neurons in the awake monkey during the last few years indicate that in the various visual areas not only different features of a stimulus are represented but also different behavioural responses to stimuli. *The different visual fields could then also be understood as a cooperative of various sensori-motor links between the eye and visually guided behaviour*. This idea is reminiscent of Foerster's (1936) metaphor where the various motor fields were compared to a working cooperative in the elaboration of motor acts. Of course, the emphasis on the behaviourally significant output of each visual area may appear to neglect the perceptual and cognitive significance of representation in these various areas. It should be realized, however, that both these aspects are complementary and may be — to some extent—interchangeable (see Creutzfeldt 1981, 1983)

The role of association fibres could be, in such a model, to reinforce mutually the responsiveness of cortical neurons to their thalamic afferent inputs. They could also provide a mechanism for coordination of activities in the various sensory fields, so that the combined output from all the visual representations becomes functionally correlated with respect to the appropriate visuomotor response. One might also consider a gating function for association fibres which allow certain afferent inputs to get through and others not, according to the state of activity in the connected field. All such suggestions must remain vague and can at best serve as models for further experimental testing. It could even be that the functional role of association fibres may become more significant in primates than in lower mammals. However, this would need experimental proof.

ACKNOWLEDGMENT

I am happy to dedicate to you, David Hubel and Torsten Wiesel—after a long period of friendship and of partial controversy—the preceding thoughts about the visual system. Some of these ideas, which I could only outline in an essayistic format, try to bring some of your important contributions into a context of neurological tradition and also try to open new aspects for alternative ways of thinking. None of these ideas claims to be highly original insofar as most of the original ideas about the functional organization of the brain have been formulated in one form or another by those who had the chance to be amongst the first to enter the adventure of scientific exploration of the brain.

REFERENCES

Albus, K., and Donate-Oliver, F. (1977). Cells of origin of the occipitopontine projection in the cat: functional properties and intracortical location. *Exp. Brain Res.*, **281**, 167–174.

Bender, D. B. (1982). Receptive field properties of neurons in the macaque inferior pulvinar. *J. Neurophysiol.*, **1**, 1–16.

Benedek, G., Norita, M., and Creutzfeldt, O. D. (1983). Electrophysiological and anatomical demonstration of an overlapping striate and tectal projection to the lateral posterior-pulvinar complex of the cat. *Exp. Brain Res.*, **521**, 157–169.

Benevento, L. A., and Davis, P. (1977). Topographical projections of the prestriate cortex to the pulvinar nucleus in the macaque monkey: an autoradiographic study. *Exp. Brain Res.*, **30**, 405–424.

Brodal, P. (1978). The cortico-pontine projection in the rhesus monkey. Origin and principles of organization. *Brain Res.*, **10**, 251–283.

Campbell, A. W. (1905). *Histological Studies in the Localization of Cerebral Function*, University Press Cambridge, Cambridge.

Creutzfeldt, O. D. (1979). *Repräsentation der visuellen Umwelt im Gehirn. Verh Dtsch Zool Ges*, Gustav Fischer Jena Verlag, Stuttgart, pp. 5–18.

Creutzfeldt, O. D. (1981). Diversification and synthesis of sensory systems across the cortical link. *Brain Mechanisms of Perceptual Awareness and Purposeful Behavior*, (Eds. O. Pompeiano and C. Ajmone-Marsan), Raven Press, New York, pp. 153–165.

Creutzfeldt, O. D. (1983). *Cortex Cerebri. Leistung, strukturelle und funktionelle Organisation der Hirnrinde*, Springer-Verlag, Berlin.

Creutzfeldt, O. D., and Ito, M. (1968). Functional synaptic organization of primary visual cortex neurons in the cat. *Exp. Brain Res.*, **61**, 324–352.

Creutzfeldt, O. D. Kuhnt, U. W., Benevento, L. A. (1974). An intracellular analysis of visual cortical neurons to moving stimuli: responses in a cooperative neuronal network. *Exp. Brain Res.*, **21**, 251–274.

Diamond, I. T. (1979) The subdivisions of neocortex: a proposal to revise the traditional view of sensory, motor, and association areas. *Progr. Psychobiol. and Physiol. Psychol.*, **8**, 2–43.

Fischer, B., Boch, R. (1981). Enhanced activation of neurons in prelunate cortex before visually guided saccades of trained rhesus monkeys. *Exp. Brain Res.*, **44**, 129–137.

Fleschsig, P. (1896). *Gehirn und Seele,* 2nd ed., Verlag Velt and Co, Leipzig.

Foerster, O. (1936). Motorische Felder und Bahnen. In *Handbuch der Neurologie*, (Eds. O. Bumke and O. Foerster), Vol. 6, Springer-Verlag, Berlin, pp. 1–357.

Gouras, P., and Krüger, J. (1979). Responses of cells in foveal visual cortex of the monkey to pure color contrast. *J. Neurophysiol*, **421**, 850–860.

Graham, J., Lin, C. S. and Kaas, J. H. (1979). Subcortical projections of six visual cortical areas in the owl monkey, *Aotus trivirgatus*. *J. comp. Neurol.*, **1871**, 557–580.

Graybiel, A. M. and Berson, D. M. (1980). Histochemical identification and afferent connections of subdivisions in the lateralis posterior-pulvinar complex and related thalamic nuclei in the cat. *Neurosci*, **5**, 115–123.

Guedes, R., Watanabe, S., Creutzfeldt, O. D. (1983). Functional role of associative fibres for a visual association area: the posterior suprasylvian sulcus of the cat. *Exp. Brain Res.*, **49**, 13–27.

Guillery, R. W., Geisert, E. E. Polley, E. M. and Mason, C. A. (1980). An analysis of the retinal afferents to the cat's medial interlaminar nucleus and to its rostral thalamic extension, the 'geniculate wing'. *J. comp. Neurol.*, **1941**, 117–142.

Hubel, D. H., Wiesel, T. N. (1962). Receptive fields binocular interaction and functional architecture in the cat's visual cortex. *J. Physiol.*, **160**, 106–154.

Hubel, D. H., and Wiesel, T. N. (1965). Receptive fields and functional architecture in two non-striate visual areas (18 and 19) of the cat. *J. Neurophysiol*, **281**, 229–289.

Hubel, D. H. and Wiesel, T. N. (1969). Visual area of the lateral suprasylvian gyrus (Clare–Bishop area) of the cat. *J. Physiol. (Lond.),* **202**, 251–260.

Itoh, K., Mizuno, N., and Kudo, M. (1983). Direct retinal projections to the lateroposterior and pulvinar nuclear complex (LP-Pul) in the cat, as revealed by the anterograde HRP method. *Brain Res*, **276**, 325–328.

Kawamura, S., Sprague, J. M., and Niimi, K. (1974). Corticofugal projections from the visual cortices to the thalamus, pretectum and superior colliculus in the cat. *J. comp. Neurol.*, **158**, 339–362.

Lin, C. S. and Kaas, J. H. (1978). The inferior pulvinar complex in the owl monkeys: architectonic subdivisions and patterns of input from the superior colliculus and subdivisions of the visual cortex. *J. comp. Neurol.*, **187**, 655–678.

Marr, D. (1982). *Vision: a Computational Investigation into the Human Representation and Processing of Visual Information*, Freeman Publishers, San Francisco.

Mason, R. (1978). Functional organization in the cat's pulvinar complex. *Exp. Brain Res.*, **31**, 51–66.

Mason, R. (1981). Differential responsiveness of cells in the visual zones of the cat's LP-pulvinar complex to stimuli. *Exp. Brain Res.*, **43**, 25–33.

Merzenich, M. M., Kaas, J. H., Sur, M., and Lin, C. S. (1978). Double representation of the body surface within cytoarchitectonic areas 3b and 1 in SI in the owl monkey (*Aotus trivirgatus*) *J. comp. Neurol.*, **181**, 41–74.

Merzenich, M. M. and Kaas, J. H. (1980). Principles of organization of sensory perceptual systems in mammals. *Progr. Psychobiol. and Physiol Psychol.* (Eds. J. M. Sprague and A. N. Epstein), **9**, 2–42.

Meynert, T. (1867–1868) Der Bau der Großhirnrinde und seine örtlichen Verschiedenheiten. *Vrtljschr. f. Psychiatr.*, **11**, 77–93 (1867) and **21**, 88–113 (1868).

Mucke, L., Norita, M., Benedek, G., and Creutzfeldt, O. D. (1982). Physiologic and anatomic investigation of a visual cortical area situated in the ventral bank of the anterior ectosylvian sulcus of the cat. *Exp. Brain Res.*, **46**, 1–11.

Munk, H. (1881). *Über die Funktionen der Großhirnrinde. Gesammelte Mitteilungen aus den Jahren 1877–1880*, A. Hirschwald, Berlin.

Poggio, G. F., and Fischer, B. (1977). Binocular interaction and depth sensitivity of striate and prestriate cortical neurons of the behaving monkey. *J. Neurophysiol.*, **40**, 1392–1405.

Prelevic, S., Burnham, W. M. and Gloor, P. (1976). A microelectrode study of amygdaloid afferents: temporal neocortical inputs. *Brain Res.*, **105**, 437–457.

Ridley, R. M. Hester, N. S., and Ettlinger, G. (1977). Stimulus- and response-dependant units from the occipital and temporal lobes of the un-anesthetised monkey performing learned visual task. *Exp. Brain Res.*, **27**, 239–552.

Robinson, D. L., Baizer, J. S. and Dows, B. M. (1980). Behavioral enhancement of visual responses of prestriate neurons of the rhesus monkey. *Invest. Ophthalmol. Vis. Sci.*, **19**, 1120–1123.

Rolls, E. T. Judge, S. J. and Sanhera, M. K. (1977). Activity of neurons in the alert monkey. *Brain Res.*, **130**, 229–238.
Sillito, A. M. (1977). Inhibitory processes underlying the directional specificity of simple, complex and hypercomplex cells in the cat's visual cortex. *J. Physiol. (Lond.),* **271**, 669–720.
Spatz, W. B. Tigges, J. and Tigges, M. (1970). Subcortical projections, cortical associations, and some intrinsic interlaminar connections of the striate cortex in the squirrel monkey (Saimiri). *J. comp. Neurol.*, **140**, 155–174.
Suga, N., O'Neill, W. E., Kujirai, K., and Manabe, T. (1983). Specificity of combination-sensitive neurons for processing of complex biosonar signals in auditory cortex of the mustached bat. *J. Neurophysiol.*, **49**, 1573–1626.
Tanaka, M., Lee, B. B., and Creutzfeldt, O. D. (1983). Spectral tuning and contour representation in area 17 of the awake monkey. In *Colour Vision. Physiology and Psychophysics* (Eds. J. D. Mollon and L. T. Sharpe), Academic Press, London, pp. 269–276.
Ungerleider, L. G. and Mishkin, M. (1980). Two cortical visual systems. In *Advances in the Analysis of Visual Behavior* (Eds. D. J. Ingle, R. J. W. Mansfield and M. A. Goodale), MIT Press, Cambridge, Mass.
Updyke, B. V. (1977). Topographic organization of the projections from cortical areas 17, 18, and 19 onto the thalamus, pretectum and superior colliculus in cat. *J. comp. Neurol*, **173**, 81–122.
Van Essen, D. C. (1979). Visual areas of the mammalian cerebral cortex. *Ann. Rev. Neurosci*, **2**, 227–263.
Vidyasagar, T. R. and Urbas, J. V. (1982). Orientation sensitivity of cat LGN neurones with and without inputs from visual cortical areas 17 and 18. *Exp. Brain Res.*, **46**, 157–169.
Vogt, C., and Vogt, O. (1919). Allgemeinere Ergebnisse unserer Hirnforschung. *J. Psychol. u. Neurol*, **25**, 279–461.
Wernicke, C. (1874). *Der aphasische Symptomenkomplex*, Cohen u. Weigert, Breslau.
Whitlok, D. G. and Nauta, J. H. W. (1956). Subcortical projections from the temporal neocortex in *Macacca mulatta. J. comp. Neurol*, **106**, 183–212.
Wurtz, R. H., and Mohler, C. W. (1976). Enhancement of visual responses in the monkey striate cortex and frontal eye fields. *J. Neurophsyiol*, **41**, 766–772.
Zeki, S. M. (1978). Uniformity and diversity of structure and function in rhesus monkey prestriate visual cortex. *J. Physiol.*, **277**, 273–290.

Models of the Visual Cortex
Edited by D. Rose and V. G. Dobson

CHAPTER 7

A model for the generation of visually guided saccadic eye movements

Peter H. Schiller
Department of Psychology, Massachusetts Institute of Technology, Cambridge, MA 02139, USA

During waking hours primates view their environment by making saccadic eye movements at the approximate rate of three per second. With each shift in eye position a different constellation of stimuli impinges on the retinal surface. The stimuli encountered can range from one or two relatively simple ones (a bird in the sky) to very complex and extended arrays (a crowd of people in the street). Following each saccade the organism faces at least two basic tasks: (a) to analyze in more detail the stimuli viewed by the fovea and (b) to select from the constellation of stimuli falling onto the extrafoveal region the one to be foveated next. These two tasks under normal conditions can be accomplished within about 300 milliseconds, which is the average fixation time between saccades. The central focus of this discussion is the processes and pathways involved in the analysis and propagation of visual information which yields saccadic eye movements. What is remarkable about this system is that in spite of its great complexity, the computations involved are carried out with considerable speed and with very little effort.

The speed of eye movement generation deserves closer scrutiny as it is crudely indicative of the computational times involved. In humans, saccadic reaction times to single targets (detection paradigm), appearing at more than one log unit above threshold, is around 200 milliseconds, and the temporal distribution of an extensive sample of saccadic latencies in most studies is unimodal. When a discrimination paradigm is used where subjects have to saccade to one of two or more stimuli, the reaction time better than doubles. Monkeys perform these tasks more rapidly. Saccades to single stimuli can be performed as fast as 100 milliseconds, and the temporal distribution of saccadic latencies is often bimodal, suggestive of two processes (Fischer and Boch, 1983; Schiller, 1984). Most of these times are not taken up by the propagation of the sensory input or of the motor output: electrical stimulation of brainstem oculomotor structures in monkeys can elicit an eye movement in 10 to 25 milliseconds, and a visual stimulus one log unit above threshold can elicit a visual response in the striate cortex in about 30 to 40 milliseconds (Malpeli, 1983; Robinson, 1972; Schiller, Finlay and Volman, 1976; Schiller and Sandell, 1983). Thus even under the most favorable conditions, much of the time involved in the generation of a saccade is what may be referred to as computational. The computational processes involve both the selection of the stimulus to be foveated next (some form of pattern analysis) and the targeting of that stimulus (based on retinal error and eye position signals).

The basic premise of the model to be presented here is that several parallel channels of information processing are involved in the computations which yield saccades. Some involve area 17, acting either through corticocortical or corticotectal outputs, and another involves direct retinal inputs to the superior colliculus.

In the primate, the superior colliculus has been extensively implicated in the generation of saccadic eye movements. The visual field in the superficial layers of this structure is laid out in a well-defined, topographic manner (Cynader and Berman, 1972). In the intermediate layers single cells discharge prior to saccades (Schiller and Koerner, 1971; Sparks and Pollack, 1977; Wurtz and Goldberg, 1972). Electrical stimulation elicits saccades at low current levels; the size and direction of saccades elicited depends on the site stimulated within the superior colliculus and not on the parameters of stimulation (Robinson, 1972; Schiller and Stryker, 1972). It has been shown that the sensory and motor maps in the colliculus are in register: stimulation of a particular collicular site elicits saccades which bring the fovea into the receptive field area of the cells which had been activated electrically (Schiller and Stryker, 1972). It appears, therefore, that the colliculus is involved, in part at least, in the computation of retinal error signals. It has been demonstrated, however, that eye position signals are also present in this structure (Sparks and Pollack, 1977).

The major problem with the hypothesis that the superior colliculus is the prime center for the generation of visually guided saccades is that its ablation in the primate produces only relatively small deficits. The frequency of spontaneous saccades decreases, saccadic reaction times increase, velocities decrease and accuracy suffers only modestly (Albano and Wurtz, 1982; Butler *et al.*, 1982; Schiller, True and Conway, 1980). These findings suggested that in addition to the superior colliculus other structures may also be involved in the production of visually guided eye movements. It has been shown that electrical stimulation of several cortical areas can elicit saccades (Robinson and Fuchs, 1969; Schiller, 1977). These include striate and extrastriate cortices, the parietal lobe and the frontal eye fields. To determine the extent to which stimulation-elicited saccades of these cortical regions are mediated by

the colliculus, studies were undertaken in which cortical areas were stimulated before and after collicular ablation. The results showed that collicular ablations eliminated saccades triggered electrically from striate cortex, extrastriate cortex and the parietal lobe, but did not interfere with saccades generated by stimulation of the frontal eye fields (Schiller, 1977; Keating, 1983). These observations then led to the hypothesis that pathways involving the frontal eye fields provide a second system for generating visually guided eye movements. This hypothesis received further support when it was found that ablation of both the frontal eye fields and of the colliculi eliminated virtually all visually guided eye movements (Schiller, True and Conway, 1979). Ablation of either structure alone produced only relatively small deficits.
parallel channels for the generation of visually guided saccades, one involving posterior cortex and the superior colliculus and the other involving posterior cortex and the frontal eye fields.

If there are two such parallel systems, the next question to address is what their functions might be. To assume that the two channels perform the same task does not seem reasonable. As the first step in considering this question, the nature of the visual input to the collicular and the frontal eye field systems will be examined. It is now a well-known fact that in most species several parallel systems of information processing originate in the retina (Cleland, Levick and Sanderson, 1973; Enroth-Cugell and Robson, 1966; Gouras, 1969; Schiller and Malpeli, 1977). In the rhesus monkey, the animal we have studied most, two systems are most prominent, the broad-band and color-opponent. The broad-band system responds transiently, it has low spatial-frequency resolution, lacks color-opponency and conducts at high velocities to the central nervous system. The color-opponent system, on the other hand, has high spatial-frequency resolution, is color selective, responds in a sustained fashion and conducts at slower velocities (Gouras, 1969; Schiller and Malpali, 1977). These two channels remain segregated, in part at least, through several stations in the visual system.

Broad-band and color-opponent cells both project to the lateral geniculate nucleus of the thalamus, but there they terminate in different laminae (Dreher, Fukuda and Rodieck, 1976; Schiller and Malpeli, 1977). The color-opponent system occupies the parvocellular laminae (layers 3 to 6), while the broad-band system occupies the magnocellular laminae (layers 1 and 2). The projection pattern from the retina to the superior colliculus, the other major recipient of retinal fibers, is quite different. This structure receives extensive inputs from broad-band cells, but not from color-opponent ones (Schiller, Malpeli and Schein, 1979). In addition, the colliculus also receives inputs from a third class of cells in the retina, the W-like cells, which in this model will not be considered further because there is insufficient knowledge about its function (Cleland and Levick, 1974; Schiller and Malpeli, 1977; Stone and Fukuda, 1974).

In the primate, the lateral geniculate nucleus projections terminate almost exclusively in the striate cortex, which thereby forms the major gateway for visual information to the rest of the cortex. The axons of the broad-band and color-opponent lateral geniculate cells terminate in different sublaminae of layer 4C (Hubel and Wiesel, 1972; LeVay and Gilbert, 1976). Studies have shown that in spite of the extensive interconnections and computations carried out in the striate cortex, a significant proportion of cells retain the segregation of the two systems identified in the retina and the lateral geniculate nucleus (Malpeli, 1983; Malpeli, Schiller and Colby, 1981; Schiller, Malpeli and Schein, 1979). This suggests that some of the striate recipient structures perform analyses based on information carried by either one or the other of these systems. That this is the case has been convincingly established for the superior colliculus, which receives extensive cortical inputs from both striate and extrastriate areas. By reversible blockage of either the color-opponent or the broad-band systems and by examination of conduction velocities, it has been shown that the corticotectal pathway is driven exclusively by the broad-band system (Hoffman, 1973; Hoffman and Sherman, 1974; Schiller, Malpeli and Schein, 1979). This pathway is essential for the function of the superior colliculus. When the corticotectal pathway is blocked, by cooling or by ablation of striate cortex, visual signals are no longer available in the intermediate and deeper layers of the colliculus (Schiller *et al.*, 1974). It appears, therefore, that the information processed by the colliculus for the generation of eye movements is derived from the broad-band channel.

Next the kind of visual information which the frontal eye fields receive will be considered. The evidence here is less formidable, but it is nevertheless suggestive. The frontal eye fields receive a prominent input from area V4 of extrastriate cortex (Van Essen, 1979). V4 cells respond in a sustained fashion, many are color specific and they receive, via area V2, a rather long latency input from the striate cortex (Malpeli, 1983). These observations suggest that the prominent system involved is the color-opponent one. A less extensive input to the frontal eye fields comes from the medial superior temporal area, which in turn is driven predominantly by the cells of the middle temporal area (Van Essen, 1979). This latter region receives a direct input from the striate cortex, and we have evidence that it is dominated by the broad-band system (Malpeli, 1983). These observations therefore suggest that there are inputs to the frontal eye fields from both systems but that the color-opponent input is perhaps more extensive.

On the basis of these data we can now extend the model. The two channels identified in the generation of visually guided eye movements, one involving

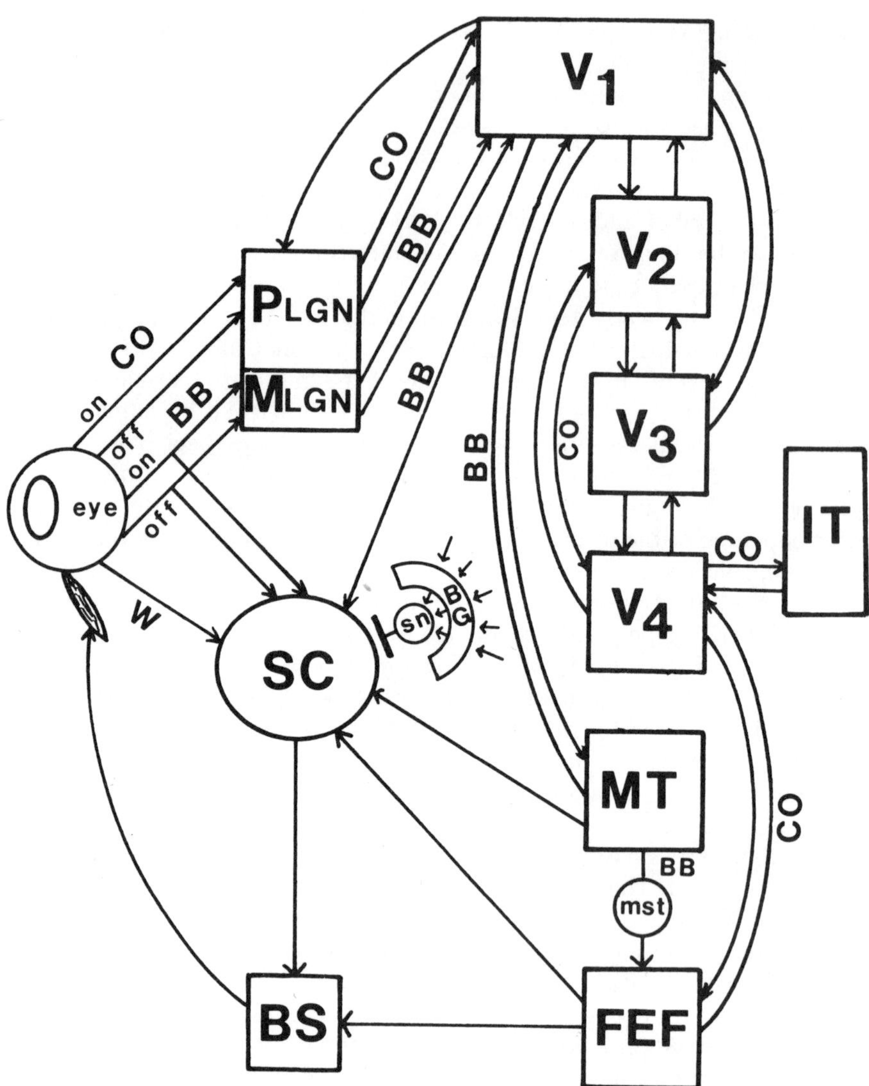

FIGURE 1 Schematic representation of the model for visually guided saccadic eye movements. The color-opponent (CO) and broad-band (BB) on and off pathways originating in the retina terminate respectively in the parvocellular (P) and magnocellular (M) portions of the lateral geniculate nucleus (LGN) of the thalamus. BB cells also project to the superior colliculus (SC), as do W-like cells. The CO and BB cell axons of the LGN terminate in striate cortex (V_1). V_1 projects to the SC and makes reciprocal connections with areas V_2, V_3 and the middle temporal area (MT). V_4 interconnects with V_2, inferotemporal cortex (IT) and with the frontal eye fields (FEF). MT sends projections to the SC and the medial superior temporal area (mst)

the superior colliculus and the other the frontal eye fields, are supplied differently by the two systems of visual information analysis. The superior colliculus receives its visual input from the broad-band system while the frontal eye fields receive an extensive input from the color-opponent system and a lesser input from the broad-band channel.

Next to consider are the functions of the two parallel visually guided saccadic eye movement generating systems. It is proposed in this model that the collicular channel is involved mainly in the generation of eye movements to simple, easily discriminable stimuli. The information is sent to the colliculus in a very rapid fashion, both from the retina and from the striate cortex, making it possible to produce quick, reflex-like responses which are essential for either prey catching or avoidance. This channel sacrifices fine detail and complex analysis for the sake of speed. The frontal eye field channel, on the other hand, undertakes a more complex, more detailed analysis of visual information for eye movement generation, such as might be necessary when complex stimuli in extensive arrays are viewed. Such analysis may require the integration of visual information with messages coming from other modalities and from memory, and may also involve the use of visual search strategies.

The evidence for the differential function of these channels at present is incomplete. It is known that collicular lesions increase saccadic reaction times and that they eliminate the recently discovered 'express' saccades observed when single stimuli are presented (Fischer and Boch, 1983; Schiller, 1984). Frontal eye field lesions, on the other hand, show a great variety of deficits, many of which disappear within two or three weeks of the ablation. In accordance with the suggestions made above are findings which show that there are deficits concerned with visual search and the temporal sequencing of eye movements when the stimuli consist of complex arrays (Butter, 1983). Our work suggests that lesions of the striate cortex in the rhesus monkey produce both sets of symptoms. In fact, striate lesions cause extremely severe deficits not only in visually guided eye movements but also, as might be expected, in pattern perception.

To round off the model we must lastly consider which signals are excitatory and which are inhibitory. It appears that the majority of pathways noted so far are excitatory. This is the case for the retinogeniculate, the geniculocort-

---

which in turn projects to the FEF. The V_1-SC connection is driven by the BB system, while the V_4-FEF connections are dominated by the CO system. The FEF and SC projections to the brainstem (BS) can generate saccadic eye movements independently. The substantia nigra (sn) makes inhibitory connections with the SC. The sn receives its projections from the basal ganglia (BG) which receives inputs from many cortical structures. Saccade generation involves both disinhibition (as of the nigral input to the SC) and excitation (as from the visual cortex to the SC).

ical and the retinocollicular pathways, and is also likely to be the case for the corticocortical connections and the corticotectal system originating in posterior cortex. Inhibitory connections, however, must also be prominent. The fact is that with each eye movement many new stimuli impinge on the retina, only one of which can become the target for the next eye movement. Therefore safeguards have to be built in to prevent the system from going haywire. This is accomplished by inhibitory networks which keep the system clamped down and need to be disinhibited to make eye movement initiation possible. Evidence for such a system has been presented recently by Hikosaka and Wurtz (1983a, 1983b, 1983c, 1983d). They have shown that the substantia nigra, which receives much of its input from the basal ganglia, makes inhibitory connections with the superior colliculus. Cells in the nigra are very active and become less so under certain conditions when visual stimuli are presented or when saccades occur. For the colliculus to generate a signal for eye movement, therefore, the nigrocollicular system, which in the absence of specific input appears to produce tonic inhibition, needs to be disinhibited. Disinhibition in itself is probably insufficient for saccade generation, and a properly synchronized excitatory signal is likely to be required.

In summary, the model presented here, as depicted in Fig. 1, has two channels for the generation of saccadic eye movements, one involving the superior colliculus and the other the frontal eye fields. The visual input to the collicular pathway is controlled by the broad-band channel, while the visual input to the frontal eye field pathway is dominated by the color-opponent channel. Eye movement generation systems are clamped down by inhibition to prevent stimulus-bound eye movement generation. Eye movements are produced by virtue of both disinhibitory and excitatory signals. The collicular pathway is involved in producing rapid, reflex-like saccades to easily discriminable stimuli. The frontal eye field system generates eye movements under more complex conditions, necessitating extensive pattern analysis and temporal sequencing. The essential gateway to all these structures for the generation of visually guided eye movements is area 17, which controls the processing of visual information to other centers for both the color-opponent and broad-band channels.

REFERENCES

Albano, J. E., and Wurtz, R. H. (1982). Deficits in eye position following ablation of monkey superior colliculus, pretectum, and posterior-medial thalamus. *J. Neurophysiol.*, **48**, 318–337.

Butter, C. M. (1983). The role of polysensory neuronal structures in spatially-directed attention and orienting responses to external stimuli. *Attention: Neural Mechanisms, Models and Clinical Application* (Eds. D. Shear and K. H. Pribram) Academic Press.

Butter, C. M., Kurtz, D., Leiby III, C. C., and Campbell Jr., A (1982). Contrasting behavioral methods in the analysis of vision in monkeys with lesions of the striate cortex or the superior colliculus. In *Analysis of Visual Behavior* (Eds. D. J. Ingle, M. A. Goodale and R. J. W. Mansfield), MIT Press, Cambridge, Mass.

Cleland, B. G., and Levick, W. R. (1974). Brisk and sluggish concentrically organized ganglion cells in the cat's retina. *J. Physiol.*, **240**, 421–446.

Cleland, B. G., Levick, W. R., and Sanderson, K. J. (1973). Properties of sustained and transient ganglion cells in the cat retina. *J. Physiol.*, **228**, 649–680.

Cynader, M., and Berman, N. (1972). Receptive-field organization of monkey superior colliculus. *J. Neurophysiol.*, **35**, 187–201.

Dreher, B., Fukuda, Y., and Rodieck, R. W. (1976). Identification, classification and anatomical segregation of cells with X-like and Y-like properties in the LGN of old world primates. *J. Physiol.*, **258**, 433–452.

Enroth-Cugell, C., and Robson, J. G. (1966). The contrast sensitivity of retinal ganglion cells of the cat. *J. Physiol.*, **187**, 517–552.

Fischer, B., and Boch, R. (1983). Saccadic eye movements after extremely short reaction times in the monkey. *Brain Res.*, **260**, 21–26.

Gouras, P. (1969). Antidromic responses of orthodromically identified ganglion cells in monkey retina. *J. Physiol.*, **204**, 407–419.

Hikosaka, O., and Wurtz, R. H. (1983a). Visual and oculomotor functions of monkey substantia nigra pars reticulata. I. Relation of visual and auditory responses to saccades. *J. Neurophysiol.*, **49**, 1230–1253.

Hikosaka, O., and Wurtz, R. H. (1983b). Visual and oculomotor functions of monkey substantia nigra pars reticulata. II. Visual responses related to fixation of gaze. *J. Neurophysiol.*, **49**, 1254–1267.

Hikosaka, O., and Wurtz, R. H. (1983c). Visual and oculomotor functions of monkey substantia nigra pars reticulata. III. Memory-contingent visual and saccade responses. *J. Neurophysiol.*, **49**, 1268–1284.

Hikosaka, O., and Wurtz, R. H. (1983d). Visual and oculomotor functions of monkey substantia nigra pars reticulata. IV. Relation of substantia nigra to superior colliculus. *J. Neurophysiol.*, **49**, 1285–1301.

Hoffman, K. P. (1973). Conduction velocity in pathways from retina to superior colliculus in the cat: a correlation with receptive-field properties. *J. Neurophysiol.*, **36**, 409–424.

Hoffman, K. P., and Sherman, S. M. (1974). Effects of early monocular deprivation on visual input to cat superior colliculus. *J. Neurophysiol.*, **37**, 1276–1286.

Hubel, D. H., and Wiesel, T. N. (1972). Laminar and columnar distribution of geniculo-cortical fibers in the macaque monkey. *J. Comp. Neurol.*, **146**, 421–450.

Keating, E. G. (1983). Removing the superior colliculus silences eye movements normally evoked from parietal and occipital eye fields. *Brain Res.*, **269**, 145–148.

LeVay, S., and Gilbert, C. D. (1976). Laminar patterns of geniculocortical projection in the cat. *Brain Res.*, **113**, 1–19.

Malpeli, J. G. (1983). Activity of cells in area 17 of the cat in absence of input from layer A of the lateral geniculate nucleus. *J. Neurophysiol*, **49**, 595–610.

Malpeli, J. G., Schiller, P. H., and Colby, C. L. (1981). Response properties of single cells in monkey striate cortex during reversible inactivation of individual geniculate laminae. *J. Neurophysiol.*, **46**, 1102–1119.

Robinson, D. A. (1972). Eye movement by collicular stimulation in the alert monkey. *Vision Res.*, **12**, 1795–1808.

Robinson, D. A., and Fuchs, A. F. (1969). Eye movements evoked by stimulation of frontal eye fields. *J. Neurophysiol.*, **32**, 637–648.

Schiller, P. H. (1977). The effect of superior colliculus ablation on saccades elicited by cortical stimulation. *Brain Res.*, **122**, 154–156.

Schiller, P. H. (1984). Parallel channels in vision and visually guided eye movements. *Assoc. Res. Vision and Ophthalmol. Abstr.*, **1984**.
Schiller, P. H., Finlay, B. L., and Volman, S. F. (1976). Quantitative studies of single-cell properties in monkey striate cortex. *J. Neurophysiol.*, **39**, 1288–1333.
Schiller, P. H., and Koerner, F. (1971). Discharge characteristics of single units in the superior colliculus of the alert rhesus monkey. *J. Neurophysiol.*, **34**, 920–936.
Schiller, P. H., and Malpeli, J. G. (1977). Properties and tectal projections of the monkey retinal ganglion cells. *J. Neurophysiol.*, **40**, 428–445.
Schiller, P. H., Malpeli, J. G., and Schein, S. J. (1979). Composition of geniculostriate input to superior colliculus of the rhesus monkey. *J. Neurophysiol.*, **42**, 1124–1133.
Schiller, P. H., and Sandell, J. H. (1983). Interactions between visually and electrically elicited saccades before and after superior colliculus and frontal eye field ablation in the rhesus monkey. *Exp. Brain Res.*, **49**, 381–392.
Schiller, P. H., and Stryker, M. (1972). Single unit recording and stimulation in the superior colliculus of the alert rhesus monkey. *J. Neurophysiol.*, **35**, 915–924.
Schiller, P. H., Stryker, M., Cynader, M., and Berman, N. (1974). The response characteristics of single cells in the monkey superior colliculus following ablation or cooling of visual cortex. *J. Neurophysiol.*, **35**, 181–194.
Schiller, P. H., True, S., and Conway, J. (1979). Paired stimulation of the frontal eye fields and the superior colliculus of the rhesus monkey. *Brain Res.*, **79**, 162–164.
Schiller, P. H., True, S. D., and Conway J. L. (1980). Deficits in eye movements following frontal eye-field and superior colliculus ablations. *J. Neurophysiol.*, **44**, 175–1189.
Sparks, D. L., and Pollack, J. G. (1977). The neural control of saccadic eye movements: the role of the superior colliculus. In *Eye Movements* (Ed. B. A. Brooks and F. J. Bajandas), Plenum Press, New York.
Stone, J., and Fukuda, Y. (1974). Properties of cat retinal ganglion cells: a comparison of W-cells with X- and Y-cells. *J. Neurophysiol.*, **37**, 7222–748.
Van Essen, D. C. (1979). Visual areas of the mammalian cerebral cortex. *Ann. Rev. Neurosci.*, **2**, 227–263.
Wurtz, R. H., and Goldberg, M. E. (1972). Activity of superior colliculus in behaving monkey: III. Cells discharging before eye movements. *J. Neurophysiol.*, **35**, 575–586.

Models of the Visual Cortex
Edited by D. Rose and V. G. Dobson

CHAPTER 8

Parallel W-, X- and Y-cell pathways in the cat: a model for visual function

S. Murray Sherman
Department of Neurobiology and Behavior, State University of New York at Stony Brook, Stony Brook, NY 11794, USA

INTRODUCTION

The model of functional significance presented here is based on studies of cats and deemphasizes the role of the striate cortex in basic-form vision. The gist of the hypothesis is outlined below; it has been described in more detail elsewhere (Sherman, 1985). It is organized around the W-, X- and Y-cell pathways (for reviews and other terminology, see Rodieck, 1979; Stone, Dreher and Leventhal, 1979; Lennie, 1980; Sherman and Spear, 1982; Sherman, 1985). These are three major, parallel pathways that start in the retina, are integrated through the lateral geniculate nucleus and innervate the various areas of the visual cortex. Figure lc,d shows that the 'visual cortex' receiving direct geniculate afferents represents some 10 to 20 separate, retinotopically mapped regions in addition to the striate cortex (Tusa, 1982; Raczkowski and Rosenquist, 1983).

It is clear that these W-, X- and Y-cell pathways are fairly independent of one another through the lateral geniculate nucleus (Cleland, Dubin and Levick, 1971; Hoffmann, Stone and Sherman, 1972; Wilson, Rowe and Stone 1976), and cells of the striate cortex seem also to belong primarily to just one of these pathways (Bullier and Henry, 1979a, 1970b, 1979c; Tanaka, 1983a, 1983b). Any such division for the extrastriate visual areas is presently unclear. In principle, any cortical cell's innervation can be traced peripherally to one or more of these pathways. Thus the use of the term 'W-cell pathway' includes all neurons having innervation that can be traced to retinal W-cells, and likewise for the X- and Y-cell pathways. Because we cannot at present delineate these pathways through all areas of cortex, the meaning of the

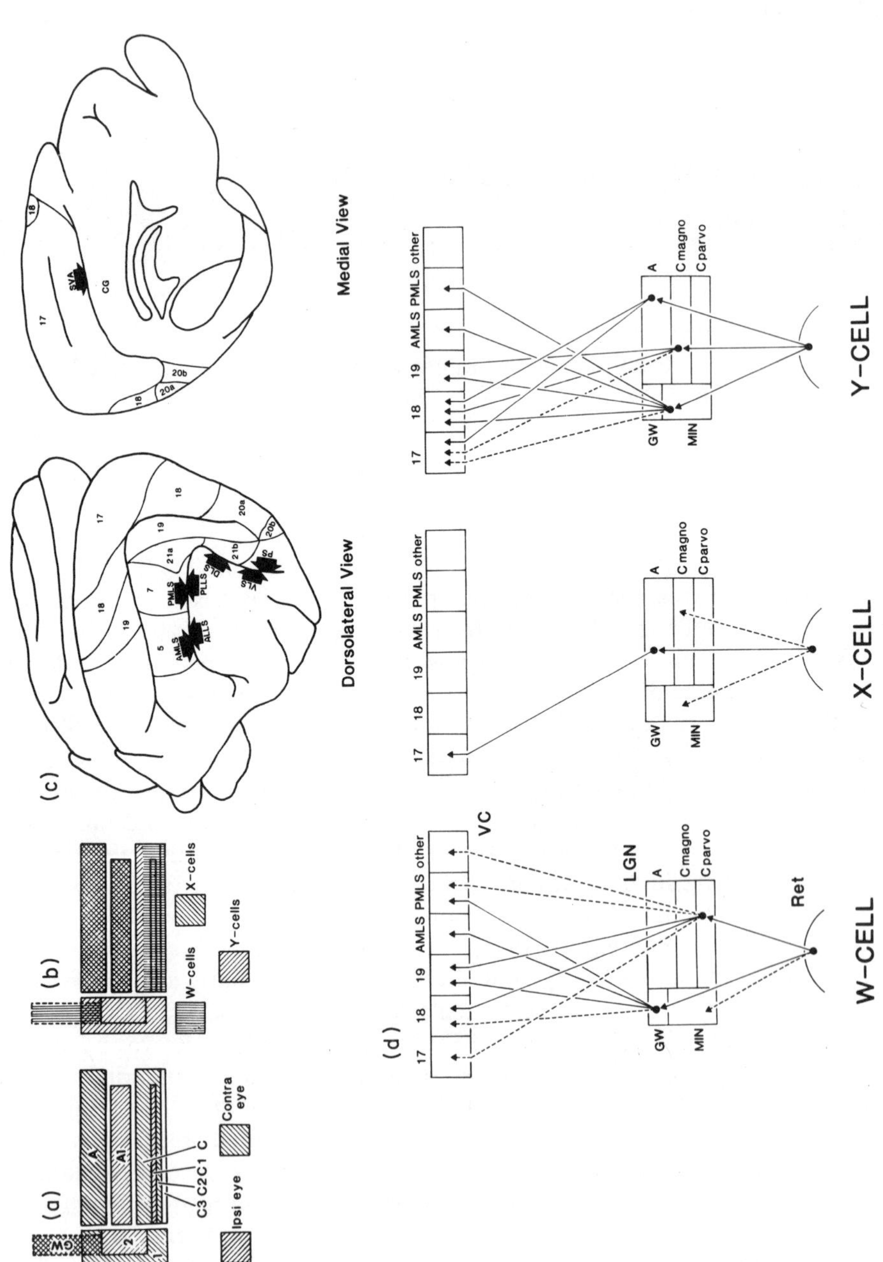
(a)
GW
2
1
A
A1
C
C3 C2 C1
Ipsi eye
Contra eye
(b)
W-cells
X-cells
Y-cells
(c)
Dorsolateral View
Medial View
17
18
19
20a
20b
21a
21b
5
7
AMLS
ALLS
PMLS
PLLS
DLS
VLS
PS
SVA
CG
(d)
17 18 19 AMLS PMLS other
VC
LGN
GW
MIN
A
C magno
C parvo
Ret
W-CELL
X-CELL
Y-CELL

FIGURE 1 Schematic representation of the W-, X- and Y-cell pathways among the divisions of the lateral geniculate nucleus and visual cortex. (a) Laminar arrangements of the right lateral geniculate nucleus according to ocular input as schematically shown in a coronal view. Illustrated are laminae A, A1, C, C1, C2, C3, the medial interlaminar nucleus (1 and 2; lamina 3, which is innervated by the contralateral temporal retina, is not shown because it occupies a more rostral position) and the geniculate wing (GW). (b) Same view of the nucleus as in (a) but showing the main locations of W-, X- and Y-cells. Not shown are the possible presence of W-cells in the medial interlaminar nucleus and rare X-cells in the medial interlaminar nucleus and C-laminae. (c) Visual cortical areas in the cat as shown in dorsolateral (left) and medial (right) views of the left hemisphere. In addition to the nine areas designated by Brodmann numbers (5, 7, 17, 18, 19, 20a, 20b, 21a, 21b) are nine additional areas (AMLS, PMLS, VLS, ALLS, PLLS, DLS, CG, PS, SVA). The abbreviations are: AMLS, anterior medial lateral suprasylvian area; PLMS, posterior medial lateral suprasylvian area; VLS, ventral lateral suprasylvian area; ALLS, anterior lateral lateral suprasylvian area; PLLS, posterior lateral lateral suprasylvian area; DLS, dorsal lateral suprasylvian area; CG, cingulate gyrus; PS, posterior suprasylvian area; SVA, splenial visual area. Thirteen of these areas (17, 18, 19, 20a, 20b, 21a, 21b, AMLS, PMLS, VLS, ALLS, PLLS and DLS) seem to be purely visual and exhibit retinotopic organization. The remaining five (5, 7, CG, PS, SVA) have some visual neurons but may not be exclusively visual, and no retinotopic organization has yet been demonstrated for any of them. (Redrawn from Tusa, 1982; Raczkowski and Rosenquist, 1983). (d) Schematic summary diagram of W-, X- and Y-cell pathways from the retina through the lateral geniculate nucleus to various areas of visual cortex as shown in (c). Abbreviations: Ret, retina; LGN, lateral geniculate nucleus; A, A-laminae; Cmagno, magnocellular lamina C; Cparvo, parvocellular C-laminae; GW, geniculate wing; MIN, medial interlaminar nucleus; VC, visual cortex; 17, area 17; 18, area 18; 19, area 19; AMLS and PMLS, as noted in (c); other, areas 20a, 20b, 21a and 21b plus ALLS, VLS, PLLS and DLS. Solid lines represent relatively dense projections and dashed lines represent relatively sparse projections. See Sherman (1985) for details of the derivation of these connections.

terms for W-, X- and Y-cell pathways is necessarily vague beyond the lateral geniculate nucleus.

It is useful to consider the functional significance of each of these parallel W-, X- and Y-cell pathways rather than the more traditional, hierarchical approach that ascribes functional roles for the lateral geniculate nucleus, striate cortex, etc. The hypothesis presented below suggests that basic form vision is largely a function of the Y-cell pathway, particularly of its extra-striate zones of innervation, whereas the X-cell pathway, including the striate cortex, is concerned chiefly with other functions, such as raising spatial resolution from that provided by the Y-cell pathway, providing for stereopsis, etc. No specific role can yet be suggested for the W-cell pathway (however, see Stone, Dreher and Leventhal, 1979).

RESPONSE PROPERTIES OF W-, X- and Y-CELLS

With few exceptions, each geniculate W-, X- and Y-cell seems to receive its innervation from a few retinal ganglion cells of the same class, and the response properties of these geniculate neurons are essentially the same as those of their retinal counterparts (Cleland, Dubin and Levick, 1971; Kaplan, Marcus and So, 1979). Details of these properties are beyond the scope of this chapter and they have already been recently reviewed (Rowe and Stone, 1977; Rodieck, 1979; Stone, Dreher and Leventhal, 1979; Lennie, 1980; Sherman and Spear, 1982; Sherman, 1985). For the purposes of the theoretical framework constructed here, only certain response properties need be elaborated. W-cells are so poorly responsive to visual stimuli (Sur and Sherman, 1982a; Thibos and Levick, 1983) that basic analysis of the visual scene most plausibly falls to the X- and Y-cell pathways. A major difference between X- and Y-cells is that the former are more responsive to higher spatial frequencies whereas the latter respond better to lower ones (Lehmkuhle *et al.*, 1980; Thibos and Levick, 1983; Troy, 1983). Thus the Y-cell pathway is likely to be most concerned with analysis of the lower spatial frequencies in the visual scene, and analysis of fine spatial details is left to the X-cell pathway. This, of course, is speculation based on response properties of neurons in anesthetized, paralyzed cats.

MORPHOLOGICAL PROPERTIES OF W-, X- AND Y-CELLS

In both the retina and the lateral geniculate nucleus, W-, X- and Y-cells are morphologically distinct from one another (for the retina: Wassle, 1982; for the lateral geniculate nucleus: Friedlander *et al.*, 1981; Stanford, Friedlander and Sherman, 1983). The details and implications of these structure/function relationships can be found elsewhere (Friedlander *et al.*, 1981; Wassle, 1982; Stanford, Friedlander and Sherman, 1983). These relationships permit the derivation by anatomical criteria of the numbers and distributions of W-, X- and Y-cells, thereby avoiding the sampling problems inherent in electrophysiology (cf. Friedlander *et al.*, 1981).

Retina

Ganglion cell density peaks near the area centralis, but this density function is steeper for X-cells than for W- or Y-cells (Fukuda and Stone, 1974; Leventhal, 1982, Wassle, 1982). While relative numbers thus depend somewhat on eccentricity, to a first approximation, X-cells represent roughly 50 to 60 per cent. of all retinal ganglion cells and Y-cells represent roughly 5 per cent. (Fukuda and Stone, 1974; Wassle, 1982; Williams and Chalupa, 1983). Every retinal X- and Y-cell projects to the lateral geniculate nucleus (Fukuda and Stone, 1974; Bowling and Michael, 1980; Illing and Wassle, 1981; Sur and Sherman, 1982b). Of the remaining retinal ganglion cells (i.e. W-cells), roughly 40 per cent. project to the lateral geniculate nucleus (Illing and Wassle, 1981; Leventhal, 1982). Thus about 15 to 20 per cent. of retinal ganglion cells participate in the W-cell pathway. By this analysis, the W- to X- to Y-cell ratio among ganglion cells that innervate the lateral geniculate nucleus is approximately 3:10:1.

Lateral geniculate nucleus

As noted in Fig. 1a,b, the cat's lateral geniculate nucleus is a laminated structure, and the distribution of ocular input as well as of W-, X- and Y-cells varies with lamination (Hickey and Guillery, 1974; Wilson, Rowe and Stone, 1976; Kratz, Webb and Sherman 1978; Guillery *et al.*, 1980). X-cells are nearly exclusively limited to the A-laminae, although these cells may rarely be found in the C-laminae and medial interlaminar nucleus. Y-cells abound in the A-laminae, magnocellular lamina C (i.e. the dorsal tier of lamina C) and the medial interlaminar nucleus. W-cells are found in the parvocellular C-laminae (i.e. those ventral to magnocellular lamina C) and possibly the geniculate wing and medial interlaminar nucleus.

The X- to Y-cell ratio of the A-laminae is roughly 2:1 (LeVay and Ferster, 1977; Friedlander *et al.*, 1981; Leventhal, 1982; Friedlander and Stanford, 1984), and when the C-laminae and medial interlaminar nucleus are considered, the geniculate X- to Y-cell ratio probably approaches 1:1. This is dramatically different from the 10:1 ratio in the retina. Estimates of geniculate W-cell numbers are not presently available. Sur and Sherman (1982b) have described a possible substrate for this amplification of the Y-cell pathway (see also Bowling and Michael, 1980): retinal X-cell axons typically innervate only lamina A or A1 in small zones, whereas those of Y-cells branch to innervate lamina A or A1 in large zones (5 to 10 times larger than those of X-cells) plus the medial interlaminar nucleus and, if from the contralateral eye, lamina C.

Visual cortex

Figure 1d summarizes the probable projections of the geniculocortical limbs of the W-, X- and Y-cell pathways (for further details, see Raczkowski and Rosenquist, 1983; Sherman, 1984). The X-cell projection is limited to the striate cortex, while those of W- and Y-cells innervate many extrastriate regions in addition to the striate cortex.

The projections of individual X- and Y-cell axons to the striate cortex in some ways resemble the patterns of their retinogeniculate counterparts (Ferster and LeVay, 1978; Gilbert and Wiesel, 1983; Humphrey, Sur and Sherman 1982). Most X-cell axons innervate layer 4 ventrally in a single, small patch that may be limited to one ocular dominance column (e.g. Shatz, Lindstrom and Wiesel 1977); Y-cell axons innervate layer 4 dorsally in several, larger patches that may represent several ocular dominance columns of the same eye. Both axon classes also sparsely innervate layer 6. In addition, most of the Y-cell axons branch to innervate area 18 (Stone and Dreher, 1973; Geisert, 1980) in extensive zones (Humphrey, Sur and Sherman, 1982). Thus the Y-cell pathway may again be relatively amplified by geniculocortical projections. However, as noted above, it is not yet clear to what extent neurons in the various visual areas are members of one or more of these W-, X- and Y-cell pathways.

Summary of X- and Y-cell pathways

Figure 2 summarizes the organizational principles for the X- and Y-cell pathways as described above. (Insufficient data are available for the W-cell pathway for it to be included in such a schematic diagram.) The pathway beginning with X-cells, which represent the majority of retinal ganglion cells, innervates only a portion of a single area of the visual cortex (i.e. striate cortex). In contrast, the pathway beginning with Y-cells, which represent a

---

FIGURE 2 Hypothetical and schematic diagram of the retino-geniculo-cortical X- and Y-cell pathways; for simplicity, only pathways from the contralateral eye are illustrated. Each retinogeniculate and geniculocortical Y-cell axon branches to innervate many more neurons than does each of the analogous X-cell axons. Also, the X-cell pathway is essentially limited to the A-laminae and area 17, whereas the Y-cell pathway occupies most regions of the lateral geniculate nucleus and visual cortex. Consequently, a small minority of retinal ganglion cells (Y-cells) come to dominate the visual cortex whereas the much greater number of retinal ganglion cells (X-cells) come to control much less cortical tissue. For details and the functional significance of this schema, see the text.

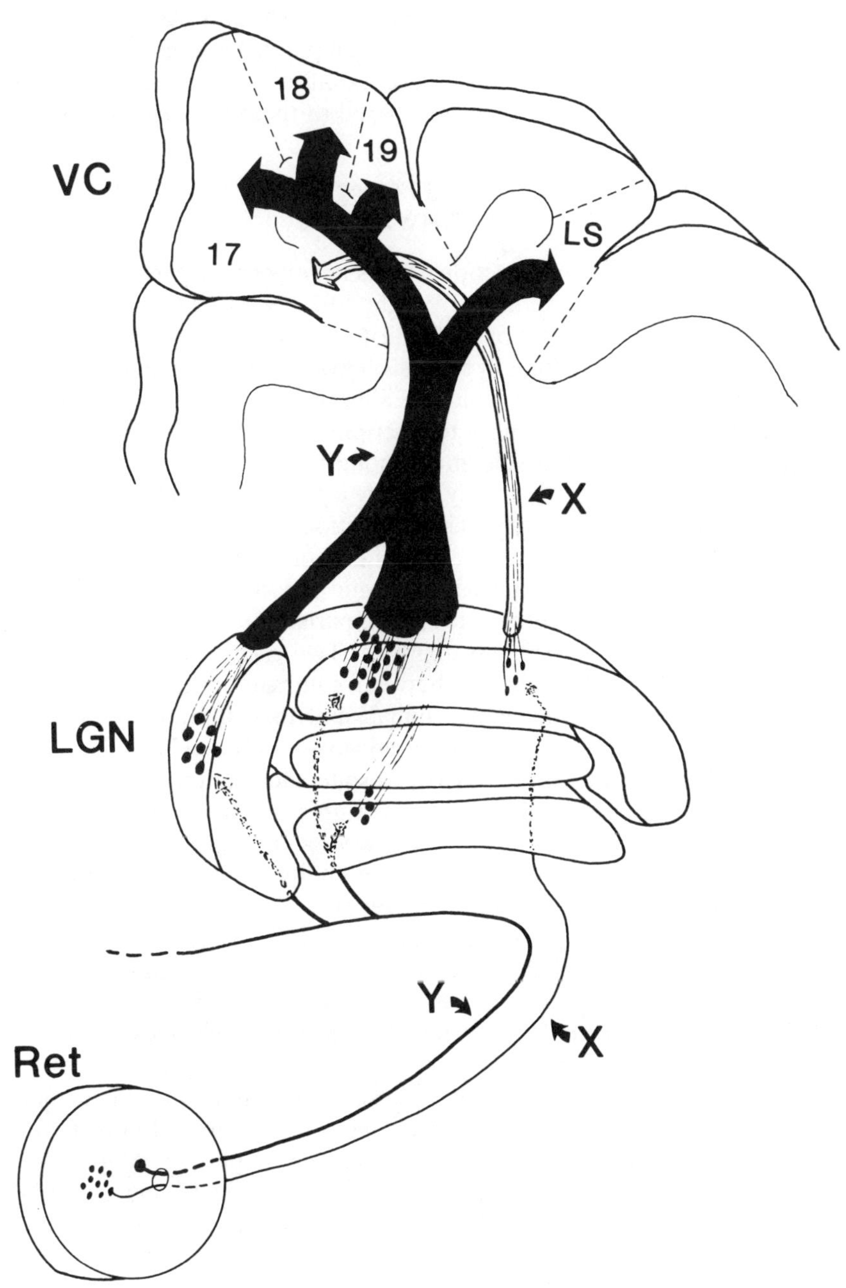
18
19
VC
LS
17
Y
X
LGN
Y
X
Ret

small minority of retinal ganglion cells, dominates geniculocortical innervation, particularly with regard to extrastriate cortical areas. The morphological basis for this relative shift in central emphasis between the X- and Y-cell pathways may be the larger, more extensively arborized innervation patterns of individual Y-cell axons compared to those of X-cell axons.

BEHAVIORAL CORRELATES FOR THE W-, X- AND Y-CELL PATHWAYS

A variety of quite different functional roles have been suggested for the W-, X- and Y-cell pathways (Ikeda and Wright, 1972; Stone, Dreher and Leventhal, 1979). However, it is most difficult to test these suggestions, because of the lack of a robust linking hypothesis between behavior and neurophysiology (for details, see the discussion in Sherman, 1985). How, for instance, does one link the animal's ability to discriminate circles from triangles with different axonal conduction velocities amongst neuronal classes?

Contrast-sensitivity functions may provide a potentially useful link between behaviour and neurophysiology. These functions can be determined psychophysically for the behaving animal and neurophysiologically for individual neurons. Comparisons between behavior and physiology are thus *relatively* straightforward. Interpretation of these comparisons is nonetheless risky and requires a number of dubious assumptions. Two such assumptions are that a straightforward relationship exists between contrast sensitivity of individual neurons and that of the behaving animal, and that different physiological conditions (i.e. anesthetized and paralyzed versus alert and behaving) do not significantly affect contrast sensitivity. Despite such serious provisos, some useful insights can be gained by these measures of contrast sensitivity. For simplicity, only spatial functions at low temporal rates of modulation (1 to 2 Hz) are considered (for other temporal parameters, see Blake and Camisa, 1977; Lehmkuhle, Kratz and Sherman, 1982).

Sprague and his colleagues (Sprague *et al.*, 1977; Berkley and Sprague, 1979) have shown that bilateral removal of striate cortex has surprisingly little effect on the cat's visual capacity beyond a 25 to 50 per cent. acuity loss. Lehmkuhle, Kratz and Sherman (1982) recently confirmed and extended this with psychophysical measures of spatial contrast sensitivity. The sensitivity losses attributed to striate cortex lesions were limited to higher spatial frequencies, and no other postoperative deficits were noted. However, Kaye, Mitchell and Cynader (1981) report that binocular depth perception may be virtually lost to cats with bilateral lesions of areas 17, 18 and part of 19, despite the small losses of visual acuity and basic form vision suffered by these cats. Much larger cortical lesions, which significantly encroach into the extrastriate areas of the visual cortex, must be made before the animal's visual performance is clearly impaired beyond a simple loss of acuity (Sprague *et al.*, 1977).

These data are most interesting in the context of Fig. 1d. An ablation limited to the striate cortex removes the X-cell pathway from cortical representation, while much of the W- and Y-cell pathways remain. It thus follows that neither the X-cell pathway nor the striate cortex is necessary for reasonable form vision, although these may be necessary for maximum visual acuity and stereopsis. It seems likely that the W-cell pathway, because of its poor responsiveness, is not responsible for the remaining visual capacity. While this disregard for W-cell function may well prove incorrect, and many important functions not yet tested in destriate cats may appear in future studies, the tentative conclusion is that the remaining Y-cell pathway is sufficient for excellent form vision. Therefore, only after sufficiently large cortical lesions that destroy the bulk of the Y-cell inputs will severe visual losses in basic form vision result.

Similar psychophysical data have also been obtained from cats reared from birth with deprivation of visual forms by eyelid suture (monocular or binocular) or dark rearing. Such cats develop serious abnormalities in their Y-cell pathway. (Sherman and Spear, 1982; Friedlander, Stanford and Sherman, 1982; Sur, Humphrey and Sherman, 1982). Recent behavioral studies of their contrast sensitivity offer interesting correlates between their ambylopia and Y-cell deficits (Blake and DiGianfillipo, 1980; Lehmkuhle, Kratz and Sherman, 1982). The visually deprived cats suffer sensitivity losses at all spatial frequencies. Thus their ambylopia is not simply an acuity loss, and indeed their sensitivity losses are greater for lower than for higher spatial frequencies. This seems to relate well to the abnormalities of the Y-cell pathway, since Y-cells are especially sensitive to the lower spatial frequencies.

Two further points can be drawn from the psychophysical data. First, spatial resolution alone may not be as good an index of visual performance as is sensitivity to lower spatial frequencies. Indeed, binocularly deprived cats have slightly better resolution than do the normally reared, destriate cats (Lehmkuhle, Kratz and Sherman, 1982), but the deprived cats exhibit a clearly inferior visual performance. Second, the observation that normally reared, destriate cats see better than do visually deprived cats implies that deficits limited to the striate cortex, no matter how severe, cannot account for the ambylopia of these visually deprived cats. A more parsimonious explanation for their ambylopia relates to the deficits among their geniculate Y-cells: as can be appreciated from Fig. 1d and 2, deficits here will produce widespread effects in many areas of the visual cortex.

COMPARISON BETWEEN CATS AND MONKEYS

These studies of cats place greater emphasis for the neural substrates of pattern vision on the extrastriate Y-cell pathways and less on the X-cell pathway and striate cortex than is usually the case. One may wonder how this conclusion relates to monkeys, especially since striate cortex lesions render monkeys practically blind (cf. Weiskrantz, 1972; Miller, Pasik and Pasik, 1980). A detailed discussion of the monkey's retino-geniculo-cortical pathways is inappropriate here, but certain points are briefly considered below.

As is the case in cats, the monkey's retino-geniculo-cortical pathways are organized into several, parallel neuronal streams (Hubel and Wiesel, 1972; Dreher, Fukuda and Rodieck, 1976; Bullier and Henry, 1980; Leventhal, Rodieck and Dreher, 1981; Blasdel and Lund, 1983; Weber *et al.*, 1983). Although many reports have emphasized basic similarities between these pathways in the monkey and the W-, X- and Y-cell pathways in the cat, the details of any such homology or analogy between species is a matter of debate (see Sherman, 1985).

It is clear in any case that the vast majority of geniculate neurons in the monkey project exclusively to the striate cortex (Tigges, Tigges and Perachio, 1977; see also Wong-Riley, 1976; Yukie and Iwai, 1981), a condition quite unlike that in the cat, for which an extensive extrastriate projection from the lateral geniculate nucleus exists. A striate cortex lesion in monkeys thus destroys nearly all of the geniculocortical input, including practically the total of any presumed Y-like pathway. As noted above, such monkeys are nearly blind without the striate cortex, while destriate cats see well. Nearly all of the geniculocortical projection must be destroyed in cats to produce an animal comparably blind to the destriate monkey. Perhaps the key to understanding the consequences of striate cortex lesions among mammalian species has less to do with striate cortex *per se* and more to do with the extent of damage to the geniculocortical innervation, and particularly to the Y-cell pathway in cats and its counterparts in other species.

CONCLUSIONS AND HYPOTHESIS

From the psychophysical studies of cats cited above, it follows that sensitivity to lower spatial frequencies may be sufficient and necessary for reasonable pattern vision. Recent psychophysical studies of human vision tend to support this: basic form information seems to be carried by the lower spatial frequencies, whereas the higher ones add detail and maximize resolution (Kabrisky *et al.*, 1970; Ginsburg, 1978; Hess and Woo, 1978).

A consideration of the spatial response properties of W-, X- and Y-cells, the visual performance of visually deprived and destriate cats and the projection patterns of the W-, X- and Y-cell pathways (Figs. 1d and 2) lead to the following hypothesis. The Y-cell pathway, with its unique sensitivity to the lower spatial frequencies, is the neural substrate for basic spatial vision. A cat with good sensitivity to lower spatial frequencies and a reasonably intact

Y-cell pathway (e.g. a normally reared cat with or without the striate cortex) displays reasonable spatial vision, whereas a cat with poor low frequency sensitivity and a deficient Y-cell pathway (e.g. a monocularly or binocularly deprived cat) exhibits serious amblyopia. The X-cell pathway (including the striate cortex), with its superior sensitivity to higher spatial frequencies, adds detail and raises acuity to the basic analysis performed by the Y-cell pathway; the X-cell pathway and striate cortex may also be essential to stereopsis. No specific role can yet be suggested for the W-cell pathway, but due to its relative unresponsiveness, it may play some unspecified but minor role in conscious visual perception (cf. Stone, Dreher and Leventhal, 1979). The above must be recognized for the speculative and incomplete hypothesis that it is.

This hypothesis can now be considered with respect to the central connections of X- and Y-cells (Fig. 2). Few Y-cells exist in retina, perhaps because the encoding of lower spatial frequencies requires relatively few peripheral elements. However, the primacy to spatial vision of these lower frequencies requires that considerable cerebral cortex be devoted to their analysis. Thus the Y-cell pathway is amplified centrally with widespread axonal connections, resulting in its relative dominance at the cortical level. The role and organization of the X-cell pathway is complementary. To extract the finest spatial detail from a visual scene requires a large number of peripheral elements to encode the higher spatial frequencies. X-cells thus abound in retina. However, the presumed secondary importance of these higher spatial frequencies to basic form vision results in an X-cell pathway that has narrowly distributed axonal connections and is channelled through a portion of a single area of the visual cortex.

Finally, this hypothesis, by playing down the role of the X-cell pathway in basic-form vision, also plays down this role for the striate cortex. For cats, this follows logically from the good spatial vision still evident after destruction of the striate cortex. Since destriate monkeys are practically blind, this secondary role for the striate cortex may seem inappropriate. However, the results of striate cortex lesions in monkeys may have more to do with a near-total destruction of geniculocortical input, including all of any monkey counterpart to the cat's Y-cell pathway, than with the loss of the striate cortex *per se*.

ACKNOWLEDGMENTS

The author's research has been supported by grants EY03038 and EYO3604 from the United States Public Health Service. The author thanks P. Palam for typing the manuscript.

REFERENCES

Berkley, M. A., and Sprague, J. M. (1979). Striate cortex and visual acuity functions in the cat. *J. comp. Neurol.*, **187**, 679–702.

Blake, R., and Camisa J. J. (1977). Temporal aspects of spatial vision in the cat. *Exp. Brain Res.*, **28**, 325–333.

Blake, R., and DiGianfillipo A. (1980). Spatial vision in cats with selective neural deficits. *J. Neurophysiol.*, **43**, 1197–1205.

Blasdel, G. G., and Lund, J. S. (1983). Terminations of afferent axons in macaque striate cortex. *J. Neurosci.*, **3**, 1389–1413.

Bowling, D. B., and Michael, R. (1980). Projection patterns of single physiologically characterized optic tract fibers in the cat. *Nature (Lond.)*, **286**, 899–902.

Bullier, J., and Henry, G. H. (1979a). Ordinal position of neurons in cat striate cortex. *J. Neurophysiol.* 42, 1251–1263.

Bullier, J., and Henry, G. H. (1979b). Neural path taken by afferent streams in striate cortex of the cat. *J. Neurophysiol.*, **42**, 1264–1270.

Bullier, J., and Henry, G. H. (1979c). Laminar distribution of first-order neurons and afferent terminals in cat striate cortex. *J. Neurophysiol.*, **42**, 1271–1281.

Bullier, J., and Henry G. H. (1980). Ordinal position and afferent input of neurons in monkey striate cortex. *J. comp. Neurol.*, **193**, 913–935.

Cleland, B. G., Dubin, M. W., and Levick, W. R. (1971). Sustained and transient neurons in the cat's retina and lateral geniculate nucleus. *J. Physiol.*, **217**, 473–496.

Dreher, B., Fukuda, Y., and Rodieck, R. W. (1976). Identification, classification and anatomical segregation of cells with X-like and Y-like properties in the lateral geniculate nucleus of old-world primates. *J. Physiol. (Lond.)*, **258**, 433–452.

Ferster, D., and LeVay S. (1978). The axonal arborizations of lateral geniculate neurons in the striate cortex of the cat. *J. comp. Neurol.*, **182**, 923–944.

Friedlander, M. J., Lin, C.-S., Stanford, L. R., and Sherman, S. M. (1981). Morphology of functionally indentified neurons in the lateral geniculate nucleus of the cat. *J. Neurophysiol.* , 46, 80–129.

Friedlander, M. J., and Stanford, L. R. (1984). The effects of monocular deprivation on the distribution of cell types in the LGNd—A sampling study with fine-tipped micropipettes. *Exp. Brain Res.*, **53**, 451–461.

Friedlander, M. J., Stanford, L. R., and Sherman, S. M. (1982). Effects of monocular deprivation on the structure/function relationship of individual neurons in the cat's lateral geniculate nucleus. *J. Neurosci.*, **2**, 321–330.

Fukuda, Y., and Stone, J. (1974). Retinal distribution and central projections of Y-, X-, and W-cells of the cat's retina. *J. Neurophysiol.*, **37**, 749–772.

Geisert, E. E. (1980). Cortical projections of the lateral geniculate nucleus in the cat. *J. comp. Neurol.*, **190**, 793–812.

Gilbert, C. D., and Wiesel, T. N. (1983). Clustered intrinsic connections in cat visual cortex. *J. Neurosci.*, **3**, 1116–1133.

Ginsburg, A. (1978). Visual information processing based upon spatial filters constrained by biological data. Ph.D. thesis, University of Cambridge.

Guillery, R. W., Geisert, E. E., Polley, E. H., and Mason, C. A. (1980). An analysis of the retinal afferents to the cat's medial interlaminar nucleus and to its rostral thalamic extension, the 'geniculate wing'. *J. comp. Neurol.*, **194**, 117–142.

Hess, R., and Woo, G. (1978). Vision through cataracts. *Invest. Ophthalmol. Visual Sci.*, **17**, 428–435.

Hickey, T. L., and Guillery, R. W. (1974). An autoradiographic study of retinogeniculate pathways in the cat and the fox. *J. comp. Neurol.*, **156**, 239–254.

Hoffmann, K.-P., Stone, J., and Sherman, S. M. (1972). Relay of receptive field properties in dorsal lateral geniculate nucleus of the cat. *J. Neurophysiol.*, **35**, 518–531.
Hubel, D. H., and Wiesel, T. N. (1972). Laminar and columnar distribution of geniculo-cortical fibers in the macaque monkey. *J. comp. Neurol.*, **146**, 421–450.
Humphrey, A. L., Sur, M., and Sherman, S. M. (1982). Cortical axon terminal arborization and soma location of single, functionally identified lateral geniculate nucleus neurons. *Soc. Neurosci. Abstr.*, **8**, 2.
Ikeda, H., and Wright, M. J. (1972). Receptive field organization of 'sustained' and 'transient' retinal ganglion cells which subserve different functional roles. *J. Physiol. (Lond.)*, **227**, 769–800.
Illing, R.-B., and Wassle, H. (1981). The retinal projection to the thalamus in the cat: a quantitative investigation and a comparison with the retinotectal pathway. *J. comp. Neurol.*, **202**, 265–285.
Kabrisky, M., Tallman, O., Day, C. M., and Radoy, C. M. (1970). A theory of pattern perception based on human physiology. *Ergonomics*, **13**, 129–142.
Kaplan, E., Marcus, S., and So, Y. T. (1979). Effects of dark adaptation on spatial and temporal properties of receptive fields in cat lateral geniculate nucleus. *J. Physiol.* (Lond.), **294**, 561–580.
Kaye, M., Mitchell, D. E., and Cynader, M. (1981). Selective loss of binocular depth perception after ablation of cat visual cortex. *Nature*, **293**, 60–62.
Kratz, K. E., Webb, S. V., and Sherman, S. M. (1978). Studies of the cat's medial interlaminar nucleus: a subdivision of the dorsal lateral geniculate nucleus. *J. comp. Neurol.*, **180**, 601–614.
Lehmkuhle, S., Kratz, K. E., Mangel, S. C., and Sherman S. M. (1980). Spatial and temporal sensitivity of X- and Y-cells in the dorsal lateral geniculate nucleus of the cat. *J. Neurophysiol.*, **43**, 520–541.
Lehmkuhle, S., Kratz, K. E., and Sherman, S. M. (1982). Spatial and temporal sensitivity of normal and ambylopic cats. *J. Neurophysiol.*, **48**, 372–387.
Lennie, P. (1980). Parallel visual pathways. *Vision Res.*, **20**, 561–594.
LeVay, S., and Ferster, D., (1977). Relay cell classes in the lateral geniculate nucleus of the cat and the effects of visual deprivation. *J. comp. Neurol.*, **172**, 563–584.
Leventhal, A. G. (1982). Morphology and distribution of retinal ganglion cells projection to different layers of the dorsal lateral geniculate nucleus in normal and Siamese cats. *J. Neurosci.*, **2**, 1024–1042.
Leventhal, A. G., Rodieck, R. W., and Dreher, B. (1981). Retinal ganglion cell classes in the old-world monkey: morphology and central projections. *Science*, **213**, 1139–1142.
Miller, M., Pasik, P., and Pasik, T. (1980). Extrageniculostriate vision in the monkey. VII. Contrast sensitivity functions. *J. Neurophysiol.*, **43**, 1510–1526.
Raczkowski, D., and Rosenquist, A. C. (1983). Connections of the multiple visual cortical areas with the lateral posterior-pulvinar complex and adjacent thalamic nuclei. *J. Neurosci.*, **3**, 1912–1942.
Rodieck, R. W. (1979). Visual pathways. *Ann. Rev. Neurosci.*, **2**, 193–225.
Rowe, M. H., and Stone, J. (1977). Naming of neurons. *Brain Behav. Evol.*, **14**, 185–216.
Shatz, C. J., Lindstrom, S., and Wiesel, T. N. (1977). The distribution of afferents representing the right and left eyes in the cat's visual cortex. *Brain Res.*, **131**, 103–116.
Sherman, S. M. (1985). Functional organization of the W-, X-, and Y-cell pathways in the cat: a review and hypothesis. In *Progress in Psychobiology and Physiological*

Psychology Volume 11, (Eds. J. M. Sprague and A. N. Epstein), Academic Press, New York, pp. 233-314
Sherman, S. M., and Spear, P. D. (1982). Organization of visual pathways in normal and visually deprived cats. *Physiol.. Rev.*, **62**, 738–855.
Sprague, J. M., Levy, J. DiBerardino, A., and Berlucchi, G. (1977). Visual cortical areas mediating form discrimination in the cat. *J. comp. Neurol.*, **172**, 441–488.
Stanford, L. R., Friedlander, M. J., and Sherman, S. M. (1983). Morphological and physiological properties of geniculate W-cells: a comparison with X- and Y-cells. *J. Neurophysiol.*, **50**, 582–608.
Stone, J., and Dreher, B. (1973). Projection of X- and Y-cells of the cat's lateral geniculate nucleus to areas 17 and 18 of visual cortex. *J. Neurophysiol.*, **36**, 551–567.
Stone, J., Dreher, B., and Leventhal, A. (1979). Hierarchical and parallel mechanisms in the organization of visual cortex. *Brain Res. Rev.*, **1**, 345–394.
Sur, M., Humphrey, A. L., and Sherman, S. M. (1982). Monocular deprivation affects X- and Y-cell retinogeniculate terminations in cats. *Nature*, **300**, 183–185.
Sur, M., and Sherman, S. M. (1982a). Linear and nonlinear W-cells in the C-laminae of the cat's lateral geniculate nucleus. *J. Neurophysiol.*, **47**, 869–884.
Sur, M., and Sherman, S. M. (1982b). Retinogeniculate terminations in cats: morphological differences between X and Y cell axons. *Science*, **218**, 389–391.
Tanaka, K. (1983a). Distinct X- and Y-streams in the cat visual cortex revealed by bicuculine application. *Brain Res.*, **265**, 143–147.
Tanaka, K. (1983b). Cross-correlation analysis of geniculostriate neuronal relationships in cats. *J. Neurophysiol.*, **49**, 1303–1318.
Thibos, L. N., and Levick, W. R. (1983). Spatial frequency characteristics of brisk and sluggish ganglion cells of the cat's retina. *Exp. Brain Res.*, **51**, 16–22.
Tigges, J., Tigges, M., and Perachio, A. A. (1977). Complementary laminar terminations of afferents to area 17 originating in area 18 and in the lateral geniculate nucleus in squirrel monkey. *J. comp. Neurol.*, **176**, 87–100.
Troy, J. B. (1983). Spatial contrast sensitivities of X and Y type neurons in the cat's dorsal lateral geniculate nucleus. *J. Physiol. (Lond.)*, **344**, 399–417.
Tusa, R. J. (1982). Visual cortex: multiple areas and multiple functions. In *Changing Concepts of the Nervous System* (Eds. A. R. Morrison and P. L. Strick), Academic Press, New York, pp. 235–259.
Wassle, H. (1982). Morphological types and central projections of ganglion cells in the cat retina. In *Progress in Retinal Research* (Eds. N. Osborne and G. Chader, Pergammon Press, pp. 125–152.
Weber, J. T., Huerta, M. F., Kaas, J. H., and Harting, J. K. (1983). The projections of the lateral geniculate nucleus of the squirrel monkey: studies of the interlaminar zones and the S layers. *J. comp. Neurol.*, **213**, 135–145.
Weiskrantz, L. (1972). Behavioural analysis of the monkey's visual nervous system. *Proc. Roy. Soc. Lond. B*, **182**, 427–455.
Williams, R. W., and Chalupa, L. M. (1983). An analysis of axon caliber within the optic nerve of the cat: evidence of size groupings and regional organization. *J. Neurosci*, **3**, 1554–1564.
Wilson, P. D., Rowe, M. H., and Stone, J. (1976). Properties of relay cells in the cat's lateral geniculate nucleus: a comparison of W-cells with X- and Y-cells. *J. Neurophysiol.*, **39**, 1193–1209.
Wong-Riley, M. T. T. (1976). Projections from the dorsal lateral geniculate nucleus to prestriate cortex in the squirrel monkey as demonstrated by retrograde transport of horseradish peroxidase. *Brain Res.*, 595–600.
Yukie, M., and Iwai, E. (1981). Direct projection from the dorsal lateral geniculate nucleus to the prestriate cortex in macaque monkeys. *J. comp. Neurol.*, **201**, 81–97.

Models of the Visual Cortex
Edited by D. Rose and V. G. Dobson

CHAPTER 9

Local log polar frequency analysis in the striate cortex as a basis for size and orientation invariance

PATRICK CAVANAGH
Département de Psychologie, Université de Montréal, Montréal, Québec, Canada, H3C 3J7

When an object moves across the visual field its image stimulates an ever-changing array of retinal receptors. If the observer is to recognize it as the same object in spite of these variations, this variable retinal input must be transformed into a unique pattern of neural activity that defines the object. There are two current models of form encoding: (a) an abstract, structural or propositional representation of the object, typically a list of the object's features and their interrelations, or (b) an analogue representation, typically a transformation onto a new set of dimensions where form information is invariant to changes in input size, position and orientation (for a further discussion of analogue versus structural models see Sutherland, 1973; Pylyshyn, 1973; Kosslyn, 1980).

A structural or propositional representation reduces the form to be identified to a list of primitive elements and the structural relations between them. If the primitives are, for example, lines (contours) and angles, the edges must first be extracted and then the relations between them determined. Pattern classification is based on structure and the descriptions of the primitive features do not necessarily specify their size, orientation or position. As a result, this type of encoding permits representations that are, in a very simple way, invariant with respect to the size, orientation and position of the overall pattern.

In the other model, stimulus patterns are transformed into analogue representations that do not vary with the size, orientation or position of the input. Following the original work of Schade (1956), Campbell and Robson (1968) proposed that the visual system performs a Fourier analysis on the spatial attributes of the retinal input. The amplitude portion of the Fourier transform is invariant with respect to the position of the input. However, the striate cortex cannot carry out a true Fourier transform because the receptive fields at this level are restricted to small local areas and do not cover the entire visual field as would be required. Various piecewise Fourier analyses have been proposed (Pollen, Lee and Taylor, 1971; Robson, 1975 Glezer and Cooperman, 1977) but even if a Fourier transform were computed, this transform has neither size nor rotation invariance (Casasent and Psaltis, 1976).

If we look at the local structure of the striate cortex, however, a very striking and potentially useful organization is revealed—that of a log polar frequency transform (Cavanagh, 1978; Maffei and Fiorentini, 1977, Berardi *et al.*, 1982). The goal of this paper is to show how this organization might help in the analysis of patterns. I will describe first the receptive fields of simple and complex cells in the striate cortex and the organization of the cortical cells into local transforms of the retinal pattern. I will show how these local transforms change the rotation and magnification of a stimulus into simple translations of an invariant pattern of cortical activity. Finally, I will describe how these local transforms might be integrated or summed to give a global transform and how a final, translation-invariant transformation will then yield a position, size, and orientation-invariant encoding.

SPATIAL FREQUENCY AND POSITION INVARIANCE

In the striate cortex, simple and complex cells respond to bars of a particular width and orientation (Hubel and Wiesel, 1962, 1968). These dimensions, size and orientation, are the basis of the form-encoding process described here.

The sensitivity profile of the receptive field of a typical simple cell shows two or more parallel, elongated excitatory and inhibitory subfields. Although the optimum stimulus for the cell is a bar, or set of bars, aligned with the excitatory subfields, the cell will actually respond to a wide variety of stimuli. In general, the cell's output can be roughly predicted from integrating the product of the stimulus intensity and the receptive field sensitivity over the entire receptive field. The responses of the set of simple cells that respond to a stimulus can be thought of as a decomposition of the stimulus pattern into localized size and orientation-specific features (Marcelja, 1980). Theoretically, the stimulus pattern can be reconstructed from the outputs of the simple cells if their receptive field locations and sensitivity profiles are known. This decomposition possesses no invariances, however. Size, orientation and position will all influence the set of cells responding.

The receptive fields of complex cells, on the other hand, exhibit no spatially distinct excitatory or inhibitory regions. These cells respond uniformly to a drifting bar whatever its position in the receptive field, as long as it has the optimal orientation and width (Hubel and Wiesel, 1962; Glezer *et al.*, 1980; Heggelund, 1981). These cells therefore show a specificity for pattern information (orientation and width of a bar or the spacing of the bars in a grating) but an indifference to position. Because the position invariance of the complex cells within their receptive fields is an important first step for the form-encoding process, it will be assumed that the output of the complex cells conveys the essential pattern information which is passed on to subsequent stages.

The position independence of the complex cell response implies that the decomposition of a stimulus by the complex cells may not be unique. Stimuli generally have a range of spatial frequency components at each orientation and those components that are sufficiently separated in frequency (about ± 1 octave) will stimulate different complex cells. Since positional information is lost within each receptive field, it is also lost for the relative locations of the frequency components detected by different cells. This implies that we should not be able to distinguish between images with the same frequency content but altered phase (position) content—e.g. a single dot would be indistinguishable from a field of random noise and a square wave (first and third harmonics in the peaks-subtract phase) indistinguishable from a triangle wave (first and third harmonics in the peaks-add phase). Since we can make these distinctions easily, phase or position information for frequency components must be encoded in some manner. In fact, the *relative* positioning or spatial phase relations between frequency components as well as the strengths of the components are together sufficient for a unique encoding of shape.

The broad bandwidth of the spatial frequency detectors may, therefore, play a part in encoding relative phase information. Simple and complex cells respond to a broad band of frequencies (about ± 1 octave; see Maffei and Fiorentini, 1973; Movshon, Thompson and Tolhurst, 1978) and may be sensitive to the relative positions or phases of the frequencies within that band. For example, simple cells with the same preferred spatial frequency but with symmetric or antisymmetric receptive fields (Stromeyer and Klein, 1974, Andrews and Pollen, 1979; Movshon, Thompson and Tolhurst, 1978) respond to similar ranges of spatial frequency but differ in their relative phase sensitivity. For a symmetrical receptive field, the frequency components of the stimulus within the range to which the cell responds must all be in the cosine phase (peaks aligned at the receptive field centre) to stimulate the cell optimally; however, for the antisymmetrical receptive field, they must be in the sine phase. Whether or not complex cells are selective for the relative phase is not so easily determined. DeValois and Tootell (1983) have shown that complex cells in cats are not sensitive to the relative phase but no work has yet been done in primates. Studies of complex cell response to drifting antisymmetric or symmetric brightness profiles could clarify the situation.

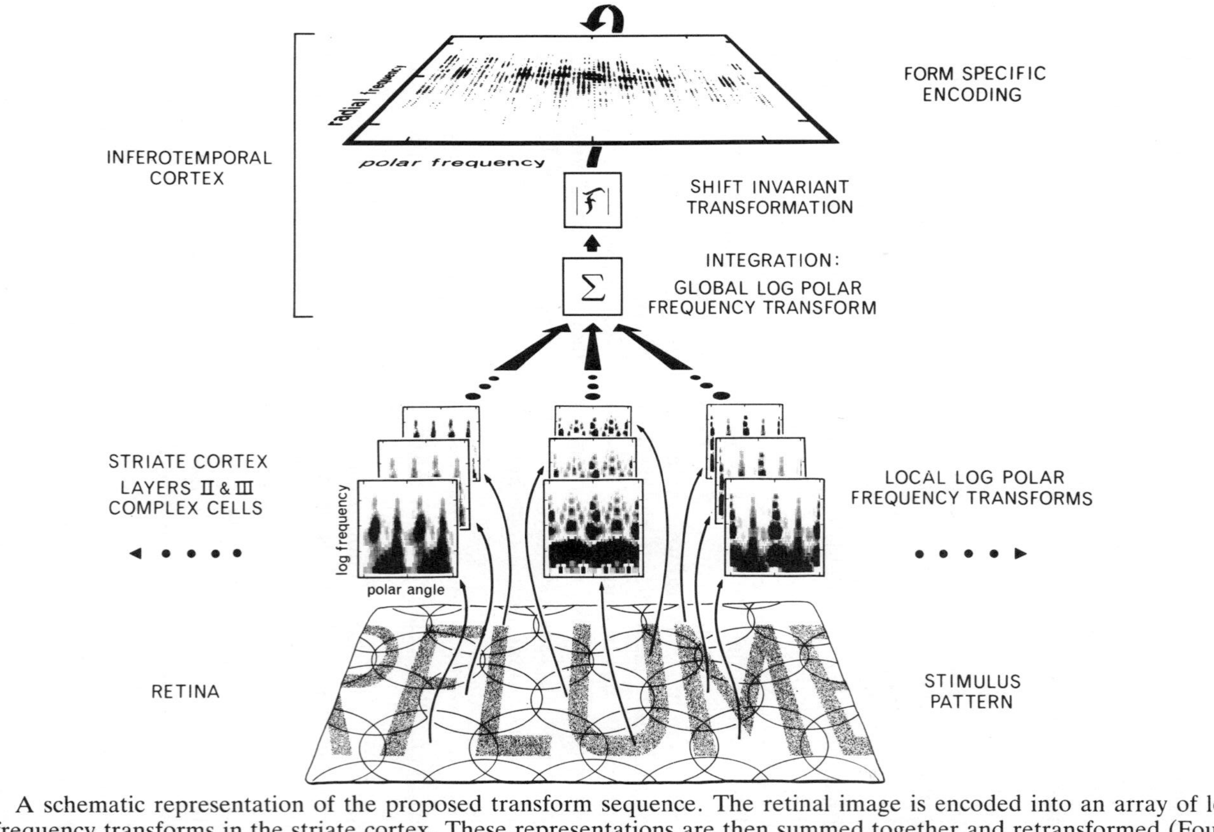

Figure 1 A schematic representation of the proposed transform sequence. The retinal image is encoded into an array of local log polar frequency transforms in the striate cortex. These representations are then summed together and retransformed (Fourier amplitude transform) to produce a position, size and orientation-invariant encoding, most likely in the inferotemporal cortex. The axes of the final transform, radial and angular frequency, are described in more detail elsewhere (Cavanagh, 1978, 1981, 1982, 1984).

LOCAL LOG POLAR FREQUENCY TRANSFORMS

Maffei and Fiorentini (1977) and Berardi *et al.* (1982) suggest that within each local region the cells in the striate cortex are organized into a two-dimensional matrix with the preferred orientation of the cells varying smoothly across the columns of the matrix and the preferred spatial frequency varying across the rows (see Fig. 1). This organization is remarkably similar to that of the log polar frequency transform. In this transform (Brousil and Smith, 1967; Casasent and Psaltis, 1976; Cavanagh, 1974, 1978; Zwicke, 1983), the spatial frequency information of the input pattern is organized in a two-dimensional array with orthogonal axes of orientation and the logarithm of the spatial frequency. This arrangement has two important properties:

(a) *Orientation*. Since one axis of the two-dimensional representation is orientation, a change in the orientation of the input shape changes the orientation of all the component features by the same amount and the whole transform pattern simply shifts left or right along the orientation axis. The shape of the transform pattern (the various blobs at the striate level of Fig. 1) is unchanged.
(b) *Log size*. When the input shape changes size by a factor x, all the component features are scaled by the same factor In order for these features to shift by a constant amount along the size axis, the axis must have a logarithmic scale. A constant multiplicative factor then becomes a constant linear shift. This constant linear shift of all components ensures that the overall pattern of the transform is retained and merely shifts up or down as a whole. A similar size invariance for a log frequency scale is seen in the uniform octave arrangement of a piano keyboard or the log frequency mapping of sounds in the auditory cortex (Tunturi, 1952). Both transform a given melody or chord into a pattern which is invariant to its absolute pitch (Altes, 1978).

This transformation converts rotation or magnification of the stimulus into translations of the activity patterns in the local transforms. However, any appreciable rotation, magnification or change in position of the stimulus will move the striate representation to a new set of local transforms. A *global* transform is essential if true position, size and orientation invariances are to be obtained. For example, a global transform could be produced by simply summing over all the local transforms. This summed transform would have the same dimensions as the local transforms but each cell's response would now represent the sum of activity at the preferred frequency and orientation over the entire visual field. This representation would (a) shift its position for size or orientation changes of the stimulus and (b) be indifferent to the stimulus position in the visual field (other than the loss of high spatial frequency information at more peripheral locations). If a further transformation were performed that was indifferent to the position shifts of the activity pattern within the global transform (Fig. 1), a final representation would be generated that would be specific to the shape of the stimulus and invariant with respect to its position, size and orientation. Local transforms in the striate cortex have been reported in various recent articles; however, the proposed integration and retransformation of these local transforms are purely hypothetical. It is quite intriguing, though, that the axes of the local transforms in the striate cortex are just those required for size and orientation invariance.

STRIATE ARCHITECTURE

Maffei and Fiorentini (1977), Tootell *et al.* (1982) and Berardi *et al.* (1982) all claim that size (or spatial frequency) and orientation comprise two orthagonal dimensions for the local representations of the visual stimulus in the striate cortex. However, they do not agree on how the local transforms are arrayed in the striate cortex. According to Maffei and Fiorentini (1977) and Berardi *et al.* (1982), the preferred orientation of cells varies along an axis parallel to the surface of the cortex while the preferred spatial frequency of cells first increases as layer II and III of the cortex are traversed and then decreases again across layers IV, V and VI. However, Tootell *et al.* (1982) claim that frequency varies parallel to the surface rather than in depth. In either case, the physical orientation of the local transforms will not change their size-invariance property. Each local transform would be sufficiently wide to cover the full range of orientations (approximately 1 mm; see Hubel and Wiesel, 1974) and sufficiently deep or broad to include a single, ordered range of spatial frequencies. There is direct evidence that preferred orientation varies linearly with distance parallel to the cortical surface (Hubel and Wiesel, 1974; Maffei and Fiorentini, 1977, Berardi *et al.*, 1982), and I have presented indirect evidence that the change of preferred frequency with cortical distance is scaled logarithmically (Cavanagh, 1978, 1984). Each local transform represents the two-dimensional frequency by orientation encoding of the pattern for the particular region of the visual field that is covered by the receptive fields of the constituent cells (see Fig. 1). The upper two cortical layers (II and III) are most suitable as the site for the local transforms for two reasons. First, the complex cells having local position invariance predominate in these layers (Hubel and Wiesel, 1962) and, second, these layers project to the prestriate cortex and from there to the inferotemporal cortex, possible sites of further processing, while the other layers (IV, V and VI) project mainly to subcortical areas (Lund *et al.*, 1975; Spatz *et al.*, 1970).

INTEGRATION AND FINAL TRANSFORMATION

Size, orientation and position information appear to have similar local organizaton in the prestriate (area 18) and striate cortices—a retinotopic organization with local transforms having size and orientation axes (Berardi *et al.*, 1982). Multiple representations appear in area 19 of the prestriate cortex (Zeki, 1978) but little is known of their local organization.

The receptive fields of the cells in the inferotemporal cortex are extremely large—up to 90 by 90 degrees—always include the fovea and typically extend into both visual hemifields (Gross, 1973). These cells must receive inputs from several cells in the striate cortex, effectively integrating across the various visual fields involved. The inferotemporal cortex is certainly a candidate area for the integration process necessary for the pattern transform sequence described here. Several studies have shown that the inferotemporal cortex plays an important role in form perception (Mishkin, 1972; Wilson and DeBauche, 1981, Dean, 1982).

The simplest way to integrate would be to sum together the local transforms of the striate cortex. Thus a cell at the global level would merely add the outputs of all complex cells preferring the same orientation, spatial frequency and relative phase. This representation is sufficient to code a stimulus pattern uniquely (Cavanagh, 1984). The resulting invariant pattern will shift in position in the log polar transform plane as a function of changes in the size and orientation of the stimulus anywhere in the visual field. (Note, however, that because of these shifts, some information will be lost or added at the borders of the transform plane. The result is a gradual drop-off in recognition performance when target and test differ in size or orientation, and this drop-off mimics that seen for human subjects under these conditions; see Cavanagh, 1978.)

To achieve a form-specific encoding, the invariant pattern on the log polar plane must be extracted and its position on that plane, which is determined by its size and orientation, ignored. The Fourier amplitude transform is, for example, position invariant. A Fourier amplitude transform of the log polar frequency plane will therefore encode the pattern of activity on that plane and ignore its position. The demonstration in Fig. 1 has used the Fourier amplitude transform at the final step but, as mentioned previously, the Fourier amplitude transform does not unambiguously encode patterns and so some other position-invariant transform would be, in fact, preferable.

Following this shift-invariant step applied to the global log polar representation, a given input shape at various different sizes, positions and orientations always produces a fixed, unchanging pattern of cell firing rates. (Size, orientation and position information are no longer represented at this level and must be processed by other means, either a more elaborate encoding transform or a parallel analysis.) Such fixed patterns could then be used as templates to identify future instances of the same pattern at new locations, sizes and orientations. Since the inferotemporal cortex is the only non-retinotopically organized visual cortex, these proposed size and orientation-invariant encodings must be located in this area as well. (Note that instead of separate and sequential steps of integration and final transformation, the two operations could be combined into a single step.)

CONCLUSIONS

An encoding transform has been described that can produce a position, size and orientation-invariant representation of form. The two essential steps in the sequence are (a) a position invariant encoding of the input that arrays the pattern information along axes of orientation and log size and (b) a position invariant encoding of this log polar representation.

Since the proposed encoding sequence represents form independently of position, size and orientation, stimuli may be classified by matching their transforms against previously stored templates—final transforms of prototype patterns such as letters, familiar faces, common shapes and familiar words. The transformed input would have to be matched in parallel against all prototypes (see Cavanagh, 1975, 1976; Anderson *et al.*, 1977, Kohonen, 1977; Murdock, 1982; Eich, 1982).

No template matching scheme could ever analyse real world scenes involving shadows, partially hidden objects and objects recognized by function (e.g. chairs). However, it seems improbable that the visual system would develop a position, size and orientation-invariant template mechanism just for a few specialized tasks. What then could be the role of such a mechanism? One possibility is that a structural analysis (e.g. Marr, 1982) might be able to take, as its data base, pattern elements identified by a transformational encoding. Rather than *having to* encode patterns and scenes as structures of simple lines and angles, the structural encoding process could start after the template mechanism had matched all elements in the scene for which stored representations were available. When the scene is totally unfamiliar, the available primitives are simply reduced to the lines and angles extracted by the receptive field profiles of the initial encoding level. The analogue and structural approaches to pattern recognition may therefore be simply two levels of a more complex process. Stored prototypes could provide a rich, high-level set of position, size and orientation-invariant primitives to serve as the basis for an intelligent structural analysis.

The proposed encoding process, if it is actually used by the visual system, would probably operate as only one of several parallel analyses of the visual input. Information from colour, depth and motion channels, as well as the brightness-based form encoding described here, would all flow into higher order structural analyses in order to build an overall representation of the visual input. Finally, the encoding sequence described here requires at least two distinct relative phase sensitivities for complex cells in order to eliminate phase ambiguities. If these ambiguities cannot be corrected, then the role of the size and orientation detectors of the striate cortex may be one of texture analysis (Robson, 1980) rather than form encoding.

ACKNOWLEDGEMENTS

This research was supported by NSERC grant A8606 and by the Ministère d'Education du Québec. The helpful comments of Stuart Anstis and Ian Howard are gratefully acknowledged.

REFERENCES

Altes, R. A. (1978). The Fourier-Mellin transform and mammalian hearing. *J. Acoust. Soc. Amer.*, **63**, 174–183.

Anderson, J. A. Silverstein, J. W., Ritz, S. A., and Jones, R. S. (1977). Distinctive features, categorical perception, and probability learning: some applications of a neural model. *Psychol. Rev.*, **84**, 413–451.

Andrews, B. W., and Pollen, D. A. (1979). Relationship between spatial frequency selectivity and receptive field profile of simple cells. *J. Physiol.*, **287**, 163–176.

Berardi, N., Bisti, S., Cattaneo, A., Fiorentini, A., and Maffei, L. (1982). Correlation between preferred orientation and spatial frequency of neurons in visual areas 17 and 18 of the cat. *J. Physiol.*, **323**, 603–618.

Brousil, J. K., and Smith, D. R. (1967). A threshold logic network for shape invariance. *IEEE Trans. Computers*, **EC-16**, 818–828.

Campbell, F. W., and Robson, J. G. (1968). Application of Fourier analysis to the visibility of gratings. *J. Physiol.*, **197**, 551–566.

Casasent, D., and Psaltis, D. (1976). Position, rotation, and scale invariant optical correlation. *App. Optics*, **15**, 1793–1799.

Cavanagh, P. (1974). A two dimensional position, size, and rotation invariant pattern transform: an electro-optical process and a neural analogue. Technical report, Département de Psychologie, Université de Montréal.

Cavanagh, P. (1975). Two classes of holographic processes realizable in the neural realm. In *Formal Aspects of Cognitive Processes*, (Eds. T. Storer and D. Winter), Springer-Verlag, Berlin, pp. 14–40.

Cavanagh, P. (1976). Holographic and trace strength models of rehearsal effects in the item recognition task. *Memory and Cognition*, **4**, 186–199.

Cavanagh, P. (1978). Size and position invariance in the visual system. *Perception*, **7**, 167–177.

Cavanagh, P. (1981). Size invariance: reply to Schwartz. *Perception*, **10**, 469–474.

Cavanagh, P. (1982). Functional size invariance is not provided by the cortical magnification factor. *Vision Res.*, **22**, 1409–1412.

Cavanagh, P. (1984). Image transforms in the visual system. In *Figural Synthesis* (Eds. P. C. Dodwell and T. M. Caelli), Lawrence Erlbaum Associates, Hillsdale, New Jersey, pp. 185–218.

Dean, P. (1982). Visual behavior in monkeys with inferotemporal lesions. In *Analysis of Visual Behavior* (Eds. D. J. Ingle, M. A. Goodale and R. J. W. Mansfield), MIT Press, Cambridge, Mass., pp. 587–628.

DeValois, K. K., and Tootell, R. B. H. (1983). Spatial-frequency-specific inhibition in cat striate cortex cells. *J. Physiol.*, **336**, 359–376.

Eich, J. M. (1982). A composite holographic associative recall model. *Psycholog. Rev.*, **89**, 609–626.

Glezer, V. D., and Cooperman, A. M. (1977). Local spectral analysis in the visual cortex. *Biolog. Cybernetics*, **28**, 101–108.

Glezer, V. D., Tsherbach, T. A., Gauselman, V. E., and Bondarko, V. M. (1980). Linear and non-linear properties of simple and complex receptive fields in area 17 of the cat visual cortex. *Biolog. Cybernetics*, **37**, 195–208.

Gross, C. G. (1973). Visual function of inferotemporal cortex. In *Handbook of Sensory Physiology* (Ed. R. Jung), Vol. VIII/3, Part B, Springer-Verlag, Berlin, pp. 451–482.
Heggelund, P. (1981). Receptive field organization of complex cells in cat striate cortex. *Exp. Brain Res.*, **42**, 99–107.
Hubel, D. H., and Wiesel, T. N. (1962). Receptive fields, binocular interaction and functional architecture in the cat's visual cortex. *J. Physiol.*, **160**, 106–154.
Hubel, D. H., and Wiesel, T. N. (1968). Receptive fields and functional architecture of monkey striate cortex. *J. Physiol.*, **195**, 215–243.
Hubel, D. H., and Wiesel, T. N. (1974). Sequence regularity and geometry of orientation columns in the monkey striate cortex. *J. Comparative Neurol.*, **158**, 267–293.
Kohonen, T. (1977). *Associative Memory*, Springer-Verlag, Berlin.
Kosslyn, S. M. (1980). *Image and Mind*, Harvard University Press, Cambridge, Mass.
Lund, J. S., Lund, R. D., Hendrickson, A. E., Bunt, A. H., and Fuchs, A. F. (1975). The origin of efferent pathways from the primary visual cortex, area 17, of the macaque monkey as shown by retrograde transport of horseradish peroxidase. *J. Comparative Neurol.*, **164**, 287–304.
Maffei, L., and Fiorentini, A. (1973). The visual cortex as a spatial frequency analyser. *Vision Res.*, **13**, 1255–1267.
Maffei, L., and Fiorentini, A. (1977). Spatial frequency rows in the striate visual cortex. *Vision Res.*, **17**, 257–264.
Marcelja, S. (1980). Mathematical description of the responses of simple cortical cells. *J. Optical Soc. Amer.*, **70**, 1297–1300.
Marr, D. (1982). *Vision*, Freeman, San Francisco.
Mishkin, M. (1972). Cortical visual areas and their interactions. In *Brain and Human Behavior* (Eds. A. G. Karczmar and J. C. Eccles), Springer-Verlag, New York, pp. 187–208.
Movshon, J. A., Thompson, I. D., and Tolhurst, D. J. (1978). Spatial summation in the receptive fields of simple cells in the cat's striate cortex. *J. Physiol.*, **283**, 53–77.
Murdoch, B. B. (1982). A theory for the storage and retrieval of item and associative information. *Psycholog. Rev.*, **89**, 609–626.
Pollen, D. A., Lee, J. R., and Taylor, J. H. (1971). How does the striate cortex begin the construction of the visual world? *Science*, **173**, 74–77.
Pylyshyn, Z. W. (1973). What the mind's eye tells the mind's brain: a critique of mental imagery. *Psycholog. Bull.*, **80**, 1–24.
Robson, J. (1975). Receptive fields: neural representation of the spatial and intensive attributes of the visual image. In *Handbook of Perception*, Vol. 5, *Seeing* (Eds. E. D. Carterette and M. D. Friedman), Academic Press, New York, pp. 81–117.
Robson, J. (1980). Neural images: the physiological basis of spatial vision. In *Visual Coding and Adaptability* (Ed. C. S. Harris), Lawrence Erlbaum Associates, Hillsdale, New Jersey, pp. 177–214.
Schade, O. H. (1956). Optical and photoelectric analog of the eye. *J. Optical Soc. Amer.*, **46**, 721–739.
Spatz, W. B., Tigges, J., and Tigges, M. (1970). Subcortical projections, cortical associations and some intrinsic interlaminar connections of the striate cortex in the squirrel monkey (*Saimiri*). *J. Comp. Neural.*, **140**, 155–174.
Stromeyer III, C. F., and Klein, S. (1974). Spatial frequency channels in human vision as asymmetrical (edge) mechanisms. *Vision Res.*, **14**, 1409–1420.
Sutherland, N. S. (1973). Object recognition. In *Handbook of Perception*, Vol. 3, Biology of Perceptual Systems (Eds. E. D. Carterette and M. P. Friedman), Academic Press, New York, pp. 157–206.

Tootell, R. B., Silverman, M. S., Switkes, E., and DeValois, R. L. (1982). Deoxyglucose analysis of retinotopic organization in primate striate cortex. *Science*, **218**, 902–904.
Tunturi, A. R. (1952). A difference in the representation of auditory signals for left and right ears in the iso-frequency contours of tight middle ectosylvian auditory cortex of the dog. *Amer. J. Physiol.*, **68**, 712–727.
Wilson, M., and DeBauche, B. A. (1981). Inferotemporal cortex and categorical perception of visual stimuli by monkeys. *Neuropsychologia*, **19**, 29–41.
Zeki, S. M. (1978). Uniformity and diversity of structure and function in rhesus monkey prestriate visual cortex. *J. Physiol.*, **277**, 273–290.
Zwicke, P. E. (1983). A new implementation of the Mellin transform and its application to radar classification of ships. *IEEE Trans. Pattern Analysis and Machine Intelligence*, **PAMI-5**, 191–199.

Models of the Visual Cortex
Edited by D. Rose and V. G. Dobson

CHAPTER 10

Representational issues and local filter models of two-dimensional spatial visual encoding

J. G. DAUGMAN
Division of Applied Sciences, Harvard University, Cambridge, MA 02138, USA

INTRODUCTION

Much of the conceptual debate in the past two decades of research on visual perception has concerned 'local' versus 'global' theories of spatial information extraction. On the one hand, the existence of visuotopic mappings to several cortical visual areas suggests some preservation of local spatial coordinates in the representation, and certainly the most salient tuning property of all cells in the striate cortex is their sharp 'tuning' for the two-dimensional spatial location of a signal, often within less than 1 deg^2 out of the 30,000 deg^2 area of visual space. On the other hand, the fundamental nature of the pattern-recognition problem for vision requires the explicit extraction of more globally defined spatial relationships; a representation which merely reproduces or eventually recovers the spatial image achieves nothing.

The aim of this chapter is to discuss several related representational issues concerning the nature of two-dimensional spatial visual encoding in the striate cortex. Some of these issues have been major foci of recent vision research but almost always in one-dimensional form, which is nonisomorphic to the obvious two-dimensionality of retinal images. The dimension in the representational analysis is crucial for many key insights, such as the intrinsic trade-offs governing resolution for variables like orientation and spatial frequency. It will be argued that both neurophysiological and psychophysical data, when cast in suitable two-dimensional form, support the view that a goal of early spatial visual encoding is to optimize the simultaneous resolution of information in both the two-dimensional space domain and the two-dimensional spatial frequency/orientation domain, subject to an inescapable four-dimensional uncertainty principle. Conceptually this leads to a 'mixed' representation, combining properties of both 'local' and 'global' representations within windows of two-dimensional visual space, functionally analogous to the speech spectrogram decomposition of the one-dimensional speech waveform into spectral formats within windows of time.

After a section raising three points of critique of the model envisioned by Marr (1982), the issue of separability of variables will be raised as a testable modelling constraint. Finally, evidence from psychophysical two-dimensional masking experiments will be presented in reference to the foregoing issues, and the two-dimensional spatial form of the putative anisotropic encoding mechanisms of local filtering will be empirically derived.

TWO-DIMENSIONAL (2D) GENERALIZED GABOR FILTERS AND JOINT ENTROPY MINIMIZATION

The need for the visual nervous system to process efficiently a vast amount of information about the spatiotemporal world requires that the incident images be represented with optimal economy. A debate became articulated in both perceptual and physiological vision research over whether local features such as edges and bars (Hubel and Wiesel, 1962) or Fourier components (Campbell and Robson, 1968; Maffei and Fiorentini, 1973; Pollen, Lee and Taylor, 1971) were the 'atoms' of visual perception. As compelling evidence was marshalled on behalf of both theories, many 'critical' experiments were pursued (Albrecht, DeValois and Thorell, 1981; DeValois, Albrecht and Thorell, 1978; DeValois, DeValois and Yund, 1979; Macleod and Rosenfeld, 1974) that would definitively resolve the issue.

Subsequently conciliatory voices (Marčelja, 1980; MacKay, 1981; Kulikowski and Bishop, 1981) affirmed that the debate was illusory because the two descriptions were complementary, and the crucial experimental results could be equally well captured by modest versions of either theory. The gradual acceptance of the complementarity and utility of both descriptions—one 'undulatory' and the other 'punctate'—is somehow reminiscent of the dissolution of the historic wave-particle debate in quantum physics. The most compelling voice of conciliation predates the spatial vision debate itself and comes from the famous 1946 paper of Dennis Gabor on the theory of communication, whose relevance to contemporary interpretations of simple cell receptive field profiles was first pointed out by Marčelja (1980).

Based on mathematical principles developed by Heisenberg two decades earlier, Gabor (1946) noted that a fundamental uncertainty principle limits the degree of specificity that a signal can have simultaneously in time and in cycles per second. He derived the family of signals which optimize this tradeoff in resolution and elaborated a quantum theory of information, in which signals (or filters) occupy regions of an abstract information diagram whose coordinates are time and frequency. The minimal area which can be occupied in this plane forms a quantal grain, which he termed a 'logon' of information. The 'Gabor signals' are the most informative because they occupy the smallest possible area, although the shape and location of their residence depend on their bandwidth, duration, frequency and time of occurrence. The analogy for cortical simple cell receptive field profiles, which are localized (tuned) in both space and spatial frequency and can be well described by Gabor functions, is now well established (Marčelja, 1980). Unfortunately, because most of the inquiry has been one-dimensional, the fundamental variable of orientation resolution has been left out of the picture.

The second dimension of a receptive field profile and its orientation resolution are clearly related to one another, and the specification of either one entails the specification of the other (Daugman, 1980). A full 2D generalization of Gabor's analysis was given for 2D spatial filters in Daugman (1985). Briefly, we define the 'occupied area' of a 2D spatial filter $f(x, y)$, i.e. the receptive field profile, as the product of its equivalent width and equivalent length, given by the normalized product of its two second moments:

$$[\int\int|f|^2x^2\mathrm{d}x\mathrm{d}y\int\int|f|^2y^2\mathrm{d}x\mathrm{d}y]^{1/2}/\int\int|f|^2\mathrm{d}x\mathrm{d}y.$$

We can thus specify its 2D spatial resolution, in analogy with the variance of a distribution. Similarly, we can measure its resolution or 'occupied area' in the 2D Fourier plane (orientation, spatial frequency) by the analogous product of the equivalent width and equivalent length of its 2D Fourier transform $F(u,v)$:

$$[\int\int|F|^2u^2\mathrm{d}u\mathrm{d}v\int\int|F|^2v^2\mathrm{d}u\mathrm{d}v]^{1/2}/\int\int|F|^2\mathrm{d}u\mathrm{d}v.$$

Without going into mathematical detail, it can be shown (Daugman, 1985) that the normalized product of these joint second moments (occupied areas) has a lower bound and that the optimal family of 2D filters $f(x,y)$ which achieve this maximum degree of joint resolution, or minimum joint entropy, have the form exemplified in Fig. 1 (from Daugman, 1980). Recall that in the Fourier plane, the distance from the origin at the center corresponds to spatial frequency and the angle around the origin corresponds to orientation.

Because the joint resolution (joint entropy) is independent of the values of any of the parameters specified in figure 1, different optimal filters may be specialized for the extraction of different information. This division of labor is captured in Fig. 2, showing isoamplitude contours of three different 2D Gabor filters in both the 2D space domain and the 2D Fourier domain. Figure 2a is a bird's-eye view of Fig. 1, showing a localized filter with unity aspect ratio (width/length) and a certain modulation spatial frequency ω_0, which in the context of its spatial dimensions gives it a certain spatial frequency bandwidth $\Delta\bar{F}$ and an orientation bandwidth $\Delta\Theta_{1/2}$ as indicated in the right-hand panel. If this filter is now simply elongated vertically as shown

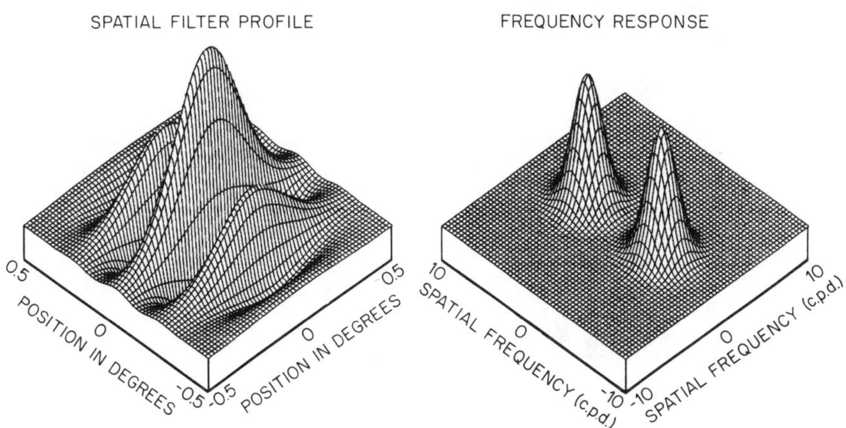

Figure 1 An even-symmetric member of the family of optimal '2D Gabor' filters, with unity width/length ratio, and its 2D Fourier transform. Members of this family have minimal joint entropy, i.e. the sharpest possible joint resolution in the two 2D domains. The spatial period and orientation of the sidelobes determine the preferred spatial frequency and orientation; the number of significant sidelobes determines spatial frequency bandwidth and their significant length determines orientation bandwidth. Different members of this optimal family are an excellent description of simple cell 2D receptive fields found in the striate cortex. The general optimal functional form in the two 2D domains is given by the following equations, for a receptive field centered on spatial coordinates (x_0, y_0) with width and length dimensions $1/a$ and $1/b$ and preferred spatial frequency $\sqrt{u_0^2 + v_0^2}$ and preferred orientation arctan (v_0/u_0):

$$f(x, y) = e^{-\pi [(x-x_0)^2 a^2 + (y-y_0)^2 b^2]} e^{-2\pi i [u_0 (x-x_0) + v_0 (y-y_0)]},$$

$$F(u, v) = e^{-\pi [(u-u_0)^2/a^2 + (v-v_0)^2/b^2]} e^{-2\pi i [x_0 (u-u_0) + y_0 (v-v_0)]}.$$

in Fig. 2b, its spatial frequency bandwidth ΔF is unaffected but its orientation bandwidth $\Delta\Theta_{1/2}$ is sharpened in proportion; this effect is indicated in the right-hand panel and the bandwidth equations are provided. Finally, Fig. 2c shows that simple horizontal elongation of the original receptive field envelope, so that it includes more excitatory/inhibitory fringes, correspondingly sharpens its spatial frequency bandwidth ΔF but has no effect on its orientation bandwidth $\Delta\Theta_{1/2}$. These consequences of the 2D similarity theorem of Fourier analysis capture not only the well-known trade-off between a receptive field's spatial resolution (in one dimension) and its spatial frequency resolution, but also capture the generally unrecognized yet equally fundamental trade-off between the other dimension of spatial resolution and the cell's orientation resolution.

Several trade-offs involving the width and length of a receptive field, its total occupied area, its preferred spatial frequency and orientation, its spatial frequency bandwidth and its orientation bandwidth were derived in the form

TWO-DIMENSIONAL GABOR OPTIMAL FILTERS

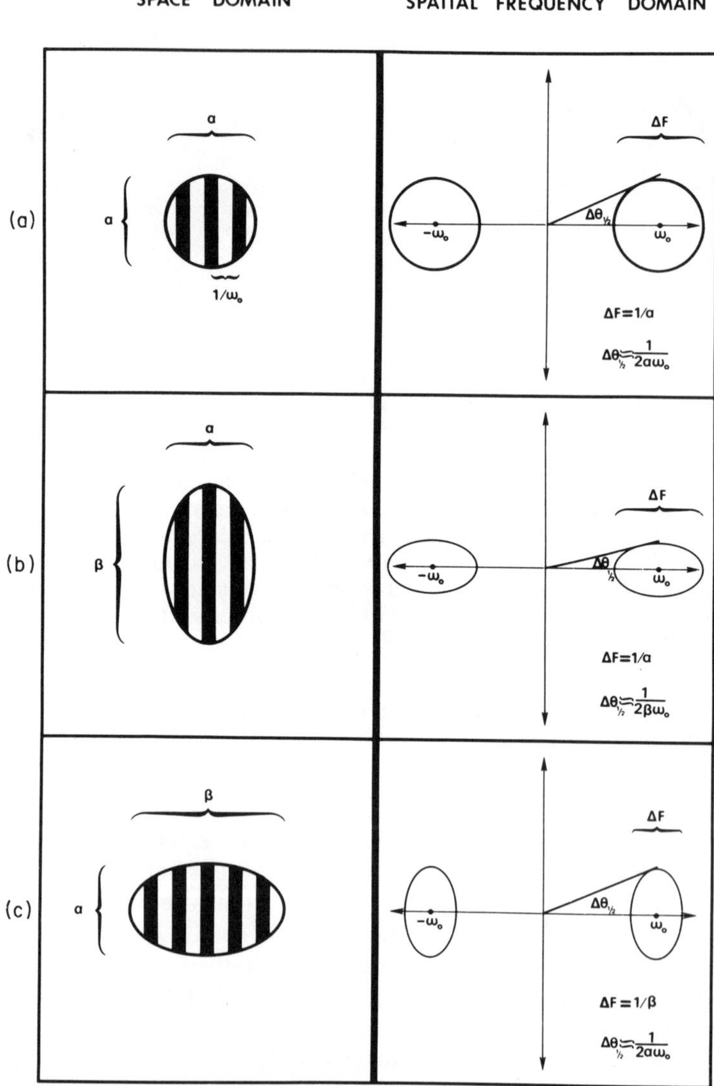

FIGURE 2 Birds-eye view of three members of the set of Gabor optimal filters, all having the same preferred spatial frequency and orientation. The panels demonstrate the dependence of the filter's spatial frequency bandwidth and orientation bandwidth on the space-domain envelope dimensions, although the *preferred* frequency and orientation are independent of those dimensions. (a) A circular filter envelope in the space domain occupies a circular region in the spatial frequency domain, with both

(*caption continued on p. 101*)

of necessary uncertainty principles in Daugman (1985). One result is that for all 2D receptive fields constrained to occupy a fixed amount of area of visual space, there is a necessary trade-off between orientation resolution and spatial frequency resolution; this competition can be understood by comparing Fig. 2b and Fig. 2c.

On the other hand, for families of receptive fields not constrained to occupy a fixed amount of spatial area but rather constrained to have a fixed width/length ratio, the situation reverses. Orientation resolution and spatial frequency resolution are no longer in competition, but rather co-vary. This appears to be the situation for neurons in the striate cortex, since a strong positive correlation has been found by Movshon (1979) in the cat and by DeValois, Albrecht and Thorell (1982) in primates. The theoretical prediction for 2D Gabor filters having a space-domain width/length ratio of λ is that their orientation resolution $\Delta\Theta_{1/2}$ and their spatial frequency bandwidth $\Delta\omega$ in octaves must be related to each other by the equation:

$$\Delta\Theta_{1/2} = \arcsin \ \lambda \frac{2^{\Delta\omega} - 1}{2^{\Delta\omega} + 1}$$

(derived in Daugman, 1985, Eq. 4). If we take as a reasonable value of the width/length ratio $\lambda = 0.6$, then this equation predicts that different simple cells should show the correlation between their orientation bandwidths and their spatial frequency bandwidths shown in Table 1.

Table 1 Predicted correlation between spatial frequency bandwidth and orientation bandwidth for 2D Gabor filters having a width/length ratio of 0.6.

Spatial frequency full-bandwidth, octaves	*Orientation half-bandwidth, degrees*
0.5	5.9
1.0	11.5
1.5	16.7
2.0	21.1
2.5	24.8

spatial frequency bandwidth and orientation bandwidth inversely related to the receptive field area. (b) Elongating the receptive field in the direction parallel to the modulation sharpens the orientation bandwidth $\Delta\Theta_{1/2}$ but has no effect on spatial frequency bandwidth ΔF. (c) Elongating the field in the perpendicular direction sharpens the spatial frequency bandwidth ΔF but has no effect on the orientation bandwidth $\Delta\Theta_{1/2}$. Thus, such filters can negotiate the necessary trade-offs for resolution in different ways, attaining, for example, sharp resolution along the *Y* direction (at the expense of orientation selectivity) or sharp resolution along the *X* direction (at the expense of spatial frequency selectivity), to favor the extraction of different kinds of information. Always, however, the product of occupied areas in the two 2D domains is the same, and is equal to the theoretical minimum.

We see that for 2D Gabor filters having spatial frequency bandwidths ranging from 0.5 to 2.5 octaves, the optimal orientation half-bandwidth increases at the rate of approximately 10° per octave.

Remarkably, this is exactly the correlation reported empirically by Movshon (1979) for simple cells in the cat visual cortex: 'orientation selectivity and spatial frequency selectivity are well correlated with one another: orientation half-widths increase by about 10 degrees for each octave increase in spatial frequency bandwidth.' Movshon's correlation coefficient was 0.7 based on 114 simple cells. This strong empirical agreement with the correlation predicted for 2D Gabor filters offers significant support for this interpretation of simple cells.

POINTS OF CRITIQUE OF $\nabla^2 G_i$ ARRAY MODELS

Much interest has developed in Marr's (1982) comprehensive treatment of visual processing, which proposed that the initial encoding of a 'primal sketch' is subserved by arrays of $\nabla^2 G_i$ filters (Gaussian-smoothed Laplacian second-derivative operators where $\nabla^2 = d^2/dx^2 + d^2/dy^2$ and $G_i = e^{-a_i^2(x^2+y^2)}$). The subscript $_i$ indicates that the smoothing Gaussians have several different sizes, corresponding to filters tuned to different spatial frequencies, thus subserving the extraction of band-limited spatial information at several different scales simultaneously. Zero crossings in the outputs of such filter arrays define edges and boundaries in an efficient manner for object segregation and pattern recognition. This attractive and elegant conceptualization has become popular in the AI community as a basis for machine vision, as well as serving as a model of the human visual system. Nonetheless, there are at least three points of critique to be noted as weaknesses of this approach:

1. As shown in Daugman (1983), such $\nabla^2 G_i$ arrays followed by zero-crossing detectors are unable to detect any subharmonic functions. An example of a subharmonic function is a sinewave grating added to a 2D paraboloid: light intensity $I(x,y) = 1 - \cos(2\pi x) + 12(x^2 + y^2)$. The human visual system is demonstrably able to see such luminance distributions.
2. The highest stage of linear filtering envisioned before the nonlinear operation which constructs the 'primal sketch' involves only isotropic (nonorientation-selective) filters, whereas in the mammalian visual system, linear filtering seems to extend to the level of cortical simple cells (Movshon Thompson and Tolhurst, 1978; Andrews and Pollen, 1979) which are fundamentally orientation selective. An interesting alternative for the Marr approach would be to replace, or combine, the isotropic Laplacian operator with directional derivatives which would confer orientation selectivity. One such proposal was offered by Daugman (1983, Fig. 1), which looks like a physiologically plausible receptive field profile and

which also has a polar separable 2D Fourier spectrum. Indeed, for any such cascade of differential operators applied in the 2D space domain, the resulting receptive field profiles would always generate independent orientation and spatial frequency tuning curves, thus preventing interaction between these two information-encoding variables.

3. Much is made of Logan's theorem (e.g. Marr, Poggio and Ullman, 1979) as supporting the utility of extracting zero crossings. Logan's theorem shows that the zero crossings of 1-octave bandwidth signals are extremely rich in information, and are generally sufficient for the full reconstruction of such waveforms within a scale factor. However, no clear account has yet been given for whether or how Logan's theorem applies to 2D signals, such as retinal images. Indeed, whereas the zero-crossings of 1D bandpass signals are countable and finite, those of 2D bandpass signals are neither.

POLAR SEPARABILITY OF SPECTRAL VARIABLES IN LOCAL 2D FILTERS

A fundamental representational issue for local 2D filter models is whether the decompositional variables are separable, i.e. have independent tuning curves. We have already seen that inescapable uncertainty principles dicate a trade-off between orientation resolution and one dimension of spatial resolution and between spatial frequency resolution and the other spatial dimension. What of the relationship between just the two variables of orientation and spatial frequency? There is no *a priori* necessary relationship between these two tuning variables, and theoretical 2D filter classes exist which have arbitrarily sharp or poor resolution in these variables simultaneously. However, among physiologically plausible organizational principles of simple cell receptive fields the situation is more constrained.

Considering just the basic *organizational principle*, regardless of the exact form of the 2D receptive field profile, reveals specific 2D spectral consequences. For example, the elementary organizational principle of concatenating LGN subunits in an aligned row to form an orientation-selective simple cell, or the principle of simply stretching an isotropic receptive field profile in one direction to form an elongated, orientation-selective receptive field, must always lead to nonseparable orientation and spatial frequency tuning properties (Daugman, 1983). That is, the cell *must* have different orientation tuning properties at different spatial frequencies, and vice versa. (In the latter case of simple elongation, the polar spectral nonseparability is so severe that the cell's maximum possible response for gratings could be obtained at *any* orientation relative to the direction of elongation, provided the right spatial frequency were used; this property is a bit counterintuitive, in view of the unambiguous elongation.) Of various theoretical organizational principles for 2D receptive fields reviewed in Daugman (1983), regardless of functional form, the only principle which was found that must always lead

2D FREQUENCY DOMAIN

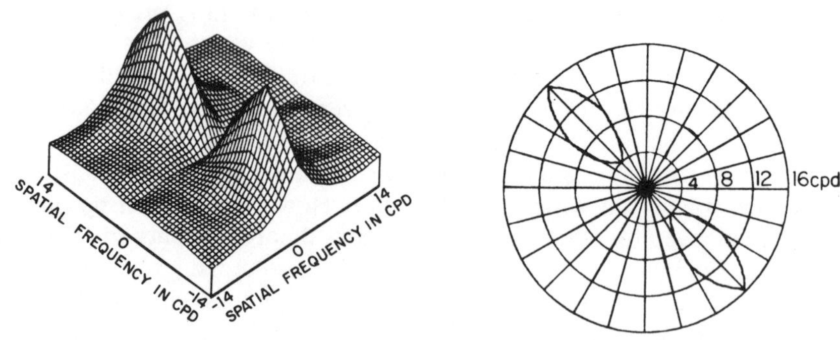

(a) THRESHOLD ELEVATION SURFACE (b) HALF-AMPLITUDE CONTOURS

2D SPACE DOMAIN

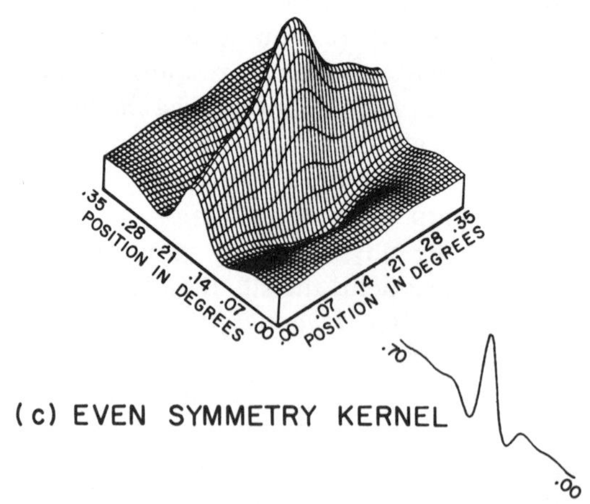

Figure 3 Psychophysical threshold elevation surface produced by an 8 c. p. d. vertical masking grating of 32 per cent contrast, interpolated over a Cartesian grid of 121 measurement points in the 2D Fourier plane for one of four subjects. The location of the peaks corresponds to 8 c.p.d., vertical, at which the threshold elevation is about 2,000 per cent. The highest spatial frequency tested, corresponding to the corners of the Cartesian-resolved coordinates, was 20 c.p.d. (b) Isoamplitude contours of the tuning surface shown in (a) sliced at the half-amplitude level over the Fourier plane. The 2D Gabor filters have elliptical contours. (c) Two-dimensional space-domain filter profile for the same subject, as inferred by numerical 2D Fourier transformation of the tuning surface in (a) and assuming even symmetry. Inset shows a 1D central cross-section at twice the spatial scale. Reproduced by permission of Pergamon Press Ltd from Daugman (1984).

to independent orientation and spatial frequency tuning properties (polar separable 2D spectra) is any arbitrary sequence of orientation-selective differential operators applied in hierarchy to any original isotropic filter, such as a center/surround retinal X-type ganglion cell. Such differential operators can be realized directly by lateral inhibitory interactions between neighbouring cells, as is well known.

PSYCHOPHYSICALLY INFERRED 2D SPATIAL CHANNEL PROFILES

Although the integration of physiological and psychophysical data is fraught with pitfalls and ambiguities, it is nonetheless perhaps useful to bring psychophysical data to bear on these representational issues. This has frequently been attempted in the past (e.g. Blakemore and Campbell, 1969) but invariably in only one-dimensional form, such as in the comparison of psychophysically inferred line-spread functions and simple cell profiles. Accordingly, a psychophysical paradigm of spatial frequency masking in 2D form was employed for the purpose of inferring the 2D spectral and spatial profiles of visual 'channels'. Four subjects, including the author, all corrected myopes aged in their twenties, participated in signal-detection masking experiments employing forced-choice trials as detailed in Daugman (1984). The percentage threshold elevation over the 2D Fourier plane induced by a vertical 8 c.p.d. mask grating of 32 per cent contrast is shown for a typical subject in Fig. 3a. Qualitatively identical results and conclusions were obtained from the other three subjects (not shown in Fig. 3).

Computationally intersecting each of these empirical tuning surfaces with upright, origin-centered concentric cylinders of different radii yielded the orientation tuning curves of the 8 c.p.d. vertical channel for different probe spatial frequencies. The obtained orientation half-widths depend strongly on the probe spatial frequency, ranging from about 30° one octave below the channel's center frequency to about 7° one octave above. Thus, for all four subjects, the 2D spectral tuning surface of the putative channel was highly polar non-separable. Such channels cannot be described simply in terms of 'a spatial frequency tuning curve' and 'an orientation tuning curve'.

The 2D spectral localization of these channels is indicated by their half-amplitude contours, a typical one of which is shown in Fig. 3b. For all four subjects these were ellipses with roughly a 0.5 width/length ratio. Presumably the 2D Fourier plane is 'paved' with such ellipses, representing selectivity for many different spatial frequencies and orientations within any local patch of visual space.

Finally, numerically computing the 2D Fourier transforms of the empirical masking data surfaces permits the 2D space-domain profiles of these filters to be inferred. That of one subject is shown in Fig. 3c within a $0.35° \times 0.35°$

patch of the fovea, and its central cross-section is also provided over the larger scale of 0.7° to indicate filter quiescence after 4 or 5 extrema. Even symmetry has been arbitrarily assumed in the 2D numerical Fourier transformation; the odd-symmetric space-domain weighting functions have also been computed but are not presented here. Any linear combination of the even and odd functions would constitute filters having the associated 2D spectral tuning properties possessed by the data surface of Fig. 3a.

With both the 2D spectral tuning surfaces and the inferred 2D space-domain filter profiles in hand, we can evaluate the efficiency of such channels in terms of the generalized joint entropy analysis developed earlier. Calculating the occupied area of the filters in each of the 2D domains by the second moment of inertia integrals defined previously, we obtain for subjects WC, JD, RF, HW the joint occupied area products 5.21, 5.37, 5.15 and 4.94 respectively (Daugman, 1984). The theoretical minimum value is 2.0, achieved by true 2D Gabor filters; other theoretical 2D filter classes such as ideal 2D bandpass (spectral disk) filters have joint occupied area products of 13.0 or more. Thus, the inferred 2D psychophysical filters come comparatively close to the theoretical limit of joint resolution imposed by the fundamental 4D uncertainty principle.

In summary, the 2D spectral characteristics as well as the computed 2D spatial profiles of the psychophysical channels inferred by masking bear close resemblance to simple cell properties, and both the psychophysical and the physiological data sets fit well within the 2D Gabor framework for encoding the 2D retinal image. Such a representation is efficient and also 'intelligent', even at this low level of encoding, since while preserving spatial locality it captures explicitly the more global relationships pertaining to form and pattern within a visual patch as an associated spectral signature.

The 'mixed' representation subserved by such an array of 2D Gabor filters in striate cortex would organize spatial visual information in a fashion analogous to the localized spectral formants of a speech spectrogram. The extraction of such local 'spectral signatures' has considerable theoretical utility in pattern-recognition algorithms, such as the Wigner approach developed in detail by Jacobson and Wechsler (1982). Physiological patterns of local sequence regularity of orientation columns and the typical distributions of orientation bandwidths and spatial frequency bandwidths for simple cells suggest that the array of 2D Gabor filters are arranged in a doubly embedded (2D space and 2D spectral) log polar sampling lattice. This entails a log polar sampling density in both the 2D Fourier domain (bandwidths spaced in octaves and in degrees) for any given patch of visual space, and similarly in the global 2D space domain (dependency of receptive field size and overlap on polar distance from the fovea). This doubly embedded log polar encoding matrix of 2D Gabor filters paves both the 2D Fourier plane locally and 2D visual space globally, and may perform the efficient and essential groundwork for higher visual processes.

REFERENCES

Albrecht, D. G., DeValois, R. L., and Thorell, L. G. (1981). Visual cortical neurons: are bars or gratings the optimal stimuli? *Science*, **207**, 88–90.

Andrews, B. W., and Pollen, D. A. (1979). Relationship between spatial frequency selectivity and receptive field profile of simple cells. *J. Physiol.*, **287**, 163–176.

Blakemore, C., and Campbell, F. W. (1969). On the existence of neurons in the human visual system selectively sensitive to the orientation and size of retinal images, *J. Physiol.*, **203**, 237–260.

Campbell, F. W., and Robson, J. G. (1968). Application of Fourier analysis to the visibility of gratings. *J. Physiol.*, **197**, 551–566.

Daugman, J. G. (1980). Two-dimensional spectral analysis of cortical receptive field profiles. *Vision Res.*, **20**, 847–856.

Daugman, J. G. (1983). Six formal properties of two-dimensional anisotropic visual filters: structural principles and frequency/orientation selectivity. *IEEE Transactions on Systems, Man, and Cybernetics*, **13**, 882–887.

Daugman, J. G. (1984). Spatial visual channels in the Fourier plane. *Vision Res.*, **24**, 891–910.

Daugman, J. G. (1985). Uncertainty relation for resolution in space, spatial frequency, and orientation optimized by two-dimensional visual cortical filters. *J. Opt. Soc. Amer.* (in press).

DeValois, K. K., DeValois, R. L., and Yund, E. Y. (1979). Responses of striate cortex cells to grating and checkerboard patterns. *J. Physiol.*, **291**, 483–505.

DeValois, R. L., Albrecht, D. G., and Thorell, L. G. (1978). Cortical cells: bar and edge detectors, or spatial frequency filters? In *Frontiers of Visual Science*, (Eds. S. J. Cool and E. L. Smith), Springer, New York.

DeValois, R. L., Albrecht, D. G., and Thorell, L. G. (1982). Spatial frequency selectivity of cells in macaque visual cortex. *Vision Research*, **22**, 545–559.

Gabor, D. (1946). Theory of communication. *J. IEEE Lond.*, **93**, 429–457.

Hubel, D., and Wiesel, T. N. (1962). Receptive fields, binocular interaction and functional architecture in the cat's visual cortex. *J. Physiol.*, **160**, 106–154.

Jacobson, L., and Wechsler, H. (1982). *A New Paradigm for Computational Vision Based on the Wigner Distribution*. University of Minnesota Digital Systems Program, Technical Report, Minneapolis.

Kulikowski, J. J., and Bishop, P. O. (1981). Fourier analysis and spatial representation in the visual cortex. *Experientia*, **37**, 160–163.

MacKay, D. (1981). Strife over visual cortical function. *Nature*, **289**, 117–118.

Macleod, I. D. G., and Rosenfeld, A. (1974). The visibility of gratings: spatial frequency channels or bar-detecting units? *Vision Res.*, **14**, 909–915.

Maffei, L., and Fiorentini, A. (1973). The visual cortex as a spatial frequency analyzer. *Vision Res.*, **13**, 1255–1267.

Marčelja, S. (1980). Mathematical description of the responses of simple cortical cells. *J. Opt. Soc. Amer.*, **70**, 1297–1300.

Marr, D. (1982). *Vision*. Freeman, San Francisco.

Marr, D., Poggio, T., and Ullman, S. (1979). Bandpass channels, zero-crossings, and early visual information processing. *J. Opt. Soc. Amer.*, **69**, 914–916.

Movshon, J. A. (1979). Two-dimensional spatial frequency tuning of cat striate cortical neurons. *Soc. Neuroscience, Abstracts*, **9**, 799.

Movshon, J. A., Thompson, I. D., and Tolhurst, D. J. (1978). Spatial summation in the receptive fields of simple cells in the cat's striate cortex. *J. Physiol.*, **283**, 53–77.

Pollen, D. A., Lee, J. R. and Taylor, J. H. (1971). How does the striate cortex begin the reconstruction of the visual world? *Science*, **173**, 74–77.

Models of the Visual Cortex
Edited by D. Rose and V. G. Dobson

CHAPTER 11

The cellular basis of perception

J. I. NELSON[1]
Department of Ophthalmology, New York University Medical Center, 550 First Avenue, New York, NY 10016, USA

This essay's thesis is that cortical columns shape intracortical interactions and selective interactions are the basis of much perceptual processing. Figural illusions (orientation contrast) and stereoscopic globality are used as examples.

LATERAL INHIBITION IN AN ORDERED CORTEX

A generation ago, Gestalt researchers such as Wolfgang Köhler and Hans Wallach surveyed the contour position distortions seen in figural after-effects. To explain their observations, they postulated electrical fields spreading through homogeneous cortical tissue and pushing contour representations around as they went (Köhler & Wallach, 1944). Today we know the tissue is not homogeneous.

Locally, single cortical neurons are selective for a particular orientation, retinal disparity (depth), velocity or other stimulus dimension. Globally, neurons are ordered according to their stimulus selectivities or *tunings*. A radial electrode penetration through the sensory cortex (perpendicular to the cortical surface) will often yield a series of neurons sharing a common stimulus tuning. This vertical aggregation was the original meaning of 'cortical column' (Mountcastle, 1978), but horizontal groupings are known as well. To avoid distortion of the meaning of 'column' (Szentagothai, 1983), horizontal progressions of tunings which can be seen with tangential electrode penetrations (Hubel and Wiesel, 1974) may be termed *slabs*. Strips of tissue containing like-tuned neurons (revealed by the 'metabolic stain' deoxyglucose; see Hubel, Wiesel and Stryker, 1977, 1978) may be termed *isoslabs* and progressions of tunings, *sequence slabs* (Hubel and Wiesel, 1974, Albus, 1975b).

[1] Current address: J. I. Nelson, Working Group in Biophysics, Dept. of Physics, Philipps University, D-3500 Marburg, F.R. Germany.

Slab organization is of more interest here than the classic column. Anatomical innervations and functional interactions can occur within sequence slabs which change as a function of position along a stimulus dimension, while separation in visuotopic space is held relatively constant, and conversely for isoslabs. I term these interactions *domain interactions*.

Originally I wished to transpose Gestalt lateral interactions to a more ordered cortex. By taking lateral interactions from the spatial domain to the orientation domain, orientation illusions could be treated as the Mach bands of visual cortex (Nelson, 1969). It was also my hypothesis that the small contour displacements which constitute sensory fusion in binocular vision must have the same basis as the equally small contour displacements of figural illusions. By transposing the concepts of lateral interactions from the orientation domain to the retinal disparity domain, a domain interaction model of sensory fusion could be created (Nelson, 1975). My goal was to emphasize the existence of instability and sensory coding errors in binocular vision, because the area had traditionally been too concerned with precise correspondence measurements. As the instabilities in binocular vision were not warmly appreciated problems, I introduced domain interactions as a means of solving the disparity detection problem in random dot stereograms, a question clearly formulated in the research of Julesz (1971).

TWO ENTRY-LEVEL PROBLEMS

Orientation illusions and the disparity detection problem in random dot stereograms are perceptual problems simple enough to be attacked with current neurophysiological techniques. These problems appear to be very different disturbances of the sensory coding process. However, they share the requirement that one must deal with interactions among responses to more than one stimulus. *Orientation* is coded well for one contour, but we make errors when judging angles formed by two contours (Zöllner, 1860; Carpenter and Blakemore, 1973). Something is 'wrong' (nonveridical) in the mechanism for coding the relationship between contours.

For *retinal disparity* the repeating nature of random dot stereograms makes extraction of disparity information impossible on the basis of point-by-point analysis—the only analysis which one neuron with two receptive fields can perform alone. The rules for random dot stereogram generation guarantee that a black with black or a white with white match will be found at 50 per cent of the positions one might test when seeking the correct retinal disparity. These matches would satisfy the stimulus requirements of binocular cortical neurons, and so false responses would occur. The response at incorrect disparity values is termed *matching noise*. Resolving ambiguity to extract the disparity signal from 'matching noise' is referred to as *globality*. The true disparity value may be identified because it is everywhere the same, or

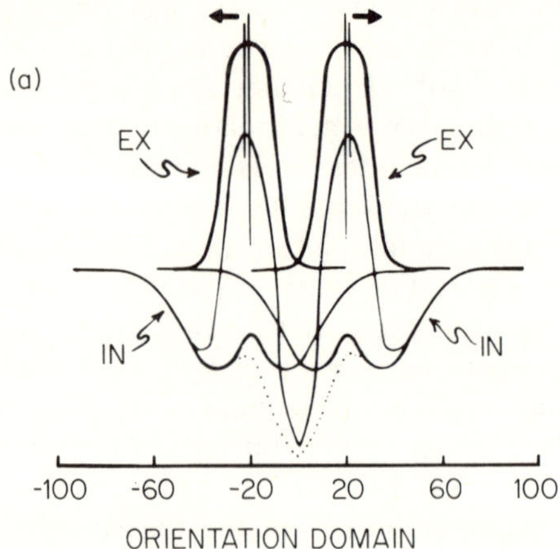

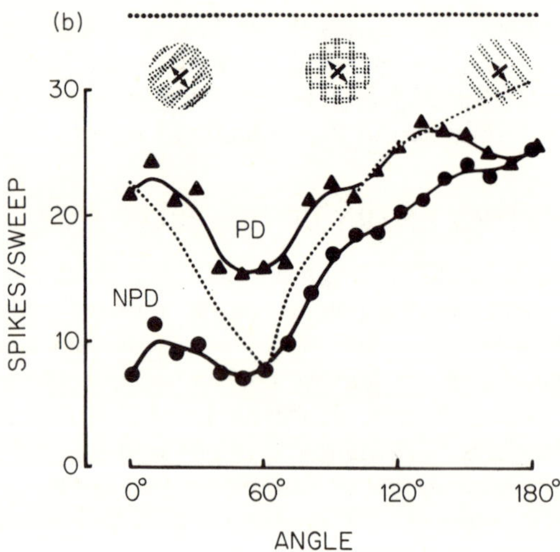

Figure 1 (a) Model for orientation contrast. EX, profiles of primary excitatory activity along the orientation domain in response to contours differing 40° in orientation; IN, orientation domain inhibition. The dotted line shows the summated orientation domain inhibition and arrows the net response after subtracting domain inhibition from the primary excitation. It displays peaks and overall activity displaced to domain positions appropriate to an angle larger than 40°. (b) Inhibition of a direction-selective simple cell in cat striate cortex by two remotely presented gratings forming various angles to each other, as schematized by the three inset 'donuts'. The cat sees a small bar well centered on the receptive field of the neuron under test and moving back and forth at the optimal orientation (center of the insets). The centered bar raises the cell's discharge to the level indicated by the dotted line. The receptive field is shielded from all other stimulation by a mask which is larger (10° diameter) than all

changes gradually, while matching noise takes on multiple and abruptly changing values. Finding this disparity requires that something in the visual system needs to respond to the relationships between disparities across multiple visual field locations.

Orientation contrast

When we view acute angles, the presence of one angle arm can distort the perceived orientation of the other; the arms appear to differ more in orientation than they do physically. This is therefore an instance of simultaneous contrast (difference enhancement), but here the contrast occurs in the orientation domain, not in the perceived brightness domain (Mach bands). Orientation contrast is a component of many figural illusions studied since psychology's earliest days (e.g. Heymans, 1897). Today, one may hope to explain the effect by mutual inhibition amongst orientation selective neurons (Carpenter and Blakemore, 1973).

Model

A typical model for orientation contrast is shown in Fig. 1. Cortical neurons are arrayed along the *X* axis according to their orientation preferences, such as they are arrayed within an orientation sequence slab. If a straight line inclined 20° counterclockwise from vertical (−20°) is presented, neurons tuned to −20° will respond best, with lesser activity in suboptimally tuned neurons, so that the overall pattern of excitatory activity might appear as portrayed in the uppermost curve of Fig. 1a (curve EX). Note that this curve represents the pattern of activity in an ordered population of neurons when a fixed stimulus is presented, not the tuning curve for one neuron when the stimulus is varied.

The model postulates that active units propagate inhibition upon their neighbors in the orientation domain and that this inhibition varies with difference in orientation, in the absence of any change in spatial separation.

---

conventional excitatory and inhibitory regions of the receptive field by a comfortable margin: hence the hole in the donuts. Large moving gratings 47° in diameter are presented by a projection system only to regions beyond the neuron's conventional receptive field and have an orientation-selective influence upon the neuron, presumably via intracortical pathways. Gratings counter-rotate; the bisector of the varying angle which they make with each other remains stationary at the test neuron's optimal orientation. Triangles, PD: inhibited responses as a function of orientation difference between the two gratings when their motion is in the preferred direction for the neuron; circles, NPD: non-preferred direction. The dotted line shows the predicted PD function for angles, derived by summing two orientation domain inhibition tuning curves, each elicited by one grating presented in isolation.

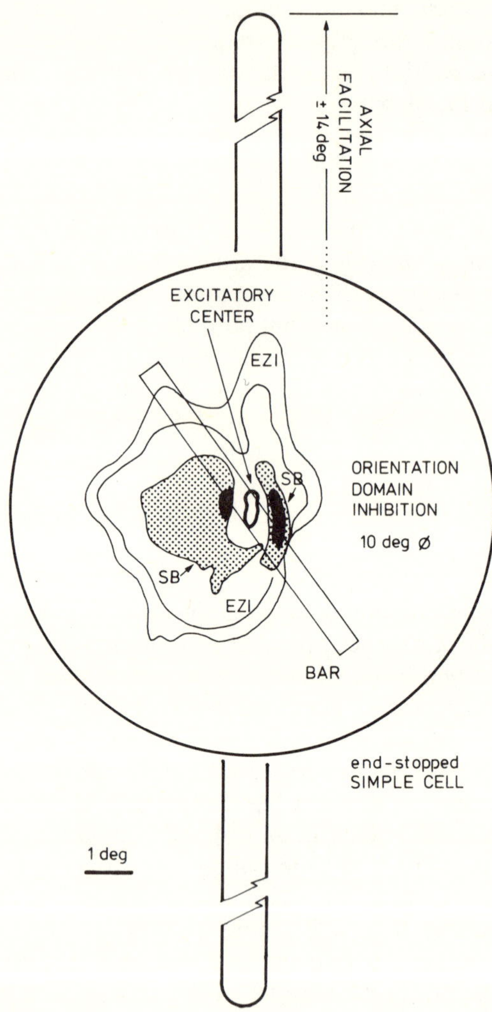

Figure 2 Topographical map showing classical receptive field components of an end-stopped ('hypercomplex') simple cell in cat striate cortex. Response is raised to over three times the background level by stimulation in the small excitatory discharge center (XDC). The cell is unimodal. SB: two flanking inhibitory sidebands. Inhibition is stronger in the left SB (black, total; shaded area and outer contour line, inhibition significant at the 1 per cent confidence level) than in the right SB (never total; 1 and 10 per cent levels shown). EZI: end-zone inhibitory regions. The end-zones confer length specificity; the optimal bar length would be about 1.6°. The sidebands convey positional specificity, making the cell selective for retinal disparity (the cue for stereoscopic depth) and orientation (response falls to half maximum for an orientation change of 18.6°). The 7 x 0.5 bar at 37° inclination falls across both inhibitory sidebands and would inhibit the cell. To optimally excite the cell, a 1.6° long bar

This inhibition is the lower curve labeled IN in Fig. 1a. If a second contour is presented, differing in orientation as would the second arm of an angle, then two distributions of excitation and inhibition will be present. Simple summation of the inhibition gives the dotted curve; linearly combining this summed inhibition with the excitation produces a distorted curve whose two peaks are displaced away from each other (arrows). This is taken to be the neurophysiological correlate of the perceptual displacement of angle arms away from each other. Perhaps the overall distribution of activity rather than the peak is important for sensory coding; either will work.

Neurophysiology: the receptive field

Conventional inhibition shapes central excitation

Orientation domain activity consists firstly of conventional *excitation* elicited by stimulation of central excitatory discharge regions in cells' receptive fields. In the view of the Canberra (Australia) group, orientation selectivity is imposed upon this broadly tuned excitation in simple cells by inhibitory regions of the conventional receptive field (Henry, Dreher and Bishop, 1974). These inhibitory areas are the inhibitory sidebands. While not themselves orientation selective (Bishop, Coombs and Henry, 1973), the inhibitory sidebands are capable of totally suppressing the neuron's response if misoriented or mispositioned contours fall across them. Inhibitory end-zones analogously confer length selectivity. These receptive field areas are illustrated in Fig. 2.

Domain inhibition

Experiments in cat striate cortex have demonstrated inhibitory inputs to most simple cells when neighbouring cells are stimulated. This inhibition of a test

---

must be vertically oriented and positioned within the XDC. Probe bar for these data, 0.5° long.

Changing to a probe bar 2° long brings out the EZI regions, which require more spatial summation (significance levels 0.1 and 10 per cent shown). At their weak outer limits, sideband and end-zone inhibitory regions merge. Direction preference and velocity selectivity for moving stimuli cannot be predicted from the spatial disposition of these excitatory and inhibitory areas.

Intracortical domain interactions can be driven from visual field areas beyond the limits of classic receptive field regions. Orientation domain inhibition arises isotropically around the receptive field (large circle). The tuning curve for orientation domain inhibition is wide; for this cell, there is a broad inhibition maximum centered around the vertical. Orientation domain facilitation spreads broadly in the spatial domain, but is confined to a region coaxially aligned with the receptive field XDC and sharply tuned in the orientation domain. This particular cell did not display facilitation.

neuron may be observed when none of the classical receptive field areas of the test neuron are themselves directly stimulated (Nelson and Frost, 1978), thus suggesting an intracortical origin. The inhibition possesses the key property required of a cellular mechanism for orientation contrast, namely it is a function of orientation. The remotely driven inhibition always falls off towards zero at orientations 90° from what is optimal for the test neuron's central excitatory response.

In some cells the inhibition never reaches zero, implying that a nonoriented and possibly noncortical component is added to the tuned component. The remotely driven, orientation selective and presumably intracortical inhibition is *orientation domain inhibition*. The summation of sharply tuned, central excitation (i.e. the conventional orientation tuning curve, as shaped by sideband inhibition) and broadly tuned, peripheral domain inhibition will give a 'Mexican hat' function to activity across the orientation domain with large, real-life stimuli.

Orientation domain inhibition is ill-suited to augment the stimulus selectivity of the central excitatory response. It is too broadly tuned, it originates too uniformly around the receptive field and it integrates too broadly across the spatial domain to respond to small intrusions of slightly misoriented contours.

Summation with angles

Granted that orientation domain inhibition exists, the summation of inhibition at middle orientations (i.e. 0° in Fig. 1a) must next be measured. We measure the strength of inhibition felt by a fixed test neuron while presenting contours with 'flanking' orientations in a broad annulus around the test neuron. Two gratings are rotated equally and oppositely to frame the test neuron's orientation in a series of ever-expanding acute angles. Orientation domain inhibition from the two gratings summates; the summated inhibition is greatest when the two gratings form an acute angle of 55° with each other. With activity within the angle relatively more reduced, activity and presumably percepts are displaced to values outside the angle, in accord with the orientation contrast (acute angle) effect. To confirm further the domain inhibition model of orientation contrast, a tuning curve for inhibition as a function of angle was derived by summing inhibition from single gratings measured separately. The theoretical curve (dotted line, Fig. 1b) agrees well with the measured angle-tuning curve.

Columns

There is selective inhibition from remotely presented gratings which is very much like that initially predicted from psychophysical observations in man.

Columnar organization of the cortex would simplify the generation of interactions which must be ordered in ways other than merely spatially.

Dendritic arbors

A neuron's receptive field is well correlated with the spread of its dendritic tree, and it is reasonable to assume greatest input at the arbor's center (Peichl and Wässle, 1983). A center-weighted dendritic arbor spreading in an orientation sequence slab would have the most connections at one central orientation and progressively fewer elsewhere. A total spread of about 1 mm, *confined to the slab*, would span 180° in the cat (Albus, 1975b, Hubel and Wiesel, 1974). Inhibitory *output* from such a neuron would give the sort of broad tuning curve we have measured for orientation domain inhibition.

Clustered axonal arbors

The inhibitory interneuron for orientation domain inhibition can have randomly arborizing dendrites. It needs no morphogenic control other than growth confinement to a sequence slab. Spread within the *spatial domain* must be considered next.

Domain inhibition has a circular spatial integration area 8° in diameter for neurons of less than 5° eccentricity in cat striate cortex. The 8° input area is too big to be achieved by the 1 mm diameter dendritic arborization which generates the 180° domain inhibitory tuning curve. It can not even be achieved within one sequence slab, whose narrowness precludes a circular area of spatial integration. One must jump to sequence slabs in other visuotopic areas, but the jump must land on the same orientation. The orientation tuning at the axonal end must match the tuning found at the center of the dendritic arbor. Such axons would be observed to arborize periodically. The periodicity is the separation of identically tuned neurons within and across orientation sequence slabs. Such a lattice-like structure has much in common with the anatomical observations of Rockland and Lund (1982, 1983). The connectivity rule for domain *inhibition* stipulates that a certain orientation be selected, but not that the spatial domain be traversed only along, or at right angles to, this orientation, as proposed in the recent Mitchison and Crick (1982) model. It is domain *facilitation* which confirms their model (below).

The above scheme can give us orientation domain inhibition, a heretofore neglected inhibitory input exhibited by at least 89 per cent of all cells in cat striate cortex and capable of inhibiting over 70 per cent of the conventional, central excitatory response in 23 per cent of the simple cells we have tested. The inhibitory interneurons mediating this input may be the cells with long, periodically arborizing axons reported by Gilbert and Wiesel (1983). The

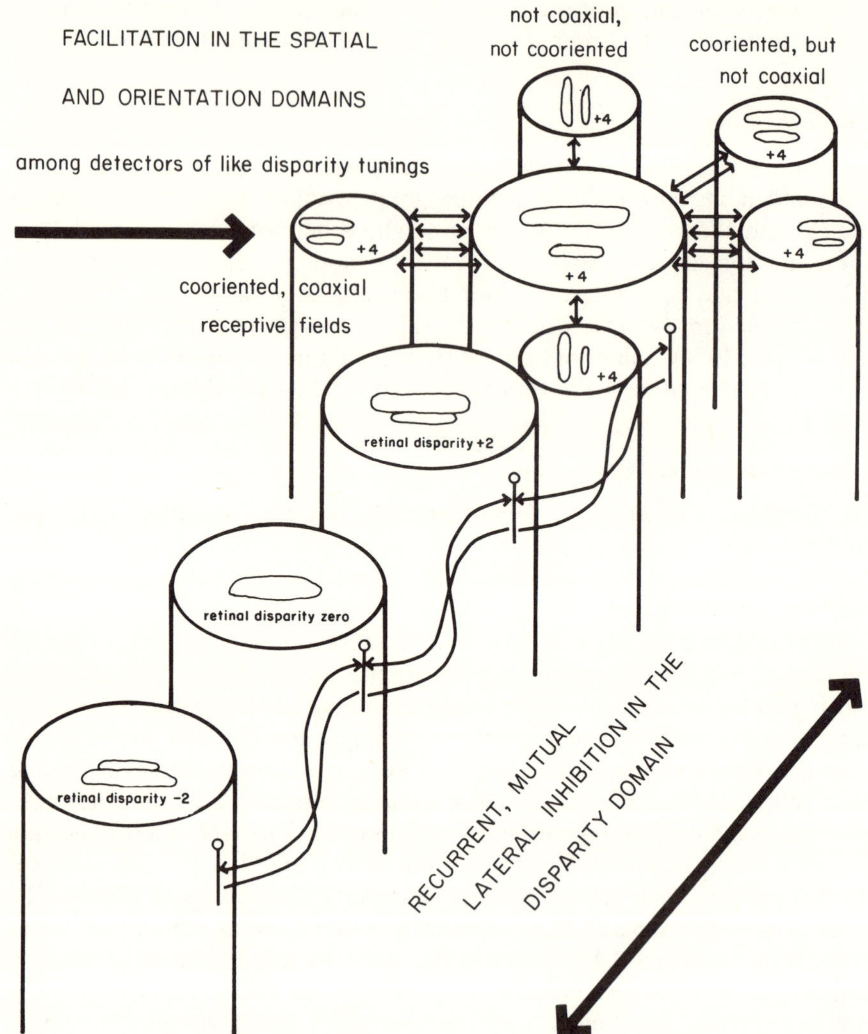

Figure 3 Cortical model for achieving globality in stereoscopic vision. Binocular neurons of similar disparity tuning are aggregated into radial columns, schematized as cylinders. These columns form arrays or sequence slabs with a regular progression of tunings (diagonal row of cylinders labeled with disparity tunings of −2, 0, +2, etc.). Neighboring neurons within a retinal disparity sequence slab are mutually inhibitory. This inhibition suppresses matching noise. Columns with matched tunings are mutually inhibitory. This inhibition suppresses matching noise. Columns with matched tunings are mutually facilitatory (top of figure), especially those matched in orientation and aligned coaxially in space, in addition to being matched in disparity. Only a few degrees of the visual field are covered by all the tissue portrayed. Inhibition, which spreads broadly along the retinal disparity domain, is confined to

assignment to these neurons of a role in the generation and propagation of domain inhibition is to be distinguished from the generation of classical receptive field regions, a role which has been suggested previously (Mitchison and Crick, 1982; Gilbert and Wiesel, 1983). An additional role for clustered axonal arbors in facilitation is considered below.

Globality

Globality is the ability of the binocular visual system to suppress local matching noise yet respond strongly to disparity values on the basis of their widespread occurrence. Globality may be achieved from mutual inhibition, spreading broadly within the disparity domain and narrowly in all directions across space (Nelson, 1975, 1977). Mutual *facilitation* confined to like-tuned disparity-selective neurons is postulated to strengthen the response of correctly tuned neurons. The facilitation spreads widely in the spatial domain, but not isotropically (see Fig. 3), in the interests of strengthening the representation of object edges. Recurrency of domain inhibition brings hysteresis to the neural response, as would seem to be required by the hysteresis in the range of the binocular sensory fusional response (Panum's area: see Fender and Julesz, 1967; Nelson, 1975; Hyson, Julesz and Fender, 1983).

This model was the first physiologically plausible one to be suggested, but it was only qualitative. Since then, more rigorous formulations have appeared (Forshaw, 1979; Grimson, 1981; Hirai and Fukushima, 1978; Marr, Palm and Poggio, 1978; Marr and Poggio, 1976; Sugi and Suwa, 1977; Trehub, 1978). Most of these models have used network concepts similar to domain interactions, although Marr and Poggio (1979) have vascillated between cooperativity and the assignment of different aspects of stereopsis to different spatial frequency channels.

The psychophysics of globality strongly suggests mutual facilitation among binocular neurons with matched disparity tunings. In monocular contour perception, neurons with matched *orientation* tunings may also be mutually facilitatory, but an analysis of the literature on figural illusions does not reveal effects which require this.

To accommodate both inhibition and facilitation, the globality model calls for sequence slabs (progressions) of disparity tunings and isoslabs. Mutual inhibition occurs within sequence slabs and mutual facilitation occurs within isoslabs. Units with cooriented, coaxially aligned receptive fields are to be

---

the visuotopic space of a few hypercolumns. Facilitation, which is confined to one retinal disparity value, spreads broadly across visuotopic space. These different spread rules for inhibition and facilitation appear to apply generally in both the orientation domain (illusions) and the disparity domain (globality). Reproduced by permission of Academic Press from Nelson (1975).

preferentially interconnected (mutually facilitated; see Fig. 3). The postulated neurophysiological interactions remain to be discovered; little is yet known about the functional architecture of the disparity domain (Blakemore, 1970; Bishop, 1973).

BEYOND THE ENTRY LEVEL

Facilitation for figure synthesis

Interactions in the disparity domain achieve globality and, perhaps beyond that, sensory fusion and a more realistic (variable) kind of retinal correspondence. Orientation domain inhibition seems a liability: with orientation contrast, percepts are less veridical. However, one can look beyond orientation contrast to find the larger goal which orientation domain interactions serve. That goal is figure synthesis.

Figure synthesis requires facilitation. Because everyday objects are large and receptive fields are small, it would serve the needs of figure synthesis if neurons with cooriented, coaxially aligned receptive fields would possess mutual facilitation (Nelson, 1975; see Fig. 3 above). In this way, a group of neurons which are active because they are responding to a common object would respond more strongly than a neuron stimulated in isolation.

Isolation of the attended figure

Facilitation of well-matched neurons coupled with inhibition spreading along domains among non-matched (differently tuned) neurons has been invoked in at least one model of *figure–ground segmentation* (Dev, 1974). This approach can achieve an invariant pattern of neural activity tolerant of spatial changes in stimulation (below). Without this invariance, we are hard pressed to deal with generalization and pattern recognition (Edelman and Reeke, 1982) or to create a stable internal perceptual environment within which to localize ourselves as we move through the world.

To understand isolation of the attended figure in focal, attentive visual perception, suppose an entire topographic area converges onto a nontopographic area. Further, thanks to the facilitation and inhibition of figure–ground segmentation processes, only responses to the constantly attended figure are ever active on these convergent pathways; the background changes, but there is never a response to it. The topographic area has now 'netted' the object of interest: in the nontopographic area, the object's representation will be constant despite visual field displacement. We are free of the tyranny of topographic representation. Let separation in tissue now be used to encode the separation between real and ideal, between what we see and what we are looking for.

Neurophysiology: axial facilitation

If a simple cell's discharge level is elevated by a small, centrally placed stimulus, facilitation can be observed when additional contours are presented beyond the classic receptive field. This facilitation most often takes the form of relief from orientation domain inhibition (disinhibition).

Recall that the inhibitory effect exhibits a tuning curve up to 180° wide and originates broadly in space. Disinhibition, contrariwise, is more selective. To disinhibit a neuron, one bar of the grating stimulus must be aligned with the excitatory discharge center of the conventional receptive field. Other experiments demonstrate an orientation selectivity for the facilitation, with half-width at half-height sometimes as narrow as that of the conventional excitatory orientation tuning curve itself. Facilitation occurs preferentially amongst neurons with cooriented, coaxially aligned receptive fields (Nelson and Frost, 1985). The spatial integration function along the length of the coaxial, cooriented strip reaches 28° for units with eccentricities less than 6°, exceeding the scatter of geniculocortical inputs several-fold (Albus, 1975a) and suggesting a lateral intracortical pathway.

Neurons possessing cooriented, coaxially aligned receptive fields need mutually facilitatory interconnections, and mutual facilitation must hop across large separations in tissue. Dendrites may arborize randomly, provided that they continue their arborizations within the walls of an isoslab; axons must carry the facilitation across as much as 14° of visuotopic space. The axons must always travel along a visuotopic direction aligned with the orientation eliciting the facilitation itself. Alas, few isoslabs of neurons tuned to X degrees will make a run across the visual field which itself has an inclination of X degrees. Therefore, the interneuron's axons will have to hop from slab to slab, arborizing at neurons of the same orientation in each one. This neuron, then, is immersed head and feet, dendritic and axonal arbors, in tissue reeking of the same orientation specialization. In sum, there are specific connectivity rules required by the neurophysiological observations, and the rules in turn suggest that each orientation tuning expresses a unique cell-surface specificity. Curiously, there is evidence that multiple sclerosis can identify and selectively attack cells according to their orientation specificity (Kupersmith *et al.*, 1984).

One class of periodically arborizing axons described by Gilbert and Wiesel (1983) provides candiates for these *facilitatory* interneurons, both because of the periodicity of their arborizations and the visuotopic orientation of the axons along the optimal orientation of the cell's receptive field. Other cells with axons at right angles to the optimal orientation have no role to play in this facilitation, but could spread orientation domain *inhibition* across the spatial domain. The overall system is similar to the proposals of Mitchison and Crick (1982), with the refinement that the system of axons *aligned* with the neuron's optimal stimulus orientation is now predicted to be facilitatory, while the system with *orthogonal* axons is predicted to be inhibitory; neither is involved in generating conventional receptive field regions.

CONCLUSION: EMERGENT PRINCIPLES OF DOMAIN INTERACTIONS

Slabs are a structural expression of intracortical interactions; they exist to make intracortical interactions easier and intracortical interactions exist to perform sensory processing. From models and limited neurophysiological data, it is a plausible conjecture that *sequence slabs support mutual inhibition and isoslabs support mutual facilitation*. Slabs support interactions by making it possible for random arborization confined to a slab to generate interactions which are selective and thus perceptually useful. For extremely important dimensions such as orientation, both global tissue ordering (slabs) and local neural targeting (clustered axonal arbors) are necessary to generate interactions with sufficient specificity.

Domain interactions provide a cellular basis for higher perceptual processing, and perception leads onward to cognition. I believe that the well-developed, modular-repeated functional architecture of the cortex makes it easier prey to a fewer number of general principles than is the case with subcortical centers. If this view is correct, some achievments which we regard as Man's highest will yield to neuroscientific enquiry first, while emotional and motivational mechanisms, which we so clearly share with animals, will be more refractory.

Science proceeds faster than we anticipate; it is important to lift one's gaze now and then to distant sights to get some grasp of where we are going. As the field matures, it will become clear that visual system modelers were every bit as hardy a band as those who stayed up all night with a cat, that many of us did both and that the efforts on all sides were worth while, for neuroscience is one of those rare enterprises by which Man may come to know himself.

ACKNOWLEDGMENT

Supported by grants from the National Institute of Mental Health, the Naval Air Systems Command and the Department of Defense-University Research Instrumentation Program, Air Force Office of Scientific Research.

REFERENCES

Albus, K. (1975a). A quantitative study of the projection area of the central and the paracentral visual field in area 17 of the cat. I. The precision of the topography. *Exp. Brain Res.*, **24**, 159–179.

Albus, K. (1975b). A quantitative study of the projection area of the central and the paracentral visual field in area 17 of the cat. II. The spatial organization of the orientation domain. *Exp. Brain Res.*, **24**, 181–202.

Bishop, P. O. (1973). Neurophysiology of binocular single vision and stereopsis. In *Handbook of Sensory Physiology*, (Ed. R. Jung) Vol. VII/3, *Central Processing of Visual Information*, Springer, New York, pp. 255–305.

Bishop, P. O., Coombs, J. S., and Henry, G. H. (1973). Receptive fields of simple cells in the cat striate cortex. *J. Physiol.*, **231**, 31–60.

Blakemore, C. (1970). The representation of three-dimensional visual space in the cat's striate cortex. *J. Physiol.*, **209**, 155–178.

Carpenter, R. H. S., and Blakemore, C. (1973). Interactions between orientations in human vision. *Exp. Brain Res.*, **18**, 287–303.

Dev, P. (1974). *Segmentation Processes in Visual Perception: A Cooperative Neural Model*, COINS Technical Report 74C–5, University of Massachusetts, Amherst.

Edelman, G. M., and Reeke Jr., G. N. (1982). Selective networks capable of representative transformation, limited generalizations, and associative memory. *Proc. Natn. Acad. Sci. USA*, **79**, 2091–2095.

Fender, D., and Julesz, B. (1967). Extension of Panum's fusional area in binocularly stabilized vision. *J. Opt. Soc. Amer.*, **57**, 819–830.

Forshaw, M. R. B. (1979). The cooperative stereo algorithm: an empirical approach. *Biolog. Cybernetics*, **33**, 143–149.

Gilbert, C. D., and Wiesel, T. N. (1983). Clustered intrinsic connections in cat visual cortex. *J. Neuroscience*, **3**, 1116–1133.

Grimson, W. E. L. (1981). A computer implementation of a theory of human stereo vision. *Phil. Trans. Roy. Soc. B*, **292**, 217–253.

Henry, G. H., Dreher, B., and Bishop, P. O. (1974). Orientation specificity of cells in cat striate cortex. *J. Neurophysiol.*, **37**, 1394–1409.

Heymans, G. (1897). Quantitative untersuchungen über die Zöllnersche und die Loebsche Täuschung. *Z. Psychol.*, **14**, 101–139.

Hirai, Y., and Fukushima, K. (1978). An inference upon the neural network finding binocular correspondence. *Biolog. Cybernetics*, **31**, 209–217.

Hubel, D. H., and Wiesel, T. N. (1974). Sequence regularity and geometry of orientation columns in the monkey striate cortex. *J. comp. Neurol.*, **158**, 267–293.

Hubel, D. H., Wiesel, T. N., and Stryker, M. P. (1977). Orientation columns in macaque monkey visual cortex demonstrated by the 2-deoxyglucose autoradiographic technique. *Nature*, **269**, 328–330.

Hubel, D. H., Wiesel, T. N., and Stryker, M. P. (1978). Anatomical demonstration of orientation columns in macaque monkey. *J. comp. Neurol.*, **177**, 361–380.

Hyson, M. T., Julesz, B., and Fender, D. H. (1983). Eye-movements and neural remapping during fusion of misaligned random-dot stereograms. *J. opt. Soc. Amer.*, **73**, 1665–1673.

Julesz, B. (1971). *Foundations of Cyclopean Perception*, University of Chicago Press, Chicago, 406 pp.

Köhler, W., and Wallach, H. (1944). Figural after-effects: an investigation of visual processes. *Proc. Amer. Phil. Soc.*, **88**, 269–357.

Kupersmith, M. J., Seiple, W. H., Nelson, J. I., and Carr, R. E. (1984). Contrast sensitivity loss in multiple sclerosis. Selectivity by eye, orientation and spatial frequency measured with the evoked potential. *Invest. Ophthal. vis. Sci.*, **25**, 632–639.

Marr, D., Palm, G., and Poggio, T. (1978). Analysis of a cooperative stereo algorithm. *Biolog. Cybernetics*, **28**, 223–239.

Marr, D., and Poggio, T. (1976). Cooperative computation of stereo disparity. *Science*, **194**, 283–287.
Marr, D., and Poggio, T. (1979). A computational theory of human stereo vision. *Proc. Roy. Soc. B*, **204**, 301–328.
Mitchison, G., and Crick, F. (1982). Long axons within the striate cortex: their distribution, orientation, and patterns of connection. *Proc. Natn. Acad. Sci. USA*, **79**, 3661–3665.
Mountcastle, V. B. (1978). An organizing principle for cerebral function: the unit module and the distributed system. In *The Mindful Brain* (Eds. G. M. Edelman and V. B. Mountcastle), MIT Press, Cambridge, Mass., pp. 7–50.
Nelson, J. I. (1969). *The Misperception of Contour*, University Microfilms, Ann Arbor, Mich. LD 00090.
Nelzon, J. I. (1975). Globality and stereoscopic fusion in binocular vision. *J. theor. Biology*, **49**, 1–88.
Nelson, J. I. (1977). The plasticity of correspondence: after-effects, illusions and horopter shifts in depth perception. *J. Theor. Biology*, **66**, 203–266.
Nelson, J. I., and Frost, B. J. (1978). Orientation-selective inhibition from beyond the classic visual receptive field. *Brain Res.*, **139**, 359–365.
Nelson, J. I., and Frost, B. J. (1985). Intracortical facilitation among co-oriented, co-axially aligned simple cells in cat striate cortex. *Exp. Brain Res.* (in press).
Peichl, L., and Wässle, H. (1983). The structural correlate of the receptive field centre of alpha-ganglion cells in the cat retina. *J. Physiol.*, **341**, 309–324.
Rockland, K. S., and Lund, J. S. (1982). Widespread periodic intrinsic connections in the tree shrew visual cortex. *Science*, **214**, 1532–1534.
Rockland, K. S., and Lund, J. S. (1983). Intrinsic laminar lattice connections in primate visual cortex. *J. comp. Neurol.*, **216**, 303–318.
Sugi, N., and Suwa, M. (1977). A scheme for binocular depth perception suggested by neurophysiological evidence. *Biolog. Cybernetics*, **26**, 1–15.
Szentágothai, J. (1983). Cortical organization—a plea for better understanding, clearer definition, and more correct use of the term column. *Behav. Brain*, **6**, 169–170.
Trehub, A. (1978). Neuronal model for stereoscopic vision. *J. theor. Biol.*, **71**, 479–486.
Zöllner, F. (1860). Ueber eine neue Art von Pseudoskopie und ihre Beziehungen zu den von Plateau und Oppel beschriebenen Bewegungsphänomenen. *Poggend. Annalen*, **20**, 500–523 and Plate VIII.

Models of the Visual Cortex
Edited by D. Rose and V. G. Dobson

CHAPTER 12

Activity-dependent self-organization of the mammalian visual cortex

W. SINGER
Neurophysiologische Abteilung, Max-Planck-Institut für Hirnforschung, Deutschordenstrasse 46, 6000 Frankfurt am Main 71, Federal Republic of Germany.

One of the factors controlling the morphogenesis of the nervous system is the electrical activity of the nerve cells. From a certain developmental stage onwards neuronal activity becomes an important shaping factor for the specification of neuronal connections. The elimination of excess climbing fibres from Purkinje cells in the cerebellum (Mariani and Changeux, 1981) and of supernumerary motor neurone axons from muscle cells (Jansen and Lomo, 1981), the retraction of supernumerary callosal fibres (Innocenti and Frost, 1979), the reorganization of the retinotectal map (Gaze *et al.*, 1979) and the segregation of the afferents from the two eyes in the tectum (Meyer, 1982) and the visual cortex (Stryker, 1981) are all activity-dependent processes. In the visual cortex of mammals these activity-dependent developmental processes persist into postnatal life and hence are influenced by visual experience (for reviews see Movshon and Van Sluyters, 1981; Sherman and Spear, 1982). In the first part of this chapter I shall review work that was aimed at elucidating the algorithms which guide these activity-dependent maturation processes. In the second part I shall attempt to derive from these developmental mechanisms some predictions on the organization of the mature cortex.

BINOCULAR INTERACTIONS

Experiments in which the cooperation between the two eyes is interfered with indicate that the efficacy of excitatory transmission from the eyes to cortical neurones can be modified by the activity conveyed in these modifiable connections (for reviews of the extensive literature see Movshon and Van Sluyters, 1981; Sherman and Spear, 1982). Both an increase and a decrease of the efficacy of excitatory transmission have been observed (Wiesel and Hubel, 1963, 1965). Evidence is available that these modifications are associated with changes of the amplitude of excitatory synaptic currents (Mitzdorf and Singer, 1980; Singer, 1977a). Changes in inhibitory mechanisms appear to be less prominent and may be secondary to modifications of excitatory pathways (Sillito, Kemp and Blakemore, 1981). A series of deprivation experiments, in which kittens were raised with reverse monocular occlusion and exposed selectively to contours of a single orientation, revealed a set of modification rules that seem to govern these activity-dependent modifications (Cynader and Mitchell, 1977; Rauschecker and Singer, 1979, 1981; Singer, 1976; Singer, Rauschecker and Werth, 1977). These rules closely resemble those postulated by Hebb (1949) and Stent (1973) for adaptive synaptic connections. The main implication of these rules is that the direction of a change—an increase or decrease of the efficacy of excitatory transmission—does depend on the correlation between the activation of interconnected pre- and postsynaptic elements, rather than on the level of activation of the afferent pathways or of the postsynaptic target cells. The experimental results indicate that a modifiable connection increases its efficacy when the probability is high that the afferent pathway is active in temporal contiguity with its postsynaptic target cell. Conversely, a connection becomes less effective when the probability is high that the afferent pathway is silent while the postsynaptic cell is active. Irrespective of the level of activation of the afferent pathway no differential changes of excitatory coupling seem to occur, however, when the postsynaptic target is inactive.

Rules for activity dependent modification of excitatory transmission

A —< ○ — C

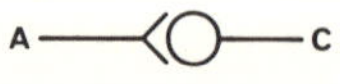

State of afferent pathway A	State of postsynaptic cell C	Efficacy of transmission A → C
active	active	increases
inactive	active	decreases
active or inactive	inactive	no change

Activity dependent selection of converging pathways

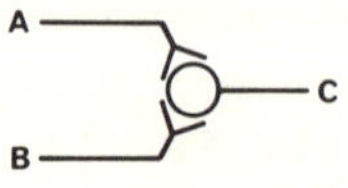

A, B → ○ — C

Statistical properties of activity in A, B, C	Resulting circuitry
A — B; A \ C; B / C	A and B consolidate on C ≙ association of A and B
A ⇸ B; A \ C; B ⇸ C	A consolidates on C on the expense of B ≙ competitive disconnection of B
A ⇸ B; A ⇸ C; B / C	B consolidates on C on the expense of A ≙ competitive disconnection of A
A — B, A ⇸ C, B ⇸ C; A ⇸ B, A ⇸ C, B ⇸ C	A and B disconnect if C is activated or A and B remain on C if C is inactive

— correlated activation
⇸ uncorrelated activation

As indicated in Table 1, such activity-dependent modifications have a dual effect when extended to conditions where several subpopulations of afferents converge onto a common target cell. Pathways which convey correlated activity and are capable of driving the follower cell will increase their gain. Hence, the modification algorithm has an associative function in that it enhances interactions between neuronal elements whose activity is correlated. However, the same algorithm also mediates competitive interactions. If converging pathways convey uncorrelated activity those afferents which drive the postsynaptic cell most effectively (or, in more general terms, which are more often active in contiguity with the postsynaptic target) will increase their efficacy at the expense of other afferent connections.

CENTRAL CONTROL OF ACTIVITY-DEPENDENT MODIFICATIONS

There is now ample evidence which indicates that retinal signals are necessary but not sufficient to induce the above-described modifications of cortical connectivity. Retinal activity fails to induce modifications under the following conditions: (a) when retinal coordinates are in gross mismatch with other sensory maps, e.g. when the eyes are rotated by 90° or more (Crewther *et al.*, 1980; Singer, Tretter and Yinon, 1982; Singer, Yinon and Tretter, 1979), (b) when the proprioceptive signals from extraocular muscles are abolished (Buisseret and Gary-Bobo, 1979; Buisseret and Singer, 1983; Trotter, Gary-Bobo and Buisseret, 1981), (c) when eye movements are suppressed by muscle relaxants (Buisseret, Gary-Bobo and Imbert, 1978; Freeman and Bonds, 1979), (d) when the kittens are anaesthetized (Freeman and Bonds, 1979; Singer, 1979; Singer and Rauschecker, 1982), (e) when ascending projections of the reticular activating system are disrupted by diencephalic lesions, a condition that is accompanied by behavioural abnormalities resembling a sensory hemi-neglect syndrome (Singer, 1982), (f) when cortical norepinephrine is depleted with 6-OH-dopamine (Kasamatsu and Pettigrew, 1979, but see Bear, Paradiso and Daniels, 1982) and (g) when the muscarinic action of acetylcholine is antagonized with locally applied scopolamine (Geiger and Singer, in preparation).

Taken together these results indicate that modifications in response to retinal activity require the presence of additional, non-retinal gating signals. A participation of ascending modulatory systems in this gating function is suggested by the evidence summarized above and is corroborated by the results of stimulation experiments. When retinal signals are conditioned with electrical stimulation of central core structures such as the mesencephalic reticular formation or the medial thalamic nuclei, long-term modifications of cortical functions can be obtained even when the kittens are anaesthetized and paralysed (Singer and Rauschecker, 1982).

Indications on the possible mechanisms of this gating process can be derived from the electrophysiological effects of central core stimulation. Activation of mes- and diencephalic central core structures facilitates transmission of retinal signals in the lateral geniculate nucleus (LGN) and in the visual cortex (for reviews, see Singer, 1977b, 1979). At least part of these facilitatory effects are mediated by the muscarinic action of acetylcholine (Francesconi, Müller and Singer, in preparation; Sillito, Kemp and Berardi, 1983). Together with the evidence that intracortically administered scopolamine abolishes modifications in response to retinal stimulation this suggests that ascending cholinergic projections which raise excitability in thalamic and cortical pathways have a permissive role in the gating of plasticity. At the cortical level the reticular facilitation of responses to retinal signals is further associated with a transient decrease of the extracellular Ca^{2+} concentration, reflecting a stimulus-dependent flux of Ca^{2+} ions from extra- to intracellular compartments (Geiger and Singer, 1982 in preparation). Interestingly, these Ca^{2+} fluxes are measurable only during the developmental stage, during which cortical functions are modifiable and they occur only under conditions of stimulation that are suitable for the induction of long-term modifications. These results suggest (a) that one prerequisite for the occurrence of activity-dependent modifications might be the activation of Ca^{2+}-dependent processes and (b) that the activation of these Ca^{2+}-dependent processes requires facilitation by non-specific modulatory projections.

Two possibilities may be considered for the action of these modulatory projections. The biochemical processes underlying long-term modifications might require the presence of certain modulatory transmitters such as, for example, norepinephrine (Kasamatsu, Pettigrew and Ary, 1979) and acetylcholine. An alternative is that the adaptive process requires that a critical threshold of neuronal activation be exceeded. This threshold might reflect, for instance, the need to activate voltage-sensitive Ca^{2+} channels in the dendrites of cortical neurones. In that case the control of adaptive changes would not be exerted by a specialized gating system but would be realized through the cooperative action of the many and distributed input systems that control the excitability of cortical neurones. I favour this latter possibility for two reasons. First, it accounts well for the evidence that numerous very different manipulations interfere with cortical plasticity. Second, the control of adaptive changes would be accomplished by the neuronal networks that are actually involved in the processing of sensory activity patterns. This overcomes the problem of postulating a specialized gating system for plasticity whose organization would probably have to be as complex as that of the processing circuits themselves.

In conclusion, the evidence reviewed above suggests that experience-dependent modifications of striate cortex connectivity are controlled both by local and by global factors. The direction of a change, i.e. whether the efficacy of a synaptic connection increases or decreases, depends on the local correlation of activity in the neuronal elements whose interconnections undergo modification. Whether a change does occur, however, is determined by the state of additional non-retinal input systems.

DEVELOPMENTAL RULES AND CORTICAL MECHANISMS

The evidence reviewed above warrants the conclusion that a functional criterion—the degree of correlation of neuronal activity—guides the specification of neuronal connections during an early phase of postnatal development. The fact that this activity-dependent specification process is gated implies further that the developing brain is able to select according to its central state the instances at which sensory activity is enabled to assume the role of a shaping factor. Such a selection algorithm is ideally suited to solve specification problems which require that neuronal connections be selected according to functional criteria. In the following paragraphs I shall give examples of such specification problems.

Binocular correspondence

The formation of binocular neurones with precisely corresponding receptive fields in the two eyes poses a specification problem that appears solvable only if the development of connections is guided by functional criteria. The pathways connecting the two eyes with the common binocular target cells in the visual cortex have to be selected so that they come from precisely matching retinal loci. Retinal correspondence depends on parameters such as the size of the eyes, the position of the eyes in the orbit, the interocular distance, etc.—parameters which are obviously under the influence of manifold epigenetic factors. Which retinal loci in the two eyes will actually be matching with each other in the mature system can therefore be anticipated only with moderate accuracy. Pathways originating from corresponding retinal loci can be identified, however, with great precision using functional criteria. These pathways convey highly correlated activity patterns when the animal is fixating a real-world object in everyday life. Since the activity-dependent modifications occurring in developing striate cortex selectively stabilize connections which convey correlated activity they are ideally suited to optimize the correspondence of binocular connections.

In this context, a crucial role can also be assigned to the gating of adaptive changes by non-retinal afferents. It is obvious that the selection process should take place only when the animal is attentively fixating a target and it should not occur at instances when the eyes are not properly aligned or are moving in an uncoordinated way. In the latter cases the signals from the two eyes are uncorrelated and the activity-dependent modifications would lead to a disruption rather than to an optimization of binocular connectivity. The gating functions mentioned above, in particular those exerted by the proprioceptive afferents from the extraocular muscles and by the ascending reticular activating system, could take care of this additional problem. They would be ideally suited to allow modifications to occur only when the conditions of attentive fixation are fulfilled.

Orientation selectivity in large-field neurones

Another specification problem is raised by cortical cells which possess large, orientation-selective receptive fields. This holds for numerous complex cells (Hubel and Wiesel, 1962) and it is particularly the case for the fraction of cortical neurones, located preferentially within layer VI, whose receptive fields are very elongated (Gilbert, 1977; Gilbert and Kelly, 1975). These cells with large receptive fields appear to integrate excitatory input from cells whose receptive fields have the same orientation preference and whose position in the cortical sheet is such that their receptive fields line up along lines parallel to the axis of preferred orientation (see Fig. 1).
From the magnification factor of the retinocortical mapping it can be inferred that the large-field neurones must sample activity from cells that are distributed over several millimetres within the cortical sheet (Hubel and Wiesel, 1974). Moreover, because of the columnar organization of the striate cortex (Hubel, Wiesel and Stryker, 1978), sampling has to be discontinuous, sparing orientation slabs whose orientation preference does not correspond to the orientation of the respective large-field neurones. The sampling problem is further complicated by the fact that the axis in the cortex along which sampling has to occur is a function of the orientation preference of the large-field neurone: cells with extended vertical receptive fields have to sample along axes that are parallel to the representation of the vertical meridian while cells with horizontally oriented fields must sample along axes parallel to the representation of the horizontal meridian, and so on. An anatomical substrate for such a discontinuous sampling process could be the far-reaching axon collaterals of cortical pyramidal cells (Lund and Boothe, 1975; Szentagothai, 1975). Particularly relevant in this context is the evidence, first, that these axon-collaterals form several discrete tufts whose spacing conforms with the spacing of columnar systems (Gilbert and Wiesel, 1979, 1983) and, second, that intracortical projections tend in general to be periodic (Rockland and Lund, 1982, 1983).

It is obvious that the development of such highly specific connections between cortical neurones requires a powerful selection algorithm that is capable of identifying the orientation preference and retinal position of each of the selectively interconnected cells. It is difficult to see how this could be accomplished by genetic instructions alone. Again selective stabilization of interconnections according to functional criteria could provide a solution to this specification problem.

Because of the continuity of contours in the outer world the probability of correlated activation is high for cortical neurones that share the same orientation preference and whose receptive fields are adjacent. This is particularly true if one considers that intracortical inhibitory interactions

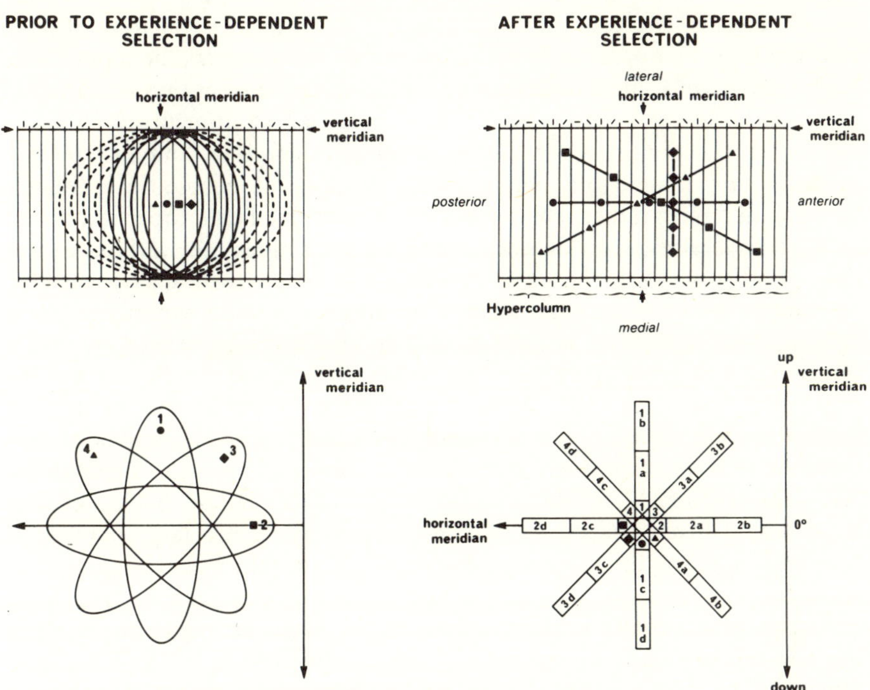

Figure 1 Schematic representation of experience-dependent selective stabilization of intracortical connections between first-order orientation selective cells and higher order orientation-selective neurones. The trajectories of intracortical connections required for the formation of the elongated receptive fields of four different neurones (1 to 4) are indicated on the cortical maps in the upper two cases. The orientation and location of the corresponding receptive fields are indicated in the visual field maps below. It is assumed that, prior to experience, higher order cells sample from numerous first-order cells which have differing orientation preferences (left pair of graphs). Correspondingly, the higher order cells have broadly tuned, poorly delineated receptive fields. After experience-dependent selective stabilization of connections, higher order cells sample selectively only from first-order cells sharing the same orientation preference (right pair of graphs). Correspondingly, the higher order cells now possess sharply tuned receptive fields. The subdivisions in the sharply tuned large receptive fields of the higher order cells are meant to reflect the receptive fields of the respective first-order cells. Their location in the visual field corresponds to the location of the first-order cells in the cortical map.

between adjacent columns have the effect of preventing simultaneous responses to nearby contours with differing orientations (Blakemore and Tobin, 1972). Thus, the very same selection algorithm that we derive from experiments on binocular convergence could serve to solve the specification problem raised by large-field, orientation-selective neurones.

If activity-dependent selection of circuits were indeed responsible for the development of orientation selectivity in higher order cortical cells with large receptive fields, two conditions must be fulfilled: (a) the population of cells, referred to as first-order cells in the following, that provides the input to the higher order cells, must develop orientation-selective receptive fields before the experience-dependent selection process starts; (b) prior to selection the intercolumnar connections should be unselective and thus coupling strength of horizontal interactions should be a function of geometrical distance between neurones rather than a function of receptive field properties. As far as experimental evidence is available these predictions are, without exception, compatible with the data. A small fraction of cortical cells develop orientation-selective receptive fields in the absence of visual experience; these cells are located in layer IV, are monocular and have simple receptive fields (Buisseret and Imbert, 1976). This makes them good candidates for the postulated first-order cells. Without experience, however, the majority of cortical cells develop only poor orientation selectivity and a number of cells continue to have surprisingly large, poorly delineated receptive fields (Buisseret and Imbert, 1976; Singer and Tretter, 1976a, 1976b). Such receptive fields would result if these cells integrate activity from a large number of cells with differing orientation preferences. This in turn would be consistent with the persistence of unselective lateral interactions. Despite the global reduction of selectivity in deprived cortex the sequence regularity of shifts in preferred orientation is maintained (Sherk and Stryker, 1976; Wiesel and Hubel, 1974). This agrees with the postulated spatial decrement of horizontal interactions. Despite unselective horizontal interactions higher order cells would continue to prefer the orientation that is preferred by the nearby first-order cells. Finally, experimental evidence indicates that the majority of cortical cells adopt a preference for the experienced orientation if exposed selectively to a single orientation (Blakemore and Cooper, 1970; Hirsch and Spinelli, 1970; Rauschecker and Singer, 1981; Stryker *et al.*, 1978). This, too, fits the hypothesis of an activity-dependent selective stabilization of horizontal intercolumnar connections. With selective exposure to a single orientation only those first-order cells are activated whose preference corresponds to the visible orientations. Hence, their axons should consolidate at the higher order target cells at the expense of axons that come from those first-order cells that are unable to respond to the visible orientations. Higher order cells should assume the orientation preference of the successfully competing first-order cells.

Deoxyglucose data are compatible with such a reorganization: orientation columns encoding the experienced orientation expand at the expense of columns encoding non-experienced orientations and this reorganization is particularly pronounced outside layer IV (Singer, Freeman and Rauschecker, 1981).

Thus, experimental evidence is compatible with the hypothesis that the same mechanisms of selective stabilization and competitive disconnection which guide the specification of binocular correspondence also guide, at another level of cortical processing, the development of higher order orientation-selective fields.

NEURONE ENSEMBLES FOR PERCEPTUAL DIMENSIONS

If the same principles of self-organization are applied to networks allowing for reciprocal recurrent excitation—and cortical circuitry provides ample opportunity for such interactions—very interesting new properties emerge.

Experience will lead to selective stabilization of excitatory connections between those elements of the network whose activation is correlated. In the striate cortex such would be the case, for example, for cells that share the same direction preference, because these cells are coactivated each time the retinal image shifts. The same should apply to cells sharing the same disparity selectivity since objects are most often confined to a particular depth plane. One likely effect of such selective coupling is that the selectively interacting units—if a sufficient number of them are activated by a sensory stimulus—mutually facilitate each other and maintain through reverberation a high level of coherent activation (see, for example, Grossberg, 1980; Marr, Palm and Poggio, 1978; Nelson, 1975; Singer, 1979). Thus, once a sufficient number of such selectively coupled cells are activated they become distinguishable as members of a cooperating neuronal assembly because of their enhanced, prolonged and coherent discharges.

As indicated in Fig. 2, one of the many interesting operations that such ensembles of selectively coupled neurones can accomplish is figure–ground separation. A process similar to the one described here has also been proposed to explain the detection of figures and depth planes in dichoptically presented random-dot stereograms. In this case it was assumed that selective cooperative interactions existed between units sharing the same disparity selectivity (see, for example, Marr, Palm and Poggio, 1978; Nelson, 1975). Obviously such ensembles of selectively coupled neurones would develop in parallel for a great number of feature domains—probably for as many as we use for figure–ground distinctions. Such feature domains could be, for example, colour, contrast, spatial frequency, orientation and, of course, velocity, direction of movement and disparity. Obviously, the coupling of such ensembles need not necessarily be realized only by circuits within area

Experience dependent selective coupling between cells encoding a coherent feature domain

Effect of activating a selectively coupled network with uncorrelated noise containing a correlated subpopulation of pattern elements.

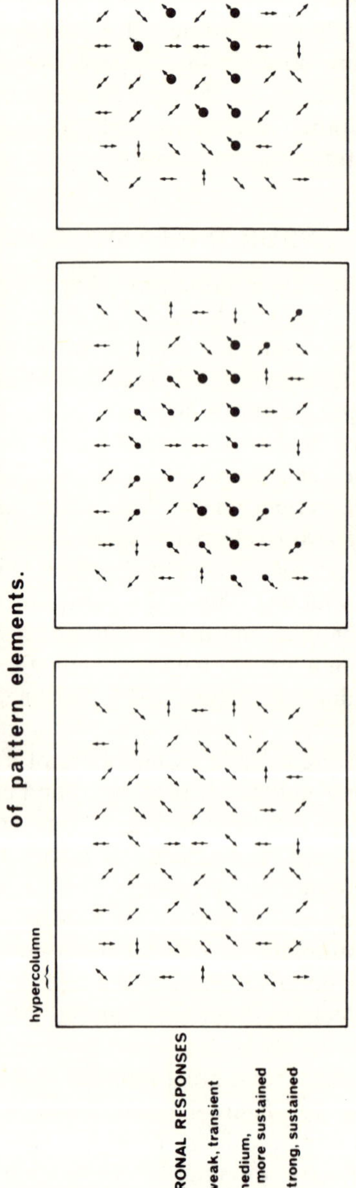

Figure 2 Schematic representation of figure–ground separation by a matrix of selectively coupled, direction-selective neurones. It is assumed that neurones which encode the same direction of movement are coupled selectively through reciprocal excitatory connections. Preferential facilitatory interactions are indicated by dotted lines (upper case). The figure consists of the dotted outlines of a triangle and moves slowly from the lower left to the upper right. It is hidden within a field of randomly moving dots which have the same size, density distribution and movement velocity as the dots of the figure. Thus, the only cue for the distinction of the figure is the coherency of the direction of dot motion. The lower three graphs show the emergence of a figure-specific activation pattern in the matrix of selectively interacting neurones. At an early stage after stimulus onset, cortical units respond similarly to all dots (left lower case). However, because of selective reciprocal facilitation, responses increase in cells which maintain selective interactions with other costimulated cells. Assuming a spatial decrement of these facilitatory interactions facilitation will be graded and depend on the local density of dots moving in the same direction. In an intermediate phase (middle case) some of the neurones encoding the dots of the triangular figure will be strongly facilitated and others weakly. If adjacent dots of the noisy background happen to move coherently, units encoding these dots will also facilitate each other's responses. Eventually, however, because of the consistent correlation in the movement of the dots of the triangular arrangement all units encoding the trajectories of the dots of the triangle will be driven into a stable resonant state. Lateral inhibition between adjacent units (columns) may further increase the difference in activation between units encoding the figure and units encoding uncorrelated noise. Since the responses to the elements of the figure are now different from the responses to the noise, the former can be treated in subsequent processing areas in the very same way as responses caused by a figure that had not been hidden in noise.

17. It could be realized as well through the reciprocal point to surface projections between striate cortex and prestriate cortical areas. In any case, however, for figure–ground distinctions the interactions must establish relations between units in a cortical sheet where retinotopy and spatial resolution is still well preserved. To my knowledge there is yet no direct physiological evidence for such selective long-range interactions in the striate cortex of normal cats. However, unusually large receptive fields consisting of multiple, spatially separated excitatory subregions have been observed in kittens with selective visual experience (Singer and Tretter, 1976b). Because these receptive fields could be related to the visual patterns which the kittens had been exposed to it was suggested that these unconventional receptive fields resulted from activity-dependent selective stabilization of long-range connections.

CORTEX AS AN ASSOCIATIVE STORAGE DEVICE

In conclusion I would propose that the major function of the cortex is to establish associations between incoming signals, using coherency as the criterion. The substrate for this function is adaptive connections stabilized during development under the guidance of the activity they convey. Coherency of activity patterns selects neuronal connectivity. This process of circuit selection seems to occur at different levels of cortical processing. Since the selective stabilization of circuits during development is gated and not left to the sensory signals alone it can be expected that the decision for the formation of particular associations (ensembles) is itself dependent on the functional state of many neurones—if not indeed on the motivational state of the organism. Thus, the criteria for the formation of particular associations (ensembles) emerge in part from the genetically determined blueprint of the wiring, from the structure of the world with which the brain interacts through its sensory and motor organs and from the actual dynamic (behavioural) state which the system maintains while it interacts with the outer world. Since these selection processes lead to long-lasting associations of neurones they can be considered as storage processes which, through structural modification, preserve information about the statistical properties of previous activation patterns.

REFERENCES

Bear, M. F., Paradiso, M. A., and Daniels, J. D. (1982), *Soc. Neurosci. Abstr.*, **8**, 4.
Blakemore, C., and Cooper, G. F. (1970). *Nature*, **228**, 477–478.
Blakemore, C., and Tobin, E. A. (1972). *Exp. Brain Res.*, **15**, 439–440.
Buisseret, P., and Gary-Bobo, E. (1979). *Neurosci. Lett.*, **13**, 259–263.
Buisseret, P., Gary-Bobo, E., and Imbert, M. (1978). *Nature*, **272**, 816–817.

Buisseret, P., and Imbert, M. (1976). *J. Physiol. (Lond.)*, **255**, 511–525.
Buisseret, P., and Singer, W. (1983). *Exp. Brain Res.*, **51**, 443–450.
Crewther, S. G., Crewther, D. P., Peck, C. K., and Pettigrew, J. D. (1980). *J. Neurophysiol.*, **44**, 97–118.
Cynader, M., and Mitchell, D. E. (1977). *Nature*, **270**, 177–178.
Freeman, R. D., and Bonds, A. B. (1979). *Science*, **206**, 1093–1095.
Gaze, R. M. Feldman, J. D., Cooke, J., and Chung, S.-H. (1979). *J. Embryol. Exp. Morphol.*, **53**, 39–66.
Geiger, H., and Singer, W. (1982). *Int. I. Develop. Neurosc.*, **Suppl.**, R328.
Gilbert, C. D. (1977). *J. Physiol. (Lond.)*, **268**, 391–421.
Gilbert, C. D., and Kelly, J. P. (1975). *J. comp. Neurol.*, **163**, 81–106.
Gilbert, C. D., and Wiesel, T. N. (1979). *Nature*, **280**, 120–125.
Gilbert, C. D., and Wiesel, T. N. (1983). *J. Neurosci.*, **3**, 1116–1133.
Grossberg, S. (1980). *Psycholog. Rev.*, **87**, 1–51.
Hebb, D. O. (1949). *The Organization of Behaviour*, Wiley, New York.
Hirsch, H. V. B., and Spinelli, D. N. (1970). *Science*, **168**, 869–871.
Hubel, D. H., and Wiesel, T. N. (1962). *J. Physiol. (Lond.)*, **160**, 106–154.
Hubel, D. H., and Wiesel, T. N. (1974). *J. comp. Neurol.*, **158**, 295–306.
Hubel, D. H., Wiesel, T. N., and Stryker, M. P. (1978). *J. comp. Neurol.*, **177**, 361–380.
Innocenti, G. M., and Frost, D. O. (1979). *Nature*, **280**, 231–234.
Jansen, J. K. S., and Lomo, T. (1981). *Trends in Neurosci.*, July **1981**, 178–181.
Kasamatsu, T., and Pettigrew, J. D. (1979). *J. comp. Neurol.*, **185**, 139–162.
Kasamatsu, T., Pettigrew, J. D., and Ary, Marylouise (1979). *J. comp. Neurol.*, **185**, 163–181.
Lund, J. S., and Boothe, R. G. (1975). *J. comp. Neurol.*, **159**, 305–334.
Mariani, J., and Changeux, J.-P. (1981). *J. Neurosci.*, **1**, 703–709.
Marr, D., Palm, G., and Poggio, T. (1978). *Biolog. Cybernetics*, **28**, 223–239.
Meyer, R. L. (1982). *Science*, **218**, 589–591.
Mitzdorf, U., and Singer, W. (1980). *J. Physiol. (Lond.)*, **304**, 203–220.
Movshon, J. A., and Van Sluyters, R. (1981). *Ann. Rev. Psychol.*, **32**, 477–522.
Nelson, J. I. (1975). *J. theor. Biol.*, **49**, 1–88.
Rauschecker, J. P., and Singer, W. (1979). *Nature*, **280**, 58–60.
Rauschecker, J. P., and Singer, W. (1981). *J. Physiol. (Lond.)*, **310**, 215–239.
Rockland, K. S., and Lund, J. S. (1982). *Science*, **214**, 1532–1543.
Rockland, K. S., and Lund, J. S. (1983). *J. comp. Neurol.*, **216**, 303–318.
Sherk, H., and Stryker, M. P. (1976). *J. Neurophysiol.*, **39**, 63–70.
Sherman, S. M., and Spear, P. D. (1982). *Physiol. Rev.*, **62**, 738–855.
Sillito, A. M., Kemp, J. A., and Berardi, N. (1983). *Brain Res.*, **280**, 249–307.
Sillito, A. M., Kemp, J. A., and Blakemore, C. (1981). *Nature*, **291**, 318–322.
Singer, W. (1976). *Brain Res.*, **118**, 460–468.
Singer, W. (1977a). *Exp. Brain Res.*, **30**, 25–41.
Singer, W. (1977b). *Physiol. Rev.*, **57**, 386–420.
Singer, W. (1979). In *The Neurosciences, Fourth Study Program*, (Eds. F. O. Schmitt and F. G. Worden), MIT Press, Cambridge, Mass., pp. 1093–1109.
Singer, W. (1982). *Exp. Brain Res.*, **47**, 209–222.
Singer, W., Freeman, B., and Rauschecker, J. (1981). *Exp. Brain Res.*, **41**, 199–215.
Singer, W., and Rauschecker, J. (1982). *Exp. Brain Res.*, **47**, 223–233.
Singer, W., Rauschecker, J., and Werth, R. (1977). *Brain Res.*, **134**, 568–572.
Singer, W., and Tretter, F. (1976a). *J. Neurophysiol.*, **39**, 613–630.
Singer, W., and Tretter, F. (1976b). *Exp. Brain Res.*, **26**, 171–184.
Singer, W., Tretter, F., and Yinon, U. (1982). *J.Physiol.*, **324**., 221–237.

Singer, W., Yinon, U., and Tretter, F. (1979). *Brain Res.*, **164**, 294–299.
Stent, G. S. (1973). *Proc. Nat. Acad. Sci.(USA)*, **70**, No. 4, 997–1001.
Stryker, M. P. (1981). *Neurosci. Abstr.*, **7**, 842.
Stryker, M. P., Sherk, H., Leventhal, A. G., and Hirsch, H. V. B. (1978). *J. Neurophysiol.*, **41**, 896–909.
Szentagothai, J. (1975). *Brain Res.*, **95**, 475–496.
Trotter, Y., Gary-Bobo, E., and Buisseret, P. (1981). *Dev. Brain Res.*, **1**, 450–454.
Wiesel, T. N., and Hubel, D. H. (1963). *J. Neurophysiol.*, **26**, 1003–1017.
Wiesel, T. N., and Hubel, D. H. (1965). *J. Neurophysiol.*, **28**, 1060–1072.
Wiesel, T. N., and Hubel, D. H. (1974). *J. comp. Neurol.*, **158**, 307–318.

Models of the Visual Cortex
Edited by D. Rose and V. G. Dobson

CHAPTER 13

Hierarchical contour coding by the visual cortex

G. Hartmann
Universität—Gesamthochschule—Paderborn, Pohlweg 47–49, 4790 Paderborn, Federal Republic of Germany

INTRODUCTION

This model starts from the basic consideration, that continuity of contours is essential for the understanding of the visual world. A cortical structure is proposed, by which all the contours of an image are completely encoded and by which continuity of contours is explicitly verified. A hierarchical contour code is introduced, describing dominant contour structures at higher levels and details at lower levels. This selection of important features makes the code well suited for the recognition strategy of the visual system.

Different types of cortical cells are necessary for the encoding procedure and detailed predictions may be made according to their function. It is very encouraging that the predicted features are in good agreement with experimental results and that also topological features may be predicted. Finally it is very surprising that in a model exclusively based on contour processing, cortical cells, behaving like spatial frequency detectors, are functionally necessary.

ENCODING OF CONTOURS BY CORTICAL NEURONS

From Hubel and Wiesel (1959) we know that a simple cortical cell will only respond if a contour of the correct type is correctly located and oriented within its receptive field. In other words, type, position and orientation of a small part of this contour are encoded by a responding cortical cell. On the other hand, the retinal projections of cortical fields can not be continuously shifted and rotated until they match all the parts of an arbitrary contour. So the basic consideration of the proposed model starts with the question of whether the unlimited number of differently running contours may be completely encoded by a finite number of cortical cells.

This question may be drastically simplified by subdividing the retinal image into overlapping islands I (see Fig. 3), cutting continuous contours into overlapping contour elements. If arbitrary contour elements, crossing an island I, may be detected and encoded by a finite set of cortical cells, arbitrary contours will also be encoded.

The area of these islands is defined by the projection of seven overlapping lateral geniculate nucleus (LGN) field centres to the retinal image (Fig. 1a).

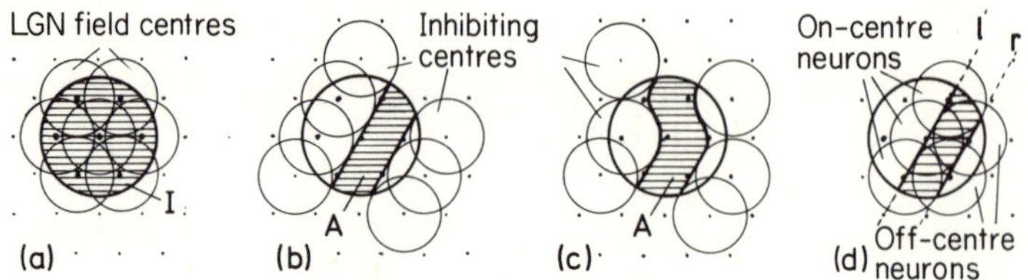

Figure 1 Differently shaped cortical fields are necessary to encode arbitrary contour elements in the region of an island I of seven overlapping LGN field centers (a). The differently shaped active areas A of tripartite cortical fields are defined by the inhibiting LGN subfields (b, c). The active areas of bipartite fields are limited by a right (r) and left (l) line, defining the location of changing subfield responses to light/dark edges (d).

This definition is very convenient, as the cortical fields can also be mapped from these retinal projections of LGN fields. This topographic mapping is, of course, a consequence of the fact that cortical neurons are driven by LGN neurons, with each geniculate cell driving hundreds of cortical cells. A hexagonally ordered structure of LGN fields with Mexican hat profiles and standard size and overlap is assumed. This is a very 'natural' arrangement, well known from closely packed coins. Preceding investigations show more details about overlap structure and relative field size (Hartmann, 1982).

Now a complete set of cortical neurons with the following features shall be assumed for all the islands I:

1 The active areas A of the receptive fields are laterally limited shape elements, running from periphery to periphery of an island I (Fig. 2).
2 Arbitrary contour elements are always enveloped by one of the different shape elements.
3 Shape elements of adjacent islands, enveloping a continuous contour, overlap each other, forming smooth sequences (Fig. 3).

It is evident that arbitrary contours will be completely encoded by the response of these cortical cells if the above set may be realized.

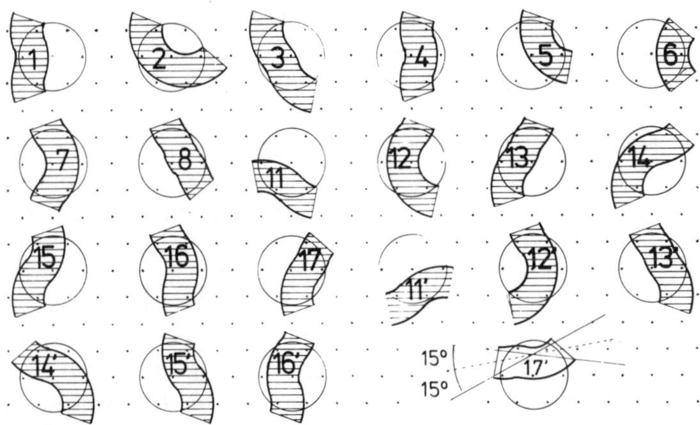

FIGURE 2 The complete set of 22 differently shaped cortical fields. Each field will be implemented in six orientations and for three contour types (bright line, dark line, edge). Orientation selectivity depends on the shape and will be approximately ±15° for most shapes.

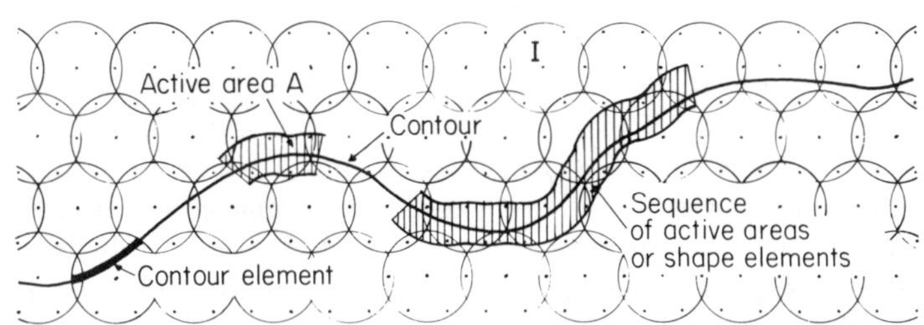

FIGURE 3 The visual field is subdivided into overlapping islands I, cutting arbitrary continuous contours into overlapping contour elements. Each contour element may be enveloped by one of the 22 differently shaped active areas of the cortical fields. These areas form smooth sequences, allowing verification of continuity.

THE SHAPE OF CORTICAL RECEPTIVE FIELDS

The shape of a cortical field will be defined by its lateral limitation. In the case of tripartite fields, the active area A may be limited by inhibiting centres of laterally arranged LGN subfields (Fig. 1b). Different arrangements of these subfields will provide 22 different shapes (Figs. 1b, c and 2) which may be arranged in six orientations, according to the hexagonal LGN field structure. Of course, the active area may be limited by LGN on-centre neurons as well as by off-centre neurons. Hence the fields of 22 x 6 cortical neurons responding to bright lines and of 22 x 6 neurons responding to dark lines will be equally structured.

Cortical fields sensitive to edges may also be discussed in this simple manner. As is known, the response of an LGN neuron to an edge changes as soon as the edge is shifted from one half of its field to the other. So in Fig. 1d the left row of on-centre LGN fields will respond if the light/dark edge is at the right side of line l and the right row of off-centre fields will respond if the edge is at the left side of line r. These lines define an active area A in the case of edges and a different arrangement of the LGN subfields provides a differently shaped area A. Again a complete set of 22 x 6 cortical cells may be realized, with all the differently shaped active areas of Fig. 2.

This proves (logically and by simulation) that an arbitrary contour element, crossing an island I, is always enveloped by one shape element of the set in Fig. 2, as long as it is less curved than the inhibiting LGN field centre. The corresponding cortical cell with the matching active area will respond and encode the contour element. So each cortical cell at this lowest level may be regarded as a specific element of a complete contour code.

Depending on the shape of the particular field, 5 to 7 geniculate neurons with inhibitory influence (Fig. 1) and an additional number of 6 to 9 geniculate neurons with excitatory influence converge to a cortical cell at this lowest detector level. More details about the coupling scheme and the responses of these cells are discussed in preceding papers (Hartmann, 1982, 1983).

From the shapes of the active areas (Fig. 2) it may be predicted that these cortical cells are sensitive to oriented contour elements and that they will only respond if the stimulus is presented within a small local region. The relation between the width and length of the particular active area A defines an orientation selectivity of approximately ± 15° for most cells which have low curvature receptive fields (Fig. 2). These predictions are in full accordance with the early results of Hubel and Wiesel (1959, 1962, 1968) about 'simple' cortical cells. 'Simple' cells with highly curved areas like numbers 2 or 12 in Fig. 2 should not be found in experiments using straight stimuli. Other cells with shapes like numbers 5, 6 or 7 should prefer curved stimuli, but they should also respond to straight stimuli, in accordance with the results of Heggelund and Hohmann (1975).

More detailed predictions, e.g. about the spontaneous discharge of cortical cells, plots of simple fields using spot stimuli, etc., depend on the special coupling scheme and are discussed in preceding work (Hartmann, 1983).

CONTINUITY OF CONTOURS AND THE LINKING HIERARCHY

A continuous contour is encoded by the set of all the responding cortical cells. This set contains all the links of a continuous and smooth chain of active areas and encodes implicitly the continuity of the enveloped contour. However, continuity is not yet verified explicitly at this lowest level of 'simple' cortical cells. To make continuity explicit, all the neurons of the responding set should converge to a higher level cell. Then this higher level cell would only respond if all the contour elements of a particular contour are encoded, in other words if the contour is continuous.

Of course, this simple coupling scheme cannot be realized, as all the different possible contours in the visual field would produce different sequences of active areas, each one requiring a specific higher level cell for verification of continuity. The tremendous number of possible combinations would lead to a 'combinatorial explosion'. On the other hand, the proposed coupling scheme will work fairly well within a small neighbourhood according to the limited number of possible combinations.

Consequently a hierarchical structure is proposed in which continuity is verified within small local regions at the first level and in which verification of continuity is extended to increasing regions from level to level by a hierarchical linking mechanism. At the first level of the linking hierarchy, verification of continuity is limited to small regions of seven overlapping islands I. These new regions are again overlapping islands of double the previous diameter, and seven of these new islands may be combined to quadruple-sized islands at the second level of the linking hierarchy. Verification of continuity will stop at level n if the 2^n-fold size of the islands is larger than the extension of the particular contour.

Consider at the first linking level a double-sized island crossed by a continuous contour, running from periphery to periphery. It will be encoded by a set of responding 'simple' cortical cells within the seven constituent islands. The active areas A of these cells will form a smooth sequence, running from periphery to periphery of the double-sized island (Fig. 4). Only a contour running through all the linked shape elements will excite all the corresponding 'simple' cells, and so by the simultaneous response continuity of the contour may be verified piecewise within the area of the double-sized island.

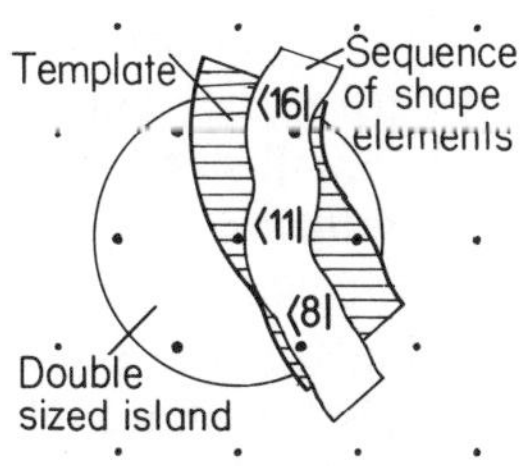

Figure 4 Sequences of shape elements which are enveloped by a double-sized template shall no longer be distinguished and shall be encoded by a common higher level neuron.

There are, however, some thousands of sequences within the area of each double-sized island, and this number would rapidly increase at higher linking levels n if all those sequences would be linked again. To avoid this combinatorial explosion, different sequences shall not be distinguished if they have approximately the same run. More exactly, all those sequences which are enveloped by a double-sized shape element shall be encoded by one common neuron at level $n = 1$ (Fig. 4). This shape element, however, is only a template, not a detector. Consequently its neuron will only respond if one of the sequences within the double-sized template is activated by a continuous contour. As different sequences with similar runs are no longer distinguished by the response of a common neuron at the first linking level, shape description is generalized but continuity is still verified. The number of neurons per island, however, is again reduced to 6 x 22 for each contour type, as the double-sized templates have the same 22 shapes and 6 orientations like the 'simple' cortical fields.

At all the higher linking levels, the same linking procedure will be repeated. First-level templates are linked to sequences within quadruple-sized islands at the second linking level. Again only one second-level cortical cell will encode all those first-level sequences which are enveloped by a quadruple-sized second-level template. By this procedure a continuous contour is simultaneously encoded by cortical cells at different levels of the linking hierarchy. From level to level, shape description is more and more generalized, but verification of continuity is more and more extended. The larger the extension of a contour, the higher will be the level to which the encoding mechanism will proceed. A dominant continuous contour will be encoded at relatively high levels of the linking hierarchy, while details will be encoded at the lower levels.

THE SPATIAL FREQUENCY HIERARCHY

The size of the island I, of the active areas A and consequently the size of the cortical receptive fields is determined by the size of the retinal and LGN receptive fields. Though retinal and cortical field size increases with retinal eccentricity, differently sized cortical fields are simultaneously necessary in all regions of the visual field. Small-sized fields are necessary to resolve closely adjacent contours or to follow contours with high curvature. On the other hand, small fields will not respond to broad lines or blurred edges.

At first sight, different coupling schemes seem to be necessary for cortical cells with differently sized fields. A more elegant cortical architecture, however, would prefer identical coupling schemes and would only change the size and distance of the basic building blocks. Just this structure may be realized: a cortical cell, receiving as excitatory inputs the spike rates of 19 LGN neurons with adjacent on-centre fields and as inhibitory inputs the spike rates of 19 LGN neurons with congruent off-centre fields, will respond with the same spike rate as an LGN neuron with a double-sized on-centre field. A corresponding cortical cell with interchanged excitatory and inhibitory inputs will behave like an LGN off-centre neuron, belonging to the antagonistic system. This surprising feature depends on the surprising fact that the weighted sum of 19 overlapping 'Mexican hats' provides a 'Mexican hat' of double the size (for more details, see Hartmann, 1982).

Those cortical neurons responding like LGN neurons with double-sized fields shall be called 'virtual LGN neurons' as they are equivalent to LGN neurons coupled to a 'virtual retina' with double-sized retinal fields. Now the above-described coupling scheme may also be applied to the 'virtual LGN neurons' themselves, providing higher level 'virtual LGN neurons' with quadruple size. This procedure may be repeated and will produce a hierarchy of virtual LGN neurons with concentric fields of 2^k-fold size, compared with the field size of the original LGN neurons.

The hierarchy of virtual LGN neurons, which must not be confused with the linking hierarchy, provides the building blocks for a hierarchy of 'simple' cortical neurons with receptive fields of 2^k-fold size. Of course, the complete set of 'simple' cortical neurons with all types, 22 shapes and 6 orientations will be implemented at all the levels of the hierarchy, and of course corresponding neurons of the sets will have the same coupling scheme, independent of the field size.

Detailed contour structures or high spatial frequency grids will be encoded by cells at the lowest hierarchical level, while gross contours or lower spatial frequency grids will be encoded at higher levels. Consequently the hierarchical levels may be interpreted as independent spatial frequency channels, encoding an image at different levels of resolution. This necessary feature of the model is in good accordance with experimental results (Braddick, Campbell and Atkinson, 1978; Maffei 1978), which confirm the existence and independence of spatial frequency channels.

Also, of course, the continuity will be verified of those contours which are encoded by large-sized 'simple' neurons in a lower resolution spatial frequency channel. For all the independent spatial frequency channels there will be an independent linking hierarchy with the known coupling scheme.

The realization of both the hierarchies also has structural consequences for the visual cortex. Assuming shortest pathways, the cells will be arranged in spatial frequency layers and in a novel arrangement of orientation columns (further details are prepared for publication).

THE HIERARCHICAL CONTOUR CODE AND PATTERN RECOGNITION

In the preceding chapters, the visual cortex was characterized as an encoding system for grey-level images, while encoding of colour, stereopsy and other well-known features was neglected. But the generation of a hierarchical contour code seems to be the most essential tool for pattern recognition.

By the proposed linking mechanism, the most extended continuous contour structures are encoded at the highest level of the linking hierarchy. It will be a good strategy to control the pursuit movements by those neurons responding at the relatively highest level and so to analyse dominant contour structures first. The 'relatively highest' level neuron may be stimulated as well by a street in a natural environment, as by a hair on a uniform background.

For the purpose of recognition, it will also be a good strategy to start the shape analysis at the relatively highest level, as generalization of shape description makes matching independent of details. If the generalized image corresponds with a generalized model, it will then be promising to analyse details of the image which are encoded at lower linking levels, perhaps making voluntary eye movements guided by the generalized model.

CONTOUR PROCESSING OR SPATIAL FREQUENCY ANALYSIS?

The proposed cortical model is exclusively based on contour processing and so it will be mentally rejected by all those scientists who are modelling the visual cortex as a spatial frequency analyser. For this reason, the response of the proposed cortical cells to grids shall be discussed.

A 'simple' cortical cell sensitive to lines will be sharply tuned to grids within a small spatial frequency interval. A low-frequency cut-off is due to increasing line width and a high-frequency cut-off is due to the neighbouring bars running through the inhibiting lateral zones. The response to a moving grid will be periodical, as no bar of the grid can enter or leave the exciting zone without moving across the inhibiting zones.

Of course, a particular sequence of active areas would also provide a periodic response to a moving grid. However, sequences are not standalone units, and the first level neuron will respond if only one of its sequences matches a bar of the grid. There will be a response for different positions of the grid per definition, as different sequences of similar run are encoded by the same neuron at the first linking level. So the response should be continuous and a periodical offset will be due to a change of response with the relative position of the bar. This periodical offset will be more and more averaged with increasing linking level, and the response will be independent of the relative position of the grid.

The orientation characteristic of the response, however, will not depend on the linking level n. It is defined by the relation between the length and width of the shape elements, which is independent of their size. Sharp spatial frequency tuning is due to the lowest level cells and it will not be influenced at all by the linking mechanism. So higher level cortical cells will respond to a moving grid like sharply tuned oriented spatial frequency detectors with unmodulated response.

This feature of the cortical cells may or may not be used in the recognition process, but it is very essential to stress that the model of these cells is based on contour processing and verification of continuity. Consequently a model based on spatial frequency analysis cannot any longer exclude a contour processing model, and vice versa.

REFERENCES

Braddick, O, Campbell, F. W., and Altkinson, J. (1978). Channels in vision: basic aspects. In *Handbook of Sensory Physiology*, Vol. III, Springer, Berlin, Heidelberg and New York, pp. 3–38.

Hartmann, G. (1982). Recursive features of circular receptive fields. *Biolog. Cybernetics*, **43**, 199–208.

Hartmann, G. (1983). Processing of continuous lines and edges by the visual system. *Biolog. Cybernetics*, **47**, 43–50.

Heggelund, P., and Hohmann, A. (1975). Responses of striate cortical cells to moving edges of different curvatures. *Exp. Brain Res.*, **23**, 211–216.

Hubel, D. H., and Wiesel, T. N. (1959). Receptive fields of single neurons in the cat's striate cortex. *J. Physiol. (Lond.)*, **148**, 574–579.

Hubel, D. H., and Wiesel, T. N. (1962). Receptive fields, binocular interaction and functional architecture in the cat's visual cortex. *J. Physiol. (Lond.)*, **160**, 106–154.

Hubel, D. H., and Wiesel, T. N. (1968). Receptive fields and functional architecture of monkey striate cortex. *J. Physiol. (Lond.)*, **195**, 215–243.

Maffei, L. (1978). Spatial frequency channels: neural mechanisms. In *Handbook of Sensory Physiology*, Vol. VIII, Springer, Berlin, Heidelberg and New York, pp. 39–66.

Models of the Visual Cortex
Edited by D. Rose and V. G. Dobson

CHAPTER 14

Local and global functional architecture in primate striate cortex: outline of a spatial mapping doctrine for perception

ERIC L. SCHWARTZ
Brain Research Laboratories, Department of Psychiatry, New York University Medical Center, 550 First Avenue, New York, and Courant Institute of Mathematical Sciences, New York University, New York, NY 10016, USA.

GLOBAL (RETINOTOPIC) MAP STRUCTURE

Characteristic scale and the concept of retinotopy

The concept of the retinotopic map implies that there is an orderly two-dimensional map of the surface of the retina onto the surface of cortical (and subcortical) areas.

More modern understanding of the architecture of the cortex has indicated that there is a substantial small-scale structure which is at some variance with the idea of retinotopy. Beginning with the classical work of Hubel and Wiesel, it has been recognized that there is in fact a substantial substructure on the scale of roughly one millimeter whose effect is to produce the local (hypercolumnar) details of cortical structure such as orientation tuning, disparity tuning, etc.

The basic unit of scale is determined by the hypercolumn (about 1 mm). For distances comparable to or smaller than this, global mapping ideas break down. In the next section, it will be argued that a local continuum approach may be applied. However, for distances on the order of several millimeters or larger, the local hypercolumn structure 'averages' out, and the cortex may be treated as a continuum map.

The problem of describing the global cortical map becomes a simple problem of continuum methods. Maps from two dimensions (retina) to two dimensions (cortex) are particularly easy to characterize. This has been done in detail in several recent references (Schwartz, 1983, 1984) and is also outlined in standard references (e.g. Segal, 1977).

All topologically stable mappings (of the plane into the place) are locally equivalent to either a regular mapping or to two possible types of singularity: a fold or a cusp. In the context of cortical mapping, we will assume that these singularities do not occur, in other words, that there is a one-to-one invertible map between the retina and cortex. Thus, the continuum approximation assumes that each point in the retina corresponds to a point in the cortex, and vice versa.

Recently, the 2-deoxyglucose technique has been available to essentially 'paint' a stimulus configuration on the cortex. Using logarithmically scaled global stimuli, this method has yielded a coarse mapping in human striate cortex, using positron tomography as an imaging technique (Schwartz, 1981b; Schwartz, Wolf and Christman, 1983). Tootel, Silverman and DeValois (1982) applied this technique to primate cortex, using autoradiography. They obtained a startlingly clear map, which is reproduced in Fig. 1.

Figure 1 indicates very good agreement between a map function of the form $\log(z + 0.3)$ where z represents the tangent plane of the visual field, and the data of Tootel, Silverman and DeValois (1982). Earlier work (Schwartz, 1977a, 1980) indicated a general form for striate topography of $\log(z + a)$, where a is a constant. The data of Dow *et al.* (1981), which provides the only detailed foveal magnification data available to date, indicates that this constant a should be 0.3° This estimate (Schwartz, *et al.* 1983) was confirmed by the data of Tootel, Silverman and DeValois (see Fig. 1).

The fact that primate striate cortex is approximately isotropic (conformal) on a global scale suggests a simple developmental rule (Schwartz, 1977b). Conformal maps are entirely determined by the following:

1. The boundary conditions at the two surfaces (i.e. the retina and cortex)
2. A corresponding point in the two surfaces and
3. An orientation through that point

Specifying these three conditions uniquely fixes the map (see Alfhors, 1966, or Schwartz, 1977b).

Thus, provided that continuity is encoded, the specific map which is formed is entirely determined by the statement that the afferent fibers isotropically fill the target area. This suggests that developmental map algorithms need to consist of two distinct mechanisms: one (fiber sorting) which imposes continuity and a second (boundary conditions) which selects the particular continuous map which is to be encoded. The arguments above indicate that selection of the specific map from the large class of continuous maps may be simply provided by a variational mechanism determined by the boundary conditions of the cortex.

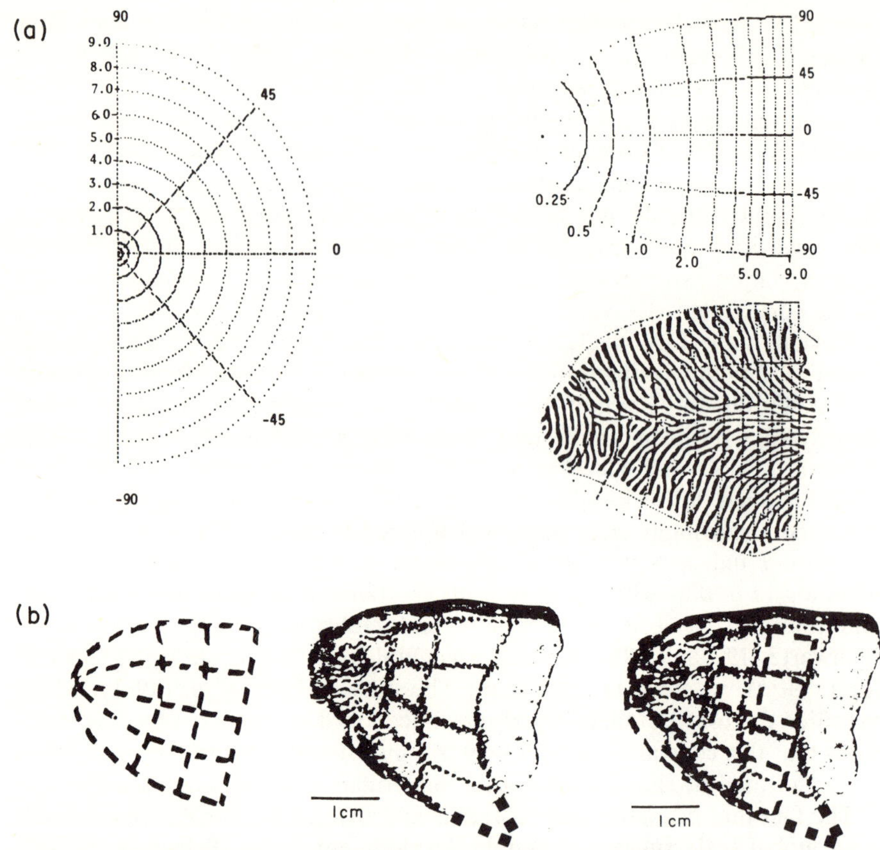

FIGURE 1 (a) On the left is shown a representation of the visual field (or the surface of the retina). The central 9° of field is shown in orthogonal projection. Polar coordinates (r, Θ) are used to represent the field. On the upper right is shown the map of this field representation using the map function log (z + 0.3). This function is derived using the recent fit of Dow *et al.* (1981). On the bottom right is shown a superposition on the anatomical data of LeVay, Hubel and Wiesel (1975). (b) At right is shown a superposition of the map function log (z + 0.3) (shown separately at left) and the corresponding 2-deoxyglucose mapping obtained by Tootel, Silverman and DeValois (1982) (shown separately at center). The operculum of the cortex is shown (as above, from LeVay, Hubel and Wiesel, 1975). Reproduced by permission of Pergamon Press from Schwartz, *Vision Res.*, **23**, 831–835 (1983).

LOCAL MAP STRUCTURE IN THE STRIATE CORTEX

As discussed above, when the scale of interest becomes comparable to the size of cortical hypercolumns, the global continuum hypothesis is no longer valid in detail.

The first effect of looking at this higher magnification is to note that there must be a significant shear on a local scale. Two sets of ocular dominance columns are interlaced to form the cortical map. The width of a set of these two columns, in primates, is about 1 mm. However, for the two sets to fit together in a way in which the global map is isotropic, then there must be a two-to-one compression in the direction perpendicular to the border of the column. Such a two-to-one compression has been observed tentatively in macaque by LeVay, Hubel and Wiesel (1975). Thus, it appears that there is a shear introduced by the ocular dominance column pattern which effectively disappears during the binocular mixing which occurs in the layers surrounding layer IV or which, in any case, is not strongly visible in the global pattern of striate cortex retinotopy. This aspect of cortical scale may be easily verified in Fig. 1, where the 'stripes' due to the ocular dominance columns are barely visible in the overall topographic map pattern of the data of Tootel, Silverman and DeValois (1982).

Local map structure and orientation columns

A second aspect of local structure in striate cortex concerns the existence of orientation columns. The global continuum hypothesis implies a point-to-point mapping. However, a salient feature of the striate cortex is that neurons respond to edges or lines (or, equivalently, spatial frequency gratings). Furthermore, oriented cells are arranged in 'columns' or 'slabs' which are roughly 0.5 x 0.05 mm in dimension (Hubel and Wiesel, 1974). Originally, it was stated that these columns were arranged in parallel slabs which filled the cortical surface with repeating 'hypercolumns' (Hubel and Wiesel, 1974) consisting of a complete 180° orientation sequence and two ocular dominance columns. More recently, it appeared that this pattern is more complicated: in the centre of the hypercolumn there seems to be an oval 'cytochrome oxidase' spot (Hubel and Livingstone, 1981) which consists of cells which stain for cytochrome oxidase (i.e. have higher than average metabolic activity) and which have little or no orientation tuning.

The experimental details of the local cortical structure are quite sketchy. Major unanswered questions are:

1. Are orientation columns parallel or radial or neither?
2. Do orientation columns intersect ocular dominance columns with any regular angle (i.e. 90°)?
3. Is the orientation column pattern discrete or continuous?
4. Is there any regular spatial frequency column structure and, if so, what is its direction with respect to the orientation columns (see Tootel, Silverman and DeValois, 1981)?
5. Most important, what is the structural origin of the orientation column pattern?

Since the actual data are unclear, it is obviously difficult to model the orientation column structure. In the following discussion, a tentative model will be presented. This model is based on the idea that the scale of hypercolumns (about 1 mm) is still macroscopic with respect to the scale of the spacing of cortical cells (about 50 μm) or the typical size of cortical dendritic trees (150 to 200 μm). However, there is only a factor of 5 to 20 separating these lower scale figures and the scale of a cortical hypercolumn. Therefore, care must be used in applying a continuum, or mapping, approach to this area. Nevertheless, a model based on using a complex logarithmic mapping to model local cortical architecture (Schwartz, 1976, 1977a, Dobson, 1980) is quite consistent with the recently observed phenomena of cytochrome oxidase spots and 'spatial frequency columns'. This model will now be reviewed.

The first basic fact concerning the functional organization of striate cortex is that neurons in this area are commonly responsive to an elongated stimulus, rather than the circularly symmetric organization of receptive fields common in the retina and lateral geniculate nucleus (LGN).

The second basic fact is that neurons which are responsive to edges are arranged in a pattern, such that a complete sequence of orientations (i.e. 180°) occurs within about 0.5 mm (in primates: Hubel and Wiesel, 1974).

Neither the details of cortical orientation tuning nor the details of the spatial arrangement of orientation tuned cells are at present understood. However, to summarize briefly a decade of work in this area, it appears that the elongated receptive field structure of cortical neurons derives from some combination of linear summation of the LGN afferents along elongated areas of cortex (in layer IV) and a lateral inhibitory operator (which is of the form of a directional derivative), which is responsible for the (elongated) inhibitory sidebands and for some component of the (elongated) excitatory field of cortical cells.

Although the precise details of the spatial structure of these cortical hypercolumns is not yet known, the original observation of Hubel and Wiesel (1974) will be followed: it will be assumed that there is an approximately *parallel* sequence of orientation columns.

Thus, the problem reduces to finding a means of generating a structure of this type, using the assumed building blocks of (dendritic) summation and (directional) lateral inhibition.

It appears that there are only two generic ways to combine (dendritic) summation and (lateral) inhibition and arrive at a series of parallel columns. The first generic way is to rely entirely on the directionality of inhibition. If the directional inhibition in the cortex were to 'rotate' appropriately along parallel strips, then a form of hypercolumn, as stated above, would result. Alternatively, one might allow the summation to 'rotate' and to keep the lateral inhibition unidirectional. This would also allow a model to be generated, as outlined above.

The complex logarithmic mapping, which has been shown earlier in this paper to provide an accurate model for global mapping, turns out to provide a useful model for generating a 'rotating' afferent structure yielding (approximately) parallel strips under this mapping. Thus, if the afferent input to the cortex were arranged in the form of *local complex logarithmic mappings, on the scale of cortical hypercolumns*, one would have a simple model of the hypercolumn.

Several details concerning this model have been discussed earlier. These details relate to the question of where the local 'origin' should be located. Two models have been suggested, one in which the local 'origin' is in the center of the hypercolumn and one in which it is on an 'edge' of the hypercolumn (Schwartz, 1977c). Dobson (1980) has studied similar models to these.

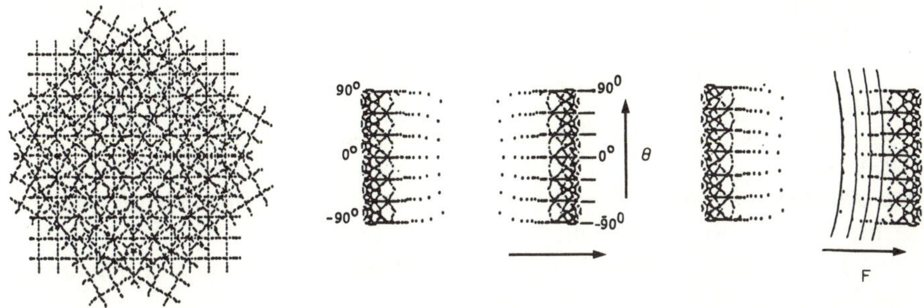

FIGURE 2 A simple simulation of the local mapping log $(z + 1)$ is shown. A series of grids of parallel lines, on the left, is mapped (center and right), using the 'moebius' strip model of the local complex log mapping (i.e. the map is folded about its midline, as in Schwartz, 1977c); cf. Fig. 1a (top right) and 1b (left). An 'orientation column' pattern is clearly visible in this figure in the form of a series of heavily overlapped 'columns', which are indexed by the angle of the grid (Θ) in the visual field. Also evident in this figure is the suggestion of a 'spatial frequency (F)' variation, which is caused by the space variant density of the underlying complex logarithmic mapping. The precise relationship of the spatial frequency columns to the orientation columns is not yet reported. This model suggests that they should be orthogonal. The existence of a cytochrome oxidase 'spot' in the center of the column is also suggested by this figure. This is essentially caused by the singularity of the complex log function, which in this model causes all orientations to collapse to the same region in the center of the column. Reproduced by permission.

Figure 2 shows a computer implementation of one of these models. The map function used is (complex) log $(z + 1)$. Note that the afferent input is arranged in roughly parallel slabs which overlap extensively. Summation of this afferent input on the dendritic trees of cells in surrounding laminae, as discussed in earlier work (Schwartz, 1977c, Dobson, 1980), would provide a roughly parallel set of diffuse orientation columns. Then, a lateral inhibitory operator running in one direction (i.e. along the direction of the change in the local orientation direction) could provide sharp orientation columns.

This model predicts two phenomena which were not part of the original Hubel and Wiesel (1974) description. First, there should be a local center, and orientation tuning should be weak about this local center. The reason for this has to do with the 'singularity' of the complex logarithm. At $z = 0$, the log function would reach minus infinity. For this reason, the function log $(z + a)$ is used, and this indeed also accounts for the curved 'shape' of the cortical surface, as seen in Fig. 1a. On the local level, near the center of the hypercolumn ($z = 0$), all orientations converge (see Fig. 2). Thus, the mechanism of the model could not provide orientation tuning near the center: all orientations would tend to be represented in some circle near the center.

Recently, Hubel and Livingstone (1981) have observed, using cytochrome oxidase, that there is a deficit of orientation tuning in the centers of hypercolumns. These 'cytochrome oxidase spots' are circular or oval, are located in the center of cortical hypercolumns and consist of striate cells which have mainly circular receptive fields. This provides interesting agreement with the requirements of the local complex log model (Schwartz, 1977c; Dobson, 1980).

A second phenomenon predicted by this model is that there should be a variation in the width of receptive fields perpendicular to the direction of the orientation change. This is because of the changing magnification factor of the complex logarithm function. Clearly, on a global level this occurs in the form of a change in magnification factor across the surface of the cortex (see Fig. 1).

Once again, recent observations have suggested that there are 'spatial frequency' columns within striate hypercolumns (Tootel, Silverman and DeValois, 1981). It is not clear at present whether these run in a direction perpendicular to the orientation columns; nor is it clear whether the 'high frequency' end should lie along the edges or the center of the hypercolumn. The local log model would suggest that the high frequency (or, more correctly, the small receptive field) cells would lie along the center, i.e. within the cytochrome oxidase spot. Determining this is complex, because spatial frequency tuning methods which have been used so far (Tootel, Silverman and DeValois, 1981) are not really appropriate to comparing different shaped fields (i.e. circular fields in the 'cytochrome oxidase' spots with elongated fields in the 'local periphery'). Furthermore, it may be that the strength of inhibitory drive may also change along these directions, as suggested by the change in receptive field structure in the region of the cytochrome oxidase spots.

Nevertheless, it appears that the existence of central regions of the hypercolumn which lack orientation tuning and the possible existence of an orderly change in receptive field size (i.e. spatial frequency tuning) are both associated with the structure of cortical hypercolumns. Since both of these phenomena are actual predictions of a model based on the existence of some form of local logarithmic mapping, it will be interesting in the next few years to view the results of more precise experimental work in this area.

SUMMARY: A SPATIAL MAPPING DOCTRINE FOR PERCEPTION

The previous parts of this paper have reviewed a descriptive model of striate cortex functional architecture. One point which emerges from this review is that a mathematical summary of striate cortex architecture has provided a surprisingly accurate model for the global structure of this area. Locally, it is uncertain at the present time whether the model reviewed here is correct or, indeed, what the experimental details are which must be modeled.

The functional implications of this work will now be briefly reviewed. At present there are, roughly, three major paradigms in use for understanding the computational basis of vision (see John and Schwartz, 1978, for a review). These are:

1. Feature extractor networks
2. Spatial frequency models
3. Spatial mapping models

Feature extractor networks propose that single cells 'extract' high-order data from the visual array and that the 'connectivity' or network structure of these arrays is the architectural basis of visual computation. Prominent examples are the 'bug detector' or 'edge' detector hypotheses in the work of Lettvin *et al.* (1959) and Hubel and Wiesel (1962). Spatial frequency models propose that the bandpass filtering properties of cortical cells are the basis of their computational function. There is a strong and a weak form of this hypothesis (Schwartz, 1981a). The strong form of this hypothesis suggests that the visual system actually performs a Fourier analysis or spatial frequency mapping. The weaker (and more reasonable!) form suggests that striate neurons merely act as bandpass filters, with no functional (as opposed to descriptive) role for Fourier analysis. (Note that this weak form of spatial frequency modeling is really a spatial mapping model: the inputs and outputs to the spatial filters are really spatial maps, i.e. are images.) Finally, spatial mapping models suggest that it is the spatial or anatomical rearrangement of visual data which is of importance to visual computation.

It is immediately obvious that these three categories of paradigms are not mutually exclusive and that, in all likelihood, all three of them are almost certainly true to some extent. However, this does not belie the importance of making these paradigm distinctions, for the following reasons. Neural modeling is in a very young and early state. The existence of competing paradigms is characteristic of beginning sciences (Kuhn, 1962), and one may learn from history that often the future progress of the science hinges very much on the dominance of one or another paradigm.

It may be argued that, in the three paradigms cited above, there is some parallel to the current state of paradigms in visual modeling. The spatial frequency hypotheses are observationally direct and correct; they are capable (because the visual system is quasi-linear, at least at threshold) of providing a fair summary of the data. However, in a fundamental sense, they may be entirely misleading. This is because vision is ultimately a spatial (not a frequency) task. The visual system is largely organized along spatial (not frequency) lines, and the underlying architecture which is currently understood (retinotopy, orientation columns) is spatially (and not frequency) referenced. The single contradiction to this statement are the 'spatial frequency' columns observed by Tootel, Silverman and DeValois (1981); as suggested earlier in this paper, these may be in fact an epiphenomenon of an underlying local logarithmic mapping.

With this historical argument outlined, concerning the importance of choosing the correct paradigm in a new field such as brain science, a brief review of the possible computational uses suggested by the spatial structure of striate cortex will be given. Because of space limitations, only a list will be provided, interested readers being referred to other reviews and papers (e.g. Schwartz, 1980, 1982, 1983, 1984).

Applications of computational anatomy in striate cortex:

1. Global topography
   (a) Size invariance
   (b) Optical flow
   (c) Data compression
2. Ocular dominance columns
   (a) Fast stereo segmentation
3. Orientation columns
   (a) Shape encoding via Fourier descriptors, applications in inferotemporal cortex
4. Visual illusions
   (a) MacKay illusion

In summary, the previous arguments suggest possible computational roles for spatial mapping in vision. Specific computational roles for ocular dominance column interlacing (Schwartz, 1982), for global patterns of topography and for orientation column patterns (Schwartz, 1980) have been proposed. Some of these have already led to correct predictions, i.e. regarding striate cortex global mapping and inferotemporal cortex trigger features (Schwartz, Wolf and Christman, 1983). In general, the question of whether neuroanatomy is computational, in the sense outlined in this paper, or whether the brain may be more profitably described by network models, spatial frequency or some other paradigm, is clearly not settled at the present time. Perhaps it is this clash of paradigms which both demonstrates the immaturity of brain science as well as provides a part of its excitement.

ACKNOWLEDGMENTS

This work was supported in part by contract No. PR 883-00663 from the AFOSR Image Understanding Program and a grant from the Systems Development Foundation.

REFERENCES

Alfhors, L. (1966). *Complex Analysis*, McGraw-Hill, New York.

Daniel, M., and Whitteridge, D. (1961). The representation of the visual field on the cerebral cortex in monkeys. *J. Physiol.*, **159**, 203–221.

Dobson, V. G. (1980). Neuronal circuits capable of generating visual cortex simple-cell stimulus preferences. *Perception*, **9**, 411–434.

Dow, B. M., Snyder, A. Z., Vautin, R. G., and Bauer, R. (1981). Magnification factor and receptive field size in foveal striate cortex of monkey. *Exp. Brain Res.*, **44**, 213–228.

Hubel, D. H., and Livingstone, M. (1981). Regions of poor orientation tuning coincide with patches of cytochrome oxidase staining in monkey striate cortex. *Neurosci. Abstr.*, **7**, 357.

Hubel, D. H., and Wiesel, T. N. (1962). Receptive fields, binocular interaction and functional architecture in the cat's visual cortex. *J. Physiol.*, **160**, 106–154.

Hubel, D. H., and Wiesel, T. N. (1974). Sequence regularity and geometry of orientation columns in the monkey striate cortex. *J. comp. Neurol.*, **158**, 267–293.

John, E. R., and Schwartz, E. L. (1978). The neurophysiology of information processing and cognition. *Ann. Rev. Psychol.*, **29**, 1–29.

Julesz, B. (1971). *Foundations of Cyclopean Perception*, University of Chicago Press, Chicago, Ill.

Kuhn, T. (1962). *The Structure of Scientific Revolutions*, University of Chicago Press, Chicago, Ill.

Lettvin, J., Maturana, H., McCulloch, W., and Pitts, W. (1959). What the frog's eye tells the frog's brain. *Proc. IRE*, **47**, 1940–1951.

LeVay, S., Hubel, D. H., and Wiesel, T. N. (1975). The pattern of ocular dominance columns in macaque visual cortex revealed by a reduced silver stain. *J. comp. Neurol.*, **159**, 559–576.

Schwartz, E. L. (1976). Analytic structure of the retinotopic mapping of visual cortex and relevance to perception. *Neurosci. Abstr.*, **1976**, No. 1636.

Schwartz, E. L. (1977a). Spatial mapping in primate sensory projection and relevance to perception. *Biolog. Cybernetics.*, **25**, 181–194.

Schwartz, E. L. (1977b). The development of specific visual connections in the monkey and the goldfish. *J. Theor. Biol.*, **69**, 655–683.

Schwartz, E. L. (1977c). Afferent geometry in primate visual cortex and the generation of neural trigger features. *Biolog. Cybernetics*, **29**, 1–24.

Schwartz, E. L. (1980). Computational anatomy and functional architecture of striate cortex: a spatial mapping approach to perceptual coding. *Vision Res.*, **20**, 645–669.

Schwartz, E. L. (1981a). Cortical anatomy and spatial frequency analysis. *Perception*, **10**, 455–468.
Schwartz, E. L. (1981b). Positron emission tomography studies of human visual cortex. *Neurosci. Abstr.*, **7**.
Schwartz, E. L. (1982). Computational anatomy and columnar architecture in primate striate cortex: segmentation and feature extraction via spatial frequency coded difference mapping. *Biolog. Cybernetics*, **42**, 157–168.
Schwartz, E. L. (1983). Spatial mapping and spatial vision from striate to inferotemporal cortex. In *Sensation, Perception, and Cognition: A Festchrift for Ivo Kohler* (Eds. L. Spillman and W. Wooten), Laurence Erlbaum Associates, Hillsdale, New Jersey, pp. 73–104.
Schwartz, E. L. (1984). Anatomical and physiological correlates of human visual perception. *IEEE Trans. (SMC)* (in press).
Schwartz, E. L., Wolf, A. W., and Christman, D. (1983). Positron emission tomography mapping of human striate cortex. *Brain Res.*, **294**, 225–230.
Schwartz, E. L., Desimone, R., Albright, T., and Gross, C. (1983). Shape recognition and inferior-temporal neurons. *Proc. Nat. Acad. Sci.*, **80**, 5776-5778.
Segal, L. A. (1977). *Mathematics Applied to Continuum Mechanics*, Macmillan, New York.
Tootel, R. B., Silverman, M. S., and DeValois, R. L. (1981). Spatial frequency columns in primary visual cortex. *Science*, **214**, 813–815.
Tootel, R. B., Silverman, M. S., and DeValois, R. L. (1982). De-oxyglucose analysis of retinotopic organization in primate striate cortex. *Science*, **218**, 902–904.

Models of the Visual Cortex
Edited by D. Rose and V. G. Dobson

CHAPTER 15

Formation of retinotopy and columnar microstructures by self-organization: a mathematical model

SHUN-ICHI AMARI
Department of Mathematical Engineering and Instrumentation Physics, University of Tokyo, Bunkyo-ku, Tokyo, Japan

INTRODUCTION

The brain has not only a highly organized structure but also the ability to modify its structure. A vast amount of experimental data accumulated so far have revealed the highly organized structure of the striate cortex: it is composed of retinotopically arranged hypercolumns, and each hypercolumn has a further microstructure in which feature detecting or ocular dominance columns are again topographically arranged. It is an interesting problem to find the mechanism of producing such fine structures. It is found that visual experiences in the critical period at an early stage of development are indispensable to the formation, refinement or fixation of the structures of the cortex. This suggests that the plasticity or modifiability of neural connections plays a fundamental role in constructing the brain structures. It seems plausible to consider the following two stages of construction. First, rough, vague and inaccurate connections are formed by a certain mechanism guided by chemoaffinity. Then, this rough and mostly random structure is refined by self-organization based on visual experiences, so that the structure of the nerve system fits well in the environmental information structure.

We present here a simple mathematical model of self-organizing nerve fields, which can explain the formation and refinement of feature-detecting cells and of topographic structures in a unified manner by a single mechanism (Amari, 1977, 1980, 1982, 1983; Amari and Takeuchi, 1978; Takeuchi and Amari, 1979). Such a mechanism is believed to play an important role in more sophisticated information processing such as association and concept formation.

BEHAVIOURS OF NEURONS WITH MODIFIABLE SYNAPSES

It is known that neurons in the visual cortex have modifiable synaptic connections. Modifiable synapses work at an early stage (critical period) of development of an animal to refine and fix the structure of the cortex. Feature detectors and their topographic arrangements are refined and fixed in this manner based on the visual experiences of an animal in this period. In order to understand the mechanism of formation and refinement of the visual cortex, we use a simple mathematical model of nerve systems and study the characteristic behaviours of the model by mathematical analysis. Let us first consider a simple mathematical model of a single neuron with modifiable synapses (Fig. 1): it receives n excitatory input signals $x_1, \ldots, x_n$ and inhibitory input signal x_0, and it emits an output signal z based on the input signals. Here, x_i and z represent the pulse rates of signals in the respective axons, taking on non-negative analogue values, and they are functions of time t. Let $s_1, \ldots, s_n$ and $-s_0$ be the synaptic efficiencies of the respective input signals. Then, by receiving signals $x_1, \ldots, x_n$ and x_0, a neuron is stimulated by the intensity $u = \Sigma s_i x_i - s_0 x_0$. It is postulated that the output firing rate z of the neuron is a monotonically increasing non-linear function $z = f(u)$ of the total synaptic input $u = \Sigma s_i x_i - s_0 x_0$. This shows the input–output behaviour of a neuron.

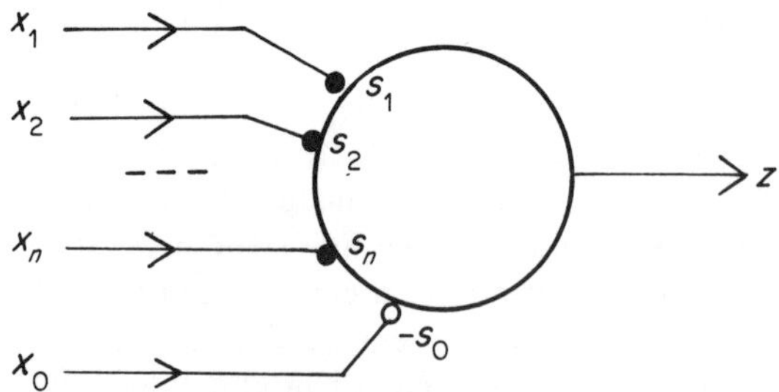

FIGURE 1 Model neuron (see text for explanation of symbols).

The behaviour of a model neuron is assumed to change slowly, depending on the input signals and the output signal: the synaptic efficiencies s_i and s_0 change in a manner as proposed by Hebb, i.e. s_i increases in proportion to the input signal x_i whenever the neuron is excited and decays slowly at the same time. Hence, the increments Δs_i and Δs_0 of the synaptic efficiencies at time t can be written as:

$$\Delta s_i = cx_i(t)r(t), \qquad \Delta s_0 = c'zx_0(t)r(t),$$

where $r(t)=1$ when the neuron fires at time t and otherwise 0, $x_i(t)$ and $x_0(t)$ are the values of the inputs at time t and c and c' are constants representing the rates of synaptic modification.

Synaptic efficiencies of a neuron change very slowly depending on the time course of the input signals $x_i(t)$ and $x_0(t)$ which represent the experiences of an animal. We call a typical combination of input signals $(x_1, x_2, \ldots, x_n, x_0)$ an input pattern, and denote it by a capital letter $X = (x_1, \ldots, x_n, x_0)$. For example, when an animal sees a vertical line, the neuron receives an input pattern X_{ver} and when it sees a horizontal line, the neuron receives another input pattern X_{hor}. We assume that an animal grows up in an environment which includes only a finite number of patterns $X_1, X_2, \ldots, X_k$ and that a neuron receives these input patterns one by one repeatedly in a random order with various relative frequencies. Then, the synaptic efficiencies of the neuron change by experiencing only these patterns. What is the final input–output behaviour of the neuron 'educated' in this environment?

We can answer this question by expressing the above considerations in the form of non-linear stochastic differential equations and by analysing them. Refer to Amari (1977) and Amari and Takeuchi (1978) for the mathematical method of analysis. A neuron is called a detector cell or representative cell of a pattern X_i, when it is excited in response to X_i but is not responsive to any other X_j's. Similarly, a neuron is a detector of a set $(X_i, X_j, \ldots)$ of patterns when it is excited in response to any X_i in the set but is not responsive to any other patterns. The mathematical analysis demonstrates that a neuron adapts to its environment by modifying the synaptic efficiencies such that it becomes a detector of one of the input patterns included in the environment. In some cases, it becomes a detector of a number of similar patterns and the resolution of pattern detectors is determined by the ratio c'/c. Hence, the modifiability of inhibitory synapses s_0 is very important in forming high-resolution detector cells. This analysis explains the experiments in which an animal comes to have very poor detectors when it is put in an abnormal environment in the critical period. A neuron is able to become potentially a detector of any pattern. However, the pattern of which detector a neuron actually becomes depends on the initial conditions and on interactions amongst neighbouring neurons. Detector cells of similar patterns come to be located at neighbouring positions by mutual interactions, when cells are arranged in a neural field with lateral inhibitory recurrent connections. It is thus believed that rough organization is formed by the use of genetic information and then it is later refined and fixed by self-organization.

SELF-ORGANIZATION OF NERVE FIELDS

The visual cortex is a multi-layered nerve field having rather a uniform structure from the macroscopic point of view. However, it has a fine microstructure composed of columns and hypercolumns, and is organized topographically and retinotopically. It is known that these structures are also affected by the visual experiences of an animal at an early stage of development. This suggests that self-organization based on experience is indispensable to the formation, refinement and fixation of these structures. We show that the same simple mechanism of synaptic modification as above accounts for the formation and refinement of the structures of nerve fields.

We consider two nerve fields R and C, where neurons in R send axons to neurons in C. (One may regard R and C as simplified mathematical models of the retina and striate cortex respectively). Let $u(\xi, t)$ be the average potential of neurons at spatial position ξ at time t of the field C, where ξ denotes the spatial coordinates. (Since the field is two-dimensional, ξ has two components $\xi = (\xi_1, \xi_2)$. However, we treat it as if ξ is one-dimensional for simplicity's sake.) The output at time t of the neurons at ξ is $z(\xi, t) = f[u(\xi, t)]$. We assume that there exist recurrent connections within the field C so that output at ξ' stimulates (or inhibits via inhibitory interneurons) neurons at ξ with the synaptic efficiency $w(\xi - \xi')$. Let η be a coordinate denoting the spatial position of neurons in the presynaptic field R and let $s(\xi, \eta)$ be the synaptic efficiencies of neural connections from neurons at position η of R to neurons at position ξ of C (Fig. 2). Then, when output signals $x(\eta)$ are emitted from the neurons at η of R, the total synaptic input arriving at the neurons at ξ of C from R is written as:

$$S(\xi) = \int s(\xi, \eta)x(\eta)\mathrm{d}\eta - s_0(\xi)x_0,$$

where x_0 is the intensity of the inhibitory signal from a common inhibitory neuron pool and $s_0(\xi)$ is its synaptic efficiency to the neurons at ξ. Here, the term $\Sigma s_i x_i$ in the previous model is replaced by integration $\int s(\xi, \eta)x(\eta)\mathrm{d}\eta$,

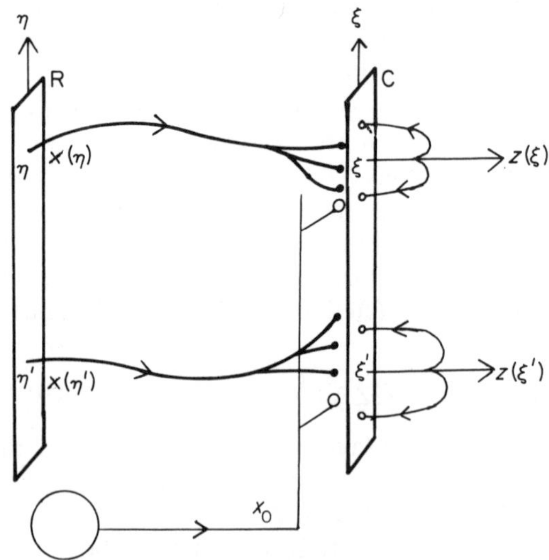

FIGURE 2 Self-organizing nerve fields (see text for explanation of symbols).

because the input signals $x(\eta)$ come from all the parts of the continuum field R. We may consider distributed inhibitory inputs. However, they are summarized into a single x_0 for the sake of simplicity.

The neurons at ξ in C receive not only the total stimulus $S(\xi)$ from R but also stimuli $w(\xi - \xi')z(\xi')$ from the neurons at position ξ' of C by the recurrent lateral connections, where $z(\xi')$ is the firing rate of the neurons at ξ'. Hence, by summing up all the recurrent signals, the total synaptic input $u(\xi)$ to the neurons at position ξ of C is expressed as:

$$u(\xi) = \int w(\xi - \xi')z(\xi')\, d\xi' + S(\xi),$$

where the firing rate $z(\xi')$ is determined from the total input $u(\xi')$ by $z(\xi') = f[u(\xi')]$. Hence, this is an integral equation determining the activity distribution $z(\xi)$ of C from the activity distribution $x(\eta)$ of R. The recurrent connections $w(\xi - \xi')$ within C are assumed to be positive where $\xi - \xi'$ is small and negative (lateral inhibitory) where $\xi - \xi'$ is large.

If the synaptic efficiencies $s(\xi, \eta)$ and $s_0(\xi)$ are modifiable, they change slowly depending on the input presynaptic activity distribution $x(\eta)$ of R and the resultant activity $z(\xi)$ of C. The equations of synaptic modification are:

$$\Delta s(\xi, \eta) = cr(\xi, t)x(\eta, t), \qquad \Delta s_0(\xi) = c'r(\xi, t)x_0.$$

where $r(\xi, t)$ is 1 when the neurons at ξ fire at t and 0 otherwise. It is easy to express these ideas in the form of exact mathematical equations but it is highly difficult to solve these equations exactly. However, by taking full advantage of the stochastic nature of the equations and the relative slowness in the synaptic modification process, we can analyse the self-organizing behaviours of nerve fields. Usually, two neighbouring neurons are excited together at the same time with a high probability, while two distant neurons are excited rather independently, in the presynaptic field R. In other words, the correlation of $x(\eta)$ and $x(\eta')$ is large when $\eta - \eta'$ is small. This shows that the activity distributions $x(\eta)$ are compatible with the topology of the two-dimensional retina R. Since the connections from R to C are modifiable, the topology of R (defined by the distance $\eta - \eta'$) is mapped to C such that the topology of the resultant activity distributions $z(\xi)$ are also compatible with that of the field C. It is proved (Amari, 1980) under some assumptions that the neural connections from R to C are rearranged retinotopically by self-organization, even when the initial connections are rough, vague and highly random. This is the neural mechanism of forming a fine and accurate topographic map from an inaccurate one. Here, the modifiability of the inhibitory synapses plays again a fundamental role in making the projection from R to C sharp. Moreover, it is proved that, when a part of R is stimulated frequently, that part comes to be mapped on a relatively large part of C so that stimuli on that part enjoy finer resolution. Even when part of R or C is destroyed, the remaining parts are topographically rearranged as a whole.

It is further proved (Takeuchi and Amari, 1979) that the equations involve more interesting characteristic features. They found that the resultant continuous map from R to C becomes unstable under a certain condition. In this case, the field C is divided into topographically arranged microregions, so that a microstructure is automatically formed in a topographically organized continuous field. This is compatible with the columnar microstructures in the actual cortex. It is rather surprising that our simple mathematical model of self-organization is able to account for so many characteristic features of nerve fields in a unified manner.

DISCUSSION

Since the work of Malsburg (1973) a number of researches have been published concerning self-organizing neural nets or the formation of feature detecting cells (Nass and Cooper, 1975; Grossberg, 1976; etc.). It was pointed out (Amari and Takeuchi, 1978) that modifiable inhibitory synapses play an important role in such systems. It was further suggested (Willshaw and Malsburg, 1976) that the self-organizing mechanism plays a role in retinotopy formation. It is through mathematical analysis (Amari, 1980; Takeuchi and Amari, 1979) that the capabilities and limitations of self-organization are elucidated from a unified point of view in self-organizing nerve fields. This leads us to the two-stage theory of the formation of neural structures (cf. Willshaw and Malsburg, 1979; Overton and Arbib, 1982, etc.). The same mechanism accounts not only for the formation of topographic structures but also for the formation in the brain of more fundamental topological information structures about the outside world (Amari, 1982; 1983; see also Kohonen, 1982).

REFERENCES

Amari, S. (1977). Neural theory of association and concept-formation. *Biolog. Cybernetics*, **26**, 175–185.

Amari, S. (1980). Topographic organization of nerve fields. *Bull. Math. Biol.*, **42**, 339–364.

Amari, S. (1982). Competitive and cooperative aspects in dynamics of neural excitation and self organization. In *Competition and Cooperation in Neural Nets* (Eds. S. Amari and M. A. Arbib), Lecture Notes in Biomathematics, Vol. 45, Springer, pp. 1–28.

Amari, S. (1983). Field theory of self-organizing neural nets. *IEEE Trans. Systems, Man and Cybernetics*, **13**, 741–748.

Amari, S., and Takeuchi, A. (1978). Mathematical theory on formation of category detecting nerve cells. *Biolog. Cybernetics*, **29**, 127–136.

Grossberg, S. (1976). Adaptive pattern classification and universal recording, I. *Biolog. Cybernetics*, **23**, 121–134.

Kohonen, T. (1982). Self-organized formation of topologically correct feature maps. *Biolog. Cybernetics*, **43**, 59–69.

Malsburg, Ch. von der (1973). Self-organization of orientation sensitive cells in the striate cortex. *Kybernetik*, **14**, 85–100.

Nass, M. M., and Cooper, L. N. (1975). A theory for the development of feature detecting cells in the striate cortex. *Biolog. Cybernetics*, **19**, 1–18.

Overton, K. J., and Arbib, M. A. (1982). The extended branch-arrow model of the formation of retino-tectal connections. *Biolog. Cybernetics*, **45**, 157–175.

Takeuchi, A., and Amari, S. (1979). Formation of topographic maps and columnar microstructures. *Biolog. Cybernetics*, **35**, 63–72.

Willshaw, D. J., and Malsburg, C. von der (1979). A marker induction mechanism for the establishment of ordered neural mappings: its application to the retinotectal connections. *Phil. Trans. Roy. Soc. (Lond.) B*. **287**, 203–243.

Willshaw, D. J., and Malsburg, C. von der (1976). How patterned neural connections can be set up by self-organization. *Proc. Roy. Soc. (Lond.) B*, **194**, 431–445.

Models of the Visual Cortex
Edited by D. Rose and V. G. Dobson

CHAPTER 16

A model for the development of neurons in the visual cortex

LEON N COOPER
Center for Neural Science and Physics Department, Brown University, Providence, RI 02912, USA.

Great attention has been lavished on the visual system in the last generation—and rightly so—for this system may provide a clue as to the means by which information is processed on its path from the sensory apparatus to 'higher' brain centers. However, as suggested in a large number of 'deprivation' experiments—experiments in which the normal development of visual cortex is modified by abnormal visual experience—the visual system may also provide a means of testing neuron learning on a single-cell level.

For the visual system, we can control the dominant input (in development as well as in testing) as well as measure the output. This gives us the possibility of measuring the change of response of individual neurons as a function of the sensory environment in which they are placed—assuming that what is observed in altered visual environments such as monocular deprivation or dark rearing are changes in responses of individual neurons rather than responses of different populations of neurons. Although it seems likely that this is indeed the case, the question will probably be settled finally by appropriate chronic experiments.

Even though such neuron learning may play only a minor role in contributing to the architecture of visual cortex, with a minimum of fortune, it could be related to learning that must take place in higher brain centers. It is this point of view that we have explored over the past few years.

Cortical neurons receive afferents from many sources. In visual cortex (layer 4, for example) the principle afferents are those from the lateral geniculate nucleus and from other cortical neurons. This leads to a complex network that we have analyzed in several stages.

In the first stage we consider a single neuron with inputs from both eyes (Fig. 1a). Here d_l, d_r, m_l, m_r are inputs and synaptic junctions from left and right eyes. The output of this neuron (in the linear region) can be written:

$$c = m_l \cdot d_l + m_r \cdot d_r. \quad (1)$$

This means that the neuron firing rate (in the linear region) is the sum of the inputs from the left eye multiplied by the appropriate left-eye synaptic weights plus the inputs from the right eye multiplied by the appropriate right-eye synaptic weights. Thus the neuron integrates signals from the left and right eyes.

According to the theory presented by Bienenstock, Cooper and Munro (1982), these synaptic weights modify as a function of local and global variables. To illustrate we consider the synaptic weight m_j (Fig. 1b). Its change in time m_j can be written very generally as:

$$\dot{m}_j = F(d_j \cdot \cdot \cdot m_j; d_k \cdot \cdot \cdot c; \bar{c} \cdot \cdot \cdot ; X, Y, Z). \quad (2)$$

Here variables such as $d_j \ldots m_j$ are designated local. These represent information (such as the incoming signal d_j and the strength of the synaptic junction m_j) available locally at the synaptic junction m_j. Variables such as $d_k \ldots c$ are designated quasi-local. These represent information (such as the firing rate c of the cell or the incoming signal d_k to another synaptic junction) that is not locally available to the junction m_j but is physically connected to the junction by the cell body itself—thus necessitating some form of internal communication between various parts of the cell and its synaptic junctions. Variables such as the time-averaged output of the cell $\bar{c}$ are averaged local or quasi-local variables. Global variables are designated $X, Y, Z, \ldots$. These latter represent information (e.g. the presence or absence of neurotransmitters such as norepinephrine or the average activity of large numbers of cortical cells) that is present in a similar fashion for all or a large number of cortical neurons (distinguished from local or quasi-local variables carrying detailed information that varies from synapse to synapse).

In a form relevant to this discussion, the Bienenstock, Cooper and Munro (BCM) modification is written:

$$\dot{m}_j = \phi\ (c, \bar{c}; X, Y, Z, \ldots) d_j, \quad (3)$$

so that the jth synaptic junction, m_j, changes its value in time as a function of quasi-local and time-averaged quasi-local variables, c and $\bar{c}$, as well as global variables X, Y, Z, through the function ϕ and a function of the local variable d_j. The crucial function, ϕ, is shown in Fig. 1c.

What is of particular significance is the change of sign of ϕ at the modification threshold θ_m and the nonlinear variation of θ_m with the average output of the cell $\bar{c}$. In a simple situation:

$$\theta_m = (\bar{c})^2 \quad (4)$$

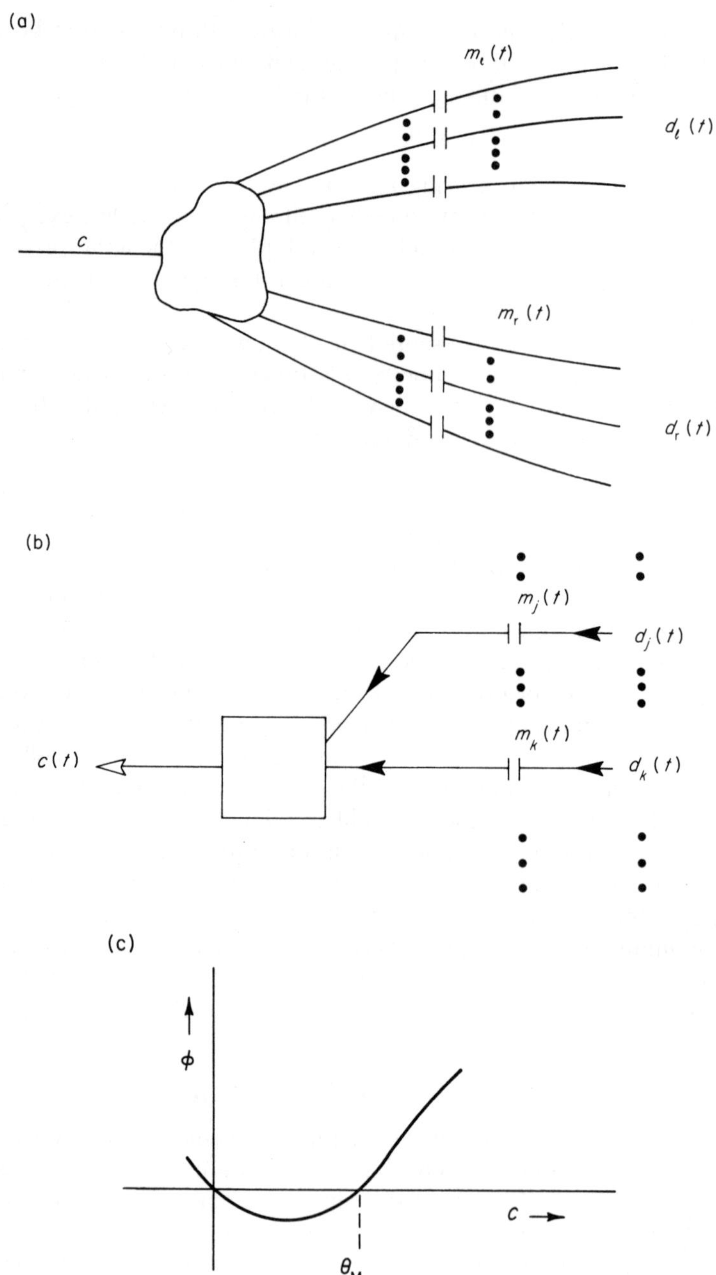

FIGURE 1 (a) A model neuron. (b) Local and quasi-local variables. (c) The BCM modification function. (see text for explanation of symbology).

The occurrence of negative and positive regions for ø drives the cell to selectivity in a 'normal' environment. This is so because the response of the cell is diminished to those patterns for which the output c is below threshold (ø negative) while the response is enhanced to those patterns for which the output c is above threshold (ø positive). The nonlinear variation of the threshold with the average output of the cell $\bar{c}$ places the threshold so that it eventually separates one pattern from all of the rest. Further, it provides the stability properties of the system.

A detailed analysis of the consequences of this form of modification is given in BCM theory. To illustrate, we consider the various final states for the cell under rearing conditions characterized as follows:

1. *Normal environment*. The inputs from the eye are a stochastic sequence of patterns possibly distorted by noise (these patterns represent the mapping by retinal and LGN cells of images or patterns viewed by the animal). In normal rearing it is presumed that certain patterns (edges, for example) are a repeated part of the environment and are viewed by the animal many times. It is assumed that such patterns may be distorted or may appear in various combinations but that for a normal environment the patterns are statistically independent.
2. *Restricted or altered environment*. The inputs are restricted to one or a few patterns that may be highly correlated. This represents the mapping in the neural space of a restricted or artificial environment (e.g. one in which only vertical lines or right angles are present).
3. *Dark reared environment*. The patterned inputs are replaced by random noise that represents the nonpatterned input due to dark discharges and spontaneous activity of the retinal cells.

Various rearing conditions can then be represented as follows:

1. *Normal binocular rearing*. Correlated patterns input from both eyes. This means that both the left and right eye see the same pattern at the same time.
2. *Strabismic rearing*. Noncorrelated patterns input from both eyes. In this case the left and right see different patterns at the same time.
3. *Monocular deprivation* (right eye deprived). Patterned input from left eye, noise from right eye. In this case the left eye sees a pattern while the right eye is closed.
4. *Binocular deprivation*. In this case, both eyes being closed (or kept in the dark), noise is received from both eyes.

The consequences of the BCM modification in these classical rearing conditions are shown in Fig. 2. These may be summarized as follows:

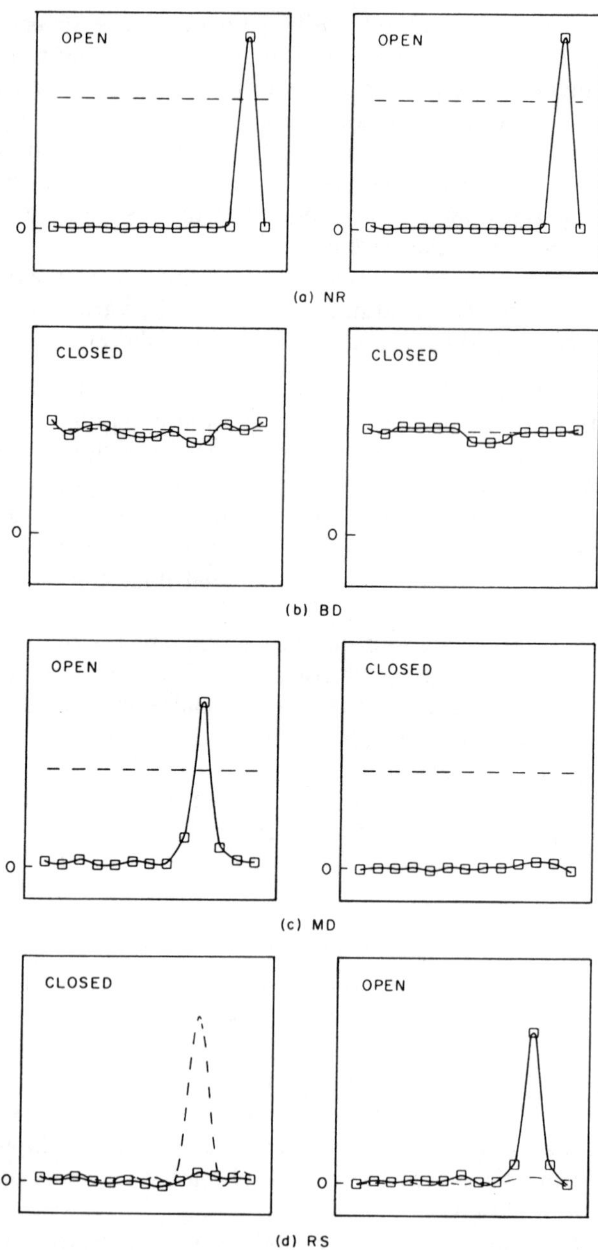

FIGURE 2 Computer simulations of various rearing conditions. Initial (dashed) and final (solid) responses to the two eyes are shown separately (left/right). NR, normally reared; BD, binocular deprivation; MD, monocular deprivation; and RS reverse suture. Ordinates are response levels and abscissae show pattern orientation (0–180).

Monocularly driven neurons:

1. A monocularly driven neuron in a 'normal' (patterned) environment becomes selective. The precise pattern to which it becomes selective is determined at random if there is no initial preference or may be biased toward a particular pattern if there is a built-in preference to this pattern.
2. This same neuron in various deprived environments evolves as follows:
   (a) Pure noise. The neuron becomes less selective but continues to be (somewhat) responsive. It may show an orientation preference, but this is relatively unstable.
   (b) Exposure to a single pattern (such as vertical lines). The neuron comes to respond preferentially to the single pattern but with less selectivity (less sharply tuned) than if all orientations were present in the environment.
3. Inhibitory synapses are required to produce maximum selectivity. If such inhibitory connections are arbitrarily set equal to zero, selectivity diminishes.

Binocularly driven neurons:

1. A binocularly driven neuron in a 'normal' (patterned) environment becomes selective and binocular. It is driven selectively by the same pattern from both eyes.
2. This same binocularly driven neuron in various deprived environments evolves as follows:
   (a) Uncorrelated patterned inputs to both eyes. The neuron becomes selective, often monocularly driven; if the neuron is binocular, it is sometimes driven by different patterns from the two eyes.
   (b) Patterned input to one eye, noise to the other (monocular deprivation). The neuron becomes selective and generally driven only by the open eye. There is a correlation between selectivity and binocularity. The more selective the neuron becomes, the more it is driven only by the open eye. A nonselective neuron tends to remain binocularly driven. This correlation is due, in part, to the fact that it is the same mechanism of synaptic change that serves to increase both the selectivity and ocular dominance of the open eye. However, there is also a subtler connection: it is the nonpreferred inputs from the open eye accompanied by noise from the closed eye that drive the neuron's response to the closed eye to zero.
   (c) Noise input to both eyes (dark rearing or binocular deprivation). The neuron remains nonselective (or loses its selectivity) and diminishes its responsiveness but remains binocularly driven (in contrast to the situation in monocular deprivation).

An unexpected consequence of this theory is a connection between selectivity and ocular dominance. The analysis given in BCM theory leads to the conclusion that the more selective a cell is, the more the state of the closed eye will be driven to zero, so that the cell will be more monocular. Thus, during cell development, pattern selectivity for the open eye is achieved before the cell becomes monocular, as shown in Fig. 3.

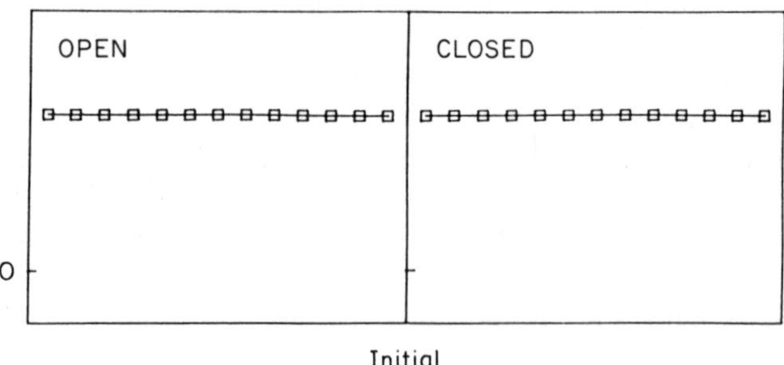

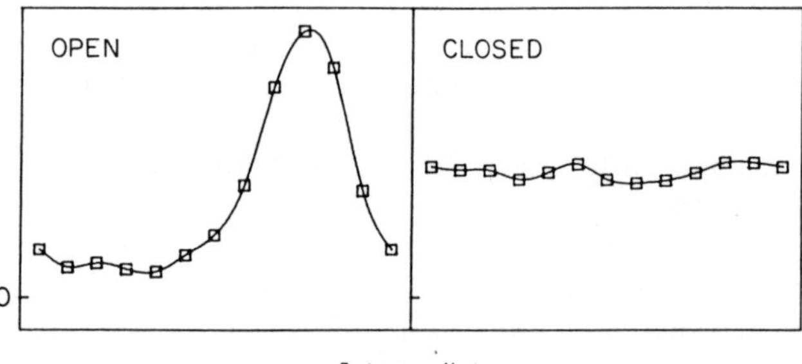

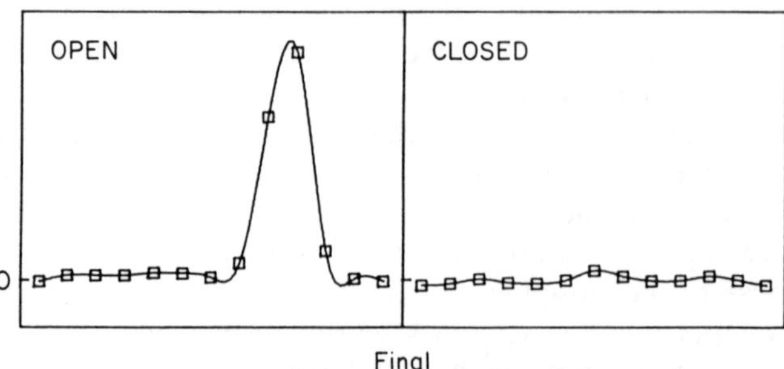

Figure 3 Progression of development of selectivity and ocular dominance. Ordinates are response levels and abscissae: show pattern orientation (0–180). Note that selectivity develops for the open eye *before* the response to the closed eye is driven to zero.

Since nonpreferred inputs presented to the open eye are a necessary part of the suppression of deprived eye responses, if inputs to the open eye are restricted to preferred patterns, selectivity may still develop but the cell will be less selective to the open eye and less monocular than it would be in a nonrestricted environment. Such effects should be experimentally observable.

To better confront these ideas with experiment a second stage of analysis is necessary. The BCM neuron must be placed in a network with the anatomical features of the visual cortex–a network in which inhibitory and excitatory cells receive input from LGN and from each other. This has been done (Scofield and Cooper, to be published). Our conclusions are similar to those of BCM with explicit further statements concerning the independent effects of excitatory and inhibitory neurons on selectivity and ocular dominance. For example, shutting off inhibitory cells lessens selectivity and alters ocular dominance (masked synapses). These inhibitory cells may be selective but there is no theoretical necessity that they be so. Further, the intracortical inhibitory synapses do not have to be very responsive to visual experience. Most of the learning processes can occur amongst the excitatory LGN–cortical synapses.

Quantitative tests of progressions such as those shown in Fig. 3 are in progress in our laboratory. We hope that such experiments can provide detailed comparisons of theory and experiment and provide us with a sensitive tool for determining synaptic modification amongst various classes of neurons—a possible entry to the process by which the nervous system organizes itself.

ACKNOWLEDGMENT

This work was supported in part by the US Office of Naval Research under contract No. N00014–81K–0136.

REFERENCES

Bienenstock, E. L., Cooper, L. N., and Munro, P. W. (1982). Theory for the development of neuron selectivity: orientation specificity and binocular interaction in visual cortex. *J. Neurosci.*,**2**, 32–48.

Scofield, C. L., and Cooper, L. N. (to be published). Selectivity and ocular dominance in visual cortex: a network theory.

Models of the Visual Cortex
Edited by D. Rose and V. G. Dobson

CHAPTER 17

Functional multicompartment models: a kinetic study of the development of orientation selectivity

YVES FRÉGNAC
Lab. de Neurobiologie du Développement, Bât. 440, Université Paris XI, Orsay 91405, France

INTRODUCTION

Twenty years of continuing debate have followed the seminal finding by Hubel and Wiesel (1963) of orientation-sensitive neurons in kitten area 17 at eye-opening. Since this milestone report, numerous studies have questioned the role of sensory experience in the development of neurons which encode specific visual attributes (see the review in Frégnac and Imbert, 1984). This had led to a profusion of models according to which sensory-evoked activity would regulate the development of cortical function. However, very few models address the kinetic aspect itself, i. e. the *timecourse* of experience-related influences on the development of neuronal selectivity.

Most theoretical schemes in fact restrict the analysis of developmental processes to self-organization mechanisms linked with associative memory. The description of 'critical' periods (Hubel and Wiesel, 1970; Daw, Berman and Ariel, 1978; Frégnac, 1979a) suggests the importance of age-dependent interactions between genetic constraints and epigenetic control. However, the few descriptions of the relative importance of both types of processes during postnatal development have so far remained quite schematic (Gottlieb, 1976; Aslin, 1981).

Apart from *maturation* which is independent of visual experience (Fig. 1a), four developmental outcomes are typically called upon; (a) *maintenance* supposes that neuronal selectivity is fully established prior to the time when experience becomes necessary for functional stabilization (Fig. 1b); (b) *facili-*

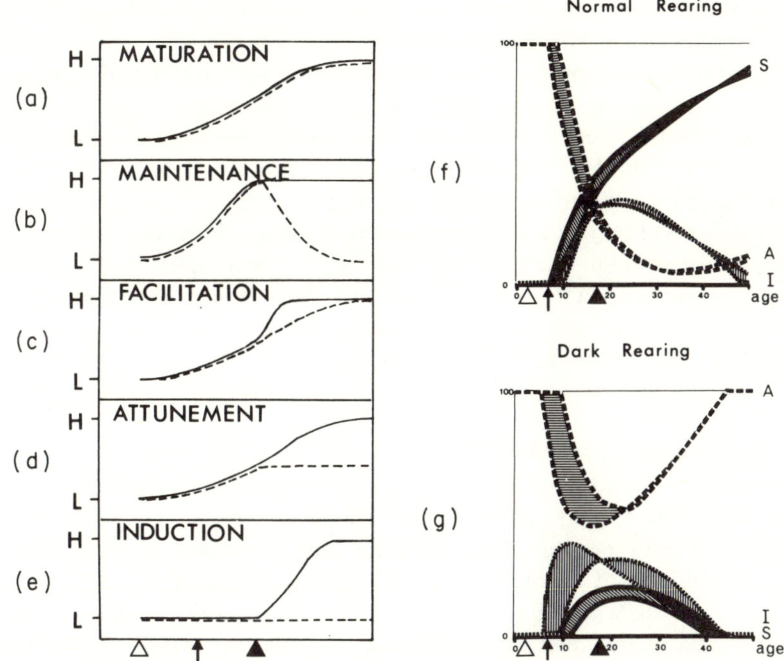

Figure 1 Roles of visual experience in the development of orientation selectivity. (a) to (e) Hypothetical developmental outcomes (adapted from Gottlieb, 1976, and Aslin, 1981 and reproduced with permission). The proportion of orientation-selective (S-type) cells is plotted with age in visually experienced (continuous line) and visually deprived (dotted line) subjects. H and L give the extrema of each graph. The open triangle represents the onset of nervous activity, the arrow gives the date of eye-opening and the filled triangle indicates the onset of the critical period. (f) and (g) Kinetics of the development of orientation tuning in area 17 of normally reared (top graph) and dark reared (bottom graph) kittens (from Figs. 3 and 4 in Frégnac, 1979a, reproduced by permission of Springer-Verlag). A, non-oriented; I, immature; S, orientation-selective cell proportions among visual cells. The age is given in days.

tation implies that at a given stage the kinetic rate is enhanced by experience (Fig. 1c); (c) *attunement* assumes that the increase in the selectivity level beyond a certain threshold is achieved with experience (Fig. 1d); (d) *induction* describes the triggering of a kinetic process by experience (Fig. 1e).

Kinetic studies of the development of functional properties of sensory neurons in experienced and deprived animals can help to separate between the various roles attributed to experience. In addition, they provide the experimental data set for constructing 'operational' models, in the sense that parameters can be extracted to fit a given kinetic without regard for a precise anatomofunctional correlate. From the identified model, predictions may be made about the effects of atypical sensory manipulation. Of particular interest is the proposal of models originally established solely on the basis of the kinetics observed with and without experience, which can account for the effects of more complex sensory manipulation such as delayed experience. The kinetic behaviour of the system when normal sensory input is restored after a deprivation period indicates a possible non-linearity which should be included in the model. This structural requirement will be referred in the text as the '*elasticity*' property (Aslin, 1981).

This paper will review a multicompartment model (Frégnac, 1979a, 1979b) which simulates the major kinetic features of the development of orientation tuning in kitten area 17. Some of its basic assumptions and predictions—which were originally introduced in 1977—are discussed in reference to more recent electrophysiological data.

EXPERIENTIAL KINETICS

The population analysis presented below compares the statistical representation of certain trigger features, established electrophysiologically over a limited sample of neurons (20 to 50 per cat), recorded in anaesthetized, paralysed subjects of different ages and rearing conditions. A characteristic in visual response which appears to be unaffected by the poor quality of the optical media in neonate kittens (Bonds, 1979) is the width of orientation tuning curves. A simple but operational classification has been introduced by Imbert and Buisseret (1975) and distinguishes three classes of visual neurons: (a) *A-type cell*: non-oriented neurons; (b) *I-type cell*: immature or broadly tuned neurons; (c) *S-type cell*: neurons which are as orientation selective as those in the adult cortex (Hubel and Wiesel, 1962).

From data obtained in 43 normally and dark-reared kittens (Buisseret and Imbert, 1976; Frégnac and Imbert, 1978), power analysis has led to the establishment of the best polynomial trends of the proportion of each cell type among visual cells (Fig. 1f, g; see Frégnac, 1979a). The extrapolation of these regressions predicts an initial cortical state where most neurons are non-oriented (A-type). From eye-opening (7 days) a process of *maturation* (increase in the proportion of S- and I-type cells) proceeds independently of visual experience during the whole of the second postnatal week. A covariance analysis shows that the S-type and A-type cell proportions evolve differently in experienced and deprived animals from the turn of the third postnatal week. From this age, a despecification process is observed if no visual experience is allowed, whereas the proportion of S-type cells increases monotonically in normally reared kittens. The kinetics of the S-cell pool appear to be minimally described by a compound process, the characteristic of which would be the product of both a *maturation* (Fig. 1a) and a *maintenance* process (Fig. 1b). Interestingly the immature compartment acts in both rearing conditions like a transit pool, reaching its peak value at the onset of the critical period and emptying completely at around the sixth postnatal week.

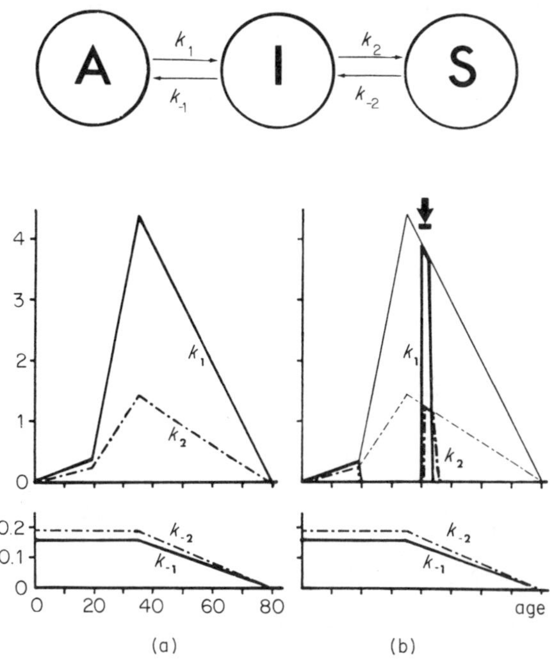

Figure 2 Expression–extinction model (from Fig. 5 in Frégnac, 1979b, reproduced by permission of Springer-Verlag). The upper flowchart symbolizes the serial structure of the three-compartment model. The lower panels show the evolutions of the exchange coefficients during (a) normal rearing and (b) dark rearing are plotted with age (in days). The values taken by k_{-1} and k_{-2} do not depend on rearing condition; k_1 and k_2 are affected by visual deprivation from the onset of the critical period (19 days). During the extinction phase following the end of delayed visual experience (underlined arrow), k_1 drops immediately to zero, while k_2 decreases linearly to zero within one day.

Similar experiential trends have been reported by numerous authors and good agreement is obtained concerning the extrapolation of the initial state from normally reared kittens older than one week of age (see Fig. 2 in Frégnac and Imbert, 1984). However, recent data obtained in kittens as young as 6 days of age (Albus and Wolf, 1984) indicate a larger proportion of oriented neurons than predicted by polynomial regressions. This could support the hypothesis of an initial sigmoid process. More controversial, since it depends on the classification scheme, is the final state of selectivity in dark-reared kittens. Kinetics established from Bonds' data (1979) are more reminiscent of an *attunement* (Fig. 1d) than a maintenance process.

Apart from these experimental discrepancies, the weakest link of this inferential analysis is that it relies on a 'constant pool' hypothesis. It is already known that the laminar development pattern is non-homogenous in area 17: layers 4, 5 and 6 exhibit orientation selectivity earlier than layers 2 and 3, although it is not yet clear which layer specializes first. (Tsumoto and Suda, 1982; Albus and Wolf, 1984). There is also a suggestion that the sampling density of visual cells is affected both by age and rearing conditions.

EXPRESSION–EXTINCTION MODEL

Compartmental Kinetics

The inferential analysis presented above suggests a catenary process of development illustrated by the flowchart in Fig. 2: the immature compartment is the transit pool between non-specific and specific pools. The kinetics, approximated by a first-order process, are given by a matricial differential equation and a conservation law:

$$\frac{\mathrm{d}}{\mathrm{d}t}\begin{pmatrix} A \\ I \\ S \end{pmatrix} = \begin{pmatrix} -k_1 & k_{-1} & 0 \\ k_1 & -(k_2 + k_{-1}) & k_{-2} \\ 0 & k_2 & -k_{-2} \end{pmatrix} \cdot \begin{pmatrix} A \\ I \\ S \end{pmatrix} \quad (1)$$

$$A(t) + I(t) + S(t) = A(\bar{o}) + I(\bar{o}) + S(\bar{o}) \quad (2)$$

Equation (1) states that, at each instant t, the rate of change of the occupancy of each pool is the net flow resulting from exchanges with its neighbours, defined by the flowchart in Fig. 2. Thus, the derivative of the occupancy of a given pool ($\mathrm{d}A/\mathrm{d}t$ or $\mathrm{d}I/\mathrm{d}t$ or $\mathrm{d}S/\mathrm{d}t$) is expressed as the algebraic sum of the flows leaving (counted negatively) or entering (counted positively) the pool. These flows are represented by arrows on the flowchart and are proportional to the occupancy of the pool from which they originate. Equation (2) states that the total number of cells (in all three pools) remains constant.

Two sequential stages occur: (a) realization of an intrinsic programme of maturation followed by (b) an epigenetic phase beginning at 19 days. The initial and final states of occupancy of each compartment are derived from the experiential kinetics. The derivative of the immature compartment occupancy is assumed initially to be zero and to reverse its sign at the onset of the critical period.

In order to fit the non-linearities of the critical period, the following assumptions have been proposed: (a) k_{-1} and k_{-2}, which are responsible for the despecification process observed during dark rearing, act independently of rearing conditions, like a 'forgetting mechanism' slowly varying with time. (b) k_1 and k_2 fit an empirical 'modifiability gradient', peaking in normally reared animals at 5 weeks of age and extinguishing around the twelfth postnatal week. As shown in Fig. 2b, during dark rearing k_1 and k_2 take a zero value from the onset of the critical period. (c) *Expression hypothesis*: if, after a period of dark rearing, visual experience is then allowed, the exchange coefficients k_1 and k_2 immediately take the value they would have if the animal had been normally reared from the beginning. Thus delayed visuomotor experience allows, at the time at which it takes place, the *expression* of a *maturation* process which has been masked up to this point by the absence of vision or eye movements. (d) *Extinction hypothesis*: at the end of the allowed period of visual experience, the *gate factor* k_1 goes instantly to zero, while the *specialization factor* k_2 drops to zero with an extinction constant (see the legend, Fig. 2). A mathematical description and numerical simulation are provided in detail elsewhere (Frégnac, 1979b).

Predictions

The predictive power of this model results from its *elasticity* properties. The non-linearity introduced by the expression hypothesis permits the restoration of a large majority of oriented (S- and I-type) cells after a few hours of visual experience following six weeks of previous dark rearing, as shown experimentally by Buisseret, Gary-Bobo and Imbert (1978). The extinction hypothesis offers an alternative to often-invoked consolidation mechanisms, according to which a return to dark may facilitate the imprinting of previous visual experience. The continuing increase in the proportion of S-type cells observed a short delay after the end of the period of normal vision (Buisseret, Gary-Bobo and Imbert, 1978) is the consequence, according to the model, of differences in decay time constants associated with each exchange coefficient k_1 and k_2.

Linear models (constant exchange coefficient kinetics in Frégnac, 1979b) or the developmental retardation hypothesis (Grobstein and Chow, 1975; Aslin, 1981) fail to show adequate rebound properties. The maximum recovery they can predict is 100 per cent of the effect which visual experience would produce at the onset of the critical period (see, for instance, Fig. 3 in Frégnac, 1979b, and case II, Fig. 2.7 in Aslin, 1981). Gating of an already built-in non-linear sensitivity to visual experience appears necessary to account for the effects of delayed visual experience.

Recent electrophysiological data have suggested a physiological correlate of the assumed *gate factor*: *extraocular proprioception*. Afferents running through the ophthalmic branch of the trigeminal nerve influence the kinetic rate of restoration of orientation selectivity and of capture of ocular dominance, following delayed visuomotor experience during the critical period (Buisseret and Gary-Bobo, 1979; Trotter, Gary-Bobo and Buisseret, 1981; Trotter, Frégnac and Buisseret, 1983; Imbert and Frégnac, 1983; Buisseret and Singer, 1983). Moreover, the time course of sensitivity to unilateral section of the ophthalmic branches of the trigeminal nerve (Trotter, Frégnac and Buisseret, 1981) fits closely with that of the 'modifiability gradient' imposed on the *gate factor* k_1 (compare Fig. 2 in the present paper or Fig. 5 in Frégnac, 1979b, and Fig. 1 in Trotter, Frégnac and Buisseret, 1981). These data strongly support the idea that extraretinal feedback of the motor act, which serves to direct gaze and to sample information in precise regions of visual space, contributes to the gating of epigenetic modifications in the primary visual cortex.

An unexpected test of the model has been raised by the recent finding that visual deprivation prolongs experience sensitivity (Cynader and Mitchell, 1980). According to equation (1), the flow rate of despecification from pool A (dA/dt)—which can be regarded as an index of experience sensitivity—is linked to the occupancy of the non-specific and to a lesser extent of the immature pool. From Figs. 1 and 2, it can be easily shown that the outgoing flow from the non-specific pool would be much higher in long-term deprived kittens (for instance 11 weeks old) than in normally reared kittens of the same age, or even than in kittens near the onset of the critical period at an age corresponding to the same value of k_1.

Also coherently related with the multicompartment serial structure of the model is the report that the last cells to lose their orientation selectivity after return to darkness (Buisseret, Gary-Bobo and Imbert, 1982) show the same orientation preference and binocularity as the first cells to acquire it during development (Frégnac and Imbert, 1978). A refinement of the model would be to divide the S-pool into a chain of specialized subcompartments underlying a slower kinetic ($S = S_1 \rightleftarrows S_2 \rightleftarrows .. \rightleftarrows S_n$). The earlier a given cell would reach the selective state, the more likely it would be to reach the end of the chain (S_n). Consequently, if the kinetic process is reversed at an intermediate stage (i.e. before all cells reach the final subcompartment S_n) such a cell would on average take more time to leave the S-pool because of the larger number of transitions it would have to cross back. From the hypothesis of serial specified subcompartments one can predict that resistance to visual deprivation depends on the time at which the cell has entered the selective state.

CONCLUDING REMARKS

In summary, the model presented here appears to fulfil the major kinetic features of orientation tuning during postnatal manipulations of the binocular visual input. It stresses the importance of *maturation* process, even during the epigenetic phase, the *expression* of which is assumed to be masked by the absence of vision and its proper *gate* context, i.e. extraocular proprioception.

However, one should keep in mind that the absence of reference in this model to orientation preference minimizes the possible extent of *attunement* processes. For instance, no prediction can be made for the effect of visual experience restricted to one orientation. Does the final state of high selectivity correspond to the attractor defined by the imposed sensory experience or to any stable attractor defined by the internal structure of the model itself (Bienenstock, Cooper and Munro, 1982; Bienenstock, 1983)? Experimental data suggests a refinement in tuning for a certain class of cells, dependent on the experienced orientation (Rauschecker, 1982; Hirsch *et al*, 1983). An improvement of the existing model would be to describe the occupancies of the orientation committed pools I and S and the exchange coefficients by functions of time *and* orientation preference. If one refers to Aslin's terminology, the first developmental period assumed by the model reveals a biased *orientation filter*, permitting the endogenous maturation of the basic grid of the orientation columnar system, probably formed by horizontally or vertically tuned cells (Frégnac and Imbert, 1978; Singer, Freeman and Rauschecker, 1981; Thompson, Kossut and Blakemore, 1983). The epigenetic phase could correspond to a flat orientation filter, which would allow visual experience to determine the precise shift in orientation preference and its columnar extension from one seed to its neighbour (see, for instance, the oblique symmetry-breaking mechanism proposed by Malsburg and Cowan, 1982). Unfortunately, a much larger amount of experimental data is needed to quantify the time dependency of the orientation filters which a proper model should associate with each exchange coefficient.

REFERENCES

Albus, K., and Wolf, R. 1984. Early postnatal development of neuronal function in the kitten's visual cortex: a laminar analysis. *J. Physiol. (Lond.)* **348,** 153–185.

Aslin, R. N. (1981). Experiental influences and sensitive periods in perceptual development: a unified model. In *Development of Perception*, Vol. 2, Academic Press, pp. 45–93.

Bienenstock, E. L. 1983. Cooperation and competition in central nervous system development: a unifying approach. In *Int. Symp. Proc. on Synergetics of Brain (Schloss Elmau, Bavaria)*, Springer-Verlag, Berlin 9 pp. 250–263

Bienenstock, E. L., Cooper, L. N., and Munro, P. (1982). Theory for the development of neuron selectivity: orientation specificity and binocular interaction in visual cortex. *J. Neurosci.*, **2**, 32–48.

Bonds, A. B. (1979). Development of orientation tuning in the visual cortex of kittens. In *Developmental Neurobiology of Vision* (Ed. R. D. Freeman), NATO ASIS, Series A: Life Sciences, Vol. 27, Plenum Press, pp. 31–41.

Buisseret, P., and Gary-Bobo, E. (1979). Development of visual cortical orientation specificity after dark-rearing: role of extraocular proprioception. *Neurosci. Lett.*, **13**, 259–263.

Buisseret, P., Gary-Bobo, E., and Imbert, M. (1978). Ocular motility may be involved in recovery of orientational properties of visual cortical neurons in dark-reared kittens. *Nature*, **272**, 816–817.

Buisseret, P., Gary-Bobo, E., and Imbert, M. (1982). Plasticity in the kitten's visual

cortex: effects of the suppression of visual experience upon the orientational properties of visual cortical cells. *Dev. Brain Res.*, **4**, 417–426.
Buisseret, P., and Imbert, M. (1976). Visual cortical cells: their developmental properties in normal and dark-reared kittens. *J. Physiol. (Lond.)*, **255**, 511–525.
Buisseret, P., and Singer, W. 1983. Proprioceptive signals from extraocular muscles gate experience-dependent modifications of receptive fields in the kitten visual cortex. *Exp. Brain Res.*, **51**, 453–460.
Cynader, M., and Mitchell, D. E. (1980). Prolonged sensitivity to binocular deprivation in dark-reared cats. *J. Neurophysiol.*, **43**, 1026–1040.
Daw, N. W., Berman, N. E. J., and Ariel, M. (1978). Interaction of critical periods in the visual cortex of kittens. *Science*, **199**, 565–567.
Frégnac, Y. (1979a). Development of orientation selectivity in the primary visual cortex of normally and dark-reared kittens. I. Kinetics. *Biolog. Cybernetics*, **34**, 187–193.
Frégnac, Y. (1979b). Development of orientation selectivity in the primary visual cortex of normally and dark-reared kittens. II. Models. *Biolog. Cybernetics*, **34**, 195–203.
Frégnac, Y., and Imbert, M. (1978). Early development of visual cortical cells in normal and dark-reared kittens: relationship between orientation selectivity and ocular dominance. *J. Physiol. (Lond.)*, **278**, 27–44.
Frégnac, Y., and Imbert, M. 1984. Development of neuronal selectivity in the primary visual cortex of the cat. *Physiol. Rev.*, **64**, 325–434.
Gottlieb, G. (1976). The roles of experience in the development of behavior and the nervous system. In *Neural and Behavioral Specificity, Studies on the Development of Behavior and the Nervous System* (Ed. G. Gottlieb), Vol. 3, Academic Press, pp. 25–54.
Grobstein, P., and Chow, K. L. (1975). Receptive field development and individual experience. *Science*, **190**, 352–358.
Hirsch, H. V. B., Leventhal, A. G., McCall, M. A., and Tieman, D. G. (1983). Effects of exposure to lines of one or two orientations on different cell types in striate cortex of cat. *J. Physiol. (Lond.)*, **337**, 241–256.
Hubel, D. H., and Wiesel, T. N. (1962). Receptive fields, binocular interaction and functional architecture in the cat's visual cortex. *J. Physiol. (Lond.)*, **160**, 106–154.
Hubel, D. H., and Wiesel, T. N. (1963). Receptive fields of cells in striate cortex of very young visually inexperienced kittens. *J. Neurophysiol.*, **26**, 994–1002.
Hubel, D. H., and Wiesel, T. N. (1970). The period of susceptibility to the physiological effects of unilateral eye closure in kittens. *J. Physiol. (Lond.)*, **206**, 419–436.
Imbert, M., and Buisseret, P. (1975). Receptive field characteristics and plastic properties of visual cortical cells in kittens reared with or without visual experience. *Exp. Brain Res.*, **22**, 25–36.
Imbert, M., and Frégnac, Y. (1983). Specification of cortical neurons by visuomotor experience. *Prog. Brain Res.*, **58**, 427–436.
Malsburg, C. von der, and Cowan, J. D. (1982). Outline of a theory for the ontogenesis of iso-orientation domains in visual cortex. *Biolog. Cybernetics*, **45**, 49–56.
Rauschecker, J. P. (1982). Instructive changes in the kitten's visual cortex and their limitation. *Exp. Brain Res.*, **48**, 301–305.
Singer, W., Freeman, B., and Rauschecker, J. (1981). Restriction of visual experience to a single orientation affects the organization of orientation columns in cat visual cortex. *Exp. Brain Res.*, **41**, 199–215.
Thompson, I. D., Kossut, M., and Blakemore, C. (1983). Development of orientation columns in cat striate cortex revealed by 2-deoxyglucose autoradiography. *Nature*, **301**, 712–715.

Trotter, Y., Frégnac, Y., and Buisseret, P. (1981). Période de sensibilité du cortex visuel primaire du chat à la suppression unilatérale des afférences proprioceptives extraoculaires. *CR Acad. Sci. (Paris)*, Série D, **293**, 245–248.
Trotter, Y., Frégnac, Y., and Buisseret, P. (1983). Synergie de la vision et de la proprioception extraoculaire dans les mécanismes de plasticité fonctionnelle du cortex visuel primaire du chaton. *CR Acad. Sci. (Paris)*, III, **296**, 665–668.
Trotter, Y., Gary-Bobo, E., and Buisseret, P. (1981). Recovery of orientation selectivity in kitten primary visual cortex is slowed down by bilateral section of ophthalmic trigeminal afferents. *Dev. Brain Res.*, **1**, 450–454.
Tsumoto, T., and Suda, K. (1982). Laminar differences in development of afferent innervation to striate neurons in kittens. *Exp. Brain Res.*, **45**, 433–446.

Models of the Visual Cortex
Edited by D. Rose and V. G. Dobson

CHAPTER 18

Decrementing associative net models of visual cortical development

VERNON G. DOBSON
Department of Psychology, Brunel University, Uxbridge, Middlesex, UK

Conventionally, the development of brain circuits and single-cell response properties has been conceptualized mainly in terms of positive processes. These are assumed to guide or join specific links, or to increment their gains selectively, on the basis of consistencies in the firing patterns or cytospecificities of their pre- and postsynaptic cells (Hebb, 1949; Sperry, 1951). An alternative conceptual strategy is to assume that links are initially formed through non-specific random growth processes followed by decrementing or selective cutting of those links where there are inconsistencies in firing patterns or cytospecificities (Dobson, 1975, 1981b, and Chapter 18 in this volume). This article will present arguments suggesting that decrementing nets may be simpler, more efficient and more consistent with the empirical evidence than the corresponding incrementing nets and that they may provide more comprehensive accounts of visual system development.

ASSOCIATIVE NETWORKS

Dobson (1975, 1976, 1985) compared the performance of the simplest incrementing and decrementing associative nets operating in 'idealized' circuits, in which all parameter values are standardized and can be characterized by binary digits or small integers. Nets are assumed to consist of two arrays of N cells each; each input array cell makes L links to each output array cell. Links are all either 'inhibitory' or 'excitatory' and can be either 'operative', with gains of G, or 'inoperative', with gains of zero. Cells can be either 'firing' at rate R or 'not-firing' or inhibited. All cells have a 'preferred' activity state, but switch to the 'non-preferred' state on receipt of a threshold level of input transmitter T.

These nets associate pairs of patterns of activity consisting of a 'figure' of P cells in one activity state and a 'ground' of $N - P$ cells in the other state. During pattern association an external teacher establishes the patterns in their arrays and gives a 'learn' signal for the modification of links between cells in the 'figures' of the two patterns.

In incrementing nets, all links are initially inoperative, but become permanently operative after experiencing 'synergistic' postsynaptic activity, i.e. postsynaptic excitation at a firing excitatory link or postsynaptic inhibition at an inhibitory link. Subsequently, reestablished input array patterns reliably retrieve their associated output patterns if output cell thresholds are set at $T = PLRG$ (Marr, 1969; Willshaw, 1981; Palm, 1982).

In decrementing nets, all links are initially operative but are made permanently inoperative by 'conflicting' postsynaptic activity, i.e. postsynaptic excitation at firing inhibitory links or postsynaptic inhibition at excitatory links. Reestablished input patterns retrieve their associated output patterns if output cells change state in response to any input, i.e. thresholds are set at $T=1$ for excitatory nets, or $T=0$ for inhibitory nets.

Despite the contrasting ways in which information is stored and retrieved in incrementing and decrementing nets, their performances in idealized circuits with the same sets of random pattern-pairs are identical. Performance is optimal when $P = \log_2 N$ (Willshaw, Buneman and Longuet-Higgins, 1969). The inhibitory and excitatory forms of each net are dual systems and give identical performance with the equivalent complementary, or inverse, output patterns. Thus, for excitatory incrementing and inhibitory decrementing nets, the number of firing cells in the output array pattern is P, but for inhibitory incrementing and excitatory decrementing nets P is the number of not-firing or inhibited output cells.

In an idealized net with $N=10^3$ cells per array, which has associated 10^3 pattern-pairs, each consisting of $P = 10$ randomly selected cells per array, the error rate in recall of output patterns, given the associated input patterns, is approximately 10^{-8} spuriously firing cells per pattern retrieved (Dobson, 1985). However, this standard of accuracy only applies to incrementing nets operating in idealized circuitries in which $T = PLRG$. If any of these precise arithmetical relationships are disturbed, the performance of the incrementing net deteriorates dramatically. For example, if the links between cells are distributed randomly, according to the Poisson distribution, so that $T = PRLG$ only on average, and the average number of links $L = 5$, then the error rate in the additive net increases by a factor of about 10^6 (Dobson, 1975, 1976, 1985); Palm, 1981, 1982). The accuracy of decrementing nets does not depend upon precise arithmetical relationships between network parameters, so that in the randomly connected circuitry described above, the performance of a decrementing net is 10^6 times more reliable than that of the corresponding incrementing net.

In terms suggested by Gregory (1968, 1974), decrementing nets operate 'synthetically' while incrementing nets operate 'analytically'. In decrementing nets information is stored in the missing links so that the numbers and gains of the remaining links can be varied, according to automatic gain control rules, to compensate for any irregularities in the circuitry left by the random growth and learning processes, without interfering with stored information. For example, if an inhibitory link is regularly required to prevent the firing of an output cell single-handedly, it could increase the gains and numbers of its synapses at strategic positions until it does prevent output cell firing. Similarly, the formation of 'cooperative' networks in the visual cortex can be explained if links between cells with similar stimulus preferences are cut and the remaining links between cells with different preferences are balanced by gain control rules so that the cells fire equal proportions of the time (Dobson, 1981a, 1981b). These nets are described in more detail below and in Dobson and Rose (Chapter 3 in this volume).

MAPPING NETWORKS

Dobson (1980, 1981a, 1981b) also compared the performance of incrementing and decrementing networks producing topographic mappings between two square mosaics of cells. In mapping nets it is assumed that, initially, each input mosaic cell distributes modifiable excitatory links randomly to the output mosaic. Cells at corresponding corners of the mosaics are connected by powerful unmodifiable links which determine the overall polarity of the map. 'Early visual experience' is simulated by 'trials' in which adjacent pairs of cells in the input mosaic are excited at random, and this excitation is projected to the output mosaic.

In incrementing nets it is assumed that cells in the output mosaic are reciprocally connected by unmodifiable short distance excitatory and longer distance inhibitory links. On each trial, excitatory inputs from the input mosaic reverberate within the local postsynaptic circuits until a stable pattern forms. The gains of the firing modifiable links to the few most excited output cells are then selectively incremented according to the products of the pre- and post-synaptic firing rates. The gains of the remaining firing links are then non-selectively decremented by similar amounts to keep total link gains constant.

In decrementing mapping nets, the only reciprocal circuits in the output mosaic are of the longer distance 'surround inhibition' type. On each trial the gains of the modifiable links firing into the few most inhibited output cells are selectively decremented, while the gains of the other firing links are non-selectively incremented by a corresponding amount. Computer simulations suggest that decrementing mapping nets are not only significantly simpler, in not requiring local excitatory links, but also generate plausible topographic mappings within a fraction of the number of trials required by the corresponding incrementing nets.

RETINOCORTICAL MAPPINGS

Dobson (1981a, 1981b) used a decrementing mapping net to construct a developmental model of the retinocortical projection-generating orientation, spatial frequency and ocular dominance columns in striate cortex. In this model it is assumed that each hemiretina consists of thousands of hemicircular regions called 'hyperfields', arranged in a sunflower seed growth pattern mosaic, which map topographically onto a corresponding square mosaic of regions in the striate cortex called hypercolumns. Each hyperfield, in turn, consists of a mosaic of X-type 'off-' and 'on-centre' retinal unit receptive field (r.f.) centres which map onto a square mosaic of 'minicolumns', or strings of cells (Mountcastle, 1978), within layer IV of the corresponding hypercolumn (Fig. 1).

It is assumed that the global retinocortical mapping, and the orientations of the local mappings within it, are guided by innate cytospecificities, while the local hyperfield–hypercolumn mappings are set up, during early experience, within decrementing mapping nets. Both local and global mappings are assumed to be characterized by the conformal logarithmic transformation shown in Fig. 1 (Schwartz, 1977, and Chapter 14 in this volume). To simplify exposition, it is assumed that adjacent hyperfields and r.f. centres do not overlap.

The hyperfield represented in Fig. 1 is assumed to consist of six concentric semicircles of X-unit r.f. centres, with diameters and spatial frequency preferences varying directly with distance from the hyperfield centre. These are assumed to map onto six parallel rows of minicolumns forming spatial frequency columns in the corresponding hypercolumn. X-unit r.f. centres are also aligned along the nine hyperfield radial bands, or wedges, at 20° intervals, which map onto nine parallel orientation columns aligned orthogonal to the spatial frequency columns. Inhibitory circuits are assumed to run along the spatial frequency columns and veto responses of cells to contour stimuli which are not aligned in orientation with the hyperfield radius projecting to them. Initially, mappings from corresponding hyperfields in the two eyes are assumed to fall 'in register' upon the same hypercolumn. These mappings then compress separately and slide over one another in opposite directions along the orientation column axes to form ocular dominance columns in layer IV. The rectilinear arrangement of the global mapping mosaic is assumed to cause columns in adjacent hypercolumns to line up over cortical distances of several hypercolumn diameters, to form striped patterns (Humphrey and Hendrickson, 1983). If hyperfields overlapped, each point in the visual field would be represented in several different orientation columns within adjacent hypercolumns.

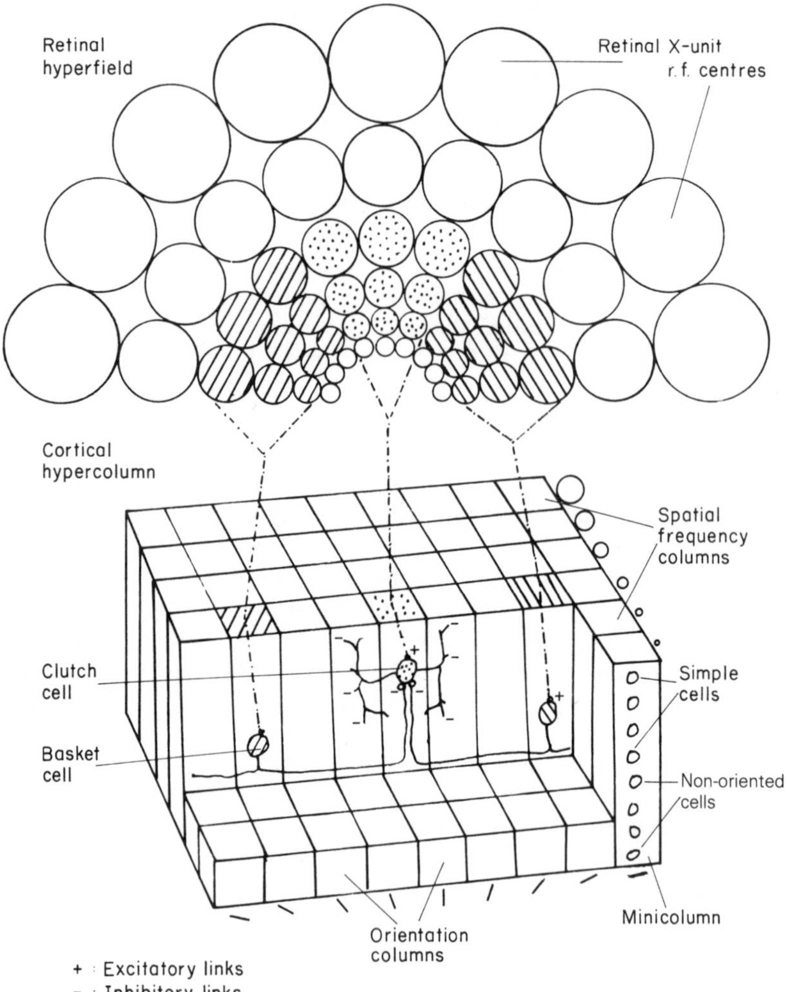

FIGURE 1 Mapping from one retinal hyperfield to the corresponding hypercolumn in the striate cortex layer IVc. For each minicolumn in the cortical hypercolumn there is an X- unit r.f. centre in the corresponding hyperfield. However, there is a local 'scatter' in the mapping of one r.f. centre diameter. Thus the 'simple cells' in the stippled minicolumn are each driven from one of the nine stippled r.f. centres and the basket cells in the striped minicolumns are driven by the clusters of striped r.f. centres. Basket-cell axons run parallel to the spatial frequency columns to inhibit simple cells three orientation columns away, so that simple cells in the stippled minicolumn are excited by a stippled r.f. centre and inhibited by the striped r.f. centres. This produces broad preferences for stimulus orientations and symmetries. However, some simple cells are also 'clutch' cells, with modifiable inhibitory links projecting to cells with different stimulus preferences in adjacent minicolumns. This produces fine-tuned stimulus selectivities and cooperativity effects.

SIMPLE-CELL RESPONSE PROPERTIES

The subtractive mapping nets described above can generate models of circuits mediating simple-cell responses which make precise quantitative predictions about the properties of the cells to be found in each minicolumn. It is assumed that competition between afferent fibres for postsynaptic sites continues until each cell in layer IVc is driven by a single lateral geniculate nucleus (LGN) cell, which, in turn, is driven by one retinal X-unit r.f. Layer IVc is conceptually subdivided into an upper sublayer of orientation-selective simple cells and a lower sublayer of non-orientation-selective cells. There are two kinds of inhibitory non-oriented cells with unmodifiable links. One of these projects isotropically to cells in both sublayers and is responsible for the 'surround inhibition' required for the production of topographic order. The other is of the 'basket-cell' type and projects asymmetrically to simple cells in the same spatial frequency column but in different orientation columns.

Consequently, each simple cell has a compound r.f. structure, with one LGN cell driving a central excitatory subunit and basket cells driving the inhibitory subunit fields, or inhibitory sidebands, on each flank. The widths of the surrounds of the X-type subunit r.f.s are assumed to be equal to the widths of the centres. Each subunit r.f. gives a standardized optimal response to edge stimuli of appropriate contrast tangential to the r.f. centre-surround border. This response declines linearly with distance until there is no response to edges outside the r.f. surround or when bisecting the r.f. centre. Assuming that all link gains are standardized and that cortical cells have no spontaneous activity, the response of a simple cell to any edge stimulus varies with the sum of the responses of its subfields, and these responses can be derived from the distance of the edge from the subfield centre-surround border (Dobson, 1980). Subtractive mapping nets such as these generate precise quantitative predictions because they do not require local reciprocal excitatory circuits which would variously blur, amplify and unpredictably transform the spatio-temporal pattern of input signals.

Simple-cell orientation tuning curves can be constructed from their responses to edges tangential to their excitatory subunit r.f. centre-surround border. Edges aligned with hyperfield radii engage the inhibitory subunit fields least of all, producing the optimal response. Edges rotated round the excitatory centre perimeter progressively engage the inhibitory subunit fields until the excitatory response is nullified. The assumption that simple-cell orientation selectivity is conferred by inhibitory circuits is consistent with the empirical evidence (Bishop, Coombs and Henry, 1971; Benevento, Creutzfeld and Kuhnt, 1972; Sillito *et al*, 1980; Barlow, 1982). The detailed shapes of these tuning curves vary with the values of key network parameters. One of these is the minimum distance D^i which is allowed between cells connected by unmodifiable inhibitory links. D^i is the internal radius of the surround inhibition determining the scatter in the topography, and is also the number of orientation column diameters separating simple cells from the basket cells determining their orientation selectivity.

If D^i is set at 1, there is no 'scatter' in the topography and all simple cells in the same minicolumn receive the same preferences for position, spatial frequency and orientation. Orientation tuning curves are symmetrical and there are no r.f. asymmetries, directional preferences or binocular disparities (other than simple cells with 'off-' and 'on-centre' excitatory subunit r.f.s in different eyes).

If D^i is set at 3 minicolumn diameters, then there is a scatter in the monocular topography like that shown in Fig. 1. Each minicolumn receives excitatory input from a cluster of LGN r.f.s with a range of positional, spatial frequency and contrast polarity preferences. Each simple cell in a minicolumn also receives inhibition from corresponding clusters of basket-cell r.f.s, which form sidebands flanking the excitatory subunits. Thus, while all simple cells in the same column have the same inhibitory subunit fields, their excitatory subunits may be at any one of nine possible positions in the space between the inhibitory sidebands. Simple cells with excitatory subunits at central positions have symmetrical r.f. properties, responding optimally to bar stimuli aligned with the corresponding radial wedge. However, cells with excitatory subunits adjacent to one sideband develop asymmetrical r.f. properties, responding only to edges stimulating, or approaching from the side of the subunit facing the more distant sideband. These cells also have a range of orientation tuning curve widths and asymmetries, with optima aligned with hyperfield radial wedge borders. At the binocular level, this monocular scatter in preferences for stimulus position, orientation and direction is translated into a range of selectivities for crossed and uncrossed disparities, and for contour stimuli tilting and moving, towards and away from the observer.

Another key network parameter is the ratio between the diameters of the r.f. centres at successive eccentricities within the same hyperfield radial wedge. In Fig. 1 this ratio is 1 : 1.4 : 2 : 2.8, forming an angle of 20°. The scatter in the mapping in Fig. 1 means that each minicolumn receives from LGN cells with 3 r.f. sizes in the ratio of 1 : 1.4 : 2. If spatial frequency preference varies inversely with r.f. diameter, then the spatial frequency preferences of cells in the same minicolumn would follow the same ratio. This is precisely consistent with empirical evidence from cat visual cortex (Thompson and Tolhurst, 1980). Figure 1 is assumed to represent a cat hyperfield, with 9 radial wedges at 20° intervals. As different simple cells in the same minicolumn may respond to opposite sides of the corresponding radial wedge, they will cover a 20° range in preferred orientation, and there will be a matching range of binocular orientation disparities similar to that found in the cat (Bishop, 1979). The orientation tuning curve half-widths generated by the mapping in Fig. 1 vary from 6 to 41°, with a mean of about 25°. This is somewhat wider than those found in the cat (Rose and Blakemore, 1974), but it is assumed that orientation selectivity is sharpened by the fine-tuning process described below.

FINE TUNING AND PSYCHOPHYSICS

The scatter in these mappings is instrumental in generating the comprehensive range of binocular stimulus preferences. However, the scatter also makes it impossible to produce optimal fine tuning in the positional selectivities of cortical cells by means of conventional unmodifiable lateral inhibitory circuits. Such conventional circuits only operate optimally within unscattered topographic mappings by means of gains which vary strictly with lateral distance according to a Mexican-hat shaped function.

However, it is possible to account for the production of optimal positional selectivities by introducing the inhibitory decrementing cells described above in the section on associative nets. These cells will be identified with 'clutch' cells (Somogyi and Martin, Chapter 54 in this volume). Clutch cells are assumed to have experientially modifiable links which grow randomly and locally to all cells in the same and immediately adjacent minicolumns (Fig. 1). The links of clutch cells obey decrementing synaptic modification rules, i.e. links between cells regularly firing synchronously are cut. Cells with the same positional preferences are assumed to cut mutual inhibitory links because they tend to fire synchronously. The remaining links between cells with different stimulus preferences are assumed to be able to adjust their gains so as to equalize the average rates of pre- and postsynaptic firing, taken over periods of minutes or hours. Thus, with experience, the cells involved would 'share out' the representation of the topography evenly amongst themselves. Such an inhibitory subtractive net could produce optimal fine tuning of positional selectivities in mappings with any degree of scatter, provided that the spread of the scatter does not exceed that of the clutch-cell axons.

Another advantage in using inhibitory decrementing nets to fine-tune stimulus preferences is that the number of dimensions of stimulus preference that can be tuned is not restricted to the three Euclidean dimensions, as is the case for conventional lateral inhibition. It is assumed that some non-oriented and simple cells in each minicolumn are clutch cells (Fig. 1). These project randomly to all other cells in the same and adjacent minicolumns (the range of the scatter in the mapping). With early experience during critical periods, inhibitory links between clutch cells and other cells with the same or similar stimulus preferences are cut by temporally contiguous firing. During normal experience, it is assumed that link gains between cells with slightly different preferences are enhanced and balanced by automatic gain control, thus producing optimal selectivities for spatial frequency and direction, and reducing orientation tuning-curve widths. The reciprocally inhibiting circuits also function to amplify and exaggerate slight, but survival-significant, spatiotemporal differences in the sensory input, such as those involved in hyperacuity (Over, 1971).

The continuous, automatic gain control modulation of the balance or 'null point' between cells with slightly differing stimulus preferences is assumed to underly colour-contingent after-effects and contrast constancies. Short-term fatigue of any set of cells is also assumed to disinhibit cells with slightly different (or opponent) stimulus preferences, producing adaptation after-effects. The mutually disinhibitory relationships between cells with similar stimulus preferences also give rise to cooperativity, hysteresis, meta-contrast masking and sensory information storage.

To account for longer distance cooperativity phenomena, activity in these local inhibitory circuits is assumed to be integrated by longer distance excitatory links from pyramidal cells with similar stimulus preferences. The excitatory links here are also assumed to develop from an initial random growth process followed by selective pruning during early experience, according to the excitatory decrementing net modification rules described above. These pyramidal cells are also reciprocally linked to columns of cells in secondary cortices with similar stimulus preferences but widely scattered positional preferences. Such columns of cells, responding to specific stimulus 'features' at all positions in the visual field, could serve as presynaptic array elements in a pattern-learning associative net. However, they could also serve as postsynaptic array elements which are 'set' by higher attentional systems to selectively enhance localized responses in the primary visual cortex to specific learned patterns of features at any point in the visual field. The cortical location of these enhanced responses would guide tracking responses by subcortical systems. The reciprocal mappings required in this model could all be generated by cytospecificities and the operation of experience in decrementing mapping nets.

ACKNOWLEDGEMENTS

This work was supported by the MRC.

REFERENCES

Barlow, H. B. (1982). David Hubel and Torsten Wiesel: their contributions towards understanding the primary visual cortex. *Trends in Neurosci.*, **55**, 145–153.

Benevento, L. A., Creutzfeldt, O. D. and Kuhnt, U. (1972). Significance of intracortical inhibition in the visual cortex. *Nature New Biol. (Lond).*,, **283**, 124–126.

Bishop, P. O. (1979). Stereopsis and the random element in the organisation of the striate cortex. *Proc. Roy. Soc. Lond. B.* , **204**, 415–434.

Bishop, P. O., Coombs, J. S., and Henry, G. H. (1971). Responses to visual contours; spatio-temporal aspects of excitation in the receptive fields of simple striate neurons. *J. Physiol. (Lond.)*, **219**, 625–657.

Dobson, V. G. (1975). Pattern learning and control of behaviour by all-inhibitory neural network hierarchies. *Perception*, **4**, 35–50.

Dobson, V. G. (1976). Towards a model of the development of adaptive behaviour in all-inhibitory network hierarchies. M. Phil. thesis, Brunel University.
Dobson, V. G. (1980). Neuronal circuits capable of generating visual cortex simple-cell stimulus preferences. *Perception*, **9**, 411–434.
Dobson, V. G. (1981a). Inhibitory circuits accounting for the development of visual cortical mappings, stimulus preferences, and psychophysical performance. *Perception*, **10**, 483–510.
Dobson, V. G. (1981b). A model of the development of functional neural organisation underlying psychophysical performance in the visual system. Ph.D. thesis, Brunel University.
Dobson, V. G. (1985). Superior accuracy of pattern retrieval in randomly connected associative nets using decrementing learning rules. *Neurosci. Lett.* Supp. **21**, S17.
Hebb, D. O. (1949). *The Organisation of Behaviour*, Wiley, New York.
Humphrey, A. L., and Hendrickson, A. E. (1983). Background and stimulus-induced patterns of high metabolic activity in the visual cortex (area 17) of the squirrel and macaque monkey. *J. Neurosci.*, **3**, 345–358.
Gregory, R. L. (1968). Perceptual illusions and brain models. *Proc. Roy. Soc. Lond. B*, **171**, 278–296.
Gregory, R. L. (1974). *Concepts and Mechanism of Perception*, Duckworth, London.
Marr, D. (1969). A theory of cerebellar function. *J. Physiol. (Lond.)*, **202**, 437–470.
Mountcastle, V. B. (1978). An organising principle for cerebral function: the unit module and the distributed system. In *The Mindful Brain* (Ed. F. O. Schmitt) MIT Press, Cambridge, Mass., pp. 7–50.
Over, R. (1971). Comparison of normalization theory and neural enhancement explanations of negative aftereffects. *Psycholog. Bull.*, **75** (4), 225–243.
Palm, G. (1981). On the storage capacity of an associative memory with randomly distributed storage elements. *Biolog. Cybernetics*, **39**, 125–134.
Palm, G. (1982). *Neural Assemblies*, Springer-Verlag, Berlin.
Rose, D., and Blakemore, C. (1974). An analysis of orientation selectivity in the cat's visual cortex. *Exp. Brain Res.*, **20**, 1–17.
Schwartz, E. L. (1977). Afferent geometry in the primate visual cortex and the generation of neuronal trigger features. *Biolog. Cybernetics*, **28**, 1–14.
Sillito, A. M., Kemp, J. A., Milson, J. A., and Beradi, N. (1980). A re-evaluation of the mechanisms underlying simple-cell orientation selectivity. *Brain Res.*, **194**, 517–520.
Sperry, R. W. (1951). Mechanisms of neural maturation. In *Handbook of Experimental Psychology* (Ed. S. S. Stevens), Wiley, New York, pp. 236–280.
Thompson, I. D., and Tolhurst, D. J. (1980). Optimal spatial frequencies of neighbouring neurons in the cat's visual cortex. *J. Physiol. (Lond.)*, **300**, 57P–58P.
Willshaw, D. J. (1981). Holography, associative memory and inductive generalisation. In *Parallel Models of Associative Memory* (Eds. G. E. Hinton and J. A. Anderson), Laurence Earlbaum Associates, Hillside, New Jersey, pp. 83–102.
Willshaw, D. J., Buneman, O. P., and Longuet-Higgins, H. C. (1969). Nonholographic associative memory. *Nature (Lond.)*, **222**, 960–962.

Models of the Visual Cortex
Edited by D. Rose and V. G. Dobson

CHAPTER 19

Monocular and binocular processes in human vision

JEREMY WOLFE
Department of Psychology, Massachusetts Institute of Technology, Cambridge, MA 02139, USA
and
RANDOLPH BLAKE
Departments of Psychology and Neurobiology/Physiology, Northwestern University, Evanston, IL 60201, USA

For years one of the foremost issues in visual science has been the nature of the mechanisms responsible for integrating information from the two eyes. Evidently the brain accomplishes this task of binocular integration very efficiently, for rarely are we aware of seeing the world through two laterally separated eyes—instead, vision is single. At the same time, of course, two-eyed vision furnishes us with stereopsis, the exquisitely precise ability to judge depth relations amongst objects. Therefore, besides integrating information from the two eyes, the brain also manages to register slight differences in the two monocular images.

In recent years neurophysiological studies of the visual cortex have provided us with glimpses of the actual neural hardware involved in the early stages of binocular vision. The landmark experiments of Hubel and Wiesel (1962, 1968) convincingly demonstrated that information from the two eyes converges onto single neurons in the striate cortex. Not all striate neurons, however, respond identically to stimulation of either eye. Many of these neurons, although receiving binocular input, are more strongly activated by stimulation of one of the two eyes than the other; other binocular neurons respond equally well to stimulation of either eye; and still others, called monocular neurons, can be activated only through one eye and not the other.

The functional significance of these variations in ocular dominance amongst striate neurons remains a debatable question (e.g. see Cynader, Gardner and Douglas, 1978; Ferster, 1981). Rather than tackle this thorny issue, we wish to describe some psychophysical evidence for the existence of neurons with varying degrees of ocular dominance in the human visual system. Besides establishing potential links between neurophysiology and perception, this evidence also reveals important information about the nature of the interaction amongst these neurons, information not available from the study of individual cortical neurons. Such information, we believe, must be considered in the formulation of models of visual cortical function.

The following section describes an inferential strategy for assessing binocular interactions psychophysically and summarizes major findings based on application of that strategy. After this, we shall consider evidence that pinpoints the site within the visual pathways where these binocular interactions occur.

MONOCULAR AND BINOCULAR ADAPTATION

One way to learn something about ocular dominance mechanisms in human vision is to examine the strength of a visual after-effect following various conditions of monocular and binocular adaptation. It is well established that prolonged exposure, or adaptation, to a visual pattern can temporarily alter the visibility or appearance of subsequently viewed test patterns. Depending on the conditions of adaptation, these so-called visual after-effects may involve an elevation in contrast threshold (Blakemore and Campbell, 1969), distortions in perceived spatial frequency (Blakemore and Sutton, 1969) or perceived orientation (Gibson and Radner, 1937) or an illusion of motion (Wohlgemuth, 1911). It is generally assumed that these after-effects result from a temporary modification of the response properties of neurons activated during the adaptation period. Following this line of reasoning, the strength of an after-effect provides an index of the extent to which the neurons activated during testing were also activated during adaptation. With this reasoning in mind, let us begin by considering the following situation.

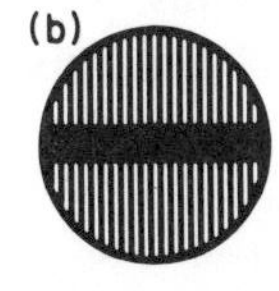

Figure 1 The tilt after-effect: the gratings in the top and bottom halves of (b) are colinear. Fixation of the center of (a) for about 1 minute will cause (b) to look like (c). If the observer adjusts (b) so that it again appears colinear, the deviation from true coliniarity will be a measure of the after-effect.

Looking at the test figure in Fig. 1b, you will probably agree that the two sets of parallel lines are aligned. Now suppose you look for a minute or so at the center of the pattern shown in Fig. 1a, where the upper set of lines is tilted anticlockwise and the lower set is tilted clockwise. Together, the two sets of lines form an arrow pointing to the right. Following this period of adaptation, you will now find that the lines in Fig. 1b, which are of course still aligned, now appear to point to the left. This is the well-known tilt after-effect. It, like several other after-effects, can be experienced even when one eye is adapted and the other eye is tested. (This, too, you can easily confirm by closing one eye during adaptation and then looking at the test figure using just your unadapted eye.) In other words, the after-effect transfers interocularly from the adapted eye to the non-adapted eye. Moreover, interocular transfer occurs even when the adapted eye is pressure-blinded immediately following adaptation, thereby eliminating any contribution from the retina of the adapted eye (see Blake and Fox, 1972, for a description of the logic of the pressure-blinding procedure). The occurrence of interocular transfer, even with the adapted eye pressure-blinded, indicates that at least some of the neurons activated during adaptation were also activated during inspection of the test figure. Because such neurons *must* receive input from both eyes, interocular transfer points to the existence of binocular neurons.

Most of the familiar after-effects, including those produced by adaptation to motion, spatial frequency and tilt, transfer interocularly. These after-effects also share something else in common: the transferred after-effect is typically weaker than the after-effect experienced when the adapted eye itself is tested. For instance, in the case of the tilt after-effect, the degree of apparent tilt is larger when the test figure is viewed using the adapted eye than it is when the figure is viewed using the unadapted eye. In other words, interocular transfer is incomplete. Based on the line of reasoning developed above, the incompleteness of interocular transfer must mean that at least some of the neurons activated during post-adaptation inspection were *not* activated during adaptation or were not stimulated as vigorously during adaptation. Such neurons evidently receive input predominantly or entirely from one eye but not the other. Thus, the incompleteness of interocular transfer points to the existence of variations in ocular dominance amongst neurons involved in adaptation. In a recent study (Blake, Overton and Lema-Stern, 1981) various adaptation procedures were used to confirm that some of the neurons involved in adaptation are genuinely monocular in nature, receiving input from one eye exclusively.

The general wiring scheme implied by these findings is summarized in Fig. 2a. To reiterate, interocular transfer points to the existence of binocular neurons and the incompleteness of interocular transfer points to the existence of monocular neurons. This conclusion, of course, dovetails nicely with the

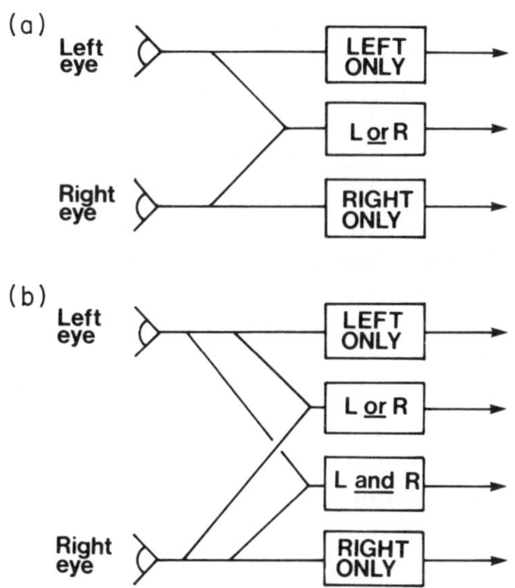

FIGURE 2 (a) Studies of interocular transfer reveal two monocular processes and one binocular process. Perception must be based in pooled responses from all active processes. (b) Further experiments reveal a second binocular process, a purely binocular process that responds only to stimulation of both the left *and* right eyes by similar input. This process is probably important for stereopsis.

variations in ocular dominance characteristic of cortical neurons. In addition, however, these psychophysical findings imply something very important about the way that activity within these monocular and binocular neurons is combined. The simple model in Fig. 2a attributes the incompleteness of interocular transfer to unadapted monocular neurons—their involvement during postadaptation testing serves to weaken the transferred after-effect. Yet at the same time, the activity of unadapted monocular neurons does not preclude the occurrence of a transferred after-effect. Therefore, the strength of a visual after-effect must be based on some form of combined activity amongst both adapted binocular neurons and unadapted monocular neurons, rather than on the most sensitive or active channel. In this respect, monocular and binocular neurons behave as if they constitute a single visual channel within which neural activity is pooled. Thinking in terms of cortical circuitry, the outputs from neurons of varying degrees of ocular dominance must be combined in a manner that reflects their total activity. To our knowledge, physiologists have yet to describe the mechanism of this pooling of monocular and binocular activity, but the results from adaptation studies would seem to require such a mechanism.

The model depicted in Fig. 2a defines binocular neurons as those that can be activated by stimulation of either eye—the conventional physiological definition of binocularity (Hubel and Wiesel, 1962). These neurons could be characterized as inclusive OR gates. Physiologists have also described another class of binocular neurons which behave like AND gates, responding only when both the left eye and the right eye receive stimulation (Bishop, 1973; Poggio and Fischer, 1977). Of course the interocular transfer paradigm just described would overlook binocular neurons of this type, since adaptation and testing always involve monocular stimulation. Consequently, we must modify the conditions of adaptation and testing in order to uncover this second possible category of binocular neurons. Towards this end, consider the following experiment.

Suppose that, following monocular adaptation, the visual after-effect is measured binocularly. According to the model in Fig. 2a, the resulting after-effect would be the product of the activity of the adapted binocular neurons, the adapted monocular neurons and the unadapted monocular neurons. Therefore, the magnitude of this after-effect should fall between that measured using just the adapted eye (direct condition) and that measured using just the unadapted eye (interocular transfer condition). This, however, is not the outcome (Wolfe and Held, 1981). Instead, the after-effect measured with binocular viewing is actually *smaller* than that measured interocularly. Apparently, then, testing both eyes together brings into play yet another set of neurons that was not active during monocular adaptation, a set besides the unadapted monocular neurons implicated by the incompleteness of interocular transfer. The process in question, it can be argued, must be a purely binocular process that is activated only if the left *and* right eyes are stimulated simultaneously. There is another piece of evidence that points to the existence of such neurons (Wolfe and Held, 1981). Imagine alternately adapting each eye to a pattern, so that both eyes receive stimulation during adaptation but never simultaneously. When testing is monocular, full-strength after-effects are found but when testing is binocular the after-effect is lessened. Again, the binocular after-effect must be reduced by a set of neurons that was not activated during adaptation of the two eyes separately but was activated during testing of both eyes together. The revised model in Fig. 2b adds this purely binocular process. Again note that the resulting after-effect must reflect the pooled activity of these various classes of neurons. Incidentally, there is psychophysical evidence (Wolfe and Held, 1982, 1983) and physiological evidence (Poggio and Fischer, 1977) which suggests that the purely binocular process may be involved in stereoscopic depth perception.

So far we have focused on evidence establishing the existence of neurons of varying degrees of ocular dominance. Several findings point to visual cortex as the site of these neurons; we shall conclude by summarizing that evidence.

THE SITE OF MONOCULAR AND BINOCULAR ADAPTATION

Because the visual after-effects described here transfer interocularly, even when the adapted eye is pressure-blinded (e.g. Blake and Fox, 1972), the neurons involved in adaptation must be postretinal. In addition, because these after-effects are orientation selective, we can safely rule out neurons in the lateral geniculate nucleus, since the activity of geniculate cells is not highly dependent on contour orientation. Moreover, tilt after-effects can be induced using an adaptation pattern created by random-dot stereoscopy (Tyler, 1975; Wolfe and Held, 1982), a procedure that effectively by-passes stimulation of precortical stages (Julesz, 1971).

Having concluded that these after-effects do not occur prior to primary visual cortex, we can further delimit the site of adaptation by capitalizing on the phenomenon of binocular rivalry (Breese, 1899). When the two eyes view dissimilar patterns, those patterns compete, or rival, for dominance, with the temporary loser suppressed from vision for several seconds at a time. The after-effects described here can be induced to full strength even when the adapting pattern is phenomenally suppressed for a substantial portion of the adaptation period (Blake and Fox, 1974; Wade and Wenderoth, 1978). In other words, suppression does not interrupt the flow of information to the site of these visual after-effects. However, pattern adaptation does alter the duration of time that an eye remains suppressed during binocular rivalry (Blake, Westendorf and Overton, 1980). This pattern of results indicates that the neural events underlying these after-effects precede and provide input to the stage of visual processing where rivalry suppression occurs. Of course, in the absence of information concerning the physiological locus of binocular rivalry (see Sloane, Chapter 21 in this volume), we cannot say with certainty that adaptation occurs in primary visual cortex. However, there is other evidence suggesting that rivalry itself occurs rather early within the visual pathways (e.g. see Blake and Overton, 1979); consequently, visual pattern adaptation must transpire at an even earlier stage of visual processing.

Thus, the monocular and binocular processes revealed in adaptation studies seem to lie early in the cortex, beyond the monocular input layers and prior to the locus of binocular rivalry. It is noteworthy, incidentally, that neurons in the visual cortex do indeed become less responsive (i.e. become adapted) following prolonged stimulation by visual patterns similar to those employed in psychophysical studies (e.g. Movshon and Lennie, 1979). It is also worth noting that the processes described here are not the only monocular and binocular processes that can be psychophysically discerned. For example, a binocular process has been implicated in the production of optokinetic nystagmus (Wolfe, Held and Bauer, 1981) and eye torsion (Wolfe and Held, 1979); this binocular process is thought to be subcortical.

SUMMARY

The findings described in this chapter illustrate how psychophysical techniques can be used to dissect the human visual system. In the examples discussed, pattern adaptation has revealed a set of monocular and binocular processes that may well correspond with classes of monocular and binocular neurons found in primary visual cortex. Furthermore, adaptation has revealed interactions between these processes that cannot yet be demonstrated by physiological methods. In particular, adaptation studies indicate that visual perception is based on the pooled responses of a number of active processes and not, for example, on the output of only the most active process.

ACKNOWLEDGMENTS

We are grateful to Julie Sandell for drawing the figures. The work described in this chapter was supported by grants from the National Institutes of Health (EY01596, EY02621 and EY04297) and the National Science Foundation (BNS 8200850). The order of authorship was determined by coin flip.

REFERENCES

Bishop, P. O. (1973). Neurophysiology of binocular single vision and stereopsis. In *Handbook of Sensory Physiology* (Ed. R. Jung), Vol. VII/3A, Springer, Berlin, pp. 255–304.

Blake, R., and Fox, R. (1972). Interocular transfer of adaptation to spatial frequency during retinal ischaemia. *Nature*, **240**, 76–77.

Blake, R., and Fox, R. (1974). Adaptation to invisible gratings and the site of binocular rivalry suppression. *Nature*, **249**, 488–490.

Blake, R., and Overton, R. (1979). The site of binocular rivalry suppression. *Perception*, **8**, 143–152.

Blake, R., Overton, R., and Lema-Stern, S. (1981). Interocular transfer of visual aftereffects. *J. Exp. Psychol; Human Perception and Performance*, **7**, 367–381.

Blake, R., Westendorf, D., and Overton, R. (1980). What is suppressed during binocular rivalry? *Perception*, **9**, 223–231.

Blakemore, C., and Campbell, F. W. (1969). On the existence of neurons in the human visual system selectively sensitive to the orientation and size of retinal images. *J. Physiol.*, **203**, 237–260.

Blakemore, C., and Sutton, P. (1969). Size adaptation: a new aftereffect. *Science*, **166**, 245–247.

Breese, B. B. (1899). On inhibition. *Psycholog. Monograph*, **3**, 1–65.

Cynader, M., Gardner, J., and Douglas, R. (1978). Neural mechanisms underlying stereoscopic depth perception in cat visual cortex. In *Frontiers in Visual Science* (Eds. J. Cool and E. L. Smith), Springer, New York, pp. 373–386.

Ferster, D. (1981). A comparison of binocular depth mechanisms in areas 17 and 18 of the cat visual cortex. *J. Physiol.*, **311**, 623–655.

Gibson, J. J., and Radner, M. (1937). Adaptation, aftereffect, and contrast in the perception of tilted lines—I. Quantitative studies. *J. Exp. Psychol.*, **20**, 453–467.

Hubel, D. H., and Wiesel, T. N. (1962). Receptive fields, binocular interaction and functional architecture in the cat's visual cortex. *J. Physiol.*, **160**, 106–154.

Hubel, D. H., and Wiesel, T. N. (1968). Receptive fields and functional architecture of monkey striate cortex. *J. Physiol.*, **195**, 215–243.

Julesz, B. (1971). *Foundations of Cyclopean Perception*, University of Chicago Press, Chicago, Ill.

Movshon, J. A., and Lennie, P. (1979). Pattern-selective adaptation in visual cortical neurons. *Nature*, **278**, 850–852.

Poggio, G., and Fischer, B. (1977). Binocular interaction and depth sensitivity in the striate and prestriate cortex of the behaving monkey. *J. Neurophysiol.*, **40**, 1392–1405.

Tyler, C. W. (1975). Stereoscopic tilt and size aftereffects. *Perception*, **4**, 187–192.

Wade, N. J., and Wenderoth, P. (1978). The influence of colour and contour rivalry on the magnitude of the tilt aftereffect. *Vision Res.*, **18**, 827–835.

Wohlgemuth, A. (1911). On the after-effect of seen movement. *Brit. J. Psychol., Monograph 1.*

Wolfe, J., and Held, R. (1979). Eye torsion and visual tilt are mediated by different binocular processes. *Vision Res.*, **19**, 917–920.

Wolfe, J., and Held, R. (1981). A purely binocular mechanism in human vision. *Vision Res.*, **21**, 1755–1759.

Wolfe, J., and Held, R. (1982). Binocular adaptation that cannot be measured monocularly. *Perception*, **11**, 287–295.

Wolfe, J., and Held, R. (1983). Shared characteristics of stereopsis and the purely binocular process. *Vision Res.*, **23**, 217–227.

Wolfe, J., Held, R., and Bauer, J. A. (1981). A binocular contribution to the production of optokinetic nystagmus in normal and stereoblind subjects. *Vision Res.*, **21**, 587–590.

Models of the Visual Cortex
Edited by D. Rose and V. G. Dobson

CHAPTER 20

Binocular interactions and their alterations resulting from abnormal visual experience

DENNIS M. LEVI
College of Optometry, University of Houston, University Park, 4901 Calhoun Blvd., Houston, TX 77004, USA

Most neurons in the visual cortex of cats and monkeys can be excited by visual stimulation of either eye. These binocular connections are functionally present at birth; however, concordant binocular visual experience is necessary for their maintenance and refinement during the months following birth (Hubel, Wiesel and LeVay, 1977; Wiesel, 1982). Thus, in animals deprived of normal binocular vision through experimentally induced strabismus or monocular form deprivation early in life, the population of neurons which can be excited by each eye is rapidly depleted and most of the remaining neurons are monocularly driven (Crawford and von Noorden, 1980; Hubel and Wiesel, 1965; Wiesel, 1982). Moreover, animals with reduced populations of binocular neurons show behavioral deficits in stereopsis and binocular summation (Blake and Hirsch, 1975; Packwood and Gordon, 1975; Crawford *et al.*, 1983). The link between the behavior and cortical physiology of cats and monkeys reared with abnormal binocular visual experience has led a number of investigators to relate these findings to clinical conditions such as strabismus and amblyopia which are quite common in humans and which are associated with a loss of stereopsis. This chapter reviews a number of recent efforts to infer binocular interaction in humans psychophysically, and attempts to relate the findings to animal models for binocular interactions in the visual cortex and their alterations resulting from abnormal visual experience.

INTEROCULAR TRANSFER OF VISUAL AFTER-EFFECTS

One of the most widely used psychophysical indices of binocularity is the ability to transfer after-effects from one eye to the other. Many simple after-effects normally exhibit interocular transfer, so that, after adapting one eye, an after-effect of somewhat reduced magnitude can be seen by the other. It is generally argued that significant interocular transfer may be interpreted as evidence for binocular neural interactions in the cortex (Day 1958; Long, 1979; Blake, Overton and Lema-Stern, 1981) since after-effects may transfer interocularly when the adapted eye is pressure blinded (Blake and Fox, 1972). Moreover, certain after-effects such as the elevation of contrast thresholds after grating adaptation are spatial frequency and orientation specific (Blakemore and Campbell, 1969). Recently Movshon and Lennie (1979) have shown in cats that the contrast thresholds of neurons in the visual cortex but not in the lateral geniculate nucleus (LGN) are elevated after adaptation to high contrast gratings.

In contrast to the robust interocular transfer shown by normal observers, stereoblind individuals with a history of abnormal binocular visual experience due to a naturally occurring strabismus or anisometropia fail to exhibit interocular transfer of the tilt after-effect (Movshon, Chambers and Blakemore, 1972; Mitchell and Ware, 1974; Mitchell, Reardon and Muir, 1975; Mann, 1978). Such results have been interpreted as evidence that stereoblind individuals lack a normal complement of binocular cortical neurons. Moreover, there is a high correlation between the stereoacuity of subjects with stereopsis and the extent of interocular transfer of the tilt after-effect (Mitchell and Ware, 1974; Mann, 1978), suggesting that the neurons involved in interocular transfer of the tilt after-effect may be linked in some way to those involved in stereopsis. While the interocular transfer of a number of suprathreshold as well as threshold after-effects is usually reduced or absent in stereoblind observers (Ware and Mitchell, 1974; Banks, Aslin and Letson, 1975; Hohman and Creutzfeld, 1975; Lema and Blake, 1977), several investigators have reported significant transfer of the threshold elevation after-effect amongst stereoblind observers (Hess, 1978; Anderson, Mitchell and Timney, 1980; Selby and Woodhouse, 1981). Most interestingly, Anderson, Mitchell and Timney (1980) reported data of several stereoblind observers who showed significant transfer of the threshold elevation after-effect, while failing to show transfer of suprathreshold after-effects. These results strongly suggest that, in such observers, some binocular connections which are incapable of mediating stereopsis (or interocular transfer of certain types of after-effect) must survive.

BINOCULAR SUMMATION

Normal observers perform better with two eyes than with one on a host of visual tasks. This superiority of binocular over monocular performance is too large to attribute to simply the statistical benefits of two-eyed viewing (i.e. probability summation), and therefore must reflect facilitative neural interactions between the two eyes (for discussions of models of binocular summation see Blake and Fox, 1973; Blake, Sloane and Fox, 1981). Binocular neural summation occurs only when the stimuli are placed on corresponding areas in the two eyes and are not separated in time by more than a hundred milliseconds or so. When these conditions are not met, simple probabilistic considerations predict the difference between monocular and binocular performance. In normal binocular vision, binocular summation is greatest when the spatial frequency and orientation of the stimuli to the two eyes are similar (Blake and Levinson, 1977; see also Fig. 1), suggesting that the neural mechanisms which mediate binocular summation are tuned to spatial frequency and orientation. In contrast, stereoblind observers (humans and monkeys) show little or no advantage of binocular over monocular viewing (Crawford *et al.*, 1983; Lema and Blake, 1977; Levi, Harwerth and Smith, 1979b; Westendorf *et al.*, 1978; Williams, 1974). The absence of binocular summation in stereoblind human observers seems to hold for a wide variety of stimulus conditions including threshold measurements using grating patterns (Lema and Blake, 1977; Levi, Harwerth and Smith, 1979; see also Fig. 1), suprathreshold measures of simple reaction time (Blake, Martens and DiGanfilippo, 1980; Harwerth, Smith and Levi, 1980; Levi, Harwerth and Manny, 1979), as well as temporal contrast sensitivity (Levi, Pass and Manny, 1982).

In summary, the facilitative binocular interactions which result in binocular summation in normal observers seem to be essentially absent in observers lacking stereopsis.

UTROCULAR DISCRIMINATION

In 1866 Helmholtz (1962) wrote 'ordinarily one is not clearly aware which eye it is with which he sees this image or the other'. The capacity to discern which eye has been stimulated, i.e. utrocular discrimination, was studied extensively in normal observers in the early 1900s (see Blake and Cormack, 1979, and Enoch, Goldmann and Sunga, 1969, for reviews). Recently, utrocular discrimination has been investigated in stereoblind and/or strabismic observers (Enoch, Goldmann and Sunga, 1969; Blake and Cormack, 1979). Blake and Cormack (1979) reasoned that eye-of-origin information would not be available from cells which could be driven by either eye, whereas monocular neurons or neurons strongly dominated by one eye would be

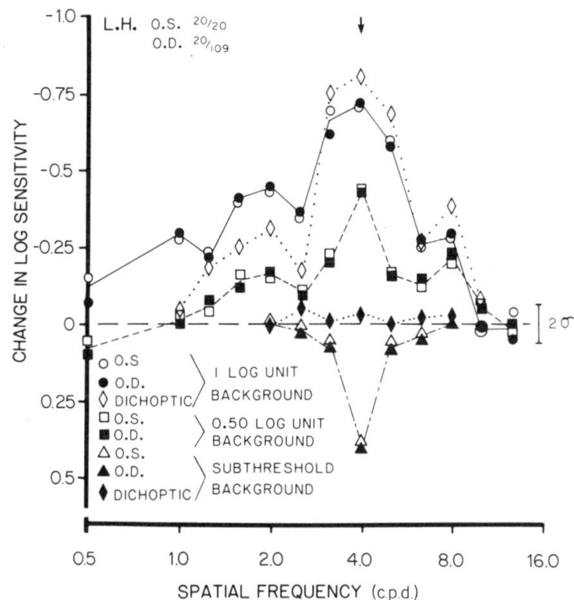

FIGURE 1 The change in log contrast sensitivity for the nonamblyopic (open symbols) and amblyopic (filled symbols) eyes of L.H. due to a 4.0 c.p.d. background 1 log unit above threshold (circles), 0.50 log unit above threshold (squares) and below threshold (triangles). The diamonds were obtained under dichoptic conditions. The open diamonds were with a 4.0 c.p.d. background about 1 log unit above threshold presented to the amblyopic eye, while the test gratings were presented to the nonamblyopic eye. The solid diamonds were also under dichoptic conditions, but the background was subthreshold. The error bar shows 2 standard deviations.

more likely to provide such information. They reported that their three stereoblind observers were able to retain eye-of-origin information under conditions in which normal observers were unable to make reliable distinctions. They concluded that the ability of stereoblind observers to discriminate utrocularly constituted evidence that the cells of the visual cortex of stereoblind observers were predominantly monocular.

There is, however, another reasonable interpretation. If the views of the two eyes are not perfectly balanced (e.g. in contrast, position, phase, duration, orientation, etc.), utrocular discrimination will be possible. Moreover, strabismics frequently show subtle suppression scotomas (Jampolsky, 1955) which can provide eye-of-origin cues. Thus utrocular discrimination may provide a strong test for the balance between the two monocular views. Loshin, Klein and Levi (1983) conducted utrocular discrimination experiments in stereoblind observers with amblyopia due to strabismus and/or anisometropia. In these experiments a number of precautions were taken to minimize potential cues. For example, the spatial frequency was randomly varied from trial to trial by 5 per cent. Phase was randomly varied. Eye-movement cues were eliminated by brief (100 ms) presentations and the use of horizontal gratings. Fusion was assured by having the observer carefully maintain alignment of horizontal and vertical fixation lines. The observers were given extensive practise with feedback. By careful attention to eliminating monocular cues, it was possible to eliminate significant utrocular discrimination near threshold in each of the observers and at high contrast levels in a number of stereoblind observers.

These results, in agreement with the data of Enoch, Goldmann and Sunga (1969), suggest that it is possible to reduce at least some of the cues for utrocular discrimination in some stereoblind observers. Our interpretation is that this finding has more bearing upon the balance between the two eye's views than it has on binocularity or its absence. Thus utrocular discrimination may provide a useful tool for studying the perceptual distortions of amblyopic eyes. If an observer could sample the input to the two eyes independently then utrocular discrimination would be easily accomplished and 100 per cent discriminability should result. Our results show that, at contrast levels near threshold, eye-of-origin information is not available to the stereoblind observer.

SPATIAL FREQUENCY SELECTIVITY

A large body of psychophysical evidence suggests that the sensitivity function of the human visual system to sinusoidal grating patterns is not that of a single detecting mechanism, but rather the envelope of the sensitivity functions of many channels, each responsive to a specific range of spatial frequencies (for reviews, see Braddick, Campbell and Atkinson, 1978; Graham, 1980). Thus, it is reasonable to ask whether the failure of binocular integration in observers deprived of normal binocular visual experience is the result of a disparity in spatial frequency selectivity between the two eyes. One method for investigating spatial frequency selectivity is simultaneous masking (Stromeyer and Julesz, 1972; Legge, 1979). In this paradigm, thresholds for detecting sine-wave gratings of various spatial frequencies are measured in the presence of simultaneously presented background gratings of a fixed spatial frequency presented to the same eye (monoptic masking).

Data from our laboratory (Fig. 1) shows how a 4.0 c/degree background grating at 1.0 log units (circles) or 0.5 log units (squares) above threshold changes the contrast sensitivity of the preferred eye (open symbols) and amblyopic eye (filled symbols) of a strabismic and anisometropic amblyope (L.H.). At each contrast level, the functions for each eye are quite similar; however, the lower contrast background elevates thresholds (decreases contrast sensitivity) to a lesser degree, and over a narrower range. If the background contrast is below threshold, detection of the test grating is facilitated (Kulikowski and King-Smith, 1973). Spatial frequency tuning functions with a subthreshold background at 4.0 c/degree are shown for each eye of L.H. (triangles). The effect of this subthreshold background was to reduce thresholds (increase sensitivity) for a narrow range of spatial frequencies in a similar fashion for each eye. Thus, we may conclude that in spite of substantial differences in contrast sensitivity between the two eyes of amblyopes, the absence of binocular interaction in humans deprived of normal binocular vision due to strabismus and/or amblyopia does not result from disturbances in the organization or specificity of the monocular spatial channels (Levi and Harwerth, 1982; Blake, 1982).

INTEROCULAR MASKING

The studies thus far have suggested that binocular interactions are either absent or weak in stereoblind observers. On the other hand, it has been suggested from clinical observations that many strabismics in fact show inhibitory binocular interactions, i.e. they suppress the input from one eye and the suppression is most marked for similar stimuli (Schor, 1977).

Levi, Harwerth and Smith (1979) investigated binocular interactions in humans deprived of normal visual experience by strabismus, amblyopia or both using an interocular masking paradigm. The results showed that a suprathreshold grating presented to one eye elevated the threshold for the discrimination of gratings similar in size and orientation presented to the corresponding retinal area in the fellow eye. The magnitude and stimulus specificity of these binocular interactions in observers with normal binocular vision were similar to those in observers deprived of normal binocular visual experience.

The open diamonds in Fig. 1 were obtained for observer L.H. under dichoptic conditions. In this experiment, the 4.0 c/degree background grating was presented to the amblyopic right eye about 1.0 log unit above threshold, while the test gratings were presented to the nonamblyopic left eye. It is of interest that the masking effects were actually slightly stronger under dichoptic conditions than under monoptic conditions, as has been reported in normal observers (Legge, 1979). In contradistinction, a subthreshold background presented to the amblyopic eye (filled diamonds) did *not* significantly alter the thresholds for discriminating the test gratings presented to the nonamblyopic eyes (or vice versa). Thus, strong suprathreshold binocular masking effects are evident in observers with abnormal binocular visual experience. These interactions result in elevated thresholds (reduced sensitivity) for stimuli similar in spatial frequency and orientation (Levi, Harwerth and Smith, 1979), suggesting that these interactions must be cortical. These interactions occur only when the stimuli are presented to corresponding areas in the two retinae, and were still evident when the visual display was limited to restricted retinal regions, including the suppression scotoma.

Since observers with abnormal binocular vision showed strong interocular masking effects, which resulted in elevated contrast thresholds but no interocular contrast summation, it was hypothesized that the effects of abnormal binocular visual experience may disrupt excitatory connections in the visual cortex while leaving the inhibitory connections more or less intact.

SINGLE CELLS AND INFERENTIAL PSYCHOPHYSICS

The pioneering studies of Hubel and Wiesel (see Wiesel, 1982, for a review) provided clear evidence that the excitatory connections between the two eyes which were present in the visual cortex of cats and monkeys at birth were rapidly disrupted when an eye misalignment or poor image quality in one eye interfered with normal binocular visual experience. It has recently been shown that monkeys reared with the visual axes of their two eyes optically dissociated early in life show a substantial loss in the proportion of neurons which can be excited through both eyes (Crawford and von Noorden, 1980) and that such animals show behaviorally both an absence of stereopsis and an absence of binocular contrast summation (Crawford *et al.*, 1983). On the other hand, several recent physiological studies suggest the presence of inhibitory binocular connections in animals reared with abnormal binocular visual experience. For example, Spear (1978) has shown an increase in the number of cortical neurons driven by the deprived eye of cats following enucleation of the experienced eye. More direct evidence for inhibitory interactions has been reported by Duffy, Burchfiel and Snodgrass (1978), who showed that intravenous administration of anti-inhibitory compounds to monocularly deprived cats resulted in substantial restoration of binocular input to the visual cortex. Furthermore, Singer (1977) has shown that electrical stimulation of the deprived eye of cats has an inhibitory influence on the responses of the experienced eye.

This chapter has reviewed a number of inferential psychophysical strategies for studying binocular interactions in humans. The main results are at least to a first approximation consistent with the animal models. Humans deprived of normal visual experience show reduced or no stereopsis and an absence of binocular summation, and generally show reduced or no interocular transfer of visual after-effects. The fact that many stereoblind observers do show some residual interocular transfer (Anderson, Mitchell and Timney, 1980) may suggest that some binocular neurons survive disruption. Whether these neurons, like the small proportion of binocular cells remaining in the visual cortex of monkeys following deprivation, are too few or are not functionally capable of supporting stereopsis and binocular summation is an open question. It is possible, too, that the residual binocular neurons evidenced in the transfer of the threshold elevation after-effect represent a different population than that which mediates stereopsis and binocular summation. In this regard, Wolfe and Held (1979) have shown that the binocular processes which mediate cyclotorsional eye movements in response to a random pattern rotating about the line of sight are apparently normal in stereodeficient subjects. Moreover, there is evoked potential evidence for binocular interactions in stereodeficient observers (Apkarian, Levi and Tyler, 1981).

The recent finding of potent, spatial frequency and orientation-specific masking in observers lacking stereopsis, binocular summation and interocular transfer of the threshold elevation after-effect (Levi, Harwerth and Smith, 1979, 1980) is of special interest, since it points out that different processes or strategies are used in masking experiments. The interpretation of these findings is not, however, straightforward. One interpretation is that interocular masking reveals inhibitory processes which survive the disruptive effects of abnormal visual experience suffered by the excitatory connections. This interpretation is consistent with the physiological studies cited. Moreover, a number of models of normal binocular vision include both excitatory and inhibitory interactions (e.g. Nelson, 1975; Marr and Poggio, 1976). On the other hand, there are alternative interpretations. For example, in masking experiments, the observers' task is to detect a low-contrast grating superimposed on a visible background (masking) grating, i.e. it is a problem of detecting a weak signal in noise. Thus, it is possible that a visual system consisting of independent (but spatially overlapping) monocular inputs would give the same masking results, so long as the two inputs were processed simultaneously and the observer could not just attend to one of the two inputs. Stereoblind observers who are unable to discriminate utrocularly stimuli near threshold are unlikely to be capable of selectively attending to the input from one eye. Thus, the definitive explanation may only be obtained by exploring the physiological responses of neurons of experimental animals under dichoptic conditions.

SUMMARY

This chapter has in a very limited way attempted to describe the psychophysicist's view of the altered binocular interactions resulting from abnormal visual experience and to relate this to animal models of the visual cortex. In the interests of brevity many important aspects of binocular interactions have not been treated, such as the question of whether strabismic suppression is simply an exaggeration of the normal binocular rivalry suppression (Smith *et al.*, 1985) or the very complex question of anomalous retinal correspondence (see Nelson, 1981, for a review). What does seem to emerge clearly is that a thorough understanding of the cortical processes involved in binocular vision and the effects of abnormal visual experience can most effectively be achieved by combining the approaches of physiology and psychophysics.

ACKNOWLEDGMENTS

I thank Ronald Harwerth, Earl Smith, Stanley Klein and Ruth Manny for their collaboration and for the exchange of ideas which led to this chapter. This work was supported by a grant from the National Eye Institute (R01 EY01728), Bethesda, Maryland.

REFERENCES

Anderson, P., Mitchell, D. E., and Timney, B. (1980). Residual binocular interaction in stereoblind humans. *Vision Res.*, **20**, 603–612.

Apkarian, P., Levi, D., and Tyler, C. (1981). Binocular facilitation in the visual-evoked potential of strabismic amblyopes. *Amer. J. Optom. Physiol. Optics.*, **58**, 820–830.

Banks, M. S., Aslin, R. N., and Letson, R. D. (1975). Sensitive period for the development of human binocular vision. *Science*, **1980**, 675–677.

Blake, R. (1982). Binocular vision in normal and stereoblind subjects. *Amer. J. Optom. Physiol. Optics*, **59**, 969–975.

Blake, R., and Cormack, R. H. (1979). Psychophysical evidence for a monocular visual cortex in stereoblind humans. *Science*, **203**, 274–275.

Blake, R., and Fox, R. (1972). Interocular transfer of adaptation to spatial frequency during retinal ischaemia. *Nature*, **240**, 76–77.

Blake, R., and Fox, R. (1973). The psychophysical inquiry into binocular summation. *Percept. Phychophys.*, **14**, 161–185.

Blake, R., and Hirsch, H. (1975). Deficits in binocular depth perception in cats after alternating monocular deprivation. *Science*, **190**, 114–116.

Blake, R., and Levinson, E. (1977). Spatial properties of binocular neurons in the human visual system. *Exp. Brain Res.*, **27**, 221–232.

Blake, R., Martens, W., and DiGanfilippo, A. (1980). Reaction-time as a measure of binocular interaction in human vision. *Invest. Ophthalmol. Vis. Sci.*, **19**, 930–941.

Blake, R., Overton, R., and Lema-Stern, S. (1981). Interocular transfer of visual aftereffects. *J. Exp. Psychol. (Hum. Percept.)*, **7**, 367–381.

Blake, R., Sloane, M., and Fox, R. (1981). Further developments in binocular summation. *Percept. Psychophys.*, **30**, 266–276.

Blakemore, C., and Campbell, F. W. (1969). On the existence of neurons in the human visual system selectively sensitive to the orientation and size of retinal images. *J. Physiol.*, **181**, 576–593.

Braddick, O., Campbell, F. W., and Atkinson, J. (1978). Channels in vision: basic aspects. In *Handbook of Sensory Physiology*, Vol. VII, *Perception* (Eds. R. Held, H. W. Leibowitz, and H. L. Teuber), Springer-Verlag, New York, pp. 3–38.

Crawford, M. L. J., and von Noorden, G. K. (1980). Optically induced concomitant strabismus in monkeys. *Invest. Ophthalmol. Vis. Sci.*, **19**, 1105–1111.

Crawford, M. L. J., von Noorden, G. K., Meharg, L. S., Rhodes, J. W., Harwerth, R. S., Smith III, E. L., and Miller, D. D. (1983). Binocular neurons and binocular function in monkeys and children. *Invest. Ophthalmol. Vis. Sci.*, **24**, 491–495.

Day, R. H. (1958). On interocular transfer and the central origin of visual aftereffects. *Amer. J. Psychol.*, **71**, 784–790.

Duffy, F. H., Burchfiel, J. L., and Snodgrass, S. R. (1978). The pharmacology of amblyopia. *Trans. Acad. Ophthalmol.*, **85**, 489–495.

Enoch, J., Goldmann, H., and Sunga, R. (1969). The ability to distinguish which eye was stimulated by light. *Invest. Ophthalmol.*, **81**, 317–331.

Graham, N. (1980). Spatial frequency channels in human vision: detecting edges without edge detectors. In *Visual Coding and Adaptability* (Ed. C. Harris), Lawrence Erlbaum Associates, Hillside, New Jersey, pp. 215–262.

Harwerth, R. S., Smith, E. L., and Levi, D. M. (1980). Suprathreshold binocular interactions for grating patterns. *Percept. Psychophys.*, **27**, 43–50.

Helmholtz, H. von (1962). In *Handbook of Physiological Optics* (Ed. J. P. C. Southall), Vol. III (translated from the third German edition, (1910), Dover Press, New York.

Hess, R. (1978). Interocular transfer in individuals with strabismic amblyopia: a cautionary note. *Perception*, **7**, 201–205.

Hohman, A., and Creutzfeldt, O. D. (1975). Squint and the development of binocularity. *Nature*, **254**, 613–614.

Hubel, D. H., and Wiesel, T. N. (1965). Binocular interaction in striate cortex of kittens reared with artificial squint. *J. Neurophysiol.*, **28**, 1041–1059.

Hubel, D. H., Wiesel, T. N., and LeVay, S. (1977). Plasticity of ocular dominance columns in monkey striate cortex. *Phil. Trans. Roy. Soc., B.*, **278**, 377–409.

Jampolsky, A. (1955). Characteristics of suppression in strabismus. *AMA Arch. Ophthal.*, **154**, 683–696.

Kulikowski, J. J., and King-Smith, P. E. (1973). Spatial arrangement of line, edge and grating detectors revealed by subthreshold summation. *Vision Res.*, **12**, 1355–1478.

Legge, G. E. (1979). Spatial frequency masking in human vision: binocular interactions. *J. Opt. Soc. Amer.*, **69**, 838–847.

Lema, S. A., and Blake, R. (1977). Binocular summation in normal and stereoblind humans. *Vision Res.*, **17**, 691–696.

Levi, D. M., and Harwerth, R. S. (1982). Psychophysical mechanisms in humans with amblyopia. *Amer. J. Optom. Physiol. Optics*, **59**, 936–951.

Levi, D. M., Harwerth, R. S., and Manny, R. E. (1979). Suprathreshold spatial frequency detection and binocular interaction in strabismic and anisometropic amblyopia. *Invest. Ophthalmol. Vos. Sci.*, **18**, 714–725.

Levi, D. M., Harwerth, R. S., and Smith, E. L. (1979). Humans deprived of normal binocular vision have binocular interactions tuned to size and orientation. *Science*, **206**, 852–854.

Levi, D. M., Harwerth, R. S., and Smith, E. L. (1980). Binocular interactions in normal and anomalous binocular vision. *Doc. Ophthal.*, **49**, 303–324.

Levi, D. M., Pass, A. F., and Manny, R. E. (1982). Binocular interactions in normal and anomalous binocular vision—effects of flicker. *Brit. J. Ophthalmol.*, **66**, 57–63.

Long, G. M. (1979). The dichoptic viewing paradigm. Do the eyes have it? *Psychol. Bull.*, **86**, 391–603.

Loshin, D. S., Klein, S. A., and Levi, D. M. (1983). Utrocular contrast discrimination—a signal detection approach. *Irvest. Ophthalmol. Vis. Sci.*, **24**, (Suppl.), 96.

Mann, V. A. (1978). Different loci suggested to mediate tilt and spiral motion aftereffects. *Invest. Ophthalmol. Vis. Sci.*, **17**, 903–909.

Marr, D., and Poggio, T. (1976). Cooperative computation of stereo disparity. *Science*, **194**, 283–287.

Mitchell, D. E., Reardon, J., and Muir, D. W. (1975). Interocular transfer of the motion aftereffect in normal and stereoblind observers. *Expl. Brain Res.*, **22**, 163–173.

Mitchell, D. E., and Ware, C. (1974). Interocular transfer of a visual aftereffect in normal and stereoblind humans. *J. Physiol. (Lond.)*, **236**, 707–721.

Movshon, J. A., Chambers, B. E. I., and Blakemore, C. (1972). Interocular transfer in normal humans and those who lack stereopsis. *Perception*, **1**, 483–490.

Movshon, J. A., and Lennie, P. (1979). Pattern selective adaptation in visual cortical neurons. *Nature*, **278**, 850–852.

Nelson, J. I. (1975). Globality and stereoscopic fusion in binocular vision. *J. Theor. Biol.*, **49**, 1–88.

Nelson, J. J. (1981). A neurophysiological model for anomalous correspondence based on mechanisms of sensory fusion. *Documenta Ophthalmol.*, **51**, 3–100.

Packwood, J., and Gordon, B. (1975). Stereopsis in normal domestic cat, siamese cat, and cat raised with alternating monocular occlusion. *J. Neurophysiol.*, **38**, 1485–1499.

Schor, C. M. (1977). Visual stimuli for strabismic suppression. *Perception*, **6**, 583–593.

Selby, S. A., and Woodhouse, J. M. (1981). The spatial frequency dependence of interocular transfer in amblyopes. *Vision Res.*, **21**, 1401–1408.

Singer, W. (1977). Effects of monocular deprivation on excitatory and inhibitory pathways in cat striate cortex. *Exp. Brain Res.*, **30**, 25–41.

Smith, E. L., Levi, D. M., Manny, R. E., Harwerth, R. S., and White, J. (1985). On the relationship between rivalry suppression and strabismic suppression. *Invest. Ophthalmol. Vis. Sci.*, **26**, 80–87.

Spear, P. D. (1978). Role of binocular interactions in visual systems development in the cat. In *Frontiers of Visual Science* (Eds. S. J. Cool and E. L. Smith), Springer-Verlag, New York.

Stromeyer III, C. F., and Julesz, B. (1972). Spatial-frequency masking in vision: critical bands and spread of masking. *J. Opt. Soc. Amer.*, **62**, 1221–1232.

Ware, C., and Mitchell, D. M. (1974). On interocular transfer of various visual aftereffects in normal and stereoblind observers. *Vision Res.*, **14**, 731–734.

Westendorf, D., Langston, A., Chambers, D., and Allegretti, C. (1978). Binocular detection by normal and stereoblind observers. *Percept. Psychophys.*, **24**, 209–214.

Wiesel, T. N. (1982). Postnatal development of the visual cortex and the influence of environment. *Nature*, **299**, 583–592.

Williams, R. (1974). The effect of strabismus on dichoptic summation of form information. *Vision Res.*, **14**, 307–310.

Wolfe, J. M., and Held, R. (1979). Eye torsion and visual tilt are mediated by different binocular processes. *Vision Res.*, **19**, 917–920.

Models of the Visual Cortex
Edited by D. Rose and V. G. Dobson

CHAPTER 21

Binocular rivalry: a psychophysics in search of a physiology

Michael E. Sloane
Department of Psychology, University of Alabama in Birmingham, Birmingham, AL 35294, USA

INTRODUCTION

Electrophysiologists have focused much attention on the neural substrate of stereoscopic depth perception. Similarly, the artificial intelligence community has devoted much effort to develop computational models of stereopsis. As a result, the neural mechanisms underlying this form of facilitative binocular interaction are well understood, and psychophysical performance of human observers on a stereoscopic task can be closely approximated by several computational algorithms. In contrast, competitive binocular interaction, as exemplified by the phenomenon of binocular rivalry (BR), has received relatively little attention. When dissimilar monocular images strike corresponding retinal areas, binocular fusion does not occur but instead the images undergo alternating phases of phenomenal suppression. Despite a wealth of psychophysical literature on the stimulus determinants of BR's temporal characteristics, little is known about the underlying neural mechanisms. This paper outlines properties of BR which need to be incorporated in any complete model and subsequently discusses plausible neural mechanisms for these properties.

SPATIAL CHARACTERISTICS OF BINOCULAR RIVALRY

In the study of BR stimuli are presented dichoptically, i.e. the conflicting stimuli are presented separately to the two eyes such that for a given region of the visual field the eyes receive conflicting information. When one views dichoptically presented gratings of orthogonal orientation which subtend visual angles of up to 1.5° at the fovea, one will see either the left eye's view or the right eye's view, i.e. suppression of one eye's view is complete. If, however, such stimuli subtend many degrees of visual angle a patchwork pattern is seen in which either grating may be dominant at a given location at a given time. It seems that BR is independent for different parts of the visual field. Since a given region of the visual field is subserved by mechanisms confined to a relatively small local area of visual cortex, discussion of the putative neural mechanisms of BR will focus on neural interactions within such a local area. The stimulus size with which unitary suppression is obtained is larger than the estimates of aggregate receptive field size (Hubel and Wiesel, 1977), suggesting that a number of orientation and ocular dominance hypercolumns are involved in mediating BR for such an area. This is in keeping with findings suggesting that roughly 2 mm movement through the cortex is required to displace receptive fields from one region of visual space to another. When considering plausible neural mechanisms, attention will be focused on local neural interactions within and between orientation and ocular dominance columns. There are additional reasons for examining local neural interactions in trying to understand BR. One can experience both stereoscopic fusion and BR simultaneously if, for example, high spatial frequency information in each eye's view is fused while low spatial frequency information is rivalrous (Julesz and Miller, 1975). Presumably only the neural mechanisms involved in processing low spatial frequency information are involved in mediating the rivalry. It should also be noted that BR can be obtained for a variety of stimulus dimensions, including color, stimulus orientation, direction of motion and spatial frequency content. For example, dichoptically presented orthogonal gratings having backgrounds of complementary colors may yield rivalry which is independent for color and orientation (Treisman, 1962). The fact that BR may be obtained along many stimulus dimensions suggests that the phenomenon may be characterized by some basic pattern of neural interactions common to the visual coding of these stimulus dimensions. Though the discussion will focus on BR obtained with the conventional dichoptic viewing of line or grating patterns having orthogonal orientations, the basic principles may be applied to BR found for other stimulus dimensions.

TEMPORAL CHARACTERISTICS

When presented with incompatible monocular images one experiences alternating phases of dominance and suppression of each eye's view. Successive dominance and suppression durations are distributed in a way which suggests that stochastic fluctuations occur in the process controlling these durations (Fox and Herrmann, 1967). The effects of increasing stimulus strength in one or both eyes have been summarized by four propositions (Levelt, 1965): (1) increasing stimulus strength in one eye increases the predominance of that eye's view; (2) increasing stimulus strength in one eye does not alter the mean duration of one period of dominance of that eye; (3) increasing stimulus strength in one eye serves to increase the rate of alternation; and (4) increasing stimulus strength in both eyes serves to increase the rate of alternation. Propositions 1 to 3 are illustrated in Fig. 1a, b. Mean suppression duration is a function of stimulus strength in the same eye, whereas mean dominance duration is inversely related to the stimulus strength in the contralateral eye (Blake and Camisa, 1979). If the stimulus strength of the suppressed grating is reduced during the suppression interval then the suppression duration will be prolonged (Leguire, 1980). If stimulus strength is increased, on the other hand, the duration will be terminated earlier.

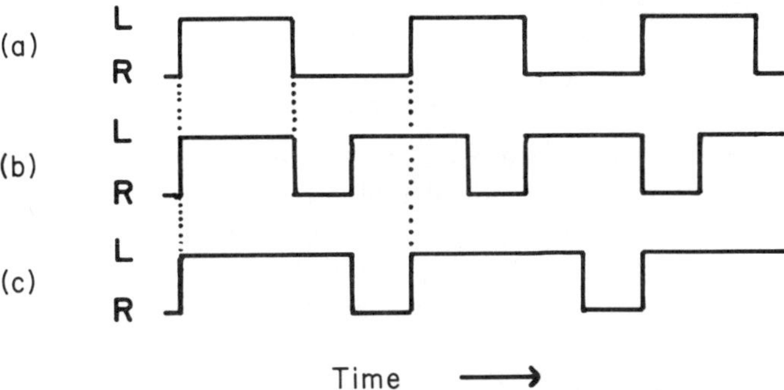

Figure 1 Schematic diagrams of alternating dominance in BR. L and R indicate dominance of the left eye and right eye respectively. (a) Baseline condition with stimuli of equal strength in the two eyes. (b) Effect of increasing stimulus strength in the left eye (illustrating Levelt's propositions 1 to 3). (c) Pattern of alternating dominance following an increase in stimulus strength in the left eye as predicted by a reciprocal inhibition-type model.

CHARACTERISTICS OF SUPPRESSION

One of the most interesting characteristics of BR suppression is its nonselectivity: once a pattern is suppressed the spatial frequency or orientation of that pattern can be changed without the change being detected for several seconds (Blake and Fox, 1974). The suppression duration will equal that normally obtained for the new stimulus, suggesting that the change has been registered on a neural level. An exception to the nonselectivity of suppression has been reported (Smith *et al.*, 1982). Incremental threshold spectral sensitivity measured during dominance reveals the three-maxima function characteristic of color-opponent channels, but when measured during suppression the one-maximum curve characteristic of the luminance channel is found. Suppression, therefore, differentially affects the color-opponent channels.

The magnitude of the suppression effect has been measured using a test probe procedure (Fox and Check, 1972). This involves the presentation of a test stimulus to be detected during either dominance or suppression. Performance during dominance is equivalent to normal monocular vision but there is a factor of 2 or 3 loss in sensitivity to the test probe during suppression. It has been reported that the loss in sensitivity to a test probe remains constant throughout the suppression duration (Fox and Check, 1972); it also appears to be independent of the relative strength and difference in orientation of the rivalrous stimuli (Blake and Lema, 1978; Blake and Camisa, 1979). Though suppression has the effect of removing the suppressed stimulus from phenomenal experience, information in the suppressed eye may interact in a facilitative way with information in the contralateral eye (Westendorf, *et al.*, 1982). The information is still available, albeit at an attenuated level, for further neural processing.

MODELS

Several lines of evidence have been used to reject, perhaps prematurely, a simplistic formulation of reciprocal inhibition as a mechanism underlying BR. Loss of sensitivity to a test probe (depth of suppression) is constant throughout the suppression duration and is also independent of the relative stimulus strength or orientation of the rivalrous stimuli. It has been argued that a reciprocal inhibition model would predict: (a) that inhibition would decrease from start to finish of a suppression period and (b) that depth of suppression would be a direct function of the relative stimulus strengths of the stimuli. Two of Levelt's propositions (numbers 2 and 3) are incompatible with a reciprocal inhibition model, namely that increasing the stimulus strength in one eye serves to increase the rate of alternation but does not alter the mean dominance duration of that eye. It has been argued that a reciprocal inhibition model would predict that dominance and suppression durations should be reciprocally related, i. e. when the dominance durations of one eye's view increase, those of the other eye's view decrease (as illustrated in Fig. 1c).

In place of a reciprocal inhibition model, a two-process model of BR has been suggested (Fox and Rasche, 1969; Blake and Camisa, 1979). This involves a central triggering mechanism controlling the alternating inhibition and a second process controlling the duration of suppression in each monocular channel. Such a model is schematized in Fig. 2a. The central triggering mechanism (T) might act like a flip-flop device initiating inhibition of the monocular channels alternately when it receives a signal from some comparison mechanism (C) signaling the presence of discrepant monocular inputs. To fit the psychophysical data this mechanism would have to be responsible for initiating a constant amount of inhibition (with some stochastic fluctuation) regardless of the stimulus strength or relative orientation of the monocular stimuli. It has been claimed that with prolonged viewing of rivalrous stimuli this mechanism, responsible for triggering alternating inhibition, is adapted, resulting in decreased amounts of complete suppression (Hollins and Hudnell, 1980). The latter argued that this was not a result of simple adaptation to the grating patterns. Such a triggering mechanism should be equally affected by adaptation to a vertical grating in the right eye and an horizontal grating in the left eye (pair A) or adaptation to a vertical grating in the left eye and a horizontal grating in the right eye (pair B). However, the temporal characteristics of BR for a test pair of rivalrous stimulus (pair A) are dramatically different, following adaptation to pair A, from those found for the same test stimuli following adaptation to pair B (Sloane, 1983).

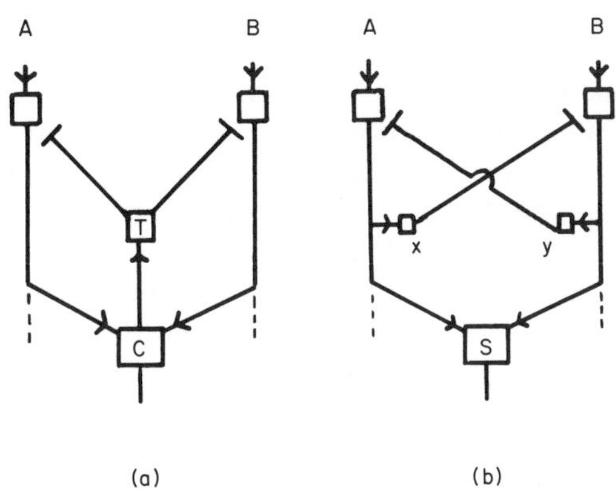

Figure 2 Schematic of two possible models of BR. Arrows indicate excitatory input; flat line terminators indicate inhibitory input. A and B represent the monocular inputs, T a triggering mechanism for inhibition, C a comparator, S a summing mechanism, and x and y inhibitory interneurons.

The second process controls suppression duration. Once inhibited, the duration of suppression would be a function of how long the excitatory pulses generated by the suppressed monocular stimulus need to summate to cancel the centrally generated inhibitory impulses. Assuming that the degree of inhibition is constant, this time will be a sole function of the stimulus strength in the suppressed eye.

One can incorporate this second process into a modified reciprocal inhibition model while eliminating the central triggering process (see Fig. 2b). One need only assume that a constant amount of inhibition is applied, i.e. inhibition is not a linear function of stimulus strength in the contralateral eye but invariant (except for stochastic fluctuations). One way to achieve this is to insert an inhibitory feedback neuron (x,y) for each monocular channel, the output of which, when triggered, is constant. Suppression would terminate when the excitatory impulses cancel the effect of this inhibition such that the postsynaptic potential of the suppressed neuron exceeds a certain level. The time required to bring this about will be shorter in the case of a stronger stimulus. When this cancellation is achieved, the excitatory impulses can then trigger the other inhibitory neuron, leading to the suppression of the contralateral monocular channel. Thus suppression will oscillate back and forth between the monocular channels. The insertion of these inhibitory neurons yielding steady feedback inhibition can account for the psychophysical data, including the predictions of Levelt's four propositions. A recent study on the temporal characteristics of BR following monocular adaptation yielded only weak support for Levelt's propositions (2 and 3), which are in conflict with a reciprocal inhibition model (Sloane, 1983). Monocular adaptation to one of the subsequently viewed rivalrous stimuli serves to reduce its stimulus strength or apparent contrast. No change in rate of alternation of the test stimuli was found. The mean dominance duration of the adapted eye was decreased but this was accompanied by an increase in the mean dominance duration of the unadapted eye (as illustrated in Fig. 1c). These findings are in keeping with some kind of reciprocal inhibition but violate propositions 2 and 3 of Levelt. Thus, even the assumption of a constant inhibitory feedback (which ensures the support of all four propositions) may not be obligatory. In a recent computer simulation of BR, Sugie (1982), using the propositions as a criteria for modeling success, incorporated inhibitory feedback of constant amplitude. He does note, however, that support for propositions 2 and 3 is only approximate. A more recent model involving nonlinear reciprocal inhibition can also be applied to the phenomenon of BR (Lehky, 1983). Much of the psychophysical data on BR can be explained within the framework of the model.

NEURAL MECHANISMS

Little is known about the neural substrate mediating the alternating phenomenal suppression of BR. Behavioral and objective assessment of BR in the behaving monkey shows it to be very similar to that of humans (Myerson, Miezin and Allman, 1981). One can therefore be optimistic about gaining an understanding of the neural mechanisms mediating BR in the near future.

The first task to be accomplished by the neural mechanisms is to assess, at some site of binocular interaction, the incompatibility of the monocular stimuli. For example, in the case of orthogonally oriented gratings viewed dichoptically, binocular neurons would be stimulated by gratings of orthogonal orientation during BR. Based on conventional orientation tuning functions one might argue that dichoptically presented orthogonal gratings would stimulate different sets of binocular neurons (Hubel and Wiesel, 1962). Orientation tuning functions show that a given neuron is triggered only by a narrow range of stimulus orientations. They are generally measured by presenting only one stimulus of a given orientation to the ipsi- or contralateral eye. Such measurements may conceal response properties dependent on simultaneous presentation of two stimuli differing in their orientations either in the same eye or dichoptically. Recent physiological studies have shown that presenting a grating orthogonal to the preferred orientation of the cell greatly reduces the response to a grating presented at that cell's preferred orientation (Morrone, Burr and Maffei, 1982). This inhibitory effect is strongest in simple cells and is thought to originate in complex cells. If this inhibition originates in complex cells, then it is more likely to come from cells in the upper and lower layers of area 17, where the majority of neurons are binocular. It has been confirmed that there are more cells in the upper and lower layers yielding this type of inhibitory effect (Bauer, *et al.*, 1983). The effect of this inhibition was not simply subtractive but changed the gain of the response versus contrast function in a way similar to that found for orientation tuning curves when GABA (the putative inhibitory transmitter) was injected extracellularly (Rose, 1977). This inhibitory effect may well be the physiological manifestation of the incompatibility of the dichoptic stimuli.

As described earlier, one of the more interesting aspects of BR is the nonselectivity of suppression. This nonselectivity may also be related to the inhibitory effect described above. The response of a cell to a grating of a given spatial frequency presented at the preferred orientation is greatly reduced when a similar grating of orthogonal orientation is simultaneously presented (Petrov, Pigarev and Zenkin, 1980). It is claimed that this inhibition reduces the responses to all spatial frequencies at the preferred orientation (Pollen and Ronner, 1982). Also, this inhibitory effect is very broadly tuned for both spatial frequency and orientation, i.e. any grating having an orientation or stimulus frequency other than the narrow range of orientations and spatial frequencies preferred by the cell can elicit this effect (Morrone, Burr and Maffei, 1982). This physiological finding may be important in explaining why suppression, assessed psychophysically, is nonselective, i.e. it is similar regardless of the relative orientations or spatial frequencies of the monocular stimuli.

There is yet another possible parallel between physiology and psychophysics. The inhibitory effect was more easily elicited by a noise grating rather than a single sinewave grating (Morrone, Burr and Maffei, 1982). This suggests that the inhibition resulted from a large number of cells sensitive to a variety of spatial frequencies. It has been shown psychophysically that the number of spatial frequency channels activated by the monocular stimuli has a profound effect on the predominance of each eye's view (Fahle, 1982). Activating a number of spatial frequency channels had a far greater effect than increasing the activity in a single spatial frequency channel (by increasing stimulus contrast).

The differential effects of suppression on color-opponent mechanisms was cited as an exception to the nonselectivity rule. Normal color opponency is disrupted in the suppressed eye during BR. Here, too, there is physiological evidence which may be related. Michael (1978a, 1978b) has recorded from double color-opponent, monocularly driven simple cells and cells with concentrically organized receptive fields in layer IV. Binocularly driven, dual color-opponent complex and hypercomplex cells were found in other layers (Michael, 1979). The monocular disruption of color-opponency may be due to the inhibition of the concentric and simple dual color-opponent coding cells in layer IV. This inhibition may originate in the binocular cells in upper and lower layers in a similar way in which the nonselectivity of suppression was explained.

At any given moment during BR, suppression tends to be monocular. How can one eye's input be suppressed? This leads us to consider the role of the ocular dominance columnar organization evident in most layers. As evidenced by 2-deoxyglucose and cytochrome oxidase studies, this is the overriding organization in the cortex. The pattern of uptake of these two labeling agents and its registration with the ocular dominance columns (Horton and Hubel, 1981) suggest that, within a given cytochrome oxidase patch, all orientations of stimuli are represented. There is psychophysical evidence that cells at either end of the ocular dominance distribution play an important role in BR. The relative effects of monocular (adapt one eye, test that eye), interocular (adapt one eye, test the other eye) and binocular adaptation to one of the rivalrous stimuli prior to tracking BR were examined (Sloane, 1983). The effect of interocular adaptation was minimal, and that of binocular adaptation was only slightly larger than that produced by monocular adaptation. To suppress one eye's input it would be optimal if the activity of cells of a particular ocular dominance could be inhibited. There is both physiological and psychophysical evidence (Blakemore, Carpenter and Georgeson, 1970; Blakemore and Tobin, 1972) for lateral inhibitory interactions between cells having different preferred stimulus orientations, but there is little physiological evidence for a similar type of inhibition across the ocular dominance dimension. Gilbert and Wiesel (1983) found recurrent lateral connections orthogonal to the orientation column border which extend quite far and are distributed in discrete repeating clusters. This clustering phenomenon may reflect excitatory or inhibitory interactions between cells with similar orientation tuning but having different ocular dominance preference. Perhaps a more direct way to achieve suppression of one eye's input is to have inhibition feeding back to a site where ocular dominance is most rigidly dichotomized (layer IV). This is in keeping with the generalized inhibitory effect described earlier, an effect more common in simple cells (biased towards monocularity) and whose source is thought to be complex cells (biased towards binocularity). To account for the alternating suppresssion of each eye's view characteristic of BR, the magnitude of this generalized inhibitory effect must vary over time. Assuming that the generalized inhibitory effect generated in a binocular neuron is due to the stimulation of one eye by a stimulus whose orientation is orthogonal to the preferred orientation of the binocular neuron, then suppression/inhibition of that eye's input would serve to reduce the magnitude of the generalized inhibitory effect. Inhibition of the monocular input would therefore provide a means of modulating the generalized suppression. It is hypothesized that the generalized inhibitory effect spreads orthogonal to the layers within an ocular dominance column.

There is evidence for within-column inhibition between upper and lower layers (Somogyi *et al*, 1981). This spread of inhibition within a column is probably augmented by lateral inhibition between ocular dominance columns. It is noteworthy that observers lacking facilitative binocular interactions reveal normal inhibitory interactions in a psychophysical task (Levi, Harwerth and Smith, 1979). These observers are thought to have a paucity of binocular neurons and may have a bimodal ocular dominance distribution with few neurons driven by both eyes. Yet binocular inhibitory interactions are normal for these observers.

LOCUS

The foregoing analysis presumes that the neural mechanisms mediating BR are located in area 17. This is unlikely to be the full picture. Other authors (e.g. Wolfe and Blake, Chapter 19 in this volume) choose to place the locus of BR beyond area 17. There are two main reasons for doing so: (a) no decrements in direct or interocular after-effects (e.g. threshold elevation, perceived shift in stimulus frequency, motion after-effect) are found after adapting to a stationary or drifting grating when these inducing patterns are phenomenally suppressed for 50 per cent. of the adapting duration (Blake and Fox, 1974); (b) prior adaptation affects the temporal characteristics of BR. It is argued that BR occurs after the site of these adaptation effects since the after-effect magnitude is determined by physical rather than phenomenal presence of the adapting stimulus. This type of evidence does not rule out that BR and these adaptation after-effects may be occurring at the same locus. To elaborate on this allow me to examine the second line of evidence—that adaptation affects the subsequently measured temporal characteristics of BR. Adaptation presumably leads to attenuated activity (whether by fatigue or prolonged inhibition) in the cells responding to the adapting stimulus. Such a reduction in activity would enhance the effect of the binocularly generated inhibitory effect due to summation of the two effects, both acting to reduce the firing rate. A reduction in activity would also prolong the time taken for this effect to be cancelled by incoming activity in the suppressed channel. The finding that there is no decrement in the after-effect magnitude when the adapting stimulus is suppressed 50 per cent. of the time is more challenging to the position taken in this paper. The after-effects in question are both orientation and spatial frequency specific, whereas during BR all orientations and all spatial frequencies are suppressed. Assuming that these after-effects reflect differential sensitivities of neurons responding to different orientations and spatial frequencies, one may explain the lack of decrement in the after-effect magnitude during BR as follows. Rivalry suppression will cause a transient attenuation of neuronal activity, which is equal in magnitude for neurons coding the different orientations and spatial frequencies. As a result of this nonselective inhibition, there will be no change in *relative* sensitivity amongst these neurons. However, relative differences in sensitivity will arise from the prolonged exposure to the adapting stimulus. The ensuing attenuation in activity will be specific to neurons sensitive to the orientation and spatial frequency of the adapting stimulus. Therefore, because BR does not eliminate the differential adaptation of neurons sensitive to orientation and spatial frequency, one would not expect the magnitude of after-effects specific to these dimensions to be affected. These arguments are, of course, speculative but may lead to a reexamination of the evidence used to place the locus of BR beyond the site of the adaptation after-effects.

If each eye's input for a given region of the visual field is to be suppressed alternately, it would seem reasonable to implement this at an early stage in visual processing where ocularity information is still readily dichotomized. I have highlighted certain aspects of the functional architecture of the visual cortex in an attempt to account for the psychophysically observed characteristics of BR. Psychophysics will continue to search for a physiology. The productivity of such a search will be a function of the degree of interfacing between psychophysics, physiology and computer simulation, a goal to which this volume is honorably directed.

ACKNOWLEDGMENT

Preparation of this paper was in part supported by grants NIH EY 04838 to the author and NIH-CORE EY 03039 to the University of Alabama in Birmingham.

REFERENCES

Bauer, R., Dow, B. M., Snyder, A. Z., and Vautin, R. (1983). Orientation shift between upper and lower layers in monkey cortex. *Exp. Brain Res.*, **50**, 133–145.

Blake, R., and Camisa, J. (1979). On the inhibitory nature of binocular rivalry suppression. *J. Exp. Psychol.: Human Perception and Performance*, **5**, 315–323.

Blake, R., and Fox, R. (1974). Binocular rivalry suppression: insensitive to spatial frequency and orientation change. *Vision Res.*, **14**, 687–692.
Blake, R., and Lema, S. A. (1978). Inhibitory effect of binocular rivalry suppression is independent of orientation. *Vision Res.*, **18**, 541–544.
Blakemore, C., Carpenter, R. H. S., and Georgeson, M. A. (1970). Lateral inhibition between orientation detectors in the human visual system. *Nature*, **228**, 37–39.
Blakemore, C., and Tobin, E. A. (1972). Lateral inhibition between orientation detectors in the cat visual cortex. *Exp. Brain Res.*, **15**, 439–440.
Fahle, M. (1982). Binocular rivalry: suppression depends on orientation and spatial frequency. *Vision Res.*, **22**, 787–800.
Fox, R., and Check, R. (1972). Independence between binocular rivalry suppression duration and magnitude of suppression. *J. Exp. Psychol.*, **93**, 283–289.
Fox, R., and Herrmann, J. (1967). Stochastic properties of binocular rivalry alternations. *Percept. Psychophys.*, **2**, 432–436.
Fox, R., and Rasche, F. (1969). Binocular rivalry and reciprocal inhibition. *Percept. Psychophys.*, **5**, 215–217.
Gilbert, C. D., and Wiesel, T. N. (1983). Clustered intrinsic projections in the cat visual cortex. *J. Neurosci.*, **3**, 1116–1133.
Hollins, M., and Hudnell, K. (1980). Adaptation of the binocular rivalry mechanism. *Invest. Ophthal. Vis. Sci.*, **19**, 1117–1120.
Horton, J. C., and Hubel, D. H. (1981). Regular patchy distribution of cytochrome oxidose staining in primary visual cortex of macaque monkey. *Nature*, **292**, 762–764.
Hubel, D. H., and Wiesel, T. N. (1962). Receptive fields, binocular interaction and functional architecture in the cat's visual cortex. *J. Physiol. (Lond.)*, **160**, 106–154.
Hubel, D. H., and Wiesel, T. N. (1977). Functional architecture of macaque visual cortex. *Proc. Roy. Soc. Lond., B*, **198**, 1–59.
Hubel, D. H., Wiesel, T. N., and Stryker, M. P. (1978). Anatomical demonstration of orientation columns in macaque monkey. *J. comp. Neurol.*, **177**, 361–380.
Julesz, B., and Miller, J. E. (1975). Independent spatial-frequency-tuned channels in binocular fusion and rivalry. *Perception*, **4**, 125–143.
Leguire, L. E. (1980). Binocular rivalry: a dual process. Paper presented to the Association for Research in Vision and Ophthalmology, Orlando, Florida.
Lehky, S. R. (1983). A model of binocular brightness and binaural loudness perception in humans with general applications to nonlinear summation of sensory inputs. *Biolog. Cybernetics*, **49**, 89.
Levelt, W. J. M. (1965). *On Binocular Rivalry*, Institute for Perception RVO-TNO, Soesterberg, The Netherlands.
Levi, D. M., Harwerth, R. S., and Smith, E. L., III (1979). Humans deprived of normal binocular vision have binocular interactions tuned to size and orientation. *Science*, **206**, 852–854.
Michael, C. R. (1978a). Color vision mechanisms in monkey striate cortex: dual opponent cells with concentric receptive fields. *J. Neurophysiol.*, **41**, 572–588.
Michael, C. R. (1978b). Color vision mechanisms in monkey striate cortex: simple cells with dual opponent-color receptive fields. *J. Neurophysiol.*, **41**, 1233–1249.
Michael, C. R. (1979). Color sensitive hypercomplex cells in monkey striate cortex. *J. Neurophysiol.*, **42**, 726–744.
Morrone, M. C., Burr, D. C., and Maffei, L. (1982). Functional complications of cross-orientation inhibition of cortical visual cells. 1. Neurophysiological evidence. *Proc. Roy. Soc. Lond., B*, **216**, 335–354.
Myerson, J., Miezin, F., and Allman, J. (1981). Binocular rivalry in macaque monkeys and humans: a comparative study in perception. *Behavior Analysis Lett.*, **1**, 149–159.

Petrov, A. P., Pigarev, I. N., and Zenkin, G. M. (1980). Some evidence against Fourier analysis as a function of the receptive fields in cat's striate cortex. *Vision Res.*, **20**, 1023–1025.

Pollen, D. A., and Ronner, S. F. (1982). Spatial computation performed by simple and complex cells in the visual cortex of the cat. *Vision Res.*, **22**, 101–118.

Rose, D. (1977). On the arithmetical operation performed by inhibitory synapses onto the neuronal soma. *Exp. Brain Res.*, **28**, 221–223.

Sloane, M. (1983). A psychophysical investigation of monocular and binocular neurons in human vision. Doctoral dissertation, Northwestern University.

Smith III, E. L., Levi, D. M., Harwerth, R. S., and White, J. M. (1982). Color vision is altered during the suppression phase of binocular rivalry. *Science*, **218**, 802–804.

Somogyi, P., Cowey, A., Halasz, N., and Freund, T. F. (1981). Vertical organization of neurons accumulating ^{3}H-GABA in visual cortex of rhesus monkey. *Nature*, **294**, 761–763.

Sugie, N. (1982). Neural models of brightness perception and retinal rivalry in binocular vision. *Biolog. Cybernetics*, **43**, 13–21.

Treisman, A. (1962). Binocular rivalry and stereoscopic depth perception. *Quart. J. Psychol.*, **14**, 23–37.

Westendorf, D. H., Blake, R., Sloane, M., and Chambers, D. (1982). Binocular summation occurs during interocular suppression. *J. Exp. Psychol.: Human Perception and Performance*, **8**, 81–90.

Models of the Visual Cortex
Edited by D. Rose and V. G. Dobson

CHAPTER 22

Inferring cortical organization from subjective visual patterns

M.A. Georgeson
Department of Psychology, University of Bristol, Bristol, BS8 1HH, UK

INTRODUCTION: SUBJECTIVE PATTERNS

A wide variety of circumstances can lead to the perception of geometric patterns, often vivid and elaborate, sometimes with colour and movement, in the absence of visual stimuli that normally correspond to those perceptions. Powerful hallucinations can be induced by various drugs (Siegel, 1977) as well as by neurological conditions such as migraine (Sacks, 1973) and viral encephalitis (Mize, 1980). In normal subjects, analogous subjective patterns can be produced by flickering uniform fields (Brown and Gebhard, 1948; Smythies, 1960; Welpe, 1975; Young *et al.*, 1975) and by prolonged pressure on the eye (Tyler, 1978). The finding of certain 'form-constants' (Klüver, 1942, 1966) common to all these conditions—honeycombs, lattices, chess-boards, cobwebs, tunnels, spirals—suggests that they have a structural basis in the visual system. In particular, the geometry of the retinocortical mapping can account for the large-scale concentric, radial and spiral forms that appear centred on the fovea, expanding in size with eccentricity (Tyler, 1978; Schwartz, 1980).

Specific spatiotemporal stimuli

All the cases cited above result from fairly non-specific activation of the visual system. Subjective patterns can also be induced during or after inspection of specific spatiotemporal stimuli—stationary, flickering or moving gratings—and here the subjective patterns depend quite precisely on the form of the stimulus. For example, adaptation to a grating of one spatial frequency can give rise to a subjective pattern after-effect ('illusory grating') of another spatial frequency about 1.5 octaves higher (Georgeson, 1976a, 1980b). The aim of this chapter is to review what is known of these more specific effects and to suggest that they arise from the local organization of the visual cortex—in particular, the dimensionally ordered arrangement and inhibitory interaction of channels within hypercolumns.

One basic notion throughout is that the specific subjective patterns are generated by disinhibition. In an ordered array of neurons, if cells A and B are mutually inhibitory (possibly via interneurons) and cells B and C are mutually inhibitory, then disinhibition can arise in two ways. If A is adapted by prolonged stimulation, then afterwards A's inhibition of B decreases below normal and B's firing rate increases. Vautin and Berkeley (1977, Fig. 7) found just such an effect in a simple cell of the cat's cortex. However, during stimulation of A, B is inhibited and so C is inhibited less than normal. C's firing rate increases by simultaneous disinhibition (Hartline and Ratliff, 1957), rather than as an after-effect. Subjective patterns are observed both during and after stimulation, and so both kinds of disinhibition may be operative in producing them.

AFTER-EFFECTS ON SINGLE DIMENSIONS

Orientation

After adapting to a vertical grating, one can observe noisy oblique gratings filling the adapted part of the visual field (Georgeson, 1976a). The perceived orientations can be reliably measured by matching the orientation of a real line, and are between 30° and 40° on either side of vertical (Georgeson, 1980b). Clockwise and anticlockwise orientations may be seen separately or superimposed. High spatial frequencies (10 c.p.d. or more) and high contrasts were optimal, in terms of the strength and duration of after-effect (Georgeson, 1976a), but any spatial frequency (2.5 to 20 c.p.d.) was capable of producing the effect.

Orientation-selective cells in the visual cortex are arranged in ordered columns (Hubel and Wiesel, 1974; Albus, 1979). If we propose that lateral inhibition operates selectively between adjacent columns, then after adaptation of cells in one column or subset of columns responsive to vertical contours, inhibition on adjacent unadapted columns would be reduced. The firing rate of cells in these columns would increase. Since the usual cause of such firing is obliquely oriented lines or gratings, that is what the subject sees as a subjective pattern. We can infer from this and other effects described below that the orientation range of such inhibition is about 40°. Schwartz (1980) has applied essentially the same explanation to the angular 'fortification illusions' of migraine.

Spatial Frequency

Adapting to a grating of one spatial frequency can produce, as a subjective after-effect, the appearance of another grating in the same orientation, having either higher or lower spatial frequency than the adapting grating. The higher spatial frequency was more easily evoked, but sometimes both components of the after-effect could be observed together (Georgeson, 1976a, 1980b; Tolhurst and Barfield, 1978). The higher component was 1 to 2 octaves above the adapting frequency, and the relation between the adapting frequency and perceived frequency of after-effect was a power function with an exponent between 0.65 and 0.85—Tyler and Nakayama (1980) obtained a much lower exponent, but I was unable to replicate their result (Georgeson, 1980b). The spatial frequency of the after-effect thus depended on the adapting spatial frequency, but did not depend on adapting temporal frequency. The frequency range of the effect, and an observed asymmetry between the higher and lower components, correspond well with results on threshold elevation after adaptation to gratings (Blakemore and Campbell, 1969). In all cases, then, the illusory gratings may be mediated by unadapted channels which lie just beyond the effective spread of adaptation and which become active when the adapted channels cease to inhibit them.

Note that the after-effects cannot be the passive consequence of 'adapting-out' some channels selectively, for then the after-effect should resemble band-stopped noise (the spontaneous activity of all the unadapted channels), whereas in fact specific bands of spatial frequency and orientation are evoked.

The importance of this after-effect lies in the implication that, along with orientation, spatial frequency selectivity forms a second locally ordered dimension in the cortex, equipped with selective inhibitory interaction between adjacent channels. The presence of spatial frequency columns intersecting the pattern of orientation columns has now been revealed anatomically by the deoxyglucose method (Thompson and Tolhurst, 1980; Tootell, Silverman and DeValois, 1981), and spatial frequency tuned inhibitory influences on cortical cells have been demonstrated physiologically. A vital result was that in some cases where a cortical cell was inhibited by the presentation of a particular grating, that cell also gave a burst of firing when the grating was removed. These off-responses occurred 'either to the cell's best frequency at an off-orientation, or to an off-frequency at the cell's best orientation' (DeValois and Tootell, 1983, p. 373). Such orientation-selective and frequency-selective after-effects at the single-cell level are a direct physiological analogue of the subjective patterns described above.

Movement

Adapting to a high-contrast grating of one orientation also produces a powerful after-effect of wavy, granular, fast, streaming movement at right angles to the adapting orientation (Pierce, 1901; MacKay, 1957). Much recent work supports the idea of a functional dichotomy between pattern-analysing mechanisms sensitive to slow velocities and spatially coarse motion-analysing mechanisms responsive to relatively fast velocities. There is no space here to review or evaluate this large field in detail. Suffice it to say that if the two mechanisms with the same preferred orientation were mutually inhibitory, then adaptation to a pattern at one orientation would disinhibit motion responses corresponding to directions orthogonal to the adapting orientation, as observed. Successful prediction of the converse after-effect (a stationary subjective pattern orthogonal to the bidirectional motion of an adapting texture) gave strong support to this model (Georgeson, 1976b).

MacKay and MacKay (1976) raised the reasonable objection that in this second case a fast-moving texture might produce its after-effect not by motion per se but via the virtual contours or 'streaks' produced by motion blur. They showed that the stationary orthogonal subjective pattern was not produced by slow adapting velocities, but became increasingly visible as adapting speed increased. However, they did not show that 'streaking' was necessary for the effect to occur, and in terms of the above model it seems straightforward to argue that fast speeds were optimal because they adapted the fast-motion channels but not the slow-pattern channels. I confirmed (unpublished observations, 1977) that the after-effect was produced by fast, superimposed counter-rotating textures (86 r.p.m., field diameter 11°) but not by slow motion (1 r.p.m.). However, I also found that when slow rotary motion of a random texture in one direction was superimposed on fast motion in the opposite direction, two superimposed patterns of radial lines and curves were observed in the after-effect, one rotating slowly, the other fast, each in a direction opposite to the corresponding adapting motion. This finding of separate, velocity-specific after-effects in opposite directions is clearly favourable to the model of slow and fast channels outlined above, though many details remain to be worked out.

Ocular Dominance

If one adapts the left and right eyes to textures rotating in opposite directions, conventional motion after-effects can be observed in opposite directions when the eyes are tested separately on stationary patterns, as if the adaptation were eye-specific (Anstis and Moulden, 1970). Testing binocularly yields no after-effect. Similarly, when the subjective after-effect of the previous section was tested on a monocular uniform field after dichoptic adaptation, the two eyes gave opposite motion after-effects (rotating radial lines and curves). Binocular testing, however, gave a strong, stationary, radial after-effect, just as after binocular adaptation (Georgeson, 1976b). These results can be explained if adaptation and disinhibition take place independently in different ocular dominance groups. The differential states of adaptation are revealed by monocular and binocular testing. Moulden (1980) gives a detailed treatment of the same idea applied to conventional after-effects.

RELATIONS BETWEEN DIMENSIONS: SIMULTANEOUSLY INDUCED EFFECTS

Orientation and Spatial Frequency

During inspection of a high-contrast grating, subjects may see a low-contrast grid superimposed on the stimulus pattern (Erb and Dallenbach, 1939). The orientation components of this grid are ± 60° from the inducing orientation and its spatial frequency is about 2 to 3 octaves below that of the inducer (Welpe, 1976; Georgeson, 1980b). Any spatial frequency is capable of inducing the effect and it is usually associated with subjective colours (Welpe, 1976).

This grid may be an example of simultaneous disinhibition at the cortical level. If channels up to 40° (orientation) away from an adapting stimulus are inhibited (see above), then channels with even more remote preferred orientations should be disinhibited, to produce a visible subjective pattern during the period of stimulation.

The fact that the subjective grid is remote in both orientation and spatial frequency suggests that the spread of inhibition from one channel to neighbouring orientations may be biased towards channels preferring lower spatial frequencies. This would also explain why the orientation after-effect (described earlier), although appearing quite broadband and noisy compared to the spatial frequency after-effect, had a mean matched frequency around 0.8 octaves lower than the adapting grating itself (Georgeson, 1980b). It is interesting that DeValois and Tootell (1983) found cortical cells to be inhibited more strongly by frequencies higher, rather than lower, than the optimum spatial frequency.

Orientation and Ocular Dominance

When a high-contrast grating is viewed monocularly, a subjective grid or lattice of fine, oblique lines can be observed 'in' the other eye (Howard, 1959). It is visible particularly when the non-viewing eye becomes dominant, and is visible weakly or not at all in binocular viewing. Its perceived orientation components are similar to the orientation after-effect—about 35° away from a vertical or horizontal inducing grating, increasing to 45° for oblique inducers (Georgeson, 1975).

It is now known that many 'monocular' cortical cells, excited only by one eye, also receive strong inhibitory influences via the other eye (Bishop, Henry and Smith, 1971; Noda, Creutzfeldt and Freeman, 1971). Ferster (1981) has shown that these cells show disparity sensitivity of the 'near' or 'far' type (Poggio and Fischer, 1977) and that the inhibitory region in the non-dominant eye has the same orientation selectivity as the excitatory region of the dominant eye.

The subjective interocular lattice may be understood as the joint effect of such interocular inhibition and the orientation-tuned inhibition discussed earlier. A grating in one eye will inhibit 'monocular' cells normally responsive to the same orientation in the contralateral eye, and hence will disinhibit those cells responsive to orientations about 40° away in the contralateral eye by reducing the normal level of orientation-domain inhibition.

Orientation, Movement, Disparity and Spatial Phase

Disinhibition as the combined result of two different, but interactive, inhibitory processes may also underlie an even more striking subjective, lattice-like effect—the 'graph-paper effect' (Georgeson, 1980a)—which involves relations between no less than four visually coded dimensions.

The 'graph-paper effect' can be induced by a variety of spatial stimuli but is optimally evoked by gratings, especially pairs of gratings superimposed at an angle of 60 to 90°. Hence, ordinary graph paper will do nicely. Tennis netting, checked shirts, etc., can also be very effective, as are rectangular-wave and sawtooth gratings, but pairs of sine- or square-wave gratings are not. The key feature turns out to be the presence of a strong second harmonic in the spatial waveform (see Georgeson, 1980a, for details). The effect requires retinal motion and can be evoked by moving the stimulus, or more simply by smoothly tracking the eyes across a stationary pattern. Tracking horizontally across an orthogonal grid of lines oriented obliquely in the frontal plane, one sees the visual field filled with roughly vertical (and less predominantly, roughly horizontal) striations moving along with the direction of gaze. Vertical tracking makes the horizontals the more dominant. Monocular and binocular stimulation can both be effective.

In simple, large-group demonstrations using graph paper as the stimulus, I find that around half the population seems to experience this effect in much the same way. With optimal conditions and appropriate instructions the proportion can be higher. Detailed experiments on small numbers of subjects also revealed: (a) that with binocular viewing the subjective pattern is stereoscopic, lying precisely in the plane of fixation, whether this was in front of or behind the stimulus plane, and (b) that the subjective pattern usually consisted partly or entirely of sharp edges with a consistent and identifiable polarity. The edge polarity depended on both the direction of movement and the polarity and waveform of the inducing grating, but was predictable from the spatial phase of the second harmonic component, with the further assumption of a movement-induced phase lag of around 45°. Put simply, the effect seems to require stimulation of movement-sensitive, phase-sensitive 'edge detectors', and then the subjective effect has the opposite edge polarity, oriented about 40° away.

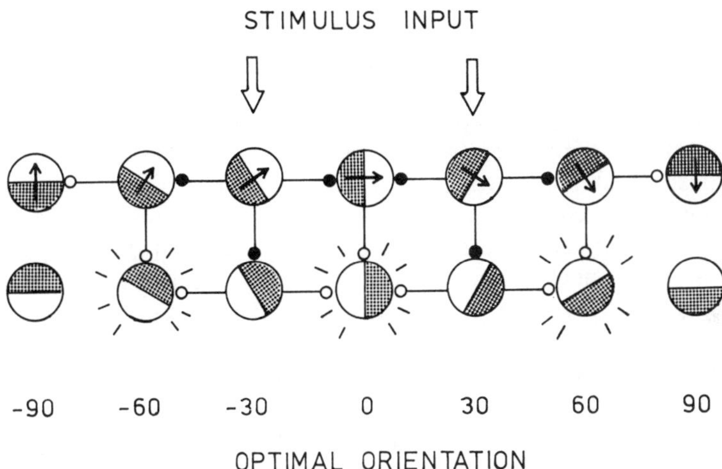

Figure 1 Cortical inhibitory model to account for major features of the 'graph-paper effect'. Large circles represent units selective for orientation, preferring zero disparity (bottom row), and units selective for direction of movement, preferring non-zero disparities (top row); half-hatching indicates preferred edge polarity. Filled circles show inhibition evoked by stimulus movement; open circles show resulting disinhibition of cells mediating the subjective pattern of illusory edges in the fixation plane. Weak disinhibitory effects (e.g. top row, ± 90 might not exceed threshold.

A tentative model to explain this elaborate effect can be constructed from the mechanisms already proposed for the other subjective patterns. Figure 1 shows parallel arrays of direction-selective and orientation-selective neurons. The ordering of orientations has been shown by Hubel and Wiesel (1974) and others; additional evidence for the ordering of preferred directions has been found by Payne, Berman and Murphy (1980) and Tolhurst, Dean and Thompson (1981). Mutual inhibition, both within and between arrays, is assumed. Movement stimulation in one direction disinhibits two orientations about 30 to 40° away. Combining two stimulus directions can produce up to four perceived orientations, but for an appropriate angle between the stimuli two of these orientations will merge into one. The 'graph-paper effect' shows just these properties. To explain the stereoscopic and phase (edge)-selective properties, it is further proposed that inhibitory interactions within the movement array operate largely between cells of the same disparity type (either 'far' cells or 'near' cells) with the same receptive-field asymmetry, giving the same spatial phase preference, and that cells in the orientation array are of the tuned excitatory kind, preferring near-zero disparities. Inhibition between the arrays would be antagonistic—i.e. between cells with opposite phase/edge preferences. With such a scheme, the disinhibited orientations would be coded with zero disparity, and hence appear to lie in the plane of fixation with an edge polarity opposite to the effective phase of the stimulus.

CONCLUSIONS

The variety of subjective patterns—gratings, lattices, edges and movements—described here are amenable to explanation in terms of the properties and interactions of cortical cells. Of course, the schemes suggested here may be criticized as *post hoc*, since they were constructed with the properties of subjective patterns in mind. I have tried to show, however, that most of the mechanisms required are already known from current neurophysiology and psychophysics. Where particular processes—such as phase-selective antagonism—are not known, the subjective patterns may serve as a rich source of hypotheses to be tested in more conventional ways. They also offer unusual insights into the relation between cortical physiology and perception. If we can accept the general proposal that the subjective orientations and spatial frequencies are indeed the consequence of activating subsets of selective cells, in the absence of a corresponding retinal pattern, then we must conclude that these cells do have quite a direct role in encoding and representing the dimensions for which they are selective.

REFERENCES

Albus, K. (1979) ^{14}C-deoxyglucose mapping of orientation subunits in the cat's visual cortical areas. *Exp. Brain Res.*, **37**, 609–613.

Anstis, S. M., and Moulden, B. P. (1970). After effect of seen movement: evidence for central and peripheral components. *Quart. J. Exp. Psychol.*, **22**, 222–229.

Bishop, P. O., Henry, G. H., and Smith, C. J. (1971). Binocular interaction fields of single units in the cat striate cortex. *J. Physiol. (Lond.)*, **216**, 39–68.

Blakemore, C., and Campbell, F. W. (1969). On the existence of neurons in the human visual system selectively to the orientation and size of retinal images. *J. Physiol. (Lond.)*, **203**, 237–260.

Brown, C. R., and Gebhard, J. W. (1948). Visual field articulation in the absence of spatial stimulus gradients. *J. Exp. Psychol.*, **38**, 188–200.

DeValois, K. K., and Tootell, R. B. H. (1983). Spatial-frequency-specific inhibition in cat striate cortex cells. *J. Physiol. (Lond.)*, **336**, 359–376.

Erb, M. B., and Dallenbach, K. M. (1939). Subjective colours from line patterns. *Amer. J. Psychol.*, **52**, 227–241.

Ferster, D. (1981). A comparison of binocular depth mechanisms in areas 17 and 18 of the cat visual cortex. *J. Physiol. (Lond.)*, **311**, 623–655.

Georgeson, M. A. (1975). Mechanisms of visual image processing. D.Phil. thesis, University of Sussex.
Georgeson, M. A. (1976a). Psychophysical hallucinations of orientation and spatial frequency. *Perception*, **5**, 99–111.
Georgeson, M. A. (1976b). Antagonism between channels for pattern and movement in human vision. *Nature*, **259**, 413–415.
Georgeson, M. A. (1980a). The graph-paper effect: subjective stereoscopic patterns induced by moving gratings. *Perception*, **9**, 503–522.
Georgeson, M. A. (1980b). The perceived spatial frequency, contrast and orientation of illusory gratings. *Perception*, **9**, 695–712.
Hartline, H. K., and Ratliff, F. (1957). Inhibitory interaction of receptor units in the eye of *Limulus*, *J. Gen. Physiol.*, **40**, 357–376.
Howard, I. P. (1959). Some new subjective phenomena apparently due to interocular transfer. *Nature*, **184**, 1516–1517.
Hubel, D. H., and Wiesel, T. N. (1974). Sequence regularity and geometry of orientation columns in the monkey striate cortex. *J. comp. Neurol.*, **158**, 267–294.
Klüver, H. (1942). Mechanisms of hallucinations. In *Studies in Personality* (Eds. Q. McNemar and M. A. Merrill), McGraw-Hill, New York.
Klüver, H. (1966). *Mescal and Mechanisms of Hallucinations*, University of Chicago Press, Chicago, Ill.
MacKay, D. M. (1957). Moving visual images produced by regular stationary contours. *Nature*, **180**, 849–850, 1145–1146.
MacKay, D. M., and MacKay, V. (1976). Antagonisms between visual channels for pattern and movement? *Nature*, **263**, 312–314.
Mize, K. (1980). Visual hallucinations following viral encephalitis: a self report. *Neuropsychologia*, **18**, 193–202.
Moulden, B. (1980). After-effects and the integration of patterns of neural activity within a channel. *Phil. Trans. Roy. Soc. (Lond.), B*, **290**, 39–55.
Noda, H., Creutzfeldt, O. D., and Freeman, R. B. (1971). Binocular interaction in the visual cortex of awake cats. *Exp. Brain Res.*, **12**, 406–421.
Payne, B. R., Berman, N., and Murphy, E. H. (1980). Organization of direction preference in cat visual cortex. *Brain Res.*, **211**, 445–450.
Pierce, A. H. (1901). *Studies in Space Perception*, Longmans, New York.
Poggio, G. F., and Fischer, B. (1977). Binocular interaction and depth sensitivity in striate and prestriate cortex of behaving rhesus monkey. *J. Neurophysiol.*, **40**, 1392–1405.
Sacks, O. (1973). *Migraine*, Faber, London.
Schwartz, E. (1980). Computational anatomy and functional architecture of striate cortex: a spatial mapping approach to perceptual coding. *Vision Res.*, **20**, 645–669.
Siegel, R. K. (1977). Hallucinations. *Sci. Amer.*, **234**, Oct.), 132–140.
Smythies, J. R. (1960). The stroboscopic patterns. III: Further experiments and discussion. *Brit. J. Psychol.*, **51**, 247–255.
Thompson, I. D., and Tolhurst, D. J. (1980). The representation of spatial frequency in cat visual cortex: a ^{14}C-2-deoxyglucose study. *J. Physiol. (Lond.)*, **300**, 58P.
Tolhurst, D. J., and Barfield, L. P. (1978). Interactions between spatial frequency channels. *Vision Res.*, **18**, 951–958.
Tolhurst, D. J., Dean, A. F., and Thompson, I. D. (1981). Preferred direction of movement as an element in the organization of cat visual cortex. *Exp. Brain Res.*, **44**, 340–342.
Tootell, R. B., Silverman, M. S., and DeValois, R. L. (1981). Spatial frequency columns in primary visual cortex. *Science*, **214**, 813–815.

Tyler, C. W. (1978). Some new entoptic phenomena. *Vision Res.*, **18**, 1633–1639.

Tyler, C. W., and Nakayama K. (1980). Grating induction: a new type of aftereffect. *Vision Res.*, **20**, 437–441.

Vautin, R. G., and Berkley, M. A. (1977). Responses of single cells in cat visual cortex to prolonged stimulus movement: neural correlates of visual after-effects. *J. Neurophysiol.*, **40**, 1051–1065.

Welpe, E. (1975). Das Schachbrettmuster, ein nur binokular anregbares Flimmermuster. *Vision Res.*, **15**, 1283–1287.

Welpe, E. (1976). Uber die durch ein Rechteckgitter induzierten subjektiven Rautenmuster. *Vision Res.*, **16**, 1337–1341.

Young, R. S. L., Cole, F. E., Gamble, M., and Rayner, M. D. (1975). Subjective patterns elicited by light flicker. *Vision Res.*, **15**, 1291–1293.

Models of the Visual Cortex
Edited by D. Rose and V. G. Dobson

CHAPTER 23

Visual motion and cortical magnification

M.J. Wright and A. Johnston
Department of Psychology, Brunel University, Kingston Lane, Uxbridge, Middlesex UB8 3PH, UK

M-SCALING AND CONTRAST SENSITIVITY

The visual field is mapped on the surface of striate cortex and in Man and other primates there is a very extensive foveal projection and a relatively small projection of peripheral parts of the visual field. The cortical magnification factor M is defined as the extent of striate cortex in millimetres corresponding to a degree of arc in visual space; M decreases progressively with eccentricity. The extrafoveal striate cortex itself however, appears to be functionally and anatomically uniform, receiving a uniform density of lateral geniculate nucleus (LGN) input fibres (Hubel and Wiesel, 1974, 1977). The linear density (cells per millimetre) of retinal ganglion cells is roughly proportional to M, provided one corrects for the displacement of ganglion cell bodies away from the fovea (Drasdo, 1977). Since there have been very few direct measurements of human M, except for a small part of the lower visual field (Cowey and Rolls, 1974), most published values are indirectly estimated from data on human ganglion cell density (Drasdo, 1977; Rovamo and Virsu, 1979).

In human psychophysical studies, Rovamo and Virsu (1979) showed that if a grating is enlarged at peripheral locations in proportion to $1/M$, the contrast at which it can be detected is equalized across the retina. This scaling procedure (M-scaling) maintains equivalence of the cortical projection of stimuli with different visual field loci. M-scaling increases the subtended area and reduces the spatial frequency of grating patches with increasing visual field eccentricity while maintaining constant their cortical area and cortical spatial frequency (cycles per millimetre). Contrast sensitivity functions measured using M-scaled gratings at different positions in the visual field superimpose when plotted against cortical spatial frequency. This is not affected by the temporal parameters of the grating (Virsu *et al.*, 1982). Contrast sensitivity for gratings is probably based on cortical mechanisms since it shows orientation- and spatial frequency-specific changes due to subthreshold summation, adaptation and masking, and these effects exhibit interocular transfer (Braddick, Campbell and Atkinson, 1978). It is necessary, for this reason, to consider *M*-scaling effects on contrast sensitivity in terms of cortical rather than retinal mechanisms.

Hubel and Wiesel (1974) found that receptive field area and scatter of extrafoveal monkey striate cortical cells is inversely related to M^2; *M*-scaling of gratings may maintain a constant relationship between spatial frequency and cortical receptive field dimensions, as well as equalizing the cortical area stimulated. Assuming that the striate cortex is composed of functionally uniform modules (Hubel and Wiesel, 1977), *M*-scaled stimuli should produce equivalent effects on striate cortical circuitry.

Contrast sensitivity for gratings increases with the total area of the stimulus but the summation function depends on spatial frequency and visual field locus (Koenderink *et al.*, 1978; Robson and Graham, 1981). It seems to be the 'cortical area' rather than the retinal area which is the critical determinant of sensitivity; contrast sensitivity increases as a power function of the calculated cortical area stimulated by the grating (Virsu and Rovamo, 1979). The simplest interpretation is that sensitivity depends on the number of cortical cells stimulated. Virsu and Rovamo have postulated a 'central integrator' which pools the activity of cortical cells over large regions of striate cortex. Probability summation across space (Robson and Graham 1981) is perhaps a more plausible mechanism. As the area of a grating is increased, it will fall on the receptive field of more independently stimulated detectors. Therefore the probability that one or more detectors will signal its presence is increased.

M-SCALING AND VISUAL MOTION

We have used *M*-scaling to investigate motion perception at suprathreshold contrasts. We noted in pilot experiments that a drifting grating appears to move more slowly with eccentric than with central viewing, and at moderate to high spatial frequencies and low velocities its motion appears to cease at a critical eccentricity. Moreover, apparent deceleration of a peripherally viewed steadily moving grating occurs with prolonged viewing. These effects indicate major differences in motion thresholds and motion adaptation, for stimuli of constant retinal size, in foveal and peripheral vision. The object of our experiments was to determine whether such differences are compensated for by scaling the stimuli to produce equivalent cortical projections.

Lower Threshold of Motion (LTM)

We measured the magnitude of the lower threshold of motion at four eccentricities (0, 1.5, 4 and 7.5°) using *M*-scaled gratings of constant contrast (0.3). LTM expressed as temporal frequency increased linearly with spatial frequency. Since velocity = temporal frequency/spatial frequency, LTM at a given visual field locus was a constant velocity which was independent of spatial frequency over a considerable range (2 to 16 c.p.d.) The LTM velocity measured as a retinal variable increased with eccentric viewing. When velocity was scaled with respect to *M*, differences in the threshold for different eccentricities disappeared; LTM was a constant in terms of cortical velocity (0.16 mm/s) (Johnston and Wright, 1983).

Threshold displacement (TD)

Westheimer (1978) showed that discrimination of direction of step displacement of line and grating targets occurred for amplitudes as low as 10 s arc and thus comes into the range of hyperacuities. We found that TD for 1 Hz square-wave displacements also fell within the hyperacuity range, and was independent of spatial frequency in the range 2 to 16 c.p.d. for foveal viewing. Displacement thresholds were constant at a given eccentricity, but increased with eccentricity. When the data were converted into cortical distances, the curves for different eccentricities were superimposed. This instantaneous displacement threshold was a constant cortical distance (0.04 mm), but for temporal frequencies of sinusoidal displacement below 1 Hz threshold is related to cortical velocity. The maximum velocity during a cycle of sinusoidal oscillation was close, at threshold, to LTM for linear motion (Wright and Johnston, 1985).

Motion After-Effect (MAE)

Following prolonged inspection of a drifting (adapt) grating, a stationary (test) grating will appear to move in the opposite direction. Although a drifting grating will appear to move more slowly when viewed peripherally, the velocity of the motion after-effect to a given grating *increases* with eccentric viewing (Wright and Johnston, 1983a, b). We quantified the motion after-effect (MAE) by nulling with real motion of the test grating. The test and adapt gratings had identical spatial frequency and contrast. The MAE to an 8 Hz adapt grating was approximately constant in terms of *cortical* velocity for *M*-scaled gratings regardless of spatial frequency (Johnston and Wright, 1983). Surprisingly, the velocity of the after-effect remained constant (for a given eccentricity) although the temporal frequency of the adapt grating was constant, so its velocity varied inversely with spatial frequency.

The temporal frequency of the adapt grating which gave rise to the greatest velocity of after-effect moreover showed very little change with eccentricity, contrast or spatial frequency.

Possible mechanisms

The 'stopped motion effect' (Campbell and Maffei, 1981) and the pronounced slowing of peripherally viewed stimuli are not mere peculiarities of peripheral vision but may be explained by reference to the calculated 'cortical velocity' of the projected stimulus (Johnston and Wright, 1983). The wide applicability to our data of the concepts of cortical velocity and cortical displacement led us to consider the idea that they might not be mere mathematical abstractions but have a concrete existence. It would, however, be anachronistic and wrong to consider such a 'cortical image' a sufficient explanation of motion thresholds. Rather, the essence of our proposal is that it is a cortical topographic map, rather than the retinal topographic map, which is sampled by motion-detecting neurons. This would seem to require a locally precise retinotopic mapping, scaled according to the cortical magnification factor, such that displacements of stimuli would give rise to changes in the location of maximum activity in an array of presynaptic terminals, and moving stimuli produce sequential activation of such terminals.

Possible candidates for such a mapping exist in layer IVc of the striate cortex which receives the major inputs from the lateral geniculate nucleus (LGN). Barlow (1981) has suggested that there are two 'interpolation layers', one receiving X-type and one Y-type inputs. These layers contain large numbers of small cells and these might be able to interpolate between the coarser sampling points of the LGN input fibres. Cells in these layers are uniform in density so that the topography of the visual field map in these layers should be scaled to M. It was originally thought that these cells were non-oriented and non-directional (Hubel, Wiesel and Stryker, 1978), but this has been disputed (Hawken and Parker, 1983). Alternatively, the terminals of the LGN input fibres themselves might constitute the retinotopic map. This map probably contains discontinuities associated with the system of ocular-dominance columns, but topographic ordering within a column is preserved (Hubel and Wiesel, 1977). It is necessary only that the ordering should be continuous over a region which is sampled by a higher order cell. Direction of motion and adaptation to motion would be determined by the higher order cells (e.g. in other layers of the striate cortex) sampling a retinotopic map whose dimensions were related to M. The inputs to motion-detecting cells would then be equivalent for central and peripheral M-scaled stimuli, and constancy of cortical motion and displacement thresholds would result, given uniform morphology of higher order cells.

At least one type of proposed motion-detecting scheme (Marr and

Ullmann, 1981) is incompatible with our findings since it gives an output depending on temporal frequency rather than velocity and is not velocity sensitive. Other proposed mechanisms for motion detection depend on the comparison of spatial and temporal variations in intensity (Reichardt, 1961; Barlow and Levick, 1965; Harris, 1980). Hildreth (1984) has pointed out a possible source of error for such mechanisms from confounding factors such as changing illumination. One possible solution to this difficulty depends on a partial or complete segregation of location from spatial contrast information *prior to* the analysis of motion.

On this view, the inputs to motion detectors should not only be *M*-scaled but also should respond very sharply to an optimum position of an intensity gradient or to an optimum phase angle of a grating, regardless of spatial frequency, intensity contrast or (within limits) velocity. Geniculate or cortical cells with spatially segregated on- and off-regions have such properties (Lee, Elepfandt and Virsu, 1981a, 1981b). With very little extrapolation beyond accepted neurophysiological findings, these properties of familiar receptive field types can be incorporated in a topographic model to account for our psychophysical results.

The constancy of LTM and TD over the middle range of spatial frequencies might appear problematic in the light of evidence that neurons in the visual pathway are tuned to spatial frequency. However, the locus of the most active cell (or presynaptic terminal) in a topographically ordered array would travel across the cortex as an edge or intensity gradient moved across the visual field. A possible complication arises if cells with different receptive field sizes and preferred spatial frequencies coexist at a given visual field locus (DeValois, Albrecht and Thorell, 1982), but as long as the cortical maps for such cells are equivalent (Hubel and Wiesel, 1977) the cortical velocity of the loci of activity would depend on the displacement or velocity of the grating, regardless of spatial frequency. Higher order directional cells sampling the array could be wired in such a way that their threshold (LTM or TD) behaviour depended on cortical displacement raher than spatial frequency (Barlow and Levick, 1965; Reichardt, 1961).

We find that LTM and TD are independent of contrast above one log unit suprathreshold, and similar for sine- and square-wave gratings, except at low spatial frequency. Some loss of performance is inevitable at low contrast and low spatial frequency due to the statistical uncertainty of locating the optimum position or phase angle in regions of gradual intensity change (Watt and Morgan, 1983). Provided that the grating contrast was sufficient for a geniculate (or simple cell) field to give a reliable indication of phase, the locus and identity of the most active cell in the hypothesized cortical array would also be independent of the contrast of the grating, and higher order directional cells sampling the array could therefore be wired to show early saturation of responses to contrast without loss of motion sensitivity.

The 'cortical constancy' of psychophysical displacement and velocity thresholds suggested an explicit topographic cortical representation of space and motion; the topography of the geniculostriate projection could be an important element in the elaboration of motion-detecting neurons. However, space and time are also *implicitly* encoded within channels or receptive fields and we have discussed elsewhere possible alternative explanations of our findings based on the spatiotemporal tuning of X- and Y-cells (Johnston and Wright, 1985). Since the visual system evidently uses *both* topographic and non-topographic encoding of motion, one of the most important problems for the future is to determine how these two kinds of spatiotemporal representation are related.

LIMITATIONS TO THE APPLICABILITY OF M-SCALING

In foveal striate cortex of the macaque monkey the parallel relationship between receptive field size, scatter and M evidently breaks down (Dow *et al* 1981; Van Essen *et al* 1984.) Comparable human data are not available. However the assumption that M is proportional to ganglion cell density in man may be questioned. Moreover psychophysical deviations from *M*-scaling have been reported for vernier acuity (Westheimer, 1982), spatial phase discrimination (Stephenson and Braddick, 1983) and detection of amplitude modulation and frequency modulation of gratings (Jamar, Kwakman and Koenderink, 1984).

A more stringent but seldom-used test of the *M*-scaling hypothesis would also take into account meridional asymmetries in the geniculostriate projection (Rovamo and Virsu, 1979). Receptive field characteristics within the striate cortex may also vary along other coordinate systems superimposed upon the general scaling with eccentricity. In the cat, the distribution of preferred orientations is related to a polar coordinate system (Payne and Berman, 1983) and ocular dominance to distance from the vertical meridian (Berman *et al.*, 1982).

The retina contains W, X and Y ganglion cells which project differentially to thalamic and midbrain nuclei. Beyond the striate cortex, interconnected both in series and in parallel are multiple visual areas which differ in the functional specialization of their cells, as well as the area and overlap of receptive fields. Major differences in the magnification function (*M* versus eccentricity) and radial anisotropy of the retinotopic projection appear in extrastriate visual areas (Van Essen and Maunsell, 1983) and in at least one projection from the striate cortex, that to the pons (Glickstein, Mercier and Whitteridge, 1983). With increasing neuroanatomical knowledge of scaling factors and anisotropies in different parts of the visual pathway, it may be possible to deduce from the goodness of fit of different scaling factors the appropriate anatomical locus for a given psychophysical task.

It is conceivable also that the morphology of cortical maps may have a functional significance; thus the exclusive lower field representation of the striatopontine projection (Glickstein, Mercier and Whitteridge, 1983) may be related to the visual control of locomotion (the limbs are usually seen in the lower visual field). Epstein (1984) has suggested that the peculiar geometry of the lower field striate projection of the cat, which differs from that of the upper field, gives equal sensitivity to small movements of prey or objects anywhere in the foreground of a typical cat's-eye view. In humans, the variation of striate M with eccentricity would tend to compensate the increase in velocities with eccentricity in the optic flow field during locomotion.

In the long term, techniques of quantitative perimetry might prove an indispensible aid to the localization of visual function, and will certainly engender theories concerning the functional significance of topographic mapping. Already these studies have indicated the need for neurophysiological explanations of visual performance based on the behaviour of assemblies of tuned neurons rather than the behaviour of single examples of neurons.

ACKNOWLEDGMENT

We thank the Medical Research Council for support (Grant no. G979/1106/N).

REFERENCES

Barlow, H. B. (1981). Critical limiting factors in the design of the eye and the visual cortex. *Proc. Roy. Soc. Lond. B*, **212**, 1–34.

Barlow, H. B., and Levick, W. R. (1965). The mechanism of directionally-sensitive units in rabbit's retina. *J. Physiol.*, **178**, 477–504.

Berman, N., Payne, B. R., Labar, D. R., and Murphy, E. H. (1982). Functional organisation of neurons in cat striate cortex; variations in ocular dominance and receptive field type with location in the visual field. *J. Neurophysiol.*, **48**, 1362–1377.

Braddick, O., Campbell, F. W., and Atkinson, J. (1978). Channels in vision: basic aspects. In *Handbook of Sensory Physiology*, Vol. 8, Springer-Verlag, pp. 3–38.

Campbell, F. W., and Maffei, L. (1981). The influence of spatial frequency and contrast on the perception of moving patterns. *Vision Res.*, **21**, 713–721.

Cowey, A., and Rolls, E. T. (1974). Human cortical magnification factor and its relation to visual acuity. *Exp. Brain Res.*, **21**, 447.

DeValois, R. L., Albrecht, D. G., and Thorell, L. G. (1982). Spatial frequency selectivity of cells in macaque visual cortex. *Vision Res.*, **22**, 545–559.

Dow, B. M., Snyder, R. G., Vautin, R. G. and Bauer, R. (1981). Magnification factor and receptive field size in foveal striate vortex of the monkey. *Expl. Brain Res.*, **44,** 213–228.

Drasdo, N. (1977). The neural representation of visual space. *Nature*, **266**, 554–556.

Epstein, L. I. (1984). An attempt to explain the differences between the upper and lower halves of the striate cortical map of the cat's field of view. *Biolog. Cybernetics*, **49**, 175–177.

Glickstein, M., Mercier, B., and Whitteridge, D. (1983). The striate cortex input to the macaque pons: selectivity and demagnification. *J. Physiol.*, **341**, 77P.
Harris, M. G. (1980). Velocity specificity of the flicker to pattern sensitivity ratio in human vision. *Vision Res.*, **20**, 687–691.
Hawken, M. J., and Parker, A. J. (1983). Contrast sensitivity and orientation tuning of neurons in layer IV of the striate cortex of Old World monkeys. *J. Physiol.*, **345**, 124P.
Hildreth, E. C. (1984). The computation of the velocity field. *Proc. Roy. Soc., B*, **221**, 189–220.
Hubel, D. H., and Wiesel, T. N. (1974). Uniformity of monkey striate cortex; a parallel relationship between field size, scatter and magnification factor. *J. comp. Neurol.*, **158**, 295–305.
Hubel, D. H., and Wiesel, T. N. (1977). Functional architecture of macaque monkey visual cortex. *Proc. Roy. Soc. Lond. B*, **198**, 1–59.
Hubel, D. H., Wiesel, T. N., and Stryker, M. P. (1978). Anatomical demonstration of orientation columns in macaque monkey. *J. comp. Neurol.*, **177**, 361–379.
Jamar, J. H. T., Kwakman, L.F.Tz., and Koenderink, J. J. (1984). The sensitivity of the peripheral visual system to amplitude-modulation and frequency-modulation of sine-wave patterns. *Vision Res.*, **24**, 243–249.
Johnston, A., and Wright, M. J. (1983). Visual motion and cortical velocity. *Nature*, **304**, 436–438.
Johnston, A., and Wright, M. J. (1985). Lower thresholds of motion for gratings as a function of eccentricity and contrast *Vision Res.*, **25**, 179–185
Koenderink, J. J., Bouman, M. A., Bueno de Mesquita, A. E., and Slappendel, S. (1978). Perimetry of contrast detection thresholds of moving sine wave patterns. I-III. *J. Opt. Soc. Amer.*, **68**, 845–860.
Lee, B. B., Elepfandt, A., and Virsu, V. (1981a). Phase of responses to moving sinusoidal gratings in cells of cat retina and lateral geniculate nucleus. *J. Neurophysiol.*, **45**, 807–817.
Lee, B. B., Elepfandt, A., and Virsu, V. (1981b). Phase of responses to sinusoidal gratings of simple cells in cat striate cortex. *J. Neurophysiol.*, **45**, 818–828.
Marr, D., and Ullmann, S. (1981). Directional selectivity and its use in early visual processing. *Proc. Roy. Soc., B.*, **211**, 151–180.
Payne, B. R., and Berman, N. (1983). Functional organisation of neurons in cat striate cortex; variations in preferred orientation and orientation selectivity with receptive field type, ocular dominance and location in the visual field map. *J. Neurophysiol.*, **49**, 1051–1072.
Reichardt, W. E. (1961). Autocorrelation, a principle for the evaluation of sensory information by the central nervous system. In *Sensory Communication*, (Ed. W. A. Rosenblith), MIT Press, Cambridge, Mass.
Robson, J. G., and Graham, N. (1981). Probability summation and regional variation in sensitivity across the visual field. *Vision Res.*, **21**, 409–418.
Rovamo, J., and Virsu, V. (1979). An estimation and application of the human cortical magnification factor. *Exp. Brain Res.*, **37**, 495–510.
Stephenson, C., and Braddick, O. J. (1983). Discrimination of relative spatial phase in fovea and periphery. *Invest. Ophthal. Vis. Sci.*, **24**, 146.
Van Essen, D. C., and Maunsell, J. H. R. (1983). Hierarchical organisation and functional streams in the visual cortex. *Trends Neurosci.*, **6**, 370–375.
Van Essen, D. C., Newsome, W. T. and Maunsell, J. H. R. (1984). The visual field representation in striate cortex of the macaque monkey: asymmetries, anisotropies and individual variability. *Vision Res.*, **24**, 429–448.

Virsu, V., and Rovamo, J. (1979). Visual resolution, contrast sensitivity and the cortical magnification factor. *Exp. Brain Res.*, **37**, 475–494.
Virsu, V., Rovamo, J., Laurinen, P., and Nasanen, R. (1982). Temporal contrast sensitivity and cortical magnification. *Vision Res.*, **22**, 1211–1217.
Watt, R. J., and Morgan, M. J. (1983). The recognition and representation of edge blur: evidence for spatial primitives in human vision. *Vision Res.*, **12**, 1465–1477.
Westheimer, G. (1978). Spatial phase sensitivity for sinusoidal grating targets. *Vision Res.*, **18**, 1073–1074.
Westheimer, G. (1982). The spatial grain of the perifoveal visual field. *Vision Res.*, **22**, 157–162 (1982).
Wright, M. J., and Johnston, A. (1983a). Spatiotemporal contrast sensitivity and visual field locus. *Vision Res.*, **23**, 983–989.
Wright, M. J., and Johnston, A. (1983b). Uniformity of motion aftereffects (MAE) and lower threshold of motion (LTM) in peripheral and central visual fields. *Perception*, **12**, A25.
Wright, M. J., and Johnston, A. (1985). The relationship of displacement thresholds for oscillating gratings to cortical magnification, spatiotemporal frequency and contrast. *Vision Res.*, **25**, 187–193.

Note added in proof

It now seems certain that the estimate of *M* used by Rovamo *et al.* (1979) and ourselves, reflects ganglion cell density and not actual cortical magnification which is higher at the fovea. Levi *et al.* (1985) argue that the former measure predicts acuity and contrast sensitivity, and the latter predicts vernier acuity and phase discrimination. There must be an unidentified *cortical* correlate of ganglion cell density, other than cortical magnification, since grating contrast sensitivity and motion thresholds depend on cortical mechanisms.

Reference

Levi, D. M., Klein, S. A., and Aitsebaomo, A. P. (1985). *Vision Res.*, **25**, 963–977.

Models of the Visual Cortex
Edited by D. Rose and V. G. Dobson

CHAPTER 24

Parallel processing of color-contrast detectors in the visual cortex

P. Gouras
Columbia University, Department of Ophthalmology, 630 West 168 Street, New York, 10032, USA

Color vision is composed of several relatively independent visual sensations, called hue, saturation and brightness. These sensations are formed by comparing the outputs of three spectrally different sets of cone photoreceptors, which subserve in parallel contiguous regions of the retina. These three sensations provide all the visual clues to the detection of objects and their movements in a two (monocular)- and three (binocular)-dimensional space. All vision depends upon color vision, so defined.

Hue, i.e. blueness, green-ness, yellowness, redness and certain admixtures of these basic hues, depends more on the spectral and less on the energy gradients across the retinal image. Saturation, i.e. whiteness, grayness or blackness, depends on both the spectral and energy gradients in this image. Brightness depends mainly on the energy gradients in an image. By combining different gradations of these three sensations more than a million different colors can be visualized. In most visual images spectral gradients tend to be greater than energy gradients and as a consequence hue and saturation often play a more important role than brightness (Gouras, 1984).

In the determination of these three visual sensations, two basic operations seem to be performed. One involves a measure of the relative outputs of the three cone mechanisms within the contours of an object, which can be considered a local process. The second involves a measure of how this information differs across the contours of an object and its background, which can be considered a more global or border-contrast phenomenon. An object is seen by virtue of gradients of wavelength and/or energy of light at its borders with its background.

Of the two operations the first or local one appears to play more of a role

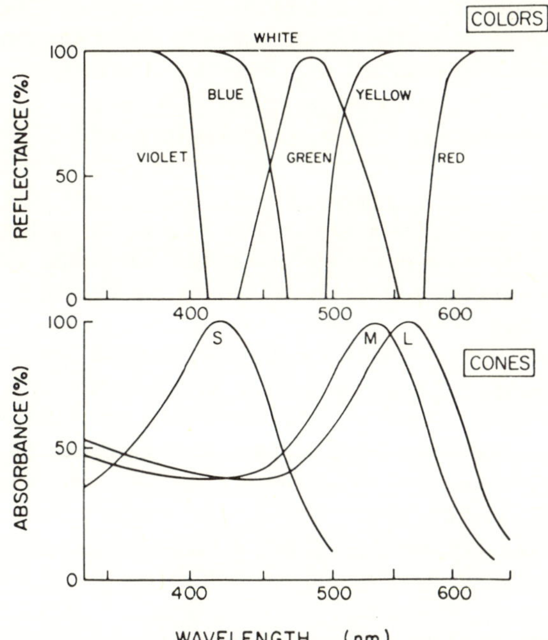

FIGURE 1 This figure shows (bottom) the absorption spectra of the three cone mechanisms of human color vision (adapted from Dartnall, Bowmaker and Mollon, 1983) and (top) the spectral reflectances associated with specific impressions of color (hue). The abscissa is the wavelength of radiation in nanometers (nm); this abscissa scale is in the fourth root of wavelength, an empirical factor for the abscissa scale which makes the cone absorption spectra have identical shapes; it also gives a striking symmetry to the spectral characteristics of the major hues.

in hue than in the sensation of brightness or saturation. The hue of an object can usually be predicted by the spectral reflectance (or transmission) of the light coming from within an object's boundaries; this enables a measurement of how it will be absorbed by each of the three cone mechanisms of human vision (Fig. 1). Nevertheless, the gradients of light at the border of an object with its background can also influence the hue of the object as exemplified by simultaneous color contrast (Land, 1977). In the determination of saturation, information must also be available about the first or local operation to establish whether an object can be white, gray or black, but whether it appears white or gray or black is determined entirely by background contrast. In the determination of brightness, background (or border) contrast plays virtually the entire role. There seems to be a descending order of importance in which gradients across the borders of an object with its background influence brightness, saturation and hue. In all three sensations, however, border contrast plays a role.

There is considerable evidence that all vision occurs by the movement of borders of contrast across the retinal mosaic by the continuous micronystagmoid movements of the normal eye. When these movements are eliminated by stabilization of the image on the retina virtually all vision is lost (Kelly, 1983). I have assumed that such border-contrast activity forms the major driving force for all cortical neurons and consequently all conscious vision.

Figure 1 (top) illustrates how the local reflectances of objects will, in general, predict their hue by virtue of how the light they reflect (or transmit) is absorbed in different proportions by the three basic cone mechanisms (Fig. 1, bottom). Objects which reflect only the long-wavelength *half* of the visible spectrum will tend to appear yellow. Objects which reflect only the short-wavelength *half* of the visible spectrum will tend to appear blue. Objects which reflect the entire visible spectrum to a similar extent will tend to appear white, gray or black. This was probably the first and most important stage in color vision (Gouras and Eggers, 1983). It is performed by comparing the outputs of the cone mechanism subserving the short-wavelength half of the spectrum (S cones) with those subserving the longer wavelength half of the spectrum (M and L cones). 'In 2 per cent' of human males and in cats, rabbits and probably other mammals, there is only one cone mechanism subserving the longer wavelength half of the spectrum, producing a divariant form of cone vision. Divariant cone vision is sufficient for the contrast which forms the sensations of blue, yellow and white. For organisms with more extensive diurnal vision, the longer wavelength half of the spectrum is further subdivided into reds and greens by employing a comparison between the M and L cones in a trivariant form of cone vision. Objects which reflect light that is absorbed more by the M than the L cones will appear greenish. Objects which reflect light that is absorbed more by the L than the M cones will appear reddish. Because the absorption spectra of the M and L cones cross again in the short-wavelength region of the spectrum, redness reappears in objects which preferentially reflect light of very short wavelengths. The reemergence of redness in the short wavelength end of the visible spectrum appears to be due to an additional facilitation of the L-cone mechanism by the S cones (Gouras, 1984). These two separate comparisons of the outputs of the three cone mechanisms in normal human and primate vision appear to occur independently of one another.

Figure 2 shows how the signals from each cone type are processed by parallel channels of bipolar cells. In this scheme each neural unit is tagged by the cone mechanism it subserves in the center of its receptive field. A plus (+) signifies that the unit is excited by light being absorbed by this central cone mechanism; a minus (−) signifies that the unit is excited by the absence or removal of light that can be absorbed by this central cone mechanism. These units can be described as on-center and off-center respectively. By this terminology the cones, themselves, are off-center units since they are excited by darkness (Tomita, 1970).

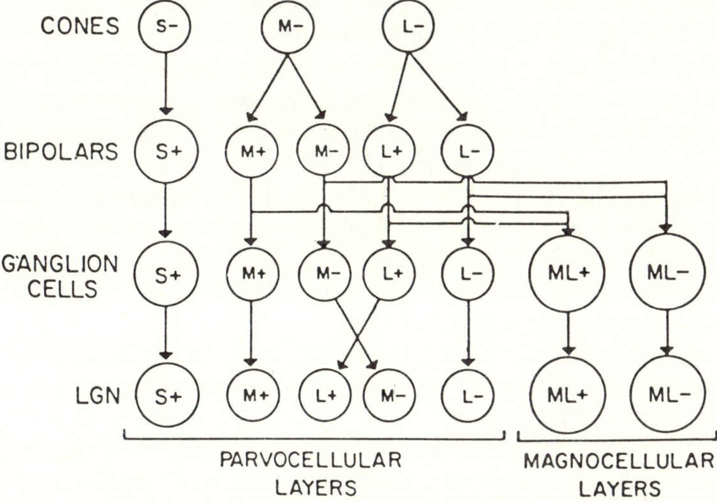

FIGURE 2 The probable flow of receptive field center signals from the cones through bipolar cells to ganglion cells and lateral geniculate cells. A plus (+) signifies an on-center and a minus (−) signifies an off-center cell. Except for the cone on-center bipolar synapse, all of the synaptic interactions are considered to be excitatory. Antagonistic (inhibitory) interactions between these parallel channels mediated mainly by horizontal, amacrine and perhaps geniculate interneurons are not illustrated. The M- and L-center cells in the parvocellular layers of the LGN are considered to have cone-opponent concentric antagonistic surrounds, i.e. L for M centers and M for L centers, which are larger than the center mechanism (type I cells of Wiesel and Hubel, 1966).

At the bipolar cell level one of the most significant changes is the creation of two separate neural channels from each M and L cone, one on-center and an off-center bipolar, a phenomenon characteristic of all vertebrate retinas (Werblin and Dowling, 1969; Kaneko, 1970). The S-cone bipolar system in primates is considered to have only on-center bipolars, with larger receptive fields than those subserving the L- and M-cone mechanisms (Gouras, 1968; DeMonasterio and Gouras, 1975; Malpeli and Schiller, 1978; Gouras and Zrenner, 1979b, 1981a, b, 1983).

At the ganglion cell level in primates, one of the most significant changes is the emergence of two parallel neural channels subserving the M-and L-cone mechanisms. These two channels have been called tonic and phasic ganglion cell systems and are undoubtedly closely related to the X and Y ganglion cell systems of cat retina (Enroth-Cugell and Robson, 1966). The unique functional difference between these two systems is that in the phasic system, both M and L cones contribute to the receptive field center response whereas in the tonic system only *one* cone mechanism contributes to the center response. The S-cone mechanism also appears to utilize a tonic system of retinal ganglion cells. These seven parallel channels form the principal input to the striate cortex where it is presumed that circuits exist which abstract hue, saturation and brightness sensations from this input.

All, or almost all, of these retinogeniculate channels, however, are basically sensitive to energy contrast, and consequently brightness contrast. A bright spot filling the center of the receptive field of each on-center cell will be more effective than a large spot of the same brightness spreading out into antagonist regions of the cell's receptive field; correspondingly, a dark spot filling the center of the receptive field of each off-center cell will be more effective than a larger spot. Spectral contrast across contours upon which hue and saturation depend does not appear to be detected by any *one* channel at this level of the visual system.

The striate cortex undoubtedly uses this input to abstract hue and saturation in addition to brightness contrast. An important clue to how this may occur comes from the reports of red/green hue contrast cells in the striate (Michael, 1978a–c, 1979) and parastriate cortex (Gouras and Kruger, 1979). These cells appear to detect the contrast between the L- and M-cone mechanisms across a contour. Michael (1981) has shown that these cells are located in columns of concentric, simple, complex and hypercomplex L/M contrast cells in the striate cortex analogous to similar columns of brightness contrast cells in cat and monkey striate cortex (Hubel and Wiesel, 1983). There has been no comparable evidence of columns selective for S- versus M- and L-cone contrast or for saturation contrast at the present time.

I am assuming the brightness contrast, red/green contrast, yellow/blue contrast and saturation contrast cells exist in the striate cortex and that they are all formed primarily by the tonic (parvocellular) input of the lateral geniculate nucleus. The existence of brightness contrast and red/green double-opponent cells in the striate cortex is well established (Hubel and Wiesel, 1968; Michael, 1978a); the existence of yellow/blue and saturation contrast cells is more speculative but a logical extension of what has been found. The assumption that all hue, saturation and brightness is determined by the tonic or parvocellular layers is also a speculation, although evidence exists that favors such an interpretation. First, the tonic cell system dominates the foveal region of the retina where these sensations are most highly developed. Second, one finds that most visual cortical cells show evidence of L- and M-cone opponent interactions (Kruger and Gouras, 1980) which is the hallmark of the tonic cell system. Figure 3 provides a scheme for reorganizing the retinogeniculate input from the parvocellular layers in a way that would create separate orientation-selective units sensitive to hue, saturation and brightness contrast. Figure 3 (top) shows the elementary forms of cone contrasts that are required to form such brightness, hue and saturation contrast detectors.

There is a high-resolution brightness contrast detecting system in the fovea which senses fine border contrast (Boynton and Kaiser, 1968) and which

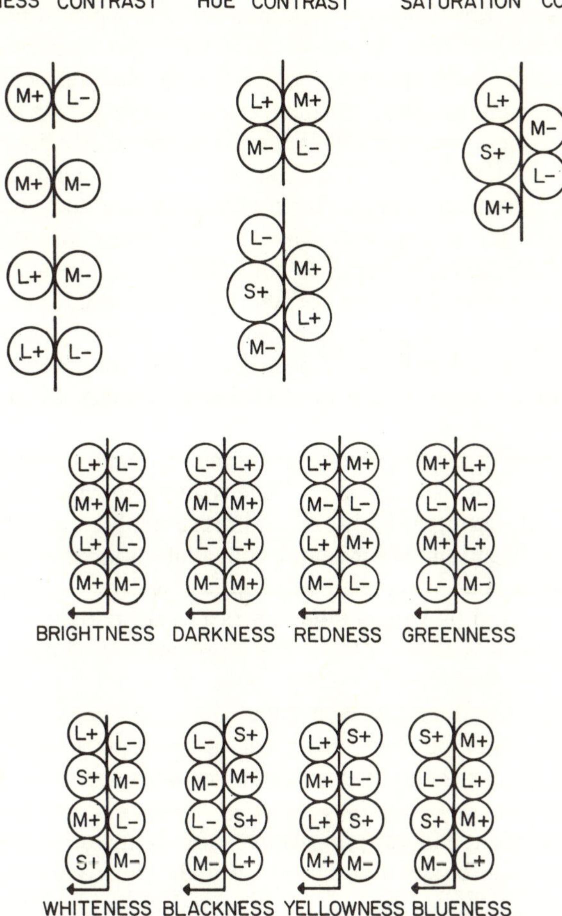

Figure 3 This figure shows (top) how the geniculate inputs of Fig. 2 can be used to form brightness contrast detectors between L and/or M cones (left column), hue contrast detectors between L and M cones (middle column, top) and between S and L with M cones (middle column, bottom), and a saturation contrast detector (right column). The two lower rows show how these spatial contrast detectors (top) can be used to form orientation-selecive edge detectors. Orientation-selective slit detectors would require a prior stage of concentrically organized color-contrast cells (top); edge detectors need not have this requirement. The arrow points to that side of the edge which has the appearance of the sensation designated below each particular contrast detector. The vertical lines represent a spatial border and the circles represent the receptive field centers of on-center (+) and off-center (−) cells.

does not receive an input from the S-cone mechanism (Tansley and Boynton, 1976). This system's response can be minimized by reducing the effective energy gradient for the L- and M-cone mechanisms across a contour. Figure

3 (top, left column) illustrates the potential ways in which units sensitive to brightness contrast between L and/or M cones could be established. By having an on-center geniculate unit of one cone and an off-center unit from an adjacent cone excite the same cortical cell, one could create a unit sensitive to brightness contrast over a gradient formed between two foveal cones.

Figure 3 (top, middle column) illustrates the way in which the same retino-geniculate inputs could be used to form red/green hue and yellow/blue hue contrast detectors. By having an L on-center geniculate unit subserving one side and an M on-center geniculate unit subserving the other side of a potential contour excite the same cortical unit, one could create a detector for L/M or so-called red/green contrast. The spatial resolution of this detector would be less than the former brightness detectors because it would not be able to take advantage of contrast across adjacent M-to-M or L-to-L cones.

The lower unit (Fig. 3, top, middle column) shows how a yellow/blue hue contrast detector could be formed. This requires having an S on-center geniculate unit subserving at least one and presumably more S cones on one side and both an L and an M on-center geniculate unit subserving the other side of a potential contour excite the same cortical unit. This cortical unit would be most excited by yellow/blue contrast. The spatial resolution of this detector would be less than both of the previous two classes of detectors since it involves S on-center geniculate cells which tend to have relatively large receptive field centers. Off-center units have also been included in these hue contrast detectors but they may not be as critical, since the S-cone mechanism contributes strongly to hue contrast but appears to use few or no off-center cells.

Figure 3 (top, right column) illustrates how a hypothetical unit which would desaturate hue contrast could be formed. This requires having S, M and L on-center geniculate units on one side and M and L off-center geniculate units subserving the other side of a potential contour excite the same cortical unit. A simple way to build such a unit is to have the brightness-contrast detector (Fig. 3, top, left column) and the S-cone mechanism feed the same cortical unit. In order to make this unit uniquely sensitive to whiteness, it would have to be a logical integrator, excited *only* when both inputs were excited. This cortical unit would be most sensitive to white/black contrast. If the unit were maximally excited, it would induce the sensation of whiteness on one side of a contour and consequently desaturate that side towards whiteness; it would induce blackness on the other side of a contour and desaturate that side towards blackness, i.e. it would make a yellow whitish on one side and a yellow brownish on the other side of the contour. When this unit is *minimally* excited, any hue contrast across the contour would be maximally saturated. Because it uniquely depends upon an S on-center geniculate unit, the spatial resolution of the cortical unit would also be poor. Such units have not been reported in the striate cortex. All these units have been arranged across a border of contrast; they could be organized concentrically by involving more units in the appropriate geometry.

Figure 3 (bottom) illustrates how this previous set of independent brightness, hue and saturation detectors could be used to form independent systems of orientation-selective contrast detectors. The orientation-selective units illustrated are designed to detect an appropriately oriented edge. Another separate system of slit detectors could be formed by the appropriate use of concentrically organized hue, saturation and brightness contrast detectors as described earlier. The principles of organization, however, are otherwise similar for the slit and edge detectors. The edge detectors could be formed directly from the geniculate input; slit detectors would seem to require a cortical stage of concentrically organized hue and saturation contrast detectors.

The edge detectors have the advantage of requiring less units for illustration and bring up the problem of polarity. Somehow visual cortex must be able to keep track of the polarity of a particular form of contrast. With orientation selectivity, the information is available to know that there is a contour of brightness and/or redness and/or yellowness and/or saturation on one side and darkness and/or green-ness and/or blueness and/or unsaturation respectively on the other side of a contour. By having all the orientation-selective contrast detectors independent one can have many combinations and their corresponding gradations on the same side of any contour. What cannot occur on the same side of any contour are the contrasting sensations—of red and green, yellow and blue, saturation and unsaturation or bright and dark. From these four contrast detectors all colors are created by basing them on different response gradations. By this system a distribution of excited cortical cells can be formed by any or all orientation-selective slit and edge contrast detectors working independently. The particular collections of orientation-selective units activated will define the form of an object and independently its color. Each orientation-selective column of cells will be reduplicated for each of the contrast-detecting systems.

The ultimate subjective sensation of brightness may not only depend upon the brightness contrast detectors depicted in Fig. 3 which involve only the L- and M-cone mechanisms. All the contrast detectors working in concert may contribute to the ultimate sensation of the brightness of an object or scene and thereby involve the S-cone mechanism as well.

The entire system would require two sets of detectors for each of the four paired forms of contrast, giving eight separate slit and edge detectors for all discriminable orientations. The slit detectors are paired for what is a slit and what is background; the edge detectors are paired for polarity or sidedness, i. e. right or left or up and down. Some of the contrast detectors, especially those dependent upon the S cones, would have an inherently lower spatial resolution and are therefore probably represented at a lower concentration than those dependent only on the L and M cones. L/M hue contrast detectors would also have a poorer spatial resolution than L/M brightness contrast detectors since the former cannot form any contrast between adjacent cones of the same type. There is simply a dilution in the number of possible cone combinations. Another factor that can influence the grain of L/M hue contrast detectors is that most retinogeniculate L and M units have a concentrically organized antagonistic surround from the opponent-cone mechanism, which extends beyond the L or M center. By making the centres contiguous as in Fig. 3, some weakening of their response will occur, because of partial activation of the opponent surrounds. Making the surrounds rather than the centers contiguous would eliminate this phenomenon but sacrifice spatial resolution. Both problems do no exist when the same cells are used for brightness contrast. Therefore, there should be more units sensitive to L/M brightness contrast and these should be preferentially affected with stimuli of high spatial frequencies. With stimuli of lower spatial frequencies, hue contrast and saturation contrast detectors will become relatively more excited. If the latter suppressed the former, as suggested by recent psychophysical results (DeValois and Switkes, 1983), this would tend to enhance hue and saturation contrast at lower spatial frequencies. This is an intuitively useful thing to do because spectral contrast would seem to be a more reliable cue to the presence of a significant contour than brightness contrast. Brightness contrast can vary enormously by the position of an object's surface relative to the illuminant. Large brightness gradients may often exist within the confines of the object whereas the spectral reflectance of the object's surface will be virtually independent of these brightness gradients. In general we perceive an object by its hue and saturation and disregard the brightness gradients across its surface.

The remapping of visual space in prestriate areas of the visual cortex is an intriguing phenomenon that so far defies explanation. There is general agreement that the size of the receptive fields of units tends to increase with distance from the primary projection area in the striate cortex. Therefore whenever stimuli are relatively small, there is little likelihood that much activity can be excited in visual areas beyond the striate cortex. As the stimuli are enlarged or elongated, activity beyond the striate cortex must ultimately come into play. It is interesting in this regard that the discrimination of color differences increases over visual areas which are enormous compared to the receptive field sizes of striate cortical cells (Brown, 1952). This could arise from the possibility that antagonistic interactions between cells and cone mechanisms increase progressively with stimulus size (Kruger and Gouras, 1980) and the entire impact of this cannot be realized in the striate cortex because here receptive field sizes are too small. By providing additional systems of cells with larger receptive field sizes that can be brought into play whenever the object of interest is large enough, the brain can take advantage of cells with an even more spectrally restrictive inputs, i.e. stronger cone-opponent interactions. This would increase the nuances of potential color contrast and the universe of possible colors. Doing this initially, at the retinogeniculate or striate cortical level, would either require larger numbers of cells subserving in parallel the same areas of visual space or require that the limited numbers of cells be more spectrally restrictive, which sacrifices potential brightness contrast.

REFERENCES

Boynton, R. M., and Kaiser, P. K. (1968). Vision: the additivity law made to work for heterochromatic photometry with bipartite fields. *Science*, **161**, 366–368.

Brown, W. R. J. (1952). The effect of field size and chromatic surroundings on colour discrimination. *J. Opt. Soc. Amer.*, **42**, 837.

Dartnall, H. J. A., Bowmaker, J. K., and Mollon, J. D. (1983). Microspectrophotometry of human photoreceptors. In *Color Vision: Physiology and Psychophysics* (Eds. J. D. Mollon and L. T. Sharpe), Academic Press, London, pp. 69–80.

DeMonasterio, F. M., and Gouras, P. (1975). Functional properties of ganglion cells of the rhesus monkey. *J. Physiol.*, **251**, 167–196.

Devalois, K. K., and Switkes, E. (1983). Simultaneous masking interactions between chromatic and luminance gratings. *J. Opt. Soc. Amer.*, **73**, 11–18.

Enroth-Cugell, C., and Robson, J. G. (1966). The contrast sensitivity of retinal ganglion cells of the cat. *J. Physiol.*, **187**, 517–552.

Gouras, P. (1968). Identification of cone mechanisms in monkey ganglion cells. *J. Physiol.*, **199**, 533–547.

Gouras, P. (1984). Color vision. Chap 8 In *Progress in Retinal Research* Vol. 3 (Eds. N. N. Osborne and G. J. Chader), Pergamon Press, Oxford.

Gouras, P., and Eggers, H. M. (1983). Responses of primate retinal ganglion cells to moving spectral contrast. *Vis. Res.,* **23**, 1175–1182.

Gouras, P., and Kruger, J. (1979). Responses of cells in foveal visual cortex of the monkey to pure color contrast. *J. Neurophysiol.*, **42**, 850–860.

Gouras, P., and Zrenner, E. (1979a). Enhancement of luminance flicker by color-opponent mechanisms. *Science*, **205**, 587–589.

Gouras, P., and Zrenner, E. (1979b). The blue sensitive cone system. *Excerpta Medica, Int. Cong. Ser.*, **450/1**, 379–384.

Gouras, P., and Zrenner, E. (1981a). Color coding in the primate retina. *Vis. Res.*, **21**, 1591–1598.

Gouras, P., and Zrenner, E. (1981b). Color vision: a review from a neurophysiological perspective. In *Progress in Sensory Physiology* (Eds. H. Autrum, D. Ottoson and E. R. Perl), Springer, Berlin, pp. 139–179.

Gouras, P., and Zrenner, E. (1983). Transient and steady state responses of ganglion cells mediating the signals of short wave sensitive cones. In *Color Vision: Physiology and Psychophysics* (Eds. J. D. Mollon and L. T. Sharpe), Academic Press, London, pp. 515–526.

Hubel, D. H., and Wiesel, T. N. (1968). Receptive fields and functional architecture of monkey striate cortex. *J. Physiol.*, **195**, 215–243.

Hubel, D. H., and Wiesel, T. N. (1983). In *Les Prix Nobel en 1981*, Nobel Foundation, Stockholm, Sweden.

Kaneko, A. (1970). Physiological and morphological identification of horizontal, bipolar and amacrine cells in goldfish retina. *J. Physiol.*, **207**, 623–633.
Kelly, D. H. (1983). Spatiotemporal variation of chromatic and achromatic contrast thresholds. *J. Opt. Soc. Amer.*, **73**, 742.
Kruger, J., and Gouras, P. (1980). Spectral selectivity of cells and its dependence on slit length in monkey visual cortex. *J. Neurophysiol.*, **43**, 1055–1069.
Land, E. H. (1977). The retinex theory of color vision. *Sci. Amer.*, **237**, 108–128.
Malpeli, J. G., and Schiller, Ph.H. (1978). Lack of blue OFF-center cells in the visual system of the monkey. *Brain Res.*, **141**, 385–389.
Michael, C. R. (1978a). Color vision mechanisms in monkey striate cortex: dual opponent cells with concentric receptive fields. *J. Neurophysiol.*, **41**, 572–588.
Michael, C. R. (1978b). Color vision mechanisms in monkey striate cortex: simple cells with dual opponent color receptive fields. *J. Neurophysiol.*, **41**, 1233–1249.
Michael, C. R. (1978c). Color-sensitive complex cells in monkey striate cortex. *J. Neurophysiol.*, **41**, 1250–1266.
Michael, C. R. (1979). Color-sensitive hypercomplex cells in monkey striate cortex. *J. Neurophysiol.*, **42**, 726–744.
Michael, C. R. (1981). Columnar organization of color cells in monkey's striate cortex. *J. Neurophysiol.*, **46**, 587–604.
Tansley, B. W., and Boynton, R. M. (1976). A line, not a space, represents visual distinctness of borders formed by different colors. *Science*, **191**, 954–957.
Tomita, T. (1970). Electrical activity of vertebrate photoreceptors. *Quart. Rev. Biophys.*, **3**, 179–222.
Werblin, F. S., and Dowling, J. E. (1969). Organization of the retina of the mudpuppy, *Necturus maculosus*. II. Intracellular recording. *J. Neurophysiol.*, **32**, 339–355.
Wiesel, T. N., and Hubel, D. H. (1966). Spatial and chromatic interactions in the lateral geniculate body of the rhesus monkey. *J. Neurophysiol.*, **29**, 1115–1156.

Models of the Visual Cortex
Edited by D. Rose and V. G. Dobson

CHAPTER 25

Contribution of striate cortical cells to pattern, movement and colour processing

J.J. KULIKOWSKI and I. J. MURRAY
Ophthalmic Optics Department, Visual Sciences Laboratory (Jackson), UMIST, P.O. Box 88, Manchester M60 1QD, UK

INTRODUCTION

Twenty-five years ago Hubel and Wiesel (1959) demonstrated that most cells in the cat striate cortex respond to contours of specific orientation. Since then orientation, contrast and hierarchical processing have dominated our conceptualization of visual coding in the primary visual area. Recent experiments described by Livingstone and Hubel (1983) force a major revision of this model: the macaque striate cortex (V1) has separate parallel projections to the secondary visual area (V2) from two distinct neuronal groups, i.e. orientation-selective (mostly non-chromatically coded) cells and non-orientation-selective (mostly chromatically coded) cells. The significance of this discovery lies in the implication that the monkey visual cortex has developed a system of separate parallel processing for the extraction of colour information, presumably because fully developed trichromatic colour vision is of particular survival value to monkeys but is not so important for subprimate mammals..

Parallel processing of visual information in mammalian vision has already attracted attention following the finding of two types of ganglion cell in the cat retina, called X (linear) and Y (non-linear) (Enroth-Cugell and Robson, 1966), alternatively described as sustained and transient respectively (Cleland, Dubin and Levick, 1971). These cell types form separate (X and Y) systems projecting along the geniculostriate pathway. Y-cells in the cat partly by-pass the primary visual area (V1) by projecting to V2), whereas in the monkey an equivalent by-passing projection is much less distinct (Wong-Riley, 1974).

Thus the striate cortex (V1) in primates is the main relay and distributor of visual signals. In other words, V1 is an essential moderator of colour and pattern vision and its damage leads to very severe visual impairment in Man and monkey (see reviews by Weiskrantz, 1972; Weiskrantz *et al.*, 1974; Cowey, 1982; Ruddock, 1982). V1 is also capable of efficiently processing information about movement of objects, although not exclusively, since even after striate lesions and severence of the optic radiation, some residual perception, especially of flicker and movement, remains.

Conversely, a patient with intact striate cortex but damaged extrastriate visual areas may have normal visual resolution for gratings (Abadi, Kulikowski and Meudell, 1981), but suffer from severe agnosia, being unable to recognize either familiar shapes, e.g. faces, or perform any complex perceptual task.

It is hoped that the models suggested may lead to the design of neurophysiological experiments to test ways in which the visual 'neural image' is resynthesized (Barlow, 1981). So far, the best-understood process is the analysis of spatial patterns carried out by simple cells, which comprise the most common class (at least 50 per cent.) of cells in the striate cortex of the cat and monkey.

STRIATE CORTICAL CELLS: GENERAL COMMENTS

Hubel and Wiesel (1962) described cells in the cat striate cortex preferring contours of specific orientations. They classified these cells into simple (S), which obey the rule of spatial summation, and complex, which do not. Simple cells have receptive fields with clear antagonistic subregions, ON (+) and OFF(−), whereas complex cells have mixed receptive fields within which both light and dark bars produce responses. In spite of many refinements and new subdivisions this classification still holds (also in the monkey cortex), and generalizations can now be made as to its functional significance: simple cells participate in linear analysis of contrast and complex cells monitor novelty, signalling movement of contours and background texture, etc.

Figure 1 illustrates receptive fields and some response characteristics of most conspicuous cell types, among which simple (S) and complex cells (subdivided into B- and C-cells by Henry, 1977) have certain common features as compared with concentric cells. Note that discrete areas of the visual field are covered by receptive fields of all types of cell. With the exception of concentric cells, most cells respond best only to specifically oriented contours if these contours move in a particular (optimal) direction. The latter property of direction preference is illustrated in Fig. 1b by assuming that for an 'average' cell the response in the non-preferred direction (opposite to the optimal) is half that obtained in the preferred direction (some cells have strong directional properties, others none at all). This directional bias may be the basis for signalling the direction of real movement,

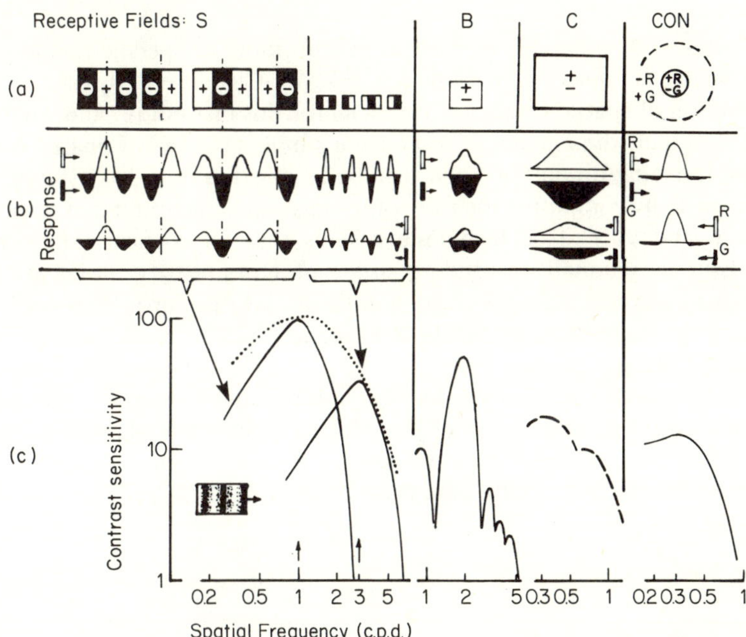

FIGURE 1 Schematic representation of typical responses of the striate cortical cells (representing 5° below the visual axis) in the macaque monkey to various stimuli: (a) Receptive field outlines obtained conventionally with stationary flashing lights; + and − represent ON and OFF responses respectively. (b) Receptive field profiles of the different cell types revealed by light and dark bars moving at optimal velocity; the upper row depicts the response profiles to movement in the preferred direction, the lower row to the opposite direction. (c) Contrast sensitivity as a function of spatial frequency of a grating drifting in the preferred direction.

The left column depicts responses of S-cells grouped in two sets of quadruplets within which the cells have symmetrical and antisymmetrical receptive fields (a, b). S-cells in each quadruplet have the same spatial frequency tuning curves (c), predictable from the spatial profiles (b): the cells with three times smaller receptive fields in the second quadruplet have three times higher optimal spatial frequency. The dotted line is an envelope of these two tuning curves.

The middle column depicts B-cells with seemingly uniform receptive fields (a). Their response profiles (b), however, show slight residual spatial periodicity which leads to very sharp spatial frequency tuning (c). C-cells do not show such periodicities, nor are their responses to drifting gratings clearly modulated (see dashed line).

The right column illustrates responses of a double cone-opponent concentric cell, responding most vigorously to a red spot or bar (b). Its responses to a red/green grating do not show tuning due to a weak low spatial frequency fall-off.

especially when similarly responding cells are grouped together (Murray, MacCana and Kulikowski, 1983; Payne, Berman and Murphy, 1980).

In general, provided stimulus parameters are carefully optimized, S-cells are distinctive in their linear response characteristics. Perhaps the most stringent test of linearity is the response to a reversing grating at a series of spatial phases. A linear cell exhibits a 'null position' or specific spatial phase (0°) at which no response is obtained (Enroth-Cugell and Robson, 1966). Optimization of spatial frequency is crucial for this procedure, and when this condition is met, the maximum response is obtained at 90° of phase. A non-linear cell, having no null position, can be assessed in terms of its departure from linearity by computing the ratio of its maximum and minimum responses (i.e. at 0 and 90°). This technique reveals that all cortical cells can be categorized in gradually ascending order of non-linearity and a pseudo-continuum emerges: linear S-cells, non-linear S-cells, through B-cells (see Fig. 1) to the most typical (highly non-linear) C-cells (Kulikowski, Bishop and Kato, 1981).

It is important to remember that for non-optimal stimuli S-cells may be non-linear and that some forms of B- and C-cells show elements of linearity. The dual linear/non-linear characteristics of some S-cells and the partial linearity of B-cells is crucially important to the role these cells play in visual processing.

SOME CHARACTERISTICS OF SIMPLE CELLS

Predictability of responses to optimally moving stimuli

The definitive criterion of classification of simple cells is whether their responses to one range of optimized stimuli predict responses to other stimuli (Movshon, Thompson and Tolhurst, 1978a; Kulikowski and Bishop, 1981). The most accurate predictions are obtained with bars, edges and gratings moving at an optimal velocity (see below, equation 1).

Figure 1b and c illustrates typical responses of S-cells to bars and gratings encountered in the striate cortex of cat and rhesus monkey (Kulikowski and Vidyasagar 1982, 1984). The shape of contrast sensitivity characteristics as a function of spatial frequency (tuning curves) is predictable as the Fourier transform of the response profiles obtained with light and dark bars (Fig. 1b; for a further explanation see Kulikowski and Bishop, 1981, and Pollen, Foster and Gaska, Chapter 28 in this volume). Note that S-cells have no spontaneous activity; hence light and dark bars must be used to reveal their ON and OFF subregions.

Association in quadruplets

There are many S-cells tuned to different orientations covering the same region of the visual field (Hubel and Wiesel, 1962, 1968). For each orientation, there are cells with receptive fields of symmetrical and antisymmetrical shapes, as is evident from the arrangements of subregions within which the cell responds to either light ON (+) or OFF(-), but not to both (Fig. 1a). Similar symmetry and antisymmetry occur in response profiles to moving light and dark bars (Fig. 1b) and these profiles clearly correspond to the maps of ON and OFF subregions outlined in Fig. 1a. In addition, the same region of visual field is represented by cells with receptive fields of different sizes (usually with a moderate number of subregions; see Kulikowski and Vidyasagar, 1984), these being 'tuned' to different ranges of spatial frequencies.

Figure 1c is an example of two tuning curves whose optimal spatial frequencies are 1:3. Earlier experiments led to the evaluation of psychophysical analogues (detectors) of S-cell receptive fields (Kulikowski and King-Smith, 1973) and of the harmonic relation between their optimal spatial frequencies (Campbell and Robson, 1968; Robson, 1975). However, these concepts should be credited to Gabor (1946) and his description of the representation of visual stimuli (see also Cowan, 1977; Marcelja, 1980; Kulikowski, Marcelja and Bishop, 1982; and Pollen, Foster and Gaska, Chapter 28 in this volume).

A set of four cells (quadruplet; see Fig. 1b, left) gives a unique pattern of responses, from which it is possible to classify any stimulus as being a light or dark bar (or a light or dark edge —note that since the responses are linear the response to an edge is the integral of the response to a bar; see Kulikowski and Bishop, 1981). Additionally, units with receptive fields of several sizes are needed in order to specify the bar width (or the spread of an edge—focused or blurred); all units respond to narrow bars and sharp edges, whereas only those with coarse receptive fields respond to blurred bars (or edges). It should be emphasized that S-cells in Fig. 1 operate as members of an assembly and not as trigger feature detectors of bars and edges (or gratings), since their individual responses do not show exclusive preferences for these stimuli. Moreover, bars, edges or gratings do not match exactly the spatial profiles of S-cells, so they cannot be regarded as optimal stimuli in each corresponding category. For a symmetrical receptive field a close match would only be achieved by using complex stimuli, e.g. a light bar flanked by two dark bars (of appropriate dimension; see Kulikowski and King-Smith, 1973), whereas for an antisymmetrical field a local edge (i.e. a local negative–positive step change in luminance; see Shapley and Tolhurst, 1973) matches its spatial profile precisely. In fact these stimuli were used in subthreshold summation psychophysical experiments to derive spatial profiles similar to those described by Gabor; the resultant (antisymmetrical/symmetrical) mechanisms were called threshold detectors of lines and edges (Kulikowski and King-Smith, 1973). Such detectors can be used as building blocks to analyse complex shapes (King-Smith and Kulikowski, 1973) or any visual scene.

Simple cells as spatiotemporal filters

In the preceding sections simple cells were described effectively as spatial filters. Now we consider their properties as temporal filters, for two reasons: (a) S-cells, even in their role of spatial filters have to respond to contours which move across the retina due to continuous eye movements; (b) S-cells are capable of signalling both transient events and direction of movement, thereby being potential processors of information about movement (see Murray, MacCana and Kulikowski, 1983).

In both cases it is necessary to know stimulus velocities over which S-cells can respond. Recent experiments (Kulikowski, 1979; Kulikowski and Bishop, 1981; and in preparation) have confirmed that a good estimate of the optimal velocity at which a cell response is maximal is determined from a simple linear model. This model, derived for a psychophysical detector (King-Smith, 1978), consists of a spatial filter of a Gabor type followed by a sequence of single-stage temporal filters, one low-pass (time constant T_1) and the other high-pass (time constant T_2). These time constants can be evaluated respectively from the rising and falling phases of responses (measured in terms of frequency of spikes) to relatively weak stationary flashing stimuli. The optimal velocity (V_0) can be estimated from the equation:

$$V_0 = \frac{1}{2\pi F_0 \sqrt{T_1 T_2}} = \frac{0.16}{F_0 \sqrt{T_1 T_2}} \qquad (1)$$

where F_0 is the optimal spatial frequency which in turn is the reciprocal of the cell spatial period (the width of two antagonistic subregions ON and OFF).

In general, the wider the receptive field subregion the higher the optimal velocity, given equal time constants. Clearly, however, some flexibility is required, as it would be inefficient to have a fixed relationship between receptive field size and optimal velocity. Linear S-cells have similar rise time constants (T_1) of about 30 ms, but the time constants of the falling phase vary between 0.1 and 2.5 s (i.e. twenty-five-fold). Thus variations of V_0 may be about a factor of 5 for similar receptive fields (see equation 1). Further extension of the range of velocities (by more than a factor of 2) is achieved in fast simple cells (Dreher, Leventhal and Hale, 1980), whose time constants (T_1 and T_2) are even shorter.

A reasonable range of operating velocities is essential for Gabor-type processing. Quadruplets of different sizes (Fig. 1b) must respond to similar velocities, thereby allowing the simultaneous processing of a range of spatial frequencies under the same dynamic conditions.

Non-linearities and their functional significance

Of particular importance is that S-cells exhibit strong, usually inhibitory, non-linearity when grossly non-optimal stimuli are introduced. This (probably active) suppressive process may be regarded as a switching mechanism, ensuring that the response is abruptly diminished, rather than decreasing gradually. This rule applies to orientation, line length, spatial frequency and velocity, sharply limiting the response of a given cell at the extremes (skirts) of its response characteristic. As a consequence, responses to complex stimuli (e.g. bars superimposed on fine texture or complex bars consisting of light and dark segments) are usually weaker than predicted by linear analysis (Hammond and MacKay, 1977, 1983). However, non-linearities are not always inhibitory (Creutzfeldt, Kuhnt and Benevento, 1974). Recent experiments (Kulikowski and Carden, in preparation) suggest the possibility of cooperative neural networks whose activity results in the enhancement of cell response to certain specific stimuli. S-cells are, therefore, adjustable spatial filters, by virtue of their flexible non-linearities (inhibitory to some complex stimuli and disinhibitory to others). Such networks may participate in resynthesis of the visual neural image.

S-cells also exhibit 'amplitude' non-linearity. 'Linearity' as described above does not imply direct proportionality of S-cell responses over the whole (perceptual) range of contrast. In fact the response range of an individual cell is limited (Tolhurst *et al.*, 1980; Albrecht and Hamilton, 1982; Kulikowski and Vidyasagar, 1983). Hence a series of S-cells must operate in cascade, thereby covering the complete range of linear contrast sensation (Kulikowski, 1976) in a piecewise manner.

There are also temporal non-linearities evident when temporal frequency characteristics obtained empirically are compared with those predicted from abrupt (pulse) presentations of high contrast stimuli (Tolhurst *et al.*, 1980). Likewise abrupt, square-wave, contrast reversals reveal more S-cells without a null position than sinusoidal temporal modulation does.

COMPLEX FAMILY CELLS: B and C

Complex cells are characteric in responding to a broad range of stimuli. Figure 1a and b shows that both types of cell (B and C) are apparently independent of contrast polarity and do not discriminate between onset and offset. However, grating stimuli reveal different classes of complex cells.

The most commonly recognized are C-cells, which exhibit marked spontaneous activity and, compared with B-cells, only a poorly modulated response to drifting gratings (some spatial frequency preference is evident; see the dashed lines in Fig. 1c). Their ratio of minimum/maximum responses to reversing gratings at phases 0 and +90° are about 0.8, thus revealing high non-linearity.

Since these cells also give a vigorous response to visual noise (Hammon and MacKay, 1977) they could be regarded as non-selective detectors of any visual event or motion (information of this kind would be needed in the superior colliculus to which some C-cells project; see Movshon, Thompson and Tolhurst, 1978b). This may be an oversimplification, since Hammond and Smith (1983) have found that C-cells respond to moving contours not only better than to background noise but also the response components of visual noise are suppressed when both are presented together.

B-cells (Henry, 1977) represent the opposite extreme in the range of complex cell types. They differ from standard C cells by having small receptive fields (Fig. 1a), no spontaneous activity, and preference for slow movement. Notable among B-cells are silent periodic cells (Kulikowski and Bishop, 1982) which respond strongly to moving gratings. These cells have atypical response characteristics. Whilst being highly sensitive and selective for a (usually high) spatial frequency, they may also respond less strongly to a broader range of parameters (fine gratings, checkerboards or visual noise). Cells acting in this way are ideally suited to monitor retinal image quality, in that they will not respond, for example, in the presence of blur. B-cells project to the Clare–Bishop area (Henry, Lund and Harvey, 1978) which has been shown to play a role in accommodation (Bando *et al.*, 1981).

The functional properties of B-cells (extraction of spatial regularities from backgrounds) make them also capable of conveying information related to other aspects of vision, e.g. the depth of objects in global stereopsis, and movement (see the discussion in Kulikowski and Bishop, 1982).

CHROMATICALLY CODED (CONCENTRIC) CELLS

Most chromatically coded cells in area V1 have concentric receptive fields and are grouped (Livingstone and Hubel, 1983) in clusters. Their projection to area V2 is clearly separated from that of oriented cells in V1, giving colour processing morphological distinction. In Fig. 1, the group of chromatically coded cells is represented by a double-opponent concentric cell that responds effectively only to a red bar on a green background moving across its excitatory centre, whereas responses to a green bar from its antagonistic surround are weak. Note that this cell has a large excitatory centre and consequently low spatial resolution. Such a cell integrates several inputs from chromatically coded cells in the parvocellular layers of the LGN, each of which may have high resolution (Hicks, Lee and Vidyasagar, 1983). These LGN cells, in turn, are supplied by a distinct group of tonic (sustained) retinal ganglion cells with thin axons (Dreher, Fukada and Rodieck, 1976). This makes colour information transfer particularly vulnerable and therefore of considerable diagnostic importance (King-Smith *et al.*, 1980).

There are also other chromatically coded cells in V1 showing less specificity than the double-opponent cells. In most of these a surround response is even weaker than in double-opponent cells, so that (unlike S-cells) they respond to diffuse flashes of light and to gratings of very low spatial frequencies (Fig. 1c).

However, since not all cells within groups of concentric cells are chromatically coded (Livingstone and Hubel, 1983) we may ask to what extent colour information is handled in complete isolation. The role of non-colour-coded concentric cells is still unclear and future experiments should establish their connections with cells in other cortical areas. One of the possible roles of achromatic concentric cells as contour processors has been discussed by Marr (1982).

CONCLUDING COMMENTS

In this chapter we assert that individual cells in the primate striate cortex (V1) are sensitive to, but not highly selective for, certain features in the visual environment. It appears that the visual cortex relies on many cells to convey information about quite simple stimuli (e.g. bars or edges), whilst a single individual visual neuron may respond to several features of a visual stimulus. S-cells, for example, can extract information about slow movement whilst performing linear analysis of contrast via a Gabor-like system.

Linear analysis of contrast could, of course, be performed in other ways, e.g. by sets of concentric cells. The attraction of the form of information-processing described by Gabor is that it is a compromise between space and spatial frequency analysis and more economical than a system based on concentric cells.

The arrangement of Gabor-like filters into symmetrical and antisymmetrical receptive field profiles means that extraction of contrast information must be confined to one axis. It is tempting to speculate that this characteristic also contributes to the emphasis on orientation specificity in the visual cortex, as compared with the preceding stages, e.g. L.G.N. (Vidyasagar and Urbas, 1982; and Vidyasagar, Chapter 42 in this volume).

The extraction of chromatic information seems to be confined to discretely grouped neurons (Livingstone and Hubel, 1983) which probably relay to equally discrete specific regions in other areas of the visual cortex; the related issues have been lucidly described by Zeki (1978, 1980, 1983).

Such precise localization of a specific visual process is generally uncharacteristic of the striate cortex. Linear S-cells, although having a clear electrophysiological (Mustari, Bullier and Henry, 1982) and morphological (Gilbert and Wiesel, 1979) identity, are well represented throughout V1. Unlike the relatively autonomous colour areas, they are, at least in primates, integrated into the overall functional organization of the primary visual cortex.

The neuronal network for extracting movement is even less specifically organized than for pattern, and may in this sense be regarded as the opposite extreme to colour. S-cells, B-cells and C-cells are all well equipped to convey information about temporal changes to such an extent that a certain amount of oversampling occurs. This incidentally may contribute to the characteristic robustness of movement detection when the visual cortex sustains damage.

REFERENCES

Abadi, R. V., Kulikowski, J. J. and Meudell, P. (1981). Visual performance in a case of visual agnosia. In *Functional Recovery from Brain Damage* (Eds. M. W. van Hof and G. Mohn), Elsevier, Amsterdam.

Albrecht, D. G., and Hamilton, D. B. (1982). Striate cortex of monkey and cat: contrast response function. *J. Neurophysiol.*, **48**, 217–237.

Bando, T., Tsukuda, K., Yamamoto, N., Maede, J., and Tsukahara, N. (1981). Cortical neurons in and around the Clare–Bishop area related with lens accommodation in the cat. *Brain Res.*, **225**, 195–199.

Barlow, H. B. (1981). Critical limiting factors in the design of the eye and visual cortex. *Proc. Roy. Soc., B,* **212**, 1–34.

Campbell, F. W., and Robson, J. G. (1968). Application of Fourier analysis to the visibility of gratings. *J. Physiol.*, **197**, 551–566.

Cleland, B. G., Dubin, M. W., and Levick, W. R. (1971). Sustained and transient neurons in the cat's retina and lateral geniculate nucleus. *J. Physiol.*, **217**, 473–496.

Cowan, J. D. (1977). Some remarks on channel bandwidth for visual contrast detection. In *Neuronal Mechanisms in Visual Perception* (Eds. R. Held, B. Dowling and H. Purple), Vol. 15, MIT Neurosciences Res. Prog. Bull., Cambridge, Mass., pp. 492–517.

Cowey, A. (1982). Sensory and non-sensory visual disorders in man and monkey. *Phil. Trans. Roy. Soc., B*, **298**, 3–13.

Creutzfeldt, O. D., Kuhnt, U., and Benevento, L. A. (1974). An intracellular analysis of visual cortical neurons to moving stimuli: responses in a co-operative neuronal network. *Exp. Brain Res.*, **21**, 251–274.

Dreher, B., Fukada, Y., and Rodieck, R. W. (1976). Identification, classification and anatomical segregation of cells with X-like and Y-like properties in the lateral geniculate nucleus of old-world primates. *J. Physiol.*, **258**, 433–452.

Dreher, B., Leventhal, A. G., and Hale, P. T. (1980). Geniculate input to cat visual cortex: a comparison of area 19 with areas 17 and 18. *J. Neurophysiol.*, **44**, 804–826.

Enroth-Cugell, C., and Robson, J. G. (1966). Contrast sensitivity of retinal ganglion cells of the cat. *J. Physiol.*, **187**, 517–552.

Gabor, D. (1946). Theory of communication. *J. IEE (Lond.)*, **13**, 429–441.

Gilbert, C. D., and Wiesel, T. N. (1979). Morphology and intracortical projections of functionally characterised neurons in the cat striate cortex. *Nature*, **280**, 120–125.

Hammond, P., and MacKay, D. M. (1977). Differential responsiveness of simple and complex cells in cat striate cortex to visual texture. *Exp. Brain Res.*, **30**, 275–296.

Hammond, P., and MacKay, D. M. (1983). Influence of luminance gradient reversal on simple cells in feline striate cortex. *J. Physiol.*, **337**, 69–89.

Hammond, P., and Smith, A. T. (1983). Directional tuning interactions between moving oriented and textured stimuli in complex cells of feline striate cortex. *J. Physiol.*, **342**, 35–49.

Henry, G. H. (1977). Receptive field classes of cells in the striate cortex of the cat. *Brain Res.*, **133**, 1–28.

Henry, G. H., Lund, J. S. and Harvey, A. R. (1978). Cells of the striate cortex projecting to the Clare–Bishop area of the cat. *Brain Res.*, **151**, 154–158.
Hicks, T. P., Lee, B. B., and Vidyasagar, T. R. (1983). The responses of cells in macaque lateral geniculate nucleus to sinusoidal gratings. *J. Physiol.*, **337**, 183–200.
Hubel, D. H., and Wiesel, T. N. (1959). Receptive fields of single neurons in the cat's striate cortex. *J. Physiol.*, **148**, 574–591.
Hubel, D. H., and Wiesel, T. N. (1962). Receptive fields, binocular interaction and functional architecture in the cat's visual cortex. *J. Physiol.*, **160**, 106–154.
Hubel, D. H., and Wiesel, T. N. (1968). Receptive fields and functional architecture of monkey striate cortex. *J. Physiol.*, **195**, 215–243.
King-Smith, P. E. (1978). Analysis of a detection of a moving line. *Perception*, **7**, 449–458.
King-Smith, P. E., and Kulikowski, J. J. (1973). Lateral interaction in the detection of composite spatial patterns. *J. Physiol.*, **234**, 5–6P.
King-Smith, P. E., Rosten, J. G., Alvarez, S., and Bhargava, S. K. (1980). Human vision without tonic ganglion cells? In *Colour Vision Deficiencies* (Ed. G. Verriest), Adam Hilger Ltd., pp. 99–105.
Kulikowski, J. J. (1976). Effective contrast constancy and linearity of contrast sensation. *Vision Res.*, **16**, 1419–1431.
Kulikowski, J. J. (1979). Neural stages of visual signal processing. In *Search and the Human Observer* (Eds. J. N. Clare and M. A. Sinclair), Taylor and Francis, London.
Kulikowski, J. J., and Bishop, P. O. (1981). Linear analysis of the responses of simple cells in the cat visual cortex. *Exp. Brain Res.*, **44**, 386–400.
Kulikowski, J. J., and Bishop, P. O. (1982). Silent periodic cells in the cat striate cortex. *Vision Res.*, **22**, 191–200.
Kulikowski, J. J., Bishop, P. O. and Kato, H. (1981). Spatial arrangement of responses by cells in the cat visual cortex to light and dark bars and edges. *Exp. Brain Res.*, **44**, 371–385.
Kulikowski, J. J., and King-Smith, P. E. (1973). Spatial arrangement of line, edge and grating detectors revealed by subthreshold summation. *Vision Res.*, **13**, 1455–1478.
Kulikowski, J. J., Marcelja, S., and Bishop, P. O. (1982). Theory of spatial position and spatial frequency relations in the receptive fields of simple cells in the visual cortex. *Biolog. Cybernetics*, **43**, 187–198.
Kulikowski, J. J., and Vidyasagar, T. R. (1982). Representation of space and spatial frequency in the macaque striate cortex. *J. Physiol.*, **332**, 10–11P.
Kulikowski, J. J., and Vidyasagar, T. R. (1983). Single-unit, multiunit and field potential responses in the striate cortex of macaque and cat as a function of contrast. *J. Physiol*, **334**, 19–20P.
Kulikowski, J. J. and Vidyasagar, T. R. (1984). Macaque striate cortex: pattern, movement and colour processing. *Ophthal. Physiol Opt.*, **4**, 77–81.
Livingstone, M. S., and Hubel, D. H. (1983). Specificity of cortico-cortical connections in monkey visual system. *Nature*, **304**, 351–354.
Marcelja, S. (1980). Mathematical description of the responses of simple cortical cells. *J. Opt. Soc. Amer.*, **70**, 1297–1300.
Marr, D. (1982). *Vision*. Freeman, San Francisco.
Movshon, J. A., Thompson, I. D., and Tolhurst, D. J. (1978a). Spatial summation in the receptive fields of simple cells in the cat's striate cortex. *J. Physiol.*, **283**, 53–77.
Movshon, J. A., Thompson, I. D., and Tolhurst, D. J. (1978b). Receptive field organisation of complex cells in the cat's striate cortex. *J. Physiol.*, **283**, 79–99.

Murray, I., MacCana, F., and Kulikowski, J. J. (1983). Contribution of two movement detecting mechanisms to central and peripheral vision. *Vision Res.*, **23**, 151–159.

Mustari, M. J., Bullier, J., and Henry, G. H. (1982). Comparison of response properties of three types of monosynaptic S-cell in cat striate cortex. *J. Neurophysiol.*, **47**, 439–454.

Payne, B. R., Berman, N., and Murphy, E. H. (1980). Organisation of directional preference in cat visual cortex. *Brain Res.*, **211**, 117–141.

Pettigrew, J. D., Nikara, T., and Bishop, P. O. (1968). Responses to moving slits by single units in cat striate cortex. *Exp. Brain Res.*, **6**, 373–390.

Robson, J. G. (1975). Receptive fields: spatial and intensive representations of the visual image. In *Handbook of Perception* (Eds. E. C. Carterette and M. P. Friedman), Vol. V, Academic Press, New York.

Ruddock, K. H. (1982). Psychophysical studies on subjects with visual defects.*J. Roy. Soc. Med.*, **75**, 315–322.

Shapley, R., and Tolhurst, D. J. (1973). Edge detectors in human vision. *J. Physiol. (Lond.)*, **229**, 165–183.

Tolhurst, D. J., Walker, N. S., Thompson, I. D., and Dean, A. F. (1980). Nonlinearities of temporal summation in neurons in 17 of the cat. *Exp. Brain Res.*, **38**, 431–435.

Vidyasagar, T. R., and Urbas, J. V. (1982). Orientation sensitivity of cat LGN neurons with and without inputs from visual cortical areas 17 and 18. *Exp. Brain Res.*, **46**, 157–169.

Weiskrantz, L. (1972). Behavioural analysis of the monkey visual system. *Proc. Roy. Soc. Lond., B*, **182**, 427–455.

Weiskrantz, L., Warrington, E. K., Sanders, M. D., and Marshall, J. (1974). Visual capacity in the hemianopic field following a restricted occipital ablation. *Brain*, **97**, 709–728.

Wong-Riley, M. (1974). Demonstration of geniculo-cortical and callosal projection neurons in the squirrel monkey by means of retrograde axonal transport of horseradish peroxidase. *Brain Res.*, **79**, 267–272.

Zeki, S. (1978). Functional specialization in the visual cortex of the rhesus monkey. *Nature*, **274**, 423–428.

Zeki, S. (1980). The representation of colours in the cerebral cortex. *Nature*, **284**, 412–418.

Zeki, S. (1983). The distribution of wavelength and orientation selective cells in different areas of monkey striate cortex. *Proc. Roy. Soc. Lond., B*, **217**, 449–470.

Models of the Visual Cortex
Edited by D. Rose and V. G. Dobson

CHAPTER 26

Spatial and spatial frequency characteristics of receptive fields of the visual cortex and piecewise Fourier analysis

V.D. GLEZER
Laboratory of Vision Physiology, I. P. Pavlov Institute of Physiology, Academy of Sciences of USSR, Leningrad, USSR

In the fundamental investigation of Hubel and Wiesel (1962) it was shown that the cortical receptive field (RF) gives maximal reaction to an elongated stimulus: an edge or a bar of definite width and orientation. It will be shown here that these properties quantitatively described as the spatial frequency and orientational characteristics of the RF in linear or quasi-linear cells are defined by a single two-dimensional weighting function (WF) of the RF.

The WFs along the orientation orthogonal to the optimal orientation were measured in the following manner. The responses of a cell in the cat's visual cortex to optimally oriented moving sinusoidal gratings of different spatial frequencies were recorded, and the amplitude–phase characteristic (APC) of the RF was derived (Fig.1a). The APC was inverse Fourier transformed to obtain the WF of the field. The calculated WFs of three cells are shown in Fig. 1b by dashed lines. The continuous lines represent the WFs derived directly from the responses to narrow light and dark bars as algebraic sums of the responses (negative response represents the response to a dark bar). The coincidence of the solid and dashed lines provides evidence that the RFs in Fig. 1b are linear systems. (Strictly considered, the RFs are quasi-linear systems. This problem is discussed elsewhere; see Glezer *et al.*, 1980).

The linearity means that an increase of luminance in the positive part of the RF evokes excitation of the cell, and a decrease of luminance evokes inhibition. In the negative part of the WF the relations are quite the opposite.

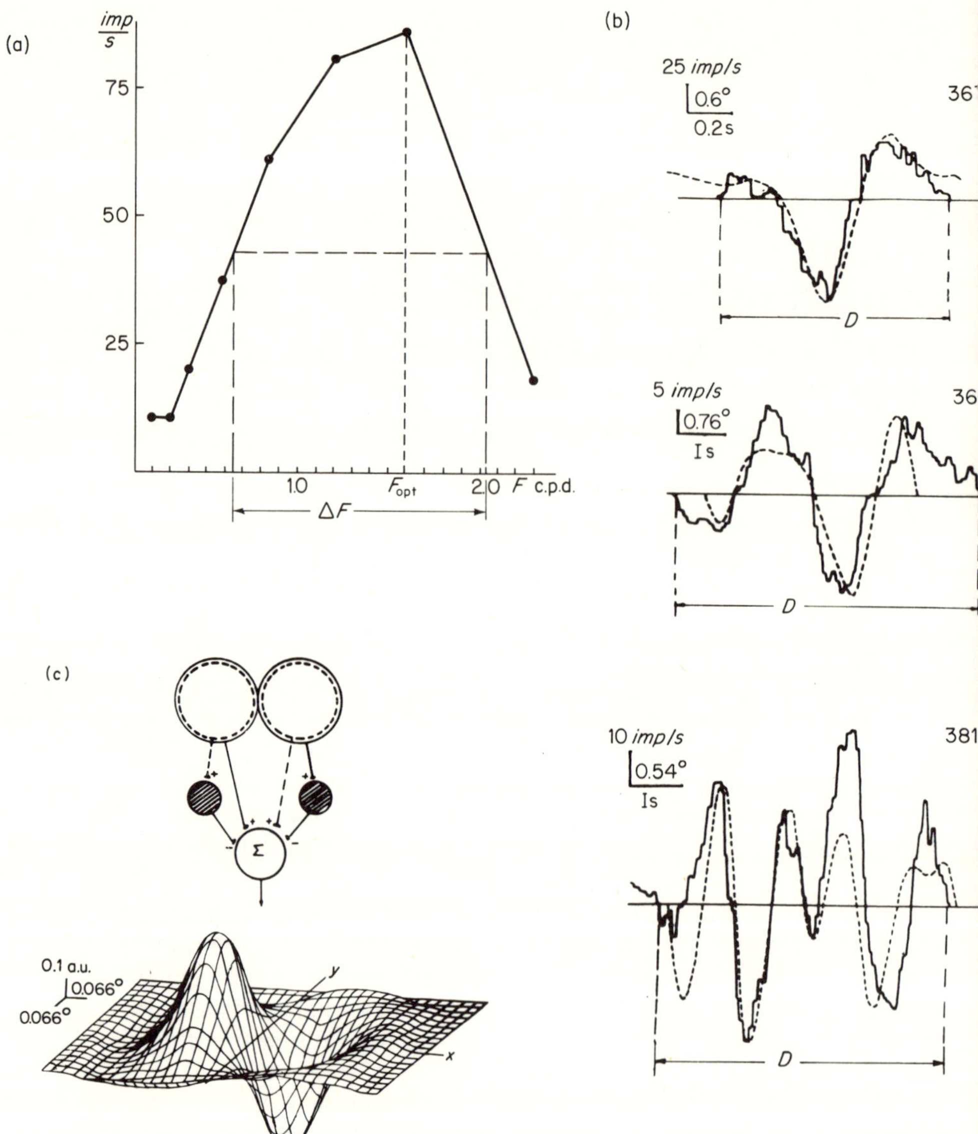

FIGURE 1 Spatial and spatial frequency characteristics of linear receptive fields (RFs) in the cat's visual cortex. (a) Amplitude characteristic of the lowest cell shown in (b). It is shown how optimal frequency F_{opt} and bandwidth ΔF were measured. (b) Weighting functions (WFs) of the RFs along the orientation orthogonal to the optimal orientation of the field. WFs received from the responses to thin 0.1 light and dark bars (continuous lines) are compared with WFs calculated from the responses to sinusoidal gratings (dashed lines). It is shown how the size of D of the RF was

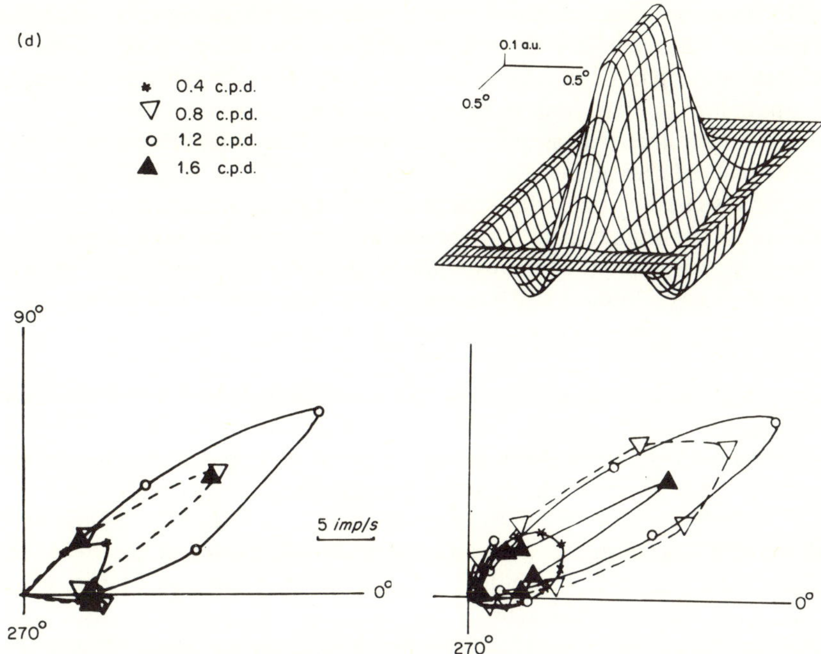

These results have been confirmed in a direct experiment. Stimulation ot the on-zone by a light line evoked excitation, by a dark line, inhibition, and vice versa in the off-zone (Glezer *et al.*, 1980, 1982; Palmer and Davis, 1981).

These facts allow us to propose a neural circuitry of the linear RF. An idealized neural scheme of the most simple RF—an edge detector—is shown in Fig. 1c (top) and also its two-dimensional WF (bottom). The on- and off-zones of the RF are composed each of an opponent pair of on- and off-subfields of LGN cell RFs. Such circuitry explains the orientation and spatial frequency characteristics of the RF. However, the orientation characteristic of the RF with the WF, such as shown in Fig. 1c, is too broad. We must therefore assume an increase in the number of subfields along the axis of the RF. In Fig. 1d the WF of such an RF is shown. The profile of the WF in the orientation orthogonal to optimal was derived by the usual procedure. The length along the optimal orientation was increased until the orientation-

---

measured. The indexes of complexity of the cells from top to bottom are equal to 0.98, 1.74, 4.53. (c) Neural circuitry of an idealized simple RF (top). Large solid circles show LGN on-fields; dashed circles, off-fields; hatched circles, inhibitory interneurones; Σ, exit cell possessing a simple RF. Two-dimensional WF (bottom). The ordinate is given in arbitrary units (a.u.). (d) Two-dimensional WF of an RF (top) and orientation characteristics (bottom) found with gratings of different SFs in the experiment (at left) and in the model (at right).

spatial frequency characteristics of the model and of the real RF coincided.

Thus orientation is not an independent and specific property of the RF and does not reflect its detector function. In the two-dimensional WF of the RF, spatial frequency and orientation characteristics are both its consequences and reflect two aspects of the spatial frequency filter. In the two-dimensional Fourier plane a two-dimensional WF filter is represented as a point which characterizes the spatial frequency and orientation properties of the filter simultaneously. This point (very narrow spatial frequency and orientation characteristics) corresponds to the case of global Fourier analysis. As the RFs have limited size we must speak not about a point but rather about an area (broad characteristics), and consequently about piecewise Fourier analysis.

The piecewise Fourier analysis hypothesis (Glezer, Ivanov and Tscherbach, 1973) suggests that the visual cortex contains modules of cells. Each module consists of RFs tuned to different orientations and spatial frequencies and projected onto the same place in the visual field.

If RFs perform piecewise Fourier analysis certain relationships should exist between the spatial and spatial frequency characteristics of the field. In a module, RFs with different indexes of complexity I must exist. I is defined as the ratio between the size of RF, D, and the period of its optimal frequency, $1/F_{\text{opt}}$, i.e. $I = DF_{\text{opt}}$. This prediction, as was shown above (Fig. 1b, dashed lines), is realized. It means that the RFs are grating filters.

The other prediction is that the RFs must be linear systems. The simple cells with $I = \pm 0.5$ are linear (or quasi-linear) systems. The complex cells have higher indices of complexity. Some complex RFs are also linear systems but other complex cells have besides linear also non-linear components (Glezer *et al.*, 1982). It can be shown that with certain limitations such RFs may participate in Fourier analysis (Glezer, Gauselman and Tscherbach, 1983). (I do not discuss in this paper the non-linear complex RFs.)

It is well known that the size of RFs increases with eccentricity, but if a module in a given area of visual field performs piecewise Fourier analysis then all the RFs of the module must have identical size D.

In the most common basis for analysing waveforms into trigonometric components, the WF must contain 1, 2, 3 or more periods of the frequency to which the RF is tuned. However, it is not certain that such a Fourier series must exist in the visual system. Many orthogonal bases are known where the number of periods in the WF is not an integer (Zhong-Di Wang, 1981). However, in all cases the interval (the size D of the RFs in the module) must be constant. The correctness of this prediction may be shown in three different experiments.

The first method consists of finding the stimulus evoking the maximal response in the RF (Glezer, Ivanov and Tscherbach, 1973). We registered the responses of cortical RFs at eccentricity about 5° to complex stimuli composed of several light bars. The number of bars, the width of the bars and the intervals between them were changed until the maximal response was attained. It was shown that for every cell under investigation, irrespective of the number of bars in the optimal stimulus, its RF width was constant and equal to 2.6°.

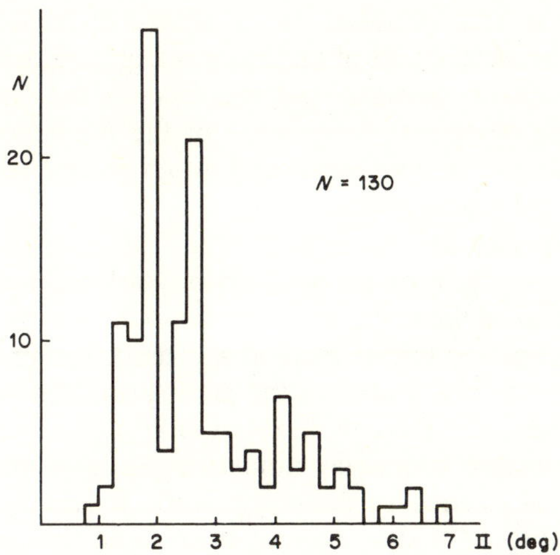

Figure 2 Distribution of sizes D of all the RFs under investigation.

In the second and third experiments we investigated RFs at eccentricities from 0 to 10°. The results are based on data received from cells of the same type as shown in Fig. 1. The second method is direct. The distribution of sizes D of a sample of RFs shows two peaks at 1.8 and 2.6° (Fig. 2), but as the number of the RFs with D greater than 2.6° was small nothing can be said about the modules with very great D. The third method, in which both independently measured parameters D and F_{opt} were used, appeared to be more sensitive. The use of both parameters allows us to eliminate the inaccuracy in their measurements. More detailed statistical analysis of the data shows five sizes of D: 1.8, 2.6, 3.6, 4.8, 6.1°.

Piecewise Fourier analysis predicts that the bandwidth ΔF of spatial frequency characteristics (Fig. 1a) diminishes inversely in proportion to the size D of the RF at a fixed index of complexity $\Delta F = K/D$, where K is a constant. Experimentally, for different indexes of complexity we found that $\Delta FD = 1.2$. The value of K provides evidence that the WFs of RFs are sinusoids or cosinusoids modulated by squarewave impulses rather than by Gabor elements.

Psychophysical experiments are in good correspondence with the predictions of the piecewise Fourier analysis hypothesis and with neurophysiological data. After the discovery of SF channels in the visual system (Blakemore and Campbell, 1969) the spatial elements comprising these channels were investigated in many works. In the overwhelming majority of papers (Kulikowski and King-Smith, 1973; Wilson, 1978; and many others) it was stated that such elements are line detectors, whose WFs have the form of the difference of two Gaussians (DOG). We came to the conclusion that an element having a DOG form is found only in the case when a subject must *detect* the presence of a stimulus. In conditions of *recognition* the spatial element serving the description of an image is a grating detector rather than a line detector (Glezer, Kostelyanets and Cooperman, 1977). It was also shown that grating detectors, irrespective of the frequency they are tuned to, have identical size, as the hypothesis about piecewise Fourier analysis predicts (Glezer and Kostelyanets, 1975).

Some new experiments have confirmed these conclusions. On the uniformly illuminated face of an oscilloscope dark vertical bars were produced. The number of bars, the width of the bars and their spacings were varied. A forced-choice staircase method was used. Under the conditions of detection the subject chose between the grating and the blank screen (Fig. 3, bottom left), and under the conditions of recognition, between the grating and the plateau equal in width and luminance energy to the grating (Fig. 3, top left). In both cases contrast thresholds were determined.

Under the conditions of detection an increase in the number of lines in the stimulus led to a fall of the threshold, providing evidence for summation in a zone 5 to 6 min in diameter (Fig. 3, bottom right). As the number of lines was increased the threshold went up because the side lines started to stimulate the inhibitory zones of the DOG. Any further increase of the number of lines did not change the threshold. Entirely different curves were found under the conditions of recognition. The zone of summation was equal to 25 to 40 min with different subjects (Fig. 3, top right). The zone of inhibition was absent. The size of the zone of summation did not change with the spatial frequency of the grating (we used 15, 20 and 27 c.p.d. gratings), as is predicted by the piecewise Fourier analysis hypothesis.

Given all the data described above it is suggested that the description of images in the visual cortex is a combination of retinotopic and local spectral analysis—piecewise Fourier analysis.

REFERENCES

Blakemore, C. B., and Campbell, F. W. (1969). On the existence in the human visual system of neurons selectively sensitive to the orientation and size of retinal images. *J. Physiol. (Lond.)*, **203**, 237–260.

Glezer, V. D., Gauselman, V. E., and Tscherbach, T. A. (1983). The dependence between spatial and spatial frequency characteristics of the receptive fields of cat's visual cortex (in Russian). *Physiol. J. SSSR*, **67**, 614–622.

Glezer, V. D., Ivanov, V. A., and Tscherbach, T. A. (1973). Investigation of complex and hypercomplex receptive fields of visual cortex of the cat as spatial frequency filters. *Vision Res.*, **13**, 1875–1904.

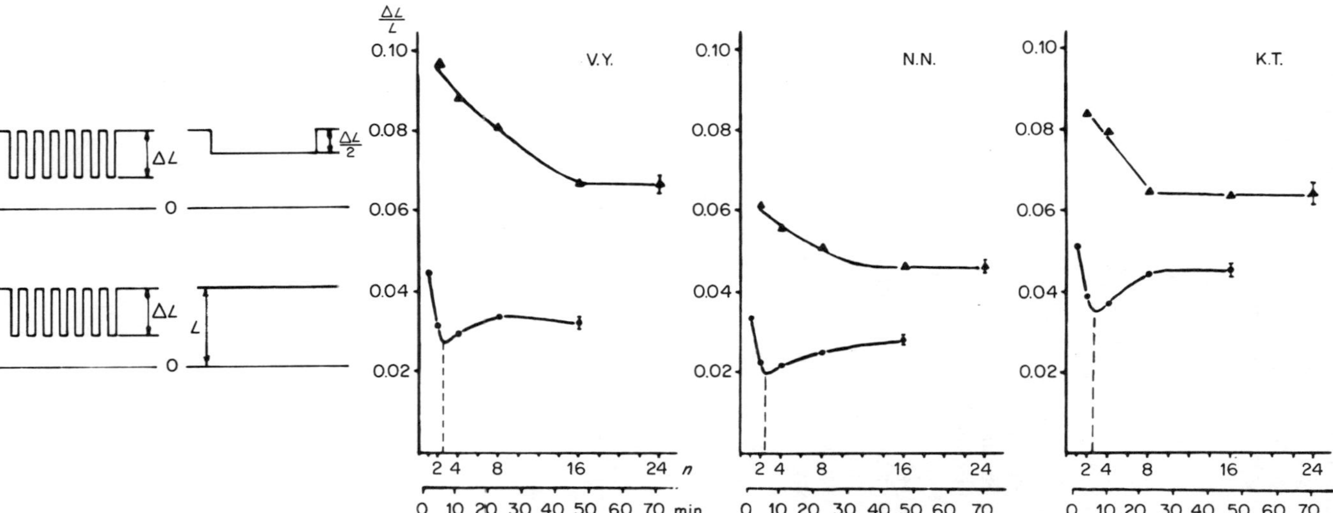

FIGURE 3 Detection and recognition are performed by different spatial elements. The dependence of the threshold contrast ($\Delta L/L$, where L is luminance) on the number of lines in the spatial pattern under the conditions of detection (bottom) and recognition (top). Abscissa: upper scale, the number of lines (1.5 ft in width) in the stimulus; lower scale, the size of the stimulus in minutes of arc. Ordinates show contrast $\Delta L/L$, where L = 5 cd/m². There are three subjects. At left the forms of spatial patterns in a pair are shown.

Glezer, V. D., and Kostelyanets, N. B. (1975). The dependence of threshold for perception of rectangular grating upon the stimulus size. *Vision Res.*, **15**, 753–756.

Glezer, V. D., Kostelyanets, N. B., and Cooperman, A. M. (1977). Composite stimuli are detected by grating detectors rather than by line detectors. *Vision Res.*, **17**, 1067–1070.

Glezer, V. D., Tscherbach, T. A., Causelman, V. E., and Bondarko, V. M. (1980). Linear and non-linear properties of simple and complex receptive fields in area 17 of the cat visual cortex. A model of the field. *Biolog. Cybernetics*, **37**, 195–208.

Glezer, V. D., Tscherbach, T. A., Gauselman, V. E., and Bondarko, V. M. (1982). Spatio-temporal organization of receptive fields of the cat striate cortex. *Biolog. Cybernetics*, **43**, 35–49.

Hubel, D. H., and Wiesel, T. N. (1962). Receptive fields binocular interaction and functional architecture in the cat's visual cortex. *J. Physiol. (Lond.)*, **160**, 106–154.

Kulikowski, J. J., and King-Smith, P. E. (1973). Spatial arrangement of line, edge and grating detectors revealed by subthreshold summation. *Vision Res.*, **13**, 1455–1478.

Palmer, L. A., and Davis, T. L. (1981). Receptive field structure in cat striate cortex. *J. Neurophysiol.*, **46**, 260–276.

Wilson, H. R. (1978). Quantitative predictions of line spread function measurements: implications for channels bandwidths. *Vision Res.*, **18**, 493–496.

Zhong-Di Wang (1981). Harmonic analysis with real function of frequency. II. Periodic and bounded cases. *Appl. Math. Comp.*, **9**, 153–163.

Models of the Visual Cortex
Edited by D. Rose and V. G. Dobson

CHAPTER 27

Functional organization of simple receptive fields

LARRY A. PALMER, JUDSON P. JONES and WALTER H. MULLIKIN
Department of Anatomy, School of Medicine, University of Pennsylvania, Philadelphia, PA 19104, USA

INTRODUCTION

Mammalian striate cortex has for some twenty years now been one of the main foci of progress in neuroscience. Testaments to this fact include the Nobel prize awarded to David Hubel and Torsten Wiesel and the publication of this book. This situation is more true today than ever and the continued application of a host of new techniques promises to yield yet further understanding of this subtle and truly impressive piece of neural machinery. In the pages to follow, we will briefly summarize our own views on some rather narrow but important issues concerning the organization of simple receptive fields.

PARALLEL X AND Y PATHWAYS IN THE STRIATE CORTEX

It is well established that the form of the neurally encoded visual message undergoes striking transformations at the level of the striate cortex. Amongst those transformations, the combining of information from on-center and off-center channels and the emergence of orientation selectivity must be considered paramount (Hubel and Wiesel, 1962). On the other hand, the fate of the W, X and Y channels in the striate cortex is not yet clear. Two recently reported experimental efforts bear heavily on this subject, however, and indicate that, for simple cells at least, the X and Y pathways remain separate in the striate cortex.

The first of these experiments is the very successful use of optic radiation stimulation to identify simple cells receiving direct lateral geniculate nucleus

(LGN) input by Bullier, Mustari and Henry (1982). These cells were found to receive input from one, two and, rarely, three pools of LGN cells. Each pool seemed to consist of a single type (X-on, X-off, Y-on, Y-off) and this input was shown to be responsible for the individual discharge zones of the simple cell. It was also found that the widths and lengths of single discharge zones and cut-off velocities were greater for those simple cells receiving Y inputs than those receiving X inputs.

We have reached rather similar conclusions regarding maintained separation of X and Y channels at the simple cell level based on a different experimental approach (Mullikin, Jones and Palmer, in press). We have found that X and Y on-center LGN cells have unique signatures in poststimulus–time (PST) response planes such as those of Stevens and Gerstein (1976). Using this method, small spots of light are turned on and then off at a sequence of points traversing the receptive field. A PST histogram is generated for each point and these are stacked to form a plane. In on-center X-cells, the center response is sustained throughout the light-on period and its boundaries in space are abrupt. The center response in Y-cells tends to be transient and exhibits arms of transient excitation which extend well beyond the center into the classically defined surround.

When the same techniques were applied to the receptive fields of simple cells, individual discharge zones showed one of two patterns—either that characteristic of X-cells or that of Y-cells. Simple cells thus identified as X-like or Y-like were found to differ in other properties as expected. Thus, X-like simple cells exhibited smaller receptive fields, more sustained responses and were tuned to lower stimulus velocities than Y-like simple cells. Further, the laminar distributions of X-like and Y-like simple cells correlated well with the recently demonstrated distribution of X and Y terminal arborizations in layer IV (Humphrey, Sur and Sherman, 1983). Thus, X-like simple cells were found throughout layer IV while Y-like simple cells were limited to layer IVab. No evidence was found for mixing X and Y inputs in the receptive fields of single simple cells.

We believe that the results of these two experiments are complementary and constitute convincing evidence that the X and Y afferent paths remain parallel up to and including simple cells in cat striate cortex. Unlike the on-center and off-center channels, convergence of X and Y channels appears to be specifically avoided, but to what ends remains unclear.

HIERARCHIES OF SIMPLE CELLS IN THE STRIATE CORTEX

Simple cells are also found outside the distribution areas of LGN axons. In particular, we find many simple cells in lamina II and III. The receptive fields of these supragranular simple cells appear to be qualitatively similar to those in layer IV, but they are found to consist of larger numbers of discrete discharge zones. The receptive fields of simple cells within layer IV consist of one, two and sometimes three excitatory regions whereas those in layers II and III consist of three to eight discrete excitatory regions. It is unlikely that these 'higher order' simple cells are driven directly by LGN inputs but, rather, that their receptive fields are synthesized by an intracortical convergence of geniculate-recipient simple cells whose receptive fields consist of only a few excitatory regions.

The data thus suggest a hierarchy of simple cells, those in layer IV deriving their receptive fields from small pools of LGN afferents and these in turn projecting onto neurons in layers II and III in such a way that receptive fields with up to eight excitatory regions result. All simple cells appear to be either X-like or Y-like and so there must really be two hierarchies, one for the X pathway and one for the Y pathway. A similar scheme was suggested by Schiller, Finlay and Volman (1976) for the organization of simple cells in monkey striate cortex based on the results of analysis with moving edges.

This scheme appears to be in general agreement with the anatomical studies of the axonal arborizations of cortical neurons by Gilbert and Wiesel (1979). They have demonstrated axonal projections of layer IV simple cells to the supragranular layers, although they appear to find very few simple cells in these superficial layers.

TWO-DIMENSIONAL ORGANIZATION OF SIMPLE RECEPTIVE FIELDS IN THE STRIATE CORTEX

For what purpose might hierarchies of X-like and Y-like simple cells be established in the cortex? Although we cannot answer such a question at present, we feel that a hint arises from a conspicuous feature of those receptive fields with four or more excitatory regions. In these cells, the width and placement of excitatory regions is such that the profile of light sensitivity as a function of distance perpendicular to the optimal orientation is obviously periodic (Mullikin, Jones and Palmer, in press). Since there is good evidence that simple cells summate spatially distributed input in a roughly linear manner (Movshon, Thompson and Tolhurst, 1978), these periodic receptive fields seem like good templates for extraction of spatial frequency information from the distribution of light in their receptive fields. That is, periodic simple receptive fields may serve as relatively narrow-band spatial frequency filters.

This suggestion was originally made by Maffei and Fiorentini (1973) and interesting theoretical support has been provided by Marcelja (1980), Kulikowski, Marcelja and Bishop (1982) and Daugman (1979, unpublished manuscript). The main argument here is that spatial response profiles of simple cells are well described mathematically by the Gabor function (Gabor, 1946) in the direction perpendicular to the optimal orientation. The Gabor function, when thought of as a linear filter, has the intriguing property that it is an optimal compromise between an ideal filter for spatial localization (an impulse function in space) and an ideal filter for localization in the spatial frequency domain (an impulse function in frequency space—a sinusoid in space). Said another way, the Gabor function minimizes the joint uncertainty in a two-dimensional information space whose axes are space (x) and spatial frequency (f). There are an infinite number of Gabor functions and they all consist of sines or cosines whose amplitude is modulated with a Gaussian.

Daugman has generalized these arguments to two dimensions and has shown that a two-dimensional Gabor function consists of a one-dimensional sine or cosine wave (free to vary in its orientation and period) modulated by a bivariate Gaussian. This function minimizes the joint uncertainty in a four-dimensional information space whose axes are the Cartesian coordinates for space (x and y) and spatial frequency (u and v).

If it could be shown that simple receptive fields are well fitted by two-dimensional Gabor functions we would gain (a) an analytic representation suitable for modeling and (b) insight into the significance of these receptive fields and the striate cortex itself insofar as their information-carrying functions are concerned. While some evidence has been given that the one-dimensional Gabor model applies to simple receptive fields perpendicular to the optimal orientation, very little has been done with the two-dimensional case.

To exploit experimentally this two-dimensional argument it will be necessary to (a) obtain two-dimensional spatial response profiles, (b) obtain two-dimensional spatial frequency response measures and (c) demonstrate spatial linearity of the receptive field mechanism. In fact, if linearity could be established (in two dimensions), either the space domain or spatial frequency domain measurements alone would suffice. Proof of linearity in two dimensions is nontrivial, however, and it seemed to us that the space and spatial frequency measurements should be made and the test for linearity would then follow automatically (two-dimensional Fourier transform).

We have undertaken a study of cat striate neurons in two spatial and spatial frequency dimensions. In the spatial frequency domain, stimuli consisted of 256 sinusoidal gratings (16×16) chosen so as to sample uniformly the responsive area of each neuron. Stimuli were drifted across the receptive field at an optimal temporal frequency which was held constant as spatial frequency and orientation were varied in random order. Five cycles of each grating crossed the receptive field and cycle histograms (64 bins) were generated from the last four cycles. From these data, amplitude of response at DC (the average spike count), the fundamental or any harmonic may be calculated and plotted.

In the space domain, small bright and dark stimuli are briefly presented (20 to 40 ms) at positions on a 16×16 grid covering the receptive field. Stimuli are presented in random order. A reverse cross-correlation is done between the spike train and the stimulus sequence, i.e. records are saved of the stimuli presented in frames preceding the action potentials.

An example of the kind of data obtained from simple cells using these methods is shown in Fig. 1. The data on the left were taken in the space domain; those on the right, in the spatial frequency domain. The cathode ray tube's face occupied 10° diameter and the region marked with a box in Fig.1a was divided into a 16 x 16 grid on which small bright and dark stimuli were presented for 20 ms each. A total of 15,363 such stimuli resulted in 2,088 action potentials and the two-dimensional spatial profile shown in Fig. 1b. The receptive field is seen to consist of two adjacent, elongated regions, one excited at light-on and the other inhibited at light-on. We have previously referred to such receptive fields as S-2 (Palmer and Davis, 1981). An ordinary response plane obtained with an elongated bar confirmed this receptive field structure. The response magnitude appears to be weighted in a roughly Gaussian manner in both the length and the width axes. The sharp separation of the two subdivisions and their elongation are evident from the contour diagram in Fig. 1c.

In the frequency domain, this cell was modestly selective in both orientation and spatial frequency. The portion of the frequency space chosen for study is indicated by the box in Fig. 1d. Sinewave gratings were drifted across the entire scope face at rates inversely proportional to their spatial frequency so that the temporal frequency was held at the near-optimal value of 2 Hz. Combinations of orientation and spatial frequency were chosen so as to sample the area within the box in a uniform Cartesian fashion with a 16 x 16 resolution. The raw data, all 256 histograms generated in response to three passes through the stimulus space, are shown in Fig. 1e. Notice the half-wave rectification and phase locking typical of simple cells.

The surface in Fig. 1f was generated by plotting the amplitude of the fundamental of the response at each orientation–frequency coordinate using an orthographic projection. The contour diagram in Fig. 1g was generated from the surface in Fig. 1f and each contour demarcates equal vertical intervals up the surface. The contours are roughly elliptical, as expected for the Fourier transform of a Gabor function, and depart only slightly from the bivariate Gaussian.

These data are very much preliminary and no definite conclusions can be drawn at this time. We are encouraged to believe that the approach is a valid one, however, by several aspects of the results obtained so far. First, we have obtained results very similar to those shown in Fig. 1 on about ten other simple cells in several animals. We have not yet encountered any of the simple cells with four or more discharge regions (Mullikin, Jones and Palmer, in press). Second, for this cell and others, there seems to be little reason for concern about two-dimensional linearity of spatial summation. Thus, data from the space and spatial frequency domains are simply related with the Fourier transform and the resulting errors are quite small.

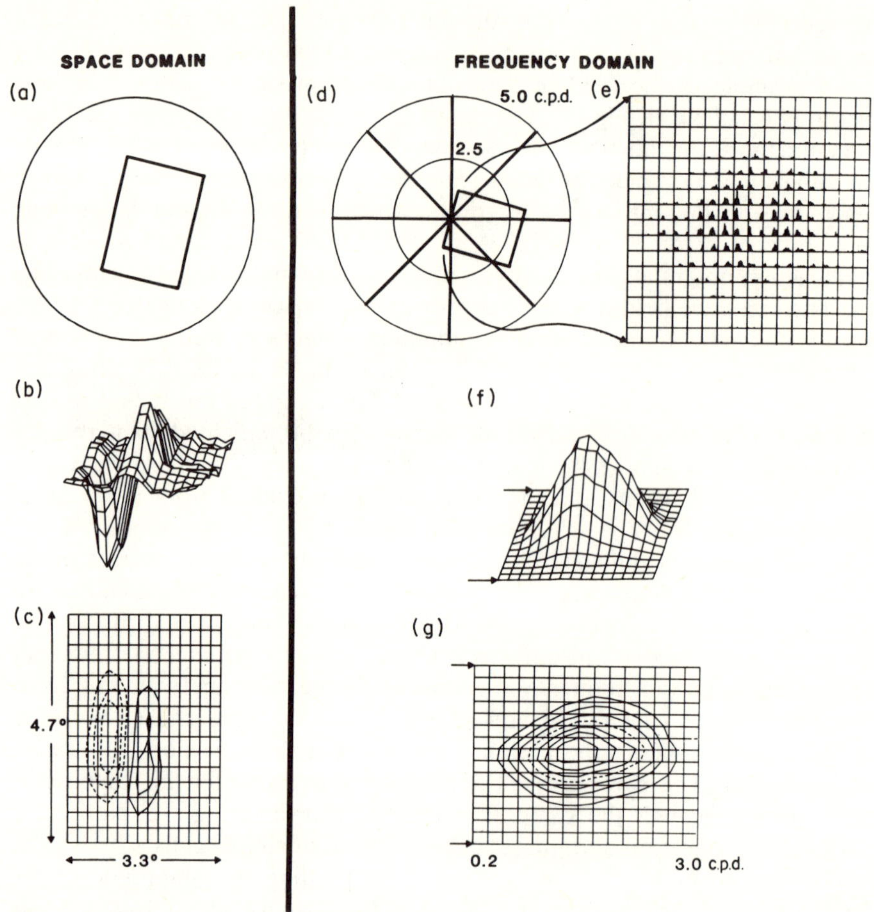

FIGURE 1 Data from a simple cell in cat striate cortex. Data to the left of the heavy line is taken in the space domain; that to the right, in the spatial frequency domain. (a) The face of the oscilloscope which subtended 10 degs and the rectangle which was centered on the receptive field. The length and width of this box were divided into 16 equal parts on which small bright and dark rectangles were presented for 20 ms each (total of 512 stimuli per pass presented in random order). Each time a bright bar preceded an action potential, the contents of a bin corresponding to the spatial location of the stimulus was incremented. Similarly, bins were decremented when a dark stimulus preceded an action potential. After roughly 15,000 stimuli, the surface in (b) emerged. This is the two-dimensional spatial impulse response of the cell. A contour diagram of the surface is shown in (c). There are two elongated discharge zones. The positive discharge zone (spikes elicited at above spontaneous levels by bright stimuli) is plotted with solid lines; the negative discharge zone (dark stimuli) is plotted with dashed lines. The stimulus space in the two-dimensional spatial frequency domain is portrayed in polar coordinates in (d). Quantitative data were acquired with 16 x 16 resolution within the box as shown. Each of the 256 histograms

In the next few months, besides collecting data from more cells, we will be applying two-dimensional curve-fitting techniques designed to measure and minimize the mean squared error between Gabor functions and data such as those of Fig. 1. We will also develop formal and quantitative tests of linearity in two dimensions. The data on hand already assures us that such further pursuit of the Gabor function hypothesis is worth while.

The need for quantitative experimental work and curve fitting results partly from the fact that there are other models of striate receptive fields which resemble Gabor functions superficially. These are different in detail, however, and are based on other fundamental assumptions, and might lead one to very different ideas about the function of the receptive fields.

One example will help to make this less abstract. Zucker and Kimia (1983) have been concerned with improving images which have been blurred by a Gaussian point spread function. The optimal inverse filter is not physically realizable but they have developed a finite approximation mathematically. Low-order approximations are very similar to the difference of a Gaussian model of retinal ganglion cell receptive fields of the X type. Higher order approximations are characterized by the emergence of extra sidelobes and resemble both Gabor functions and the profiles of simple striate receptive fields. They differ from Gabor functions in the two-dimensional placement of the sidelobes, however, and it will be important to develop methods which can distinguish these two functions convincingly. Zucker and Kimia suggest, based on the (presently superficial) similarity of striate receptive fields and their deblurring filters, that one function of the striate cortex is related to image deblurring. This is not necessarily incompatible with the hypothesis that the function of striate receptive fields is to minimize jointly the entropy in space and spatial frequency domains, which derives from the information theory of Gabor.

The study of the striate cortex appears to be poised on the brink of major revelations about the function and design of the receptive fields of at least one class of neurons. The very successful application of theory borrowed from the physical sciences has led to the design of better and more specific experiments. We are perhaps about to learn what is conserved in the encoding of visual information in this tissue and about trade-offs made in the process. This will no doubt lead to new levels of inquiry and will lead to similar efforts with complex cells and other cells in the many extrastriate areas.

---

in (e) show the averaged response to one cycle of the drifting sinusoidal grating. The spatial frequency and orientation of each grating is given by the position of the histogram in the array. The amplitude of the response in each histogram at the fundamental stimulus frequency was plotted as a surface (orthographic projection) in (f). The corresponding contour diagram is shown in (g). All axes are linear.

REFERENCES

Bullier, J., Mustari, M. J., and Henry, G. H. (1982). Receptive-field transformations between LGN neurons and S-cells of cat striate cortex. *J. Neurophysiol.*, **47**, 417–438.

Daugman, J. G. (1979). Two dimensional spectral analysis of cortical receptive field profiles. *Vision Res.*, **20**, 847–856.

Daugman, J. G. (unpublished manuscript). Uncertainty relation for resolution in space, spatial frequency, and orientation optimizing by two-dimensional visual cortical filters (submitted for publication).

Gabor, D. (1946). Theory of communication. *J. IEE*, Pt. 3, **93**, 429–457.

Gilbert, C. D., and Wiesel, T. N. (1979). Morphology and intracortical projections of functionally characterized neurons in the cat visual cortex. *Nature*, **280**, 120–125.

Hubel, D. H., and Wiesel, T. N. (1962). Receptive fields, binocular interaction and functional architecture in the cat's visual cortex. *J. Physiol.*, **160**, 106–156.

Humphrey, A. L., Sur, M., and Sherman, S. M. (1983). Relationships between cell body size and axon terminal fields in single X- and Y-cells of the retina and lateral geniculate nucleus. *Soc. Neurosci. Abstr.*, **9**, 813.

Kulikowski, J. J., Marcelja, S., and Bishop, P. O. (1982). Theory of spatial position and spatial frequency relations in the receptive fields of simple cells in the visual cortex. *Biolog. Cybernetics*, **43**, 187–198.

Maffei, L., and Fiorentini, A. (1973). The visual cortex as a spatial frequency analyzer. *Vision Res.*, **13**, 1255–1267.

Marcelja, S. (1980). Mathematical description of the responses of simple cortical cells. *J. Opt. Soc. Amer.*, **70**, 1297–1300.

Movshon, J. A., Thompson, I. D., and Tolhurst, D. J. (1978). Spatial summation in the receptive fields of simple cells in the cat's striate cortex. *J. Physiol.*, **283**, 53–77.

Mullikin, W. H., Jones, J. P., and Palmer, L. A. (1984). Periodic simple cells in cat area 17. *J. Neurophysiol.*, **52**, 372–387.

Mullikin, W. H., Jones, J. P., and Palmer, L. A. (1984). Receptive field properties and laminar distributions of X-like and Y-like simple cells in cat area 17. *J. Neurophysiol.*, **52**, 350-371.

Palmer, L. A., and Davis, T. L. (1981). Receptive field structure in cat striate cortex. *J. Neurophysiol.*, **46**, 260–276.

Schiller, P. H., Finlay, B. L., and Volman, S. F. (1976). Quantitative studies of single cell properties in monkey striate cortex. I. Spatio-temporal organization of receptive fields. *J. Neurophysiol.*, **39**, 1288–1319.

Stevens, J. K., and Gerstein, G. L. (1976). The spatio-temporal distribution of excitation and inhibition of cat LGN receptive fields. *J. Neurophysiol.*, **39**, 219–238.

Zucker, S., and Kimia, B. (1983). Deblurring the visual image: a functional interpretation of receptive field structure. Tech. Rep. 83–4R, Computer Vision and Robotics Lab., McGill University, Montreal.

Models of the Visual Cortex
Edited by D. Rose and V. G. Dobson

CHAPTER 28

Phase-dependent response characteristics of visual cortical neurons

Daniel A. Pollen, Kent H. Foster and James P. Gaska
Department of Neurology, University of Massachusetts Medical School, Worcester, MA 01605, USA

INTRODUCTION

The use of sinewave gratings as a visual stimulus (Enroth-Cugell and Robson, 1966) has provided important information about the receptive field properties of visual neurons. Numerous studies of the spatial-frequency-dependent response characteristics of visual cortical neurons have been published (for a recent review, see Pollen and Ronner, 1983). In this paper we will examine the phase-dependent response characteristics of visual neurons to one-dimensional sinewave gratings. In particular, we will consider possible ways in which the receptive field properties of cortical neurons and their neighbors may be used to encode phase information over a small subdomain or patch of visual space. Since the phase-dependent response properties of simple and complex cells to sinewave grating stimulation are quite different, we will discuss the results obtained from the two cell types in separate sections.

SIMPLE CELLS

Hubel and Wiesel (1962) defined simple cells as those orientation-selective visual cortical neurons for which the receptive field could be subdivided into regions which produced pure 'on' or pure 'off' responses when small spots or elongated slits of light were flashed at appropriate positions within the visual field. In addition, they qualitatively demonstrated that a simple cell's response to bars and edges could be predicted from mapping the receptive field with small spots of light. The degree to which a simple cell's response to bars could be predicted from its response to sinewave gratings was quantitatively examined by several investigators (Movshon, Thompson and Tolhurst, 1978a; Andrews and Pollen, 1979; Maffei *et al.*, 1979; DeValois, DeValois and Yund, 1979; Kulikowski and Bishop, 1981) who showed that the response could be predicted assuming linear summation across the receptive field. More stringent tests of linear summation such as the null-phase test have shown that the majority of simple cells exhibit linear summation across the receptive field (Movshon, Thompson and Tolhurst, 1978a).

The approximately linear spatial summation of simple cells predicts that the response amplitude should change sinusoidally as the phase of a sinewave grating is changed. Therefore, the amplitude of the cell's response could be used to provide information about the phase of the grating. However, the response of a single cell is not sufficient to completely specify the phase of the grating because, except for the positive and negative peaks, there are two points on a sinewave which have the same amplitude. A pair of cells whose phase response to gratings are offset by 90°, e.g. sine and cosine phases, would, however, be sufficient to specify the phase of the grating (Pollen and Ronner, 1975). Robson (1975) suggested that a bipartite (odd-symmetric) and a tripartite (even-symmetric) simple cell with the same axis of symmetry could be utilized for this purpose.

Pollen and Ronner (1981) investigated the phase issue by simultaneously recording from pairs of simple cells using a single electrode. They found that cell pairs whose members preferred the same spatial frequency, orientation and direction had a phase response to gratings which differed by approximately 90° (Fig. 1a, b). However, because the responses of simple cells to a drifting sinewave grating are often truncated due to low spontaneous firing rates, Pollen and Ronner (1981) hypothesized the existence of a second pair of simple cells with receptive field polarities opposite to those of the first pairs which would preserve information which would otherwise be lost due to response truncation. Subsequent experiments (Foster *et al.*, 1983) confirmed this prediction. In addition to finding more examples of pairs whose members had a 90° phase offset, they found cell pairs whose members had an 180° phase offset (Fig. 1c, d). Analysis of the response of these 180° pairs to moving narrow light and dark bars indicated that the cells had odd-symmetric receptive fields of opposite polarity sharing a common axis of symmetry.

We have not simultaneously recorded from cells pairs with on-center and off-center even-symmetric receptive fields. It is possible that cells with pure on-center and pure off-center receptive fields, which are segregated by lamina in the LGN (Schiller and Malpeli, 1978), may also be segregated in the cortex and, therefore, cannot be recorded simultaneously using a single electrode. Nevertheless, since Pollen and Ronner (1981) found simple cells with either on-center or off-center even-symmetric receptive fields paired

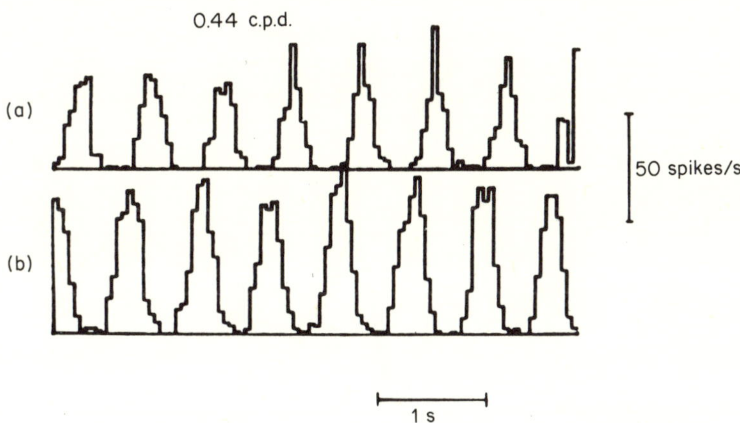

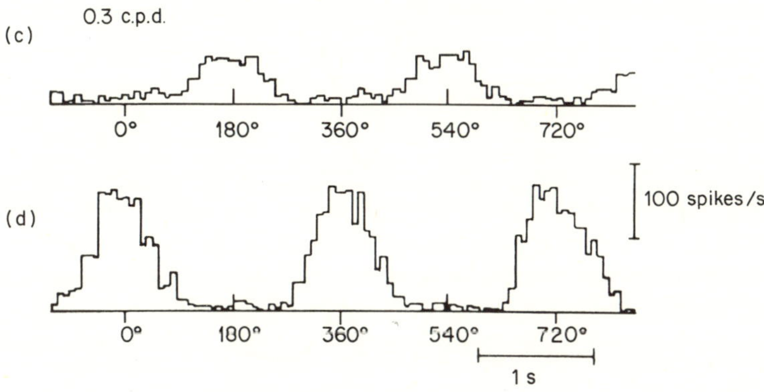

Figure 1 (a) and (b) Responses of two adjacent simple cells recorded simultaneously from the same microelectrode placement to a drifting sinewave grating at the preferred orientation and spatial frequency. The responses of the paired cells differ by approximately 90. (Reprinted from Pollen and Ronner 1981, *Science*, Vol. 212, p. 19, Fig. 3. Reproduced by permission of *Science*. Copyright 1981 by the AAAS.) (c) and (d) Responses from another pair of simple cells, similarly studied; however, in this case, the responses of the two cells differ by 180. This is a previously unpublished record from Foster *et al*. (1983).

with one or the other polarity of odd-symmetric receptive fields, it is implied that there exist four elemental receptive field types which share the same axis of symmetry. The idealized one-dimensional receptive field profiles of the four simple cell types are shown in Fig. 2 as Gaussian-attenuated sinusoids and cosinusoids, following Marcelja (1980).

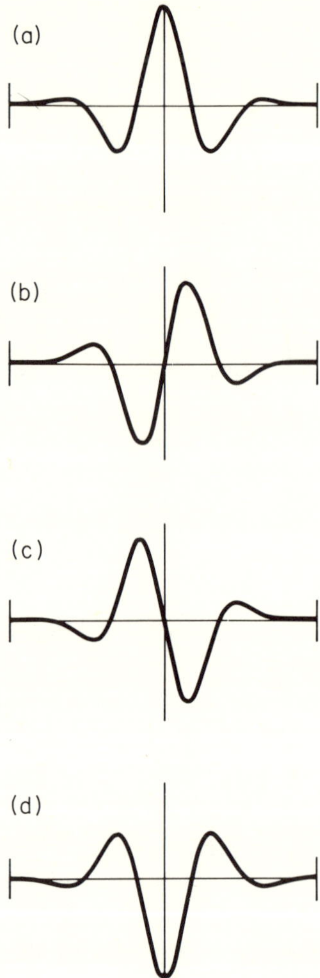

Figure 2 Four elementary one-dimensional Gabor signals sharing a common axis of symmetry are shown. The 'on-center' even-symmetric function $e^{-(1/2)(x/\sigma)^2}\cos 2\pi f_0 x$ is shown in (a); the odd-symmetric functions $e^{-(1/2)(x/\sigma)^2}\sin 2\pi f_0 x$ and $-e^{-(1/2)(x/\sigma)^2}\sin 2\pi f_0 x$ are shown respectively in (b) and (c); and the 'off-center' even-symmetric function $-e^{-(1/2)(x/\sigma)^2}\cos 2\pi f_0 x$ is shown in (d). The value of f_0 is 2.8 c.p.d., σ is 0.137°, the extent of the horizontal axis is 1° and the full bandwidth at half-amplitude of the Fourier transform of these functions is, in all cases, 1.54 octaves.

Marcelja (1980) noted that the response properties of simple cells closely correspond to the basis set of a representational scheme known in physics since the early 1920s and proposed by Gabor (1946) as eminently suitable for communications. In the original Gabor scheme, a function is expanded in terms of odd-symmetrical and even-symmetrical elementary signals which consist of a Gaussian-attenuated sinusoid and a Gaussian-attenuated cosinusoid. In fact the pair of elementary signals need not be purely even and purely odd-symmetric functions as long as the phase offset of the pair members is 90°. The elementary signals of the Gabor representation are at the same time centered at a given position and a given frequency. They have the property that the product of the uncertainty in the localization of the signal in position and frequency is equal to the theoretical minimum (Gabor, 1946). Therefore, to the extent that the spatial filtering properties of a simple cell resemble those of Gabor elementary signals, the analysis of the retinal image by simple cells is the finest grain sampling that a filter set can achieve simultaneously in spatial position and spatial frequency.

If the density of positions and frequencies of a set of Gabor elementary signals is high enough to satisfy the requirements of the sampling theorem, then an expansion in terms of Gabor signals is complete; i.e. an arbitrary function can be exactly represented (reviewed by Marcelja, 1980). A model of a cortical hypercolumn using Gabor elementary signals which is capable of representing the spatial information in the visual system (as determined from human psychophysics) has been outlined (Sakitt and Barlow, 1981). This model not only provides insight into how physiologically plausible filters can be arranged to represent visual information but also shows that the number of elements necessary is consistent with the number of cortical neurons actually present. This model stresses the efficiency of the Gabor scheme. At present we do not know if the cortical representation of the retinal image is efficiently sampled or perhaps oversampled (Kulikowski, Marcelja and Bishop, 1982).

The similarities between Gabor elementary signals and simple cells suggests that the Gabor scheme provides a useful framework for investigating the neuronal processing of spatial information. Other functions such as directional derivatives of Gaussians (Wilson, McFarlane and Phillips, 1983) may resemble the receptive field profile of even-symmetric simple cells. However, the presence of odd-symmetric and even-symmetric simple cells with a common axis of symmetry recommends the Gabor scheme. This does not imply that a Gabor analysis is performed in the cortex, but only that Gabor elementary signals and simple cells share many properties and that these similarities, if judiciously used, can allow a more quantitative examination of how neurons act together to provide a representation of the retinal image.

COMPLEX CELLS

Hubel and Wiesel (1962) found that the receptive fields of complex cells cannot be subdivided into pure on- or pure off-regions. A stationary flashed bar of either increased or decreased luminance produces both an 'on' and 'off' response when tested at most positions within the receptive field. When drifting narrow slits of either increased or decreased luminance are used as stimuli, the cell increases its firing rate as either stimulus crosses the field.

In the first study of the spatial frequency selectivity of complex cells, Maffei and Fiorentini (1973) reported that response modulation occurred only to spatial frequencies well below the cell's preferred spatial frequency. Subsequently, Pollen, Andrews and Feldon (1978) found that some complex cells in the cat also show a modulated component above the unmodulated level when tested at the preferred spatial frequency. These results have been confirmed by Glezer *et al.* (1980), Kulikowski and Bishop (1982) and Dean and Tolhurst (1983). Modulated components sometimes appear only within a narrow temporal frequency range and generally show a considerable second harmonic distortion (Pollen, Andrews and Feldon, 1978).

Movshon, Thompson and Tolhurst (1978b) demonstrated that many of the response properties of complex cells are consistent with the idea that they are built up from rectified responses of preceding subgroups which perform an approximately linear convolution with the visual stimulus. However, the spatial arrangement of these inputs is unknown. Glezer *et al.* (1982) investigated this issue in the cat by 'windowing out' either the left half or the right half of the receptive field of the complex cell and tested the response to sinewave gratings of different spatial frequencies. In some cases they could show that the responses within the hemifields became much more modulated and, in some cases, the response patterns in the hemifields were out of phase with each other.

Following this lead, we have used this 'windowing' technique to study the contribution from preceding subunits to the overall response of the complex cell in the striate cortex of the macaque monkey. We tested several such cells with 'complexity indices' of two or greater, as defined by Glezer *et al.* (1982). This index represents the product of the preferred spatial frequency and the receptive field width and is an indication of the envelope of the minimum number of cycles of a grating of the preferred spatial frequency that would be needed to completely cover the receptive field width. We first determined the preferred orientation, direction, spatial frequency and temporal frequency preferences of the cell. Having specified the cell's preferences and receptive field characteristics, we windowed a grating in the width dimension and stimulated the cell with an integral number of cycles at the preferred spatial and temporal frequency. The use of an integral number of cycles is essential to ensure that the mean luminance remains constant so that response modulation cannot be attributed solely to changes in the mean luminance of the stimulus. The grating patches were centered at the receptive field center.

The results of such an experiment are shown in Fig. 3. When a one-cycle patch is tested, the response occurs at double the drift frequency of the test grating, and the minima between successive peaks sometimes reach zero

(Fig. 3a). The result precludes an input to this subregion of the field from all four subunits with common axes of symmetry as shown in Fig. 2, because there then would have been four peaks per cycle and an elimination of the null response. The phase offset between two successive peaks actually observed is 180°, which is the same offset as seen in the simple cell pair of Fig. 1c and d. However, such data cannot tell us whether the receptive field of the complex cell is formed from such preceding simple cells pairs or *de novo* from geniculate axons.

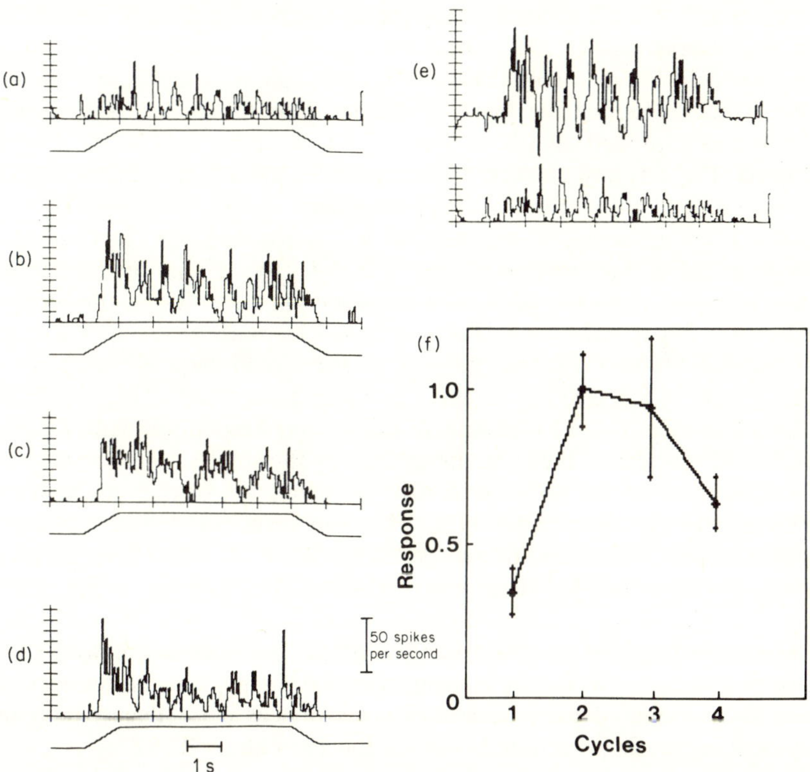

FIGURE 3 The responses of a complex cell to 'windowed' gratings of 1, 2, 3 and 4 full cycles are shown respectively in (a) to (d). The cell was tested at its preferred spatial frequency of 2.8 c.p.d. and at a temporal frequency of about 0.8 Hz. The receptive field width was 1.0; so the 3 cycle grating fully covered the receptive field core. The vertical notches along the *x* axis indicate one second intervals and the ramp functions below each histogram indicate the gradual increase, plateau and gradual decrease of contrast which peaked at 0.35. Diagram (e) (top) represents the subtraction of histogram (a) from histogram (b) plus a DC level to show the 90 phase shift of the new 'modulated' component compared to that in (a), shown again as the bottom trace. The mean response density and the standard error of the mean are shown for tests with 1, 2, 3 and 4 cycles in (f). (See text for further details.)

We then widened the window to include two full cycles, still centered around the same receptive field center, and found the response shown in Fig. 3b. It appears as if a new, but still frequency-doubled, response appears within the same regions that were previously occupied by response minima. The overall response pattern looks very similar to that often found in those complex cells that show both an unmodulated and some modulated response to extended drifting sinewave gratings.

We then widened the window to include three full cycles and found the response pattern shown in Fig. 3c. Here, except for a couple of response decreases at 3 and 4.5 s, the response pattern shows a generally unmodulated response characteristic. The pattern now resembles those of complex cells which show little or no modulation to drifting sinewave gratings.

We further widened the patch to include four full cycles. The response pattern now extended slightly beyond the full receptive field width. The new response (Fig. 3d) now showed a slightly decreased response density and the reappearance of a slight modulated component.

After the experiment was completed we 'subtracted' the response of the histogram in Fig. 3a from that of Fig. 3b in order to see what the contribution to the response from the additional half-cycle on each side of the receptive field center might have been. The subtracted response (Fig. 3e, top) shows a frequency-doubled response pattern, phase shifted about 90° with respect to the peaks in Fig. 3a, shown again as the bottom trace in Fig. 3e. It is as if the subunits enlisted by extending the grating responded with 90° phase offsets with respect to the subunits excited by the single-cycle grating patch.

We also plotted the response density as a function of the number of full cycles in the grating patches (Fig. 3f). The response density appears to summate linearly from one to two cycles but levels off when three cycles are tested. We have not yet determined whether the response leveling off at three cycles indicates a saturated response or another type of nonlinearity. (This should be determinable when we repeat the experiment with stimuli of lower contrast.) When the grating extended slightly beyond the receptive field, the response density decreased somewhat as if to suggest that there may be a suppressive surround (Maffei and Fiorentini, 1976).

These experiments provide evidence for a precisely organized pattern of inputs to the complex cell. Such an organization could produce the phase-independent response of the complex cell utilizing a small number of preceding subunits.

Glezer *et al.* (1980) and Pollen and Ronner (1980, 1982) have suggested that the response of most complex cells could represent the amplitude of a spatially localized spatial frequency coefficient, but that phase information is either not conveyed at all or not conveyed with much precision in the action potential code of the complex cell. DeValois and Tootell (1983) showed that some simple cells which are selective to low or middle ranges of spatial frequency are inhibited by gratings of higher spatial frequency in a manner independent of the relative phase of the two gratings. Complex cells, as we have described above, could mediate the phase-independent cross-frequency inhibition of this subset of simple cells. DeValois and Tootell (1983) also proposed that this cross-frequency inhibition might provide the mechanism for preserving contrast constancy, as described by Georgeson and Sullivan (1975).

Such complex cells could also participate in the phase-independent cross-orientation inhibition of simple cells at all orientations outside their orientation-selective band, as shown by Morrone, Burr and Maffei (1982). Such cross-orientation inhibition is also very broadly tuned for spatial frequency. Thus, in effect, the simple cell is inhibited at all spatial frequencies at all orientations outside its preferred orientation band. We suggest that this same mechanism may be at least partially responsible for maintaining a neuron's independent preferences for spatial frequency and orientation (Movshon, 1979; Glezer *et al.* 1982), which Daugman (1980) realized would not have resulted had simple cells been simply constructed from aligned rows of center/surround LGN cells. A phase-independent inhibitory mechanism would also have the property that it could affect the mean firing rate without phase shifting the response of a simple cell.

Thus, within the striate cortex there are several very useful functions for complex cells which do not require their conveyance of precise phase information. However, complex cells also provide major projections to the extrastriate visual cortices. It is hard to believe that the precise phase representation found at the simple cell stage is lost as one proceeds to higher cortical areas. In recent work (Foster *et al.*, 1985), we did find the presence of simple cells in V2 of the macaque monkey, but these simple cells, which could carry precise phase information, represented a lower spatial frequency range than those of simple cells found in V1.

Some additional positional information from complex cells might be encoded at higher levels assuming that the 'coordinates' of the receptive field centers of all complex cells are retained in their anatomical projections; however, it seems difficult to understand how an assembly of complex cells can abstract the absolute phase of a drifting grating if the phase information is not precisely specified within each member of the class. However, *absolute* phase information need not be of great importance in spatial pattern recognition, for as Cavanagh (1981) has noted: '. . . the *relative* phase offset between components together with their amplitudes are sufficient to uniquely specify the pattern.' Thus, the breadth of the spatial frequency selectivity of visual neurons may actually be advantageous in permitting such interactions between adjacent spatial frequency components.

Nevertheless, it remains unknown as to whether absolute phase information for medium and high spatial frequencies is represented beyond the striate cortex. Moreover, little is known about the mechanisms of the encoding of relative phase information at the single-cell level. One aspect of our future work will be to try to determine whether there exist additional mechanisms for the encoding of absolute and/or relative phase information by neurons in and beyond the striate cortex.

ACKNOWLEDGEMENT

We thank Carol Humphrey for excellent technical assistance. This research was supported by grant EY05156 from the National Eye Institute.

REFERENCES

Andrews, B. W., and Pollen, D. A. (1979). Relationship between spatial frequency selectivity and receptive field profile of simple cells. *J. Physiol.*, **287**, 163–176.

Cavanagh, P. (1981). Size invariance: Reply to Schwartz. *Perception*, **7**, 167–177.

Daugman, J. G. (1980). Two-dimensional spectral analysis of cortical receptive field profiles. *Vision Res.*, **10**, 847–856.

Dean, A. F., and Tolhurst, P. J. (1983). On the distinctiveness of simple and complex cells in the visual cortex of the cat. *J. Physiol.*, **344**, 305–325.

DeValois, K. K., DeValois, R. L., and Yund, E. W. (1979). Responses of striate cortex cells to grating and checkerboard patterns. *J. Physiol.*, **291**, 483–505.

DeValois, K. K., and Tootell, R. B. H. (1983). Spatial frequency-specific inhibition in cat striate cortex cells. *J. Physiol.*, **336**, 359–376.

Enroth-Cugell, C., and Robson, J. G. (1966). The contrast sensitivity of the retinal ganglion cells of the cat. *J. Physiol.*, **187**, 517–552.

Foster, K. H., Gaska, J. P., Marcelja, S., and Pollen, D. A. (1983). Phase relationships between adjacent simple cells in the feline visual cortex. *J. Physiol.*, **345**, 26P.

Foster, K. H., Gaska, J. P., Nagler, M., and Pollen, D. A. (1985). Spatial and temporal frequency selectivity of neurons in V1 and V2 of the macaque monkey. *J. Physiol.*, *In press*.

Georgeson, M. A., and Sullivan, G. D. (1975). Contrast constancy deblurring in human vision by spatial frequency channels. *J. Physiol.*, **252**, 627–656.

Glezer, V. D., Tscherbach, T. A., Gauselman, V. E., and Bondarko, V. M. (1980). Linear and non-linear properties of simple and complex receptive fields in area 17 of the cat visual cortex. *Biolog. Cybernetics*, **37**, 195–208.

Glezer, V. D., Tscherbach, T. S., Gauselman, V. E., and Bondarko, V. M. (1982). Spatio-temporal organization of receptive fields of the cat striate cortex. *Biolog. Cybernetics*, **43**, 35–49.

Gabor, D. (1946). Theory of communication. J.IEE, **93**, 429–459.

Hubel, D. H., and Wiesel, T. N. (1962). Receptive fields, binocular interaction and functional architecture in the cat's visual cortex. *J. Physiol*, **160**, 106–154.

Kulikowski, J. J., and Bishop, P. O. (1981). Fourier analysis and spatial representation in the visual cortex. *Experientia*, **37**, 160–163.

Kulikowski, J. J., and Bishop, P. O. (1982). Silent periodic cells in the cat striate cortex. *Vision Res.*, **22**, 191–200.

Kulikowski, J. J., Marcelja, S., and Bishop, P. O. (1982). Theory of spatial position and spatial frequency relations in the receptive fields of simple cells in the visual cortex. *Biolog. Cybernetics*, **43**, 187–198.

Maffei, L., and Fiorentini, A. (1973). The visual cortex as a spatial frequency analyzer. *Vision Res.*, **13**, 1255–1267.

Maffei, L., and Fiorentini, A. (1976). The unresponsive regions of visual cortical receptive fields. *Vision Res.*, **16**, 1131–1139.

Maffei, L., Morrone, C., Pirchio, M., and Sandini, G. (1979). Responses of visual cortical cells to periodic and nonperiodic stimuli. *J. Physiol.*, **296**, 24–47.

Marcelja, S. (1980). Mathematical description of the responses of simple cortical cells. *J. Opt. Soc. Amer.*, **70**, 1297–1300.

Morrone, M. Concetta, Burr, D. C., and Maffei, L. (1982). Functional implications of cross-orientation inhibition of cortical visual cells. I. Neurophysiological evidence. *Proc. Roy. Soc. Lond. B.*, **216**, 335–354.

Movshon, J. A. (1979). Two-dimensional spatial frequency tuning of cat striate cortical neurons. *Soc. Neurosci. Abstr.*, **5**, 799.

Movshon, J. A., Thompson, I. D., and Tolhurst, D. J. (1978a). Spatial summation in the receptive fields of simple cells in the cat's striate cortex. *J. Physiol.*, **283**, 53–77.

Movshon, J. A., Thompson, I. D., and Tolhurst, D. J. (1978b). Receptive field organization of complex cells in the cat's striate cortex. *J. Physiol.*, **283**, 79–99.

Pollen, D. A., Andrews, B. W., and Feldon, S. E. (1978). Spatial frequency selectivity of periodic complex cells in the visual cortex of the cat. *Vision Res.*, **18**, 665–682.

Pollen, D. A., and Ronner, S. F. (1975). Periodic excitability changes across the receptive fields of complex cells in the striate and parastriate cortex of the cat. *J. Physiol.*, **245**, 667–697.

Pollen, D. A., and Ronner, S. E. (1980). Spatial computation performed by simple and complex cells in the cat visual cortex. *Exp. Brian Res.*, **41**, A14–15.

Pollen, D. A., and Ronner, S. E. (1981). Phase relationships between adjacent simple cells in the visual cortex. *Science*, **212**, 1409–1411.

Pollen, D. A., and Ronner, S. F. (1982). Spatial computation performed by simple and complex cells in the visual cortex of the cat. *Vision Res.*, **22**, 101–118.

Pollen, D. A., and Ronner, S. F. (1983). Visual cortical neurons as localized spatial frequency filters. *IEEE Trans. SMC*, **13**, 907–915.

Robson, J. G. (1975). Receptive fields: neural representation of the spatial and intensive attributes of the visual image. In *Handbook of Perception* (Eds. E. C. Carterette and M. P. Friedman), Vol. 5, Academic Press, New York, pp. 81–116.

Sakitt, B., and Barlow, H. B. (1982). A model for the economic cortical encoding of the visual image. *Biolog. Cybernetics*, **43**, 97–108.

Schiller, P. H., and Malpeli, J. G. (1978). Functional specificity of lateral geniculate nucleus laminae of the rhesus monkey. *J. Neurophysiol.*, **41**, 788–797.

Wilson, H. R., McFarlane, D. K., and Phillips, G. C. (1983). Spatial frequency tuning of orientation selective units estimated by oblique masking. *Vision Res.*, **23**, 893–882.

Models of the Visual Cortex
Edited by D. Rose and V. G. Dobson

CHAPTER 29

Inhibition and spatial selectivity in the visual cortex: the cooperative neuronal network revisited

A.B. Bonds and E. J. DeBruyn
Department of Electrical and Biomedical Engineering, Vanderbilt University, Nashville, TN 37235, USA

Single visual cortical cells are perhaps best known for the vast array of visual stimuli, varied over both space and time, to which they do not respond. Cortical activity is relatively sedate in comparison with that of retinal and lateral geniculate cells, which react vigorously to virtually any visual stimulus falling within their receptive fields. The reduction of the abundance of spikes delivered by the lateral geniculate nucleus (LGN) is apparently a fair trade for the high stimulus specificity of cortical cells, which (arguably) plays an important role in the processing of visual information.

This discussion will center on how intracortical inhibition is associated with stimulus specificity in the striate cortex. We will review the current state of support, on both physiological and anatomical grounds, for the hypothesis that cortical excitation is predominantly undifferentiated, stemming primarily from LGN afferents, and that both spatial and temporal selectivity is added (or, more appropriately, subtracted) through specific intracortical inhibition. The shaping of cortical response properties by aggregates rather than by selective combination of LGN afferents has been described as the action of a cooperative neuronal network, in which the response of a single neuron is not an independent event but rather an indicator of changes throughout the network (Creutzfeldt, Kuhnt and Benevento, 1974).

We will examine mechanisms that support selectivity for spatial frequency, orientation, direction of motion and temporal frequency. Unless otherwise specified, examples and specific values used in this discussion are from the cat. There is a general tendency for selectivity for each of these properties to become more marked as one moves up from the retina through the LGN to area 17. Even retinal ganglion cells show some response limits in the domains of spatial frequency (Enroth-Cugell and Robson, 1966), orientation (Levick and Thibos, 1980) and temporal frequency (Frishman *et al.*, 1983), although one suspects that since the primary job of ganglion cells is efficient information compression and transmission, the latter two limits result from physical constraints of the system rather than by design. Some refinement, especially in low spatial frequency attenuation, is added in the LGN (DeValois, Albrecht and Thorell, 1977). Orientation selectivity is also enhanced (Vidyasagar and Urbas, 1982). The average LGN afferent to the cortex thus has a spatial frequency bandwidth of ± 2+ octaves, a response dependence on orientation averaging 2:1 (maximally 7:1), virtually no direction selectivity and the ability to respond to temporal frequencies of up to 40 to 60 Hz (DeValois, Albrecht and Thorell, 1977; Vidyasagar and Urbas, 1982; Kaplan, Marcus and So, 1979). The typical cortical cell reduces these figures to a spatial frequency bandwidth of ± 0.9 octaves (Movshon, Thompson and Tolhurst, 1978b), absolute orientation selectivity of ± 18.4° (Rose and Blakemore, 1974), more or less complete direction selectivity and a temporal frequency cut-off of about 16 Hz (Movshon, Thompson and Tolhurst, 1978b). In addition, overall responsiveness is reduced; retinal and LGN units can easily achieve discharge rates of 300 to 400 imp/s, while complex cells rarely exceed 200 imp/s and simple cells are usually limited to below 100 imp/s. The resting discharge is also reduced from 50 to 100 imp/s in retina to a few (or even negative) imp/s in the cortex (Movshon, Thompson and Tolhurst, 1978a).

Clearly, the cortex possesses a powerful mechanism that shapes the response characteristic of individual cells. The reduction of general responsiveness is the first hint that inhibition must play a major role. With this introduction, let us examine the process in more detail. The hypothesis first states that photic excitation derives primarily from LGN afferent signals. Exclusively excitatory feed from LGN to cortical cells is well supported on anatomical (Garey and Powell, 1971) and physiological (Creutzfeld and Ito, 1968) grounds. A recent cross-correlation study (Toyama, Kimura and Tanaka, 1981) has demonstrated an abundance of common excitation from LGN afferents, with only limited intracortical excitation restricted to complex–complex and complex–hypercomplex II (i.e. end-stopped with otherwise complex characteristics) pathways. The general consensus seems to be that excitation of a given cortical cell derives from a very few (1 to 4) LGN fibers and that the excitatory response field is very nearly circular rather than elongate. This conclusion is supported by intracellular recording (Creutzfeldt, Kuhnt and Benevento, 1974), simultaneous retina–cortical recording (Lee, Cleland and Creutzfeldt, 1977), response field mapping by stimulus interaction (Heggelund, 1981a, 1981b) and pharmacological studies in which inhibition is blocked (Sillito, 1975, 1977).

The remainder of the premise—that inhibition is responsible for the observed selectivities—is more elusive. Inhibition is usually manifested in something that is not there, and the natural reticence of cortical cells makes the observation of its effects doubly difficult. The usual approach has been to resort to some artifice to elevate (either randomly or by design) a cell's discharge to serve as a backdrop against which the impact of inhibition can be viewed. In the earliest such work, an unsynchronized moving bar (Bishop, Coombs and Henry, 1971) or iontophoretic injection of glutamate (Sillito, 1975) raised the resting discharge enough to demonstrate clearly the presence of direction- or orientation-specific inhibition.

More recent and informative work has used the strategy of dual stimuli—either two flashing spots (Heggelund, 1981a, 1981b), a sinewave grating with a noise pattern (Morrone, Burr and Maffei, 1982) or two gratings (Morrone, Burr and Maffei, 1982; DeValois and Tootell, 1983; Bonds, 1983). One stimulus (the base) is usually chosen to excite the cell optimally, presumably minimizing (but not eliminating—this will be discussed later) inhibition. This provides a baseline activity against which the effects of inhibition are apparent. The second stimulus (the mask) is then situated to produce that inhibition or, in some cases, excitation. All of the stimulus properties we are interested in have been studied in this way.

Dean, Hess and Tolhurst (1980) report that a grating drifting in the null direction for a direction-selective cell reduces the gain (slope) but does not change the sensitivity (intercept) of the response plotted versus the log of the (base) stimulus contrast. This implies a divisive inhibitory mechanism which could result from increased potassium conductance in the postsynaptic membrane. The net effect would be a shunting or leakage of excitatory postsynaptic potentials (EPSPs) which would reduce responsiveness proportionally.

A similar gain reduction is reported by DeValois and Tootell (1983) for a sinewave grating drifting in the optimal direction but of a spatial frequency that inhibits the cell. The remainder of their results are not quite so clear. They saw strong inhibition from the addition of a mask grating in 97 per cent of the simple cells, but in only 38 per cent of the complex cells studied. Also, the revealed inhibition was usually insufficient to account for the tight spatial frequency tuning of cortical cells; while inhibition was usually strong for high spatial frequencies, it was less apparent for spatial frequencies below the optimal for a given cell. Cortical cells differ most from LGN cells by having a more profound low spatial frequency cut-off (DeValois, Albrecht and Thorell, 1977), so inhibition alone is clearly insufficient to account for all spatial frequency selectivity.

The shaping of orientation tuning by inhibition is somewhat more clear-cut. Again, superposition of a mask grating with nonoptimal orientation reduces the gain of the contrast-response function of the base stimulus

(Morrone, Burr and Maffei, 1982; Bonds, 1983) without significant change of the sensitivity. In some instances, the tuning function measured in the usual way by varying the orientation of a single grating was matched by the function produced by varying the orientation of an inhibitory grating; here, orientation selectivity can be accounted for entirely by inhibition. This behavior was more evident in simple than complex cells (Bonds, 1983; Morrone, Burr and Maffei, 1982). We found inhibition to be strongest from masks oriented just at the limits of the excitatory tuning function, decreasing slightly as one moves further away from the optimal orientation. While Morrone, Burr and Maffei (1982) report that inhibition was uniform over all orientations outside the excitatory region, there is also an inhibitory peak visible at the excitatory limit in their Fig. 6. Such an arrangement has a definite advantage in tightening orientation selectivity, and is in keeping with the observation of Toyama, Kimura and Tanaka (1981) that intracortical inhibition seemed to be strongest from the same or neighbouring orientation columns.

Temporal frequency has also been studied in dual-grating experiments (Bonds, 1983). Inhibition clearly results from mask gratings with moderately high (6 to 16 Hz) drift frequencies, although it is much reduced or nonexistent if drift rates are increased (20 to 32 Hz). At temporal frequencies effective for inhibition, the spatial configuration does not seem to matter much; 8 Hz inhibits effectively even if it has optimal spatial qualities.

Where does this inhibition come from? It is clearly generated within the visual cortex. Intracellular cortical records of shocks to the optic radiation show that inhibition always follows (mono- or disynaptic) excitation by a delay of one synapse (Toyama *et al.*, 1974). It is probably postsynaptic, since a visually evoked suppression is apparent when the resting discharge is elevated by iontophoretic application of glutamate, but is blocked when bicuculline is added (Sillito, 1975). We are less certain about the types of cells in which the inhibition originates. Both Morrone, Burr and Maffei (1982) and DeValois and Tootell (1983) agree that complex cells are a likely source, since inhibition is largely independent of the spatial phase of the mask. However, apparent phase independence could result from the combination of signals from several (e.g. greater than five) simple cells. This alternative is reasonable since at least that many cells with average orientation tuning widths would be required to provide orientation tuning through inhibition to a unidirectional simple cell. Moreover, Toyama, Kimura and Tanaka (1981) report inhibitory cross-correlation only between eon/off cells (with exclusively ON or OFF fields, often classed as simple by other workers), eon/off onto simple cells, eon/off onto complex cells and simple onto complex cells.

The inhibition that mediates the temporal frequency bandpass of cortical cells is least specific, since it seems to be independent of the mask configuration. All intracellular studies report that excitation is always accompanied by inhibition regardless of the stimulus configuration (e.g. Creutzfeldt and Ito, 1968; Toyama *et al.*, 1974). This principle of 'ubiquitous inhibition' is probably reflected in cortical response dynamics, where saturation of the contrast-response function occurs at much lower contrasts (ca. 30 per cent.; e.g. Dean, 1981) than it does for LGN cells, even when the cortical stimulus is optimal. The impact of inhibition on the temporal frequency response most likely stems from the observation that inhibitory post synaptic potentials (IPSPs) seem to decay more slowly (20 to 40 ms) than do EPSPs (10 ms; see Creutzfeldt and Ito, 1968). Repetitive stimulation of the cortical receptive field thus causes collision of excitatory bursts with inhibition lingering from the last burst, with a concomitant decrease in responsiveness as the stimulus frequency is increased. At the highest stimulus frequencies the impact of inhibition is lessened since the cells which provide it are themselves less responsive. Cortical response saturation may itself be a consequence of an equilibrium between excitation and lingering inhibition. The slow decay of IPSPs could also account for the apparent spatial phase independence of intracortical inhibition.

Examination of the sources of inhibition that governs spatial selectivity should acknowledge the physical organization of the visual cortex. From the studies of the cortical representation of spatial frequency, this parameter seems to be less well organized than is orientation. It is thus not too surprising that spatial frequency inhibition is less readily understood than orientation inhibition. We know that prolonged stimulation with gratings of a single spatial frequency (but all orientations) leads to columnar labeling (with 2-deoxyglucose, 2-DG) throughout the visual cortex (Thompson and Tolhurst, 1980; Silverman, Tootell and DeValois, 1980; Tootell, Silverman and DeValois, 1981). In contrast, however, single-unit studies show only marginally organized spatial frequency subunits. Maffei and Fiorentini (1977) and to a lesser extent Berardi *et al.* (1982) have reported that spatial frequency is organized in a laminar rather than columnar fashion, while Tolhurst and Thompson (1982) conclude that the organization is neither laminar nor columnar. These authors report that cells with similar spatial frequency tuning are grouped in clusters that extend for approximately 200 μm in both laminar and columnar planes, and suggest that the columns seen in the 2-DG experiments may represent several spatial frequencies. They further suggest that a 2-DG column could be composed of several 200 μm clusters, each representing one (harmonically) related spatial frequency. The 2-DG label would be found throughout the entire column due to 'labelling both of discrete clusters of neuronal somata and also of the terminals of vertically running axonal connections . . .' Some anatomical support for this model may be found in the work of Rockland and Lund (1983) who report that intracortical injections of horeradish peroxidase into squirrel and macaque monkeys produces a lattice of intrinsic connections consisting of labeled walls surrounding unlabeled central lacunae in layers II, III and IVb. Within the walls are a series of regularly arranged loci of particularly dense label, in essence a column containing smaller clusters. In any event, the patchy organization for spatial frequency that is now apparent is consistent with the 'patchy' nature of spatial frequency inhibition.

In the same vein, the regular organization of orientation columns and inhibition is also consistent. In cat (Albus, 1975) and monkey (Hubel and Wiesel, 1974) as well as other species such as the tree shrew (Humphrey and Norton, 1980), orientation columns can be conceptualized as slabs of tissue approximately 50 μm thick, with each column representing a change of about 10° in orientation. If, as stated above, inhibition is strongest at the limits of the excitatory tuning curve (within a range of about ± 20° to 30° of the optimum orientation), then one might expect a given cell to be most strongly inhibited by cells lying within about 100 to 150 μm of itself. If the density of inhibitory synapses is a useful indicator of the strength of inhibition, then one might expect to find a high density of cellular processes within a distance of about 150 μm from the cell soma. Recent studies in which cells were intracellularly filled with horseradish peroxidase (e.g. Gilbert and Wiesel, 1979, 1983) suggest that this is, in fact, true. Most cells thus illustrated show both local and widespread intrinsic connections, with the local projections appearing considerably more dense than the widespread processes. The majority of these local connections are located within 150 μm of the soma when measured in a plane parallel to the cortical surface. We feel that these figures are more than coincidental in supporting the role of inhibition in specifying orientation selectivity.

It should be clear at this point that, despite the abundance of attractive evidence in support of the basic hypothesis, several fundamental and frustrating inconsistencies remain unexplained. For example, if inhibition stems from eon/off and simple cells, as suggested by the cross-correlation studies, what force is responsible for the reduction or elimination of their own maintained discharge? Certainly not other cells of the same type—because of their low maintained discharge. Also, with regard to spatial frequency and, to a lesser extent, orientation selectivity, inhibition alone is insufficient to account for observed tuning. One could invoke 'shaped' (intracortical) excitation as an alternative, but according to the cross-correlation studies this only impinges onto complex and hypercomplex II cells, and the inadequacies of inhibition are also seen in simple cells. Perhaps most fundamental to the comprehension of the cooperative neuronal network, we are confronted with a classic 'chicken and egg' problem. If neuron A receives its spatial specificity through inhibition from neuron B, and vice versa, where does the process begin? It may be necessary to propose the existence of some *intrinsically specified* neurons that rely on mechanisms other than network inhibition (e.g.

the classical model of selective combination of LGN afferents or simply specification in the LGN itself; see Vidyasagar and Urbas, 1982) to control their own selectivity. These 'ur-cells' (primitive or basic cells), which may constitute the population for which inhibition alone is insufficient to explain selectivity, could serve as templates for the organization of orientation columns or spatial frequency clusters. The observation of a few cells (about 20 per cent.) with mature orientation selectivity in very young kittens (Bonds, 1978) where synaptic density is probably insufficient to support much of a cooperative network lends some credibility to the existence of such cells.

The most important message to be taken from this discussion is that, as agreed by all of the contributors in this area, there is no such thing as a cortical single unit. Cortical response properties are not the exclusive domain of the unit under study, but belong to an elaborate local syncytium. Apparently ineffective stimuli may be of more importance to the network than those which cause the unit under study to fire. Attempts at understanding cortical response mechanisms through either mapping of discrete areas of the receptive field or by treating the neuron as an independent linear analyzer thus cannot tell the whole story.

REFERENCES

Albus, K. (1975). A quantitative study of the projection area of the central and paracentral visual field in the cat: the spatial organization in the orientation domain. *Exp. Brain Res.*, **24**, 181–202.

Berardi, N., Bisti, S., Cattaneo, A., Fiorentini, A., and Maffei, L. (1982). Correlation between the preferred orientation and spatial frequency of neurons in visual areas 17 and 18 of the cat. *J. Physiol.*, **323**, 603–618.

Bishop, P. O., Coombs, J. S., and Henry, G. H. (1971). Interaction effects of visual contours on the discharge frequency of simple striate neurons. *J. Physiol.*, **219**, 659–687.

Bonds, A. B. (1978). Development of orientation tuning in the visual cortex of kittens. In *Developmental Neurobiology of Vision* (Ed. Freeman R.D.), Plenum Press, New York, pp. 31–42.

Bonds, A. B. (1983). Inhibitory contribution to the spatial selectivity of single cells in cat striate cortex. *Inv. Ophthal. Vis. Sci. Suppl. (ARVO Spring Meeting)*, **1983**, 229.

Creutzfeldt, O. D., and Ito, M. (1968). Functional synaptic organization of primary visual cortex neurons in the cat. *Exp. Brain Res.*, **6**, 324–352.

Creutzfeldt, O. D., Kuhnt, U., and Benevento, L. A. (1974). An intracellular analysis of visual cortical neurons to moving stimuli: responses in a cooperative neuronal network. *Exp. Brain Res.*, **21**, 251–274.

Dean, A. F. (1981). The relationship between response amplitude and contrast for cat striate cortical neurons. *J. Physiol.*, **318**, 413–427.

Dean, A. F., Hess, R. F., and Tolhurst, D. J. (1980). Divisive inhibition involved in directional selectivity. *J. Physiol.*, **308**, 84P.

DeValois, K. K., and Tootell, R. B. H. (1983). Spatial-frequency specific inhibition in cat striate cortex cells. *J. Physiol.*, **336**, 359–376.

DeValois, R. L., Albrecht, D. G., and Thorell, L. G. (1977). Spatial tuning of LGN and cortical cells in the monkey visual system. In *Spatial Contrast* (Eds. H. Spekreijse and L. H. van der Tweel), North-Holland, Amsterdam, pp. 60–63.

Enroth-Cugell, C., and Robson, J. G. (1966). The contrast sensitivity of retinal ganglion cells of the cat. *J. Physiol.*, **187**, 517–552.

Frishman, L. J., Schweitzer-Tong, D. E., Ronson, J. G., Troy, J. B., and Enroth-Cugell, C. (1983). Temporal frequency tuning of cat retinal ganglion cells. *Inv. Ophthal. Vis. Sci. Suppl. (ARVO Spring Meeting)*, **1983**, 264.

Garey, L. J., and Powell, T. P. S. (1971). An experimental study of the termination of the lateral geniculo-cortical pathway in the cat. *Proc. Roy. Soc., B*, **179**, 41–63.

Gilbert, C. D., and Wiesel, T. N. (1979). Morphology and intracortical projections of functionally identified neurons in cat visual cortex. *Nature*, **280**, 120–125.

Gilbert, C. D. and Wiesel, T. N. (1983). Clustered intrinsic connections in cat visual cortex. *J. Neurosci.*, **3**, 1116–1133.

Heggelund, P. (1981a). Receptive field organization of simple cells in cat striate cortex. *Exp. Brain Res.*, **42**, 89–98.

Heggelund, P. (1981b). Receptive field organization of complex cells in cat striate cortex. *Exp. Brain Res.*, **42**, 99–107.

Hubel, D. H., and Wiesel, T. N. (1974). Uniformity of monkey striate cortex: a parallel relationship between field size, scatter and magnification factor. *J. comp. Neurol.*, **158**, 295–306.

Humphrey, A. L. and Norton, T. T. (1980). Topographic organization of the orientation column system in the striate cortex of the tree shrew (*Tupaia glis*). I. Microelectrode recording. *J. comp. Neurol.*, **192**, 531–548.

Kaplan, E., Marcus, S., and So, Y. T. (1979). Effects of dark adaptation on spatial and temporal properties of receptive fields in cat lateral geniculate nucleus. *J. Physiol.*, **294**, 561–580.

Lee, B. B., Cleland, B. G., and Creutzfeldt, O. D. (1977). The retinal input to cells in area 17 of the cat's cortex. *Exp. Brain Res.*, **30**, 527–538.

Levick, W. R., and Thibos, L. N. (1980). Orientation bias of cat retinal ganglion cells. *Nature*, **286**, 389–390.

Maffei, L., and Fiorentini, A. (1977). Spatial frequency rows in the striate visual cortex. *Vision Res.*, **17**, 257–264.

Morrone, M. C., Burr, D. C., and Maffei, L. (1982). Functional implications of cross-orientation inhibition of cortical visual cells. *Proc. Roy. Soc. (Lond.), B*, **216**, 335–354.

Movshon, J. A., Thompson, I. D., and Tolhurst, D. J. (1978a). Spatial summation in the receptive fields of simple cells in the cat's striate cortex. *J. Physiol.*, **283**, 53–77.

Movshon, J. A., Thompson, I. D., and Tolhurst, D. J. (1978b). Spatial and temporal contrast sensitivity of neurons in areas 17 and 18 of the cat's visual cortex. *J. Physiol.*, **283**, 101–120.

Rockland, K. S., and Lund, J. S. (1983). Intrinsic laminar lattice connections in primary visual cortex. *J. comp. Neurol.*, **216**, 303–318.

Rose, D., and Blakemore, C. B. (1974). An analysis of orientation selectivity in the cat's visual cortex. *Exp. Brain Res.*, **20**, 1–17.

Sillito, A. M. (1975). The effectiveness of bicuculline as an antagonist of GABA and visually evoked inhibition in the cat's striate cortex. *J. Physiol.*, **250**, 287–304.

Sillito, A. M. (1977). Inhibitory processes underlying the directional specificity of simple, complex and hypercomplex cells in the cat's visual cortex. *J. Physiol.*, **271**, 699–720.

Silverman, M. S., Tootell, R. B., and DeValois, R. L. (1980). Deoxyglucose mapping of orientation and spatial frequency in the cat visual cortex. *Inv. Ophth. Vis. Sci. Suppl.* (*ARVO Spring Meeting*), **1980**, 225.

Thompson, I. D., and Tolhurst, D. J. (1980). The representation of spatial frequency in cat visual cortex: a ^{14}C-2-deoxyglucose study. *J. Physiol.*, **300**, 58P–59P.

Tolhurst, D. J., and Thompson, I. D. (1982). Organization of neurons preferring similar spatial frequencies in cat striate cortex. *Exp. Brain Res.*, **48**, 217–227.

Tootell, R. B., Silverman, M. S., and DeValois, R. L. (1981). Spatial frequency columns in primary visual cortex. *Science*, **214**, 813–815.

Toyama, K., Kimura, M., and Tanaka, K. (1981). Cross-correlation analysis of interneural activity in cat visual cortex. *J. Neurophysiol.*, **46**, 191–201.

Toyama, K., Matsunami, K., Ohno, T., and Tokashiki, S. (1974). An intracellular study of neural organization in the visual cortex. *Exp. Brain Res.*, **21**, 45–66.

Vidyasagar, T. R., and Urbas, J. V. (1982). Orientation sensitivity of cat LGN neurons with and without inputs from visual cortical areas 17 and 18. *Exp. Brain Res.*, **46**, 157–169.

Models of the Visual Cortex
Edited by D. Rose and V. G. Dobson

CHAPTER 30

Serial processing of color in the monkey's striate cortex

CHARLES R. MICHAEL
Department of Physiology, Yale Medical School, 333 Cedar Street, New Haven, CT 06510, USA

PROPERTIES OF CORTICAL COLOR CELLS

In 1962 David Hubel and Torsten Wiesel first described simple and complex cells in the cat's striate cortex. They detailed the response properties and receptive fields of the two cell types and proposed a serial hierarchical scheme for the wiring diagram of these neurons. They suggested that geniculate axons synapse on simple cells and they, in turn, terminate on complex cells. The first stage leads to the properties of orientation selectivity and the second to a positional generalization of that property. In a later paper (Hubel and Wiesel, 1965) they described the attributes of hypercomplex cells and suggested that they received information from complex cells. Subsequent research has, in some cases, supported the original scheme but in other corners the concept of serial processing has been criticized.

A growing school of thought projects the belief that parallel processing is the important issue. For instance, it has been reported that geniculate axons end directly on complex cells and perhaps even on hypercomplex cells, that simple cells cannot be the input to complex neurons and that hypercomplex cells are really extreme forms of simple cells or complex ones. For a comprehensive review of this subject, see Stone, Dreher and Leventhal (1979).

I believe that the really interesting question is not whether integration is serial or parallel but just how does it occur between one cell and the next. By definition, there must be serial processing. When cell A synapses on cell B and it, in turn, contacts cell C, it is clearly a case of serial processing. It does not follow that there must be parallel processing but, indeed, there is considerable evidence that it does exist. The X- and Y-cell classes first enunciated by Enroth-Cugell and Robson (1966) are now thought of as the classical parallel systems in the visual pathway. However, even before the discovery of the X- and Y-cell types, the existence of on- and off-center cells was taken as evidence that there were two separate systems conveying opposite information from the eye to the brain.

In my work on the physiology of color vision in the monkey I have found that luminance-contrast information and color-contrast information are processed along separate, parallel channels at the cortical level. Within the confines of the color cell population information appears to be processed in a serial manner. In the rhesus monkey's striate cortex I have studied four classes of color-sensitive cells: concentric, simple, complex and hypercomplex (Michael, 1978a–c, 1979). All four types are generally unresponsive to white stimuli, strongly preferring monochromatic ones. I have suggested that these cells are connected in serial order. Geniculate fibers may make contact with concentric cells which then synapse on simple neurons. They might serve as the inputs to complex cells which, in turn, may end on hypercomplex neurons.

The concentric cells have circular center surround receptive fields with one red/green opponent color system in the center and the opposite in the surround. They are most sensitive to the simultaneous presentation of a two-color stimulus, one color covering the field center and the other illuminating the surround. The best evidence for the proposed geniculate/concentric connection comes from multiple-unit recordings in which the stimulus requirements, response properties and receptive field organization of two units were studied simultaneously with a single tungsten microelectrode (Michael, 1978a). Sometimes in such dual recordings one of the units had the characteristics of a concentric cortical cell while the other had the properties of a geniculate afferent. The units always had field centers which were in register and of the same diameter. Both were always monocular and driven by the same eye. Subsequent histological reconstructions revealed that they were limited to layer 4. Early studies (Michael, 1978a) indicated that the afferents in these multiple-unit recordings were often of the type II variety. Geniculate type II neurons do not have a center/surround receptive field organization (Wiesel and Hubel, 1966). Instead, they receive opponent inputs from two sets of cones with identical spatial distributions over the retina. More recently, however, experiments with improved tungsten electrodes (unpublished observations) have revealed that the geniculate member of these paired recordings was usually of the type I class. Type I geniculate cells have the classical center/surround field arrangement with red or green cones in the center and the other type in the surround (Wiesel and Hubel, 1966). In the dual recordings the center sign response of the afferent was always the same as that of the corresponding spectral element of the concentric cell's field center, i.e. a red on-center geniculate unit was paired with a red on-green off-center concentric cell.

There is general agreement amongst those who have studied them (Dow, 1974; Dow and Gouras, 1973; Gouras, 1974; Michael, 1978a) that the concentric cells are probably the first stage in the integration of color-contrast information in the striate cortex. Geniculate afferents which do not convey color information probably synapse directly on simple cells (Hubel and Wiesel, 1962, 1968). It appears, then, that in color analysis there is an extra stage, the concentric neuron, but it is not clear why an additional step should be necessary.

The simple cells are also double-opponent color cells but in this case they have rectangular receptive fields. They usually consist of a central oblong area containing one red/green system flanked by two parallel regions with the opposite organization. Because of their field organization these cells are most responsive to bar or edge stimuli with the proper orientation. In some multiple-unit recordings from pairs of cells, one unit was a simple neuron while the other member had the characteristics of a concentric cell. In all such cases the concentric unit's field center was in register with the central region of the simple cell's field and they were the same width. The field center of the concentric cell never occupied one of the flanks of the simple unit. Furthermore, the two units had the same red/green opponent color organization in the central elements of their receptive fields, as well as the same spectral sensitivities. Finally, these monocular cells were always activated by the same eye.

The third class of neurons, the complex cells, differed significantly from the first two types. They had roughly square receptive fields which did not have excitatory/inhibitory subdivisions and consequently could not be mapped with stationary stimuli. They responded only to moving monochromatic bars or edges of the proper orientation. Their spectral sensitivity curves consisted of single, relatively narrow bands which resembled either the short- or the long-wavelength component for the double-opponent color cells. Chromatic adaptation studies revealed that the complex cells received excitatory connections from either the red or the green cones and 'silent' antagonistic inputs from the other class. In other words, they had an opponent color input but only the on excitatory component manifested itself in the responses of the cell.

Paired multiple-unit recordings from simple and complex cells yielded information which may relate to their possible synaptic contact. In such recordings the fields of the two cells were always in register, with their lengths being equal. The complex field was always wider than that of the simple unit. Both units had the same axis orientation, directional selectivity and stimulus preference. The complex cells were often binocular and when one eye dominated, it was the same one that influenced the simple unit exclusively. Finally, the spectral sensitivity curve for the complex cell closely matched that of the on-center component for the simple neuron. It never coincided with the off portion of the simple cell's spectral sensitivity.

Hypercomplex color cells, like complex ones, were excited only by movement of a specifically oriented monochromatic bar or edge of light. However, in this case, it had to be limited in its length at both ends. The rectangular receptive field of a hypercomplex cell consisted of a central activating area and two 'silent' antagonistic flanks, all three of which had the same axis-orientation requirements and the same spectral sensitivities.

Multiple-unit recordings were sometimes made from a complex/-hypercomplex pair of cells. The complex unit's field was always the same size as and in register with either the activating area or one of the two antagonistic flanks. Both cells always had the same axis orientation, directional selectivity, spectral sensitivity, similar stimulus preferences and usually belonged to the same ocular dominance group.

COLUMNAR ORGANIZATION OF COLOR CELLS

The color cells are restricted to columnar regions which are separated from other areas containing cells that respond to both white and color stimuli (Michael, 1981). A color column, by definition, contains cells which are excited only by monochromatic stimuli. The cells within a single column have the full range of receptive field types and may vary in their axis orientation or in their eye preference, but they always have a strong preference for quite similar wavelengths. For instance, within a column the concentric and simple cells have the same double-opponent color organization, e. g. all red on-, green off-center. The complex and hypercomplex cells in the same column have single narrow-band spectral sensitivity curves which always peak at about the same locus as that of the on-center component of the double-opponent cells. The columns extend through the entire thickness of the gray matter and their walls appear to be perpendicular to the layers. Surface projections of multiple electrode tracks have revealed that the color columns are really shaped like slabs (100 to 250 μm wide), as are the ocular preference and axis orientation columns (Hubel and Wiesel, 1974, LeVay, Hubel and Wiesel, 1975). The boundaries of the three columnar systems appear to have no relationship to one another.

The center of Fig. 1 depicts the reconstruction of an electrode track which passed through the striate cortex at an angle nearly perpendicular to the surface and to the layers. All sixteen cells encountered in the penetration were exclusively responsive to color stimuli and were members of a single color column. With the exception of the five concentric units, all of the cells had the same axis orientation. The concentric and simple cells were driven exclusively by the contralateral eye while the binocular complex and hypercomplex neurons were driven preferentially by it. The complex and hypercomplex cells were found only above or below layer 4 while simple and concentric cells were confined to it.

To the right of the electrode track are examples of the four receptive field classes. All of the concentric and simple cells in this column, as illustrated by cells 7 and 10, were double-opponent color units with green on-, red off-center elements and the opposite organization in the surround or flanks, as the case may be. The spectral sensitivity curves of both the concentric and the simple cells had basically the same maxima and shape, namely 'on' peaks at about 480 nm and 'off' peaks at around 620 nm (far right of Fig. 1). All of the complex and hypercomplex cells responded to moving green bars and had single spectral sensitivity curves peaking at about 480 nm, coinciding with the short-wavelength green on component of the simple and concentric curves (Fig. 1, cells 4 and 15). None of the complex or hypercomplex cells in this penetration were sensitive to long-wavelength stimuli.

To the left side of the electrode track in Fig. 1 is a 'wiring diagram' of the first stages of cortical integration (Blasdel and Lund, 1983; Lund and Boothe, 1975). Nearly all of the opponent color cells in the monkey's lateral geniculate nucleus are confined to the parvocellular layers. They project primarily to layer 4Cß of the striate cortex while a somewhat lesser number, and an apparently separate population, go to layer 4A (Blasdel and Lund, 1983). In contrast, the magnocellular layers of the geniculate, which contain mostly luminance-sensitive cells, send axons almost exclusively to layer 4Cα which is sandwiched in between the two parvocellular-recipient layers. The parvocellular/magnocellular division appears to represent a separation of color and noncolor processing, the color cells described here presumably being part of the parvocellular pathway. In this regard, the laminar distributions of the color cells were most interesting. The concentric and simple double-opponent cells were restricted to 4A and 4Cß while the complex and hypercomplex color cells were confined to the layers above and below lamina 4. In other words, the cells presumed to represent earlier stages in the cortical integration of color were found in laminae where most of the parvocellular afferents terminate. In contrast, the more complicated cell types were found distant from the geniculate inputs, implying that their afferents probably come from cells within the cortex itself.

The suggested serial synaptic scheme is based mainly on the multiple-unit extracellular recordings described earlier. To prove definitively that one of these cell types synapses on another class would require intracellular recordings combined either with horseradish peroxidase injections or electron microscopy. Ultimately it will be necessary to determine the relationship between the morphological characteristics of these various cells and their functional properties. For the time being, however, the evidence from the multiple-unit recordings, from the laminar positions of the cells and from their columnar organization suggests an exclusive color system processing wavelength information in a serial fashion. One should remember, however, that it represents only one of a number of separate, parallel pathways into

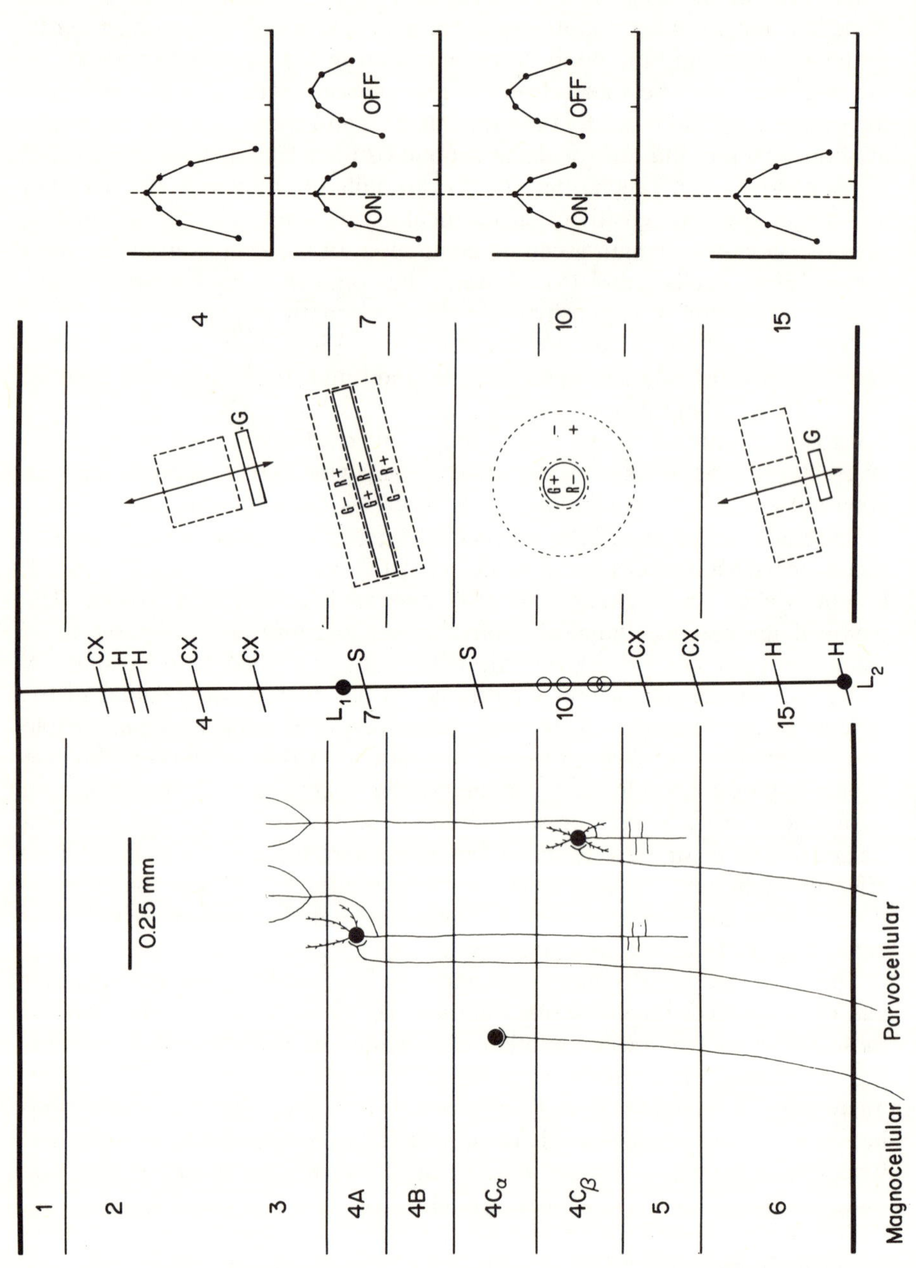
1
2
3
4A
4B
4Cα
4Cβ
5
6
Magnocellular
Parvocellular
0.25 mm
CX
H
S
L1
L2
4
7
10
15
G
ON
OFF

Figure 1 (Center) Reconstruction of a nearly perpendicular electrode penetration made through the layers of the monkey's striate cortex. All of the units encountered were responsive only to color stimuli. The positions of the sixteen cells studied along the track are indicated by circles or short lines intersecting the electrode's path (circles, concentric cells; S, simple; CX, complex; H, hypercomplex). The angles of the short lines represent the cells' axis orientations. L_1 is a lesion made at the site of the first concentric (nonoriented) cell encountered. L_2 marks the last unit in the track. To the right of the electrode track are examples of the four types of receptive fields and (far right) their corresponding spectral sensitivity curves. The interrupted vertical line emphasizes the close correspondence of the short-wavelength component of the concentric and simple cells with the spectral sensitivities of the complex and hypercomplex cells. The ordinates for the four curves represent the logarithm of the relative sensitivities of the cells (logarithm of the reciprocal of the threshold intensity, threshold taken as the intensity which produced a criterion number of spikes in response to 50 per cent of the test runs). The abscissa represents the spectral range 400 to 700 nm in 100 nm increments. To the left of the track is a diagram of the major laminar distributions of the parvocellular (color: 4A and $4C_\beta$) and magnocellular (luminance contrast: $4C_\alpha$) geniculate inputs to the striate cortex, as well as the axon projections of the spiny stellate cells in 4A and $4C_\beta$ to lower 3 and upper 5 (adapted from Lund and Boothe, 1975 by permission of Alan R. Liss, Inc.).

and within the cortex which deal with color contrast, luminance contrast, orientation selectivity, eye preference and spatial frequency (Movshon, Thompson and Tolhurst, 1978).

ACKNOWLEDGEMENTS

This research was generously supported by The National Eye Institute (grant number EY00568).

REFERENCES

Blasdel, G. G., and Lund, J. S. (1983). Termination of afferent axons in macaque striate cortex. *J. Neurosci.*, **3**, 1389–1413.

Dow, B. M. (1974). Functional classes of cells and their laminar distribution in monkey visual cortex. *J. Neurophysiol.*, **37**, 927–946.

Dow, B. M., and Gouras, P. (1973). Color and spatial specificity of single units in rhesus monkey foveal striate cortex. *J. Neurophysiol.*, **36**, 79–100.

Enroth-Cugell, C., and Robson, J. G. (1966). The contrast sensitivity of retinal ganglion cells of the cat. *J. Physiol.*, **187**, 517–552.

Gouras, P. (1974). Opponent-colour cells in different layers of foveal striate cortex. *J. Physiol.*, **238**, 583–602.

Hubel, D. H., and Wiesel, T. N. (1962). Receptive fields, binocular interaction and functional architecture in the cat's visual cortex. *J. Physiol.*, **160**, 106–154.

Hubel, D. H., and Wiesel, T. N. (1965). Receptive fields and functional architecture in two nonstriate visual areas (18 and 19) of the cat. *J. Neurophysiol.*, **28**, 229–289.

Hubel, D. H., and Wiesel, T. N. (1968). Receptive fields and functional architecture of monkey striate cortex. *J. Physiol.*, **195**, 215–243.

Hubel, D. H., and Wiesel, T. N. (1974). Sequence regularity and geometry of orientation columns in the monkey striate cortex. *J. comp. Neurol.*, **158**, 267–294.

LeVay, S., Hubel, D. H., and Wiesel, T. N. (1975). The pattern of ocular dominance columns in macaque visual cortex revealed by a reduced silver stain. *J. comp. Neurol.*, **159**, 559–576.

Lund, J. S., and Boothe, R. G. (1975). Interlaminar connections and pyramidal neuron organization in the visual cortex, area 17, of the macaque monkey. *J. comp. Neurol.*, **159**, 305–334.

Michael, C. R. (1978a). Color vision mechanisms in monkey striate cortex: dual-opponent cells with concentric receptive fields. *J. Neurophysiol.*, **41**, 572–588.

Michael, C. R. (1978b). Color vision mechanisms in monkey striate cortex: simple cells with dual opponent-color receptive fields. *J. Neurophysiol.*, **41**, 1233–1249.

Michael, C. R. (1978c). Color-sensitive complex cells in monkey striate cortex. *J. Neurophysiol.*, **41**, 1250–1266.

Michael, C. R. (1979). Color-sensitive hypercomplex cells in monkey striate cortex. *J. Neurophysiol.*, **42**, 726–744.

Michael, C. R. (1981). Columnar organization of color cells in monkey's striate cortex. *J. Neurophysiol.*, **46**, 587–604.

Movshon, J. A., Thompson, I. D., and Tolhurst, D. J. (1978). Spatial and temporal contrast sensitivity of neurons in areas 17 and 18 of the cat's visual cortex. *J. Physiol.*, **283**, 101–120.

Stone, J., Dreher, B., and Leventhal, A. (1979). Hierarchical and parallel mechanisms in the organization of visual cortex. *Brain Res. Rev.*, **1**, 345–394.
Wiesel, T. N., and Hubel, D. H. Spatial and chromatic interactions in the lateral geniculate body of the rhesus monkey. *J. Neurophysiol.*, **29**, 1115–1156.

Models of the Visual Cortex
Edited by D. Rose and V. G. Dobson

CHAPTER 31

Visual cortex and retinal eccentricity: uniformity or non-uniformity?

G.A. Orban
KUL Laboratorium voor Neuro- en Psychofysiologie, Campus Gasthuisberg, Herestraat, B-3000 Leuven, Belgium

Hubel and Wiesel (1974, 1977) have formulated the notion that the visual cortex is a stereotype uniform structure in which the functional subunit—the hypercolumn—is repeated over and over again, almost in a crystal-like fashion. The only change with eccentricity is a quantitative one: the number of hypercolumns per visual degree decreases with eccentricity. These studies involved mainly mapping experiments in which receptive field (RF) dimensions and RF scatter were measured. Recently we (Orban and Kennedy, 1981; Orban, Kennedy and Maes 1981a, 1981b; Duysens *et al.*, 1982a 1982b; and also Duysens, Orban and Cremieux, in preparation) have conducted systematic studies of areas 17, 18 and 19 of the cat as a function of eccentricity. In these studies neuronal properties other than just RF dimension were investigated: RF organization and functional properties such as orientation selectivity, binocularity, direction selectivity, velocity sensitivity and duration sensitivity. One of the major outcomes of these studies is that almost every neuronal property we have looked at changed significantly as a function of eccentricity in at least one of the three cortical areas investigated. This led us to propose that the cortex does not only undergo quantitative changes with eccentricity but also qualitative changes: not only do the hypercolumns vary in number but their content changes with eccentricity (Orban, 1984). Also, work from other laboratories (Albus, 1975a; Berman *et al.*, 1982; Leventhal, 1983; Poggio *et al.*, 1975; Wilson and Sherman, 1976; Zeki, 1983) has confirmed the functional non-uniformity of cat and monkey visual cortex. The purpose of the present review is to evaluate the hypothesis of a structural uniformity of the visual cortex in view of the many functional non-uniformities of the visual cortex, which likely underlie the changes in visual perception with eccentricity. First, I shall review the evidence for functional non-uniformity obtained in our laboratory and, second, the implications of these non-uniformities for the structural organization of the visual cortex will be evaluated.

Standard electrophysiological procedures were followed to record from single cells in areas 17, 18 and 19 of 64 cats. Cells were subjected both to a qualitative study with hand-held stimuli and a quantitative study with a computer-controlled stimulator. The study with hand-held stimuli yielded the following information: RF dimensions evaluated from the minimum discharge field (Kato, Bishop and Orban, 1978), ocular dominance (degree of excitatory input from both eyes), optimal orientation and orientation tuning width (range of orientations to which the cell responded), the presence and degree of end-stopping (decrease of response for lengths over an optimum) and RF organization. For RF classification we use the ABCS scheme (Henry, Harvey and Lund, 1979; Orban and Kennedy, 1981; Orban, 1984) which we feel improves upon the initial simple-complex scheme of Hubel and Wiesel (1962) by better defining the classification criteria.

The computer-controlled stimuli were high-contrast, narrow light slits of optimal length and orientation. Twenty velocities ranging from 0.3 to 700°/s were tested with the multihistogram technique which controls for cortical variability (Henry *et al.*, 1973; Maes and Orban, 1980). Velocity-response (VR) curves using the maximum firing rate as the measure were compiled. Comparison of VR curves for opposite directions of motion yielded a quantitative estimate of direction selectivity. Electrolytic lesions were made during and at the end of the penetration, allowing histological reconstruction of the electrode tracks. The positions of the 17–18 and 18–19 borders was determined by a set of criteria including cytoarchitectonics, retinotopy and velocity sensitivity (Orban, Kennedy and Maes, 1980; Duysens *et al.*, 1982a). The data base includes 849 cells which were hand plotted, of which 555 were quantitatively investigated. Eccentricities explored ranged from 0 to 45° in area 17 and 0 to 35° in areas 18 and 19.

As already shown by Hubel and Wiesel (1974) and others (Albus, 1975b; Wilson and Sherman, 1976; Dow *et al.*, 1981), the RF size increases with eccentricity. It is noteworthy, however, that the slope is about 2 to 3 times steeper in areas 18 and 19 than in area 17 (Table 1).

Any cell of which the response reliably decreased with increasing slit length was considered as end-stopped. According to Kato, Bishop and Orban (1978), this corresponds to an end-zone inhibition of at least 40 per cent. The proportion of end-stopped cells decreased with eccentricity in all three areas, this being clearest in area 19 and least in area 18 (Table 2). S-family cells are cells which have a narrow RF as plotted with a moving light slit and non-overlapping subregions for light onset and offset. It includes both end-stopped cells (HS-cells) and end-free cells (S-cells). The proportion of S-family cells decreased slightly with eccentricity in areas 17 and 18 (Table 2). With respect to orientation selectivity, the width of tuning changed little or not at all with eccentricity except for one cell class in area 17, the S-cells. This cell class is remarkable by having the narrowest width of tuning of all cell types of the three areas (Orban, 1984). The median orientation tuning width of area 17 S-cells increased on average from 42° below 10° eccentricity to 60° over 10° eccentricity. This cell class is also the only type to prefer orientations around the vertical or horizontal. This preference disappeared for eccentricities greater than 10°. Other changes in orientation preferences with eccentricity were also observed. Area 17 end-stopped cells, with RFs within 10° from the fixation point, preferred vertical orientations. This preference disappeared over 10° eccentricity. Finally, fewer area 18 cells preferred one of the oblique meridians. This could be the expression of a radial organization (Leventhal, 1983) as the oblique was orthogonal to the line linking RF to the fixation point. If one defines binocular cells as belonging to classes 3, 4 and 5 in the seven-point scale of Hubel and Wiesel (1962), their proportion increased with eccentricity in areas 17 and 18 and decreased in area 19 (Table 2).

Table 1 Slopes RF size (hand-plotted RF diameter of LGN cells and RF width of cortical cells) — eccentricity.

LGN	0.01
Area 17	0.09
Area 18	0.28
Area 19	0.21

Table 2 Changes with eccentricity in three cortical areas. (Numbers between brackets indicate the sample size on which percentages were calculated.)

	Area 17		Area 18		Area 19	
Eccentricity	≤ 10°	<10°	≤ 10°	< 10°	≤ 10°	< 10°
% S family	59(330)	51(72)	50(129)	44(102)	15(67)	23(43)
% end-stopped	38(375)	24(76)	27(150)	18(109)	76(76)	49(43)
% binocular	36(370)	57(76)	34(125)	57(105)	51(78)	23(59)

The sample of cortical cells tested quantitatively was divided into the same three eccentricity classes for the three areas: 0 to 5°, 5 to 15° and over 15°.

The range of velocities to which cells are sensitive changed with eccentricity (Fig. 1a, b). The lower end of this range is measured by 'response to slow

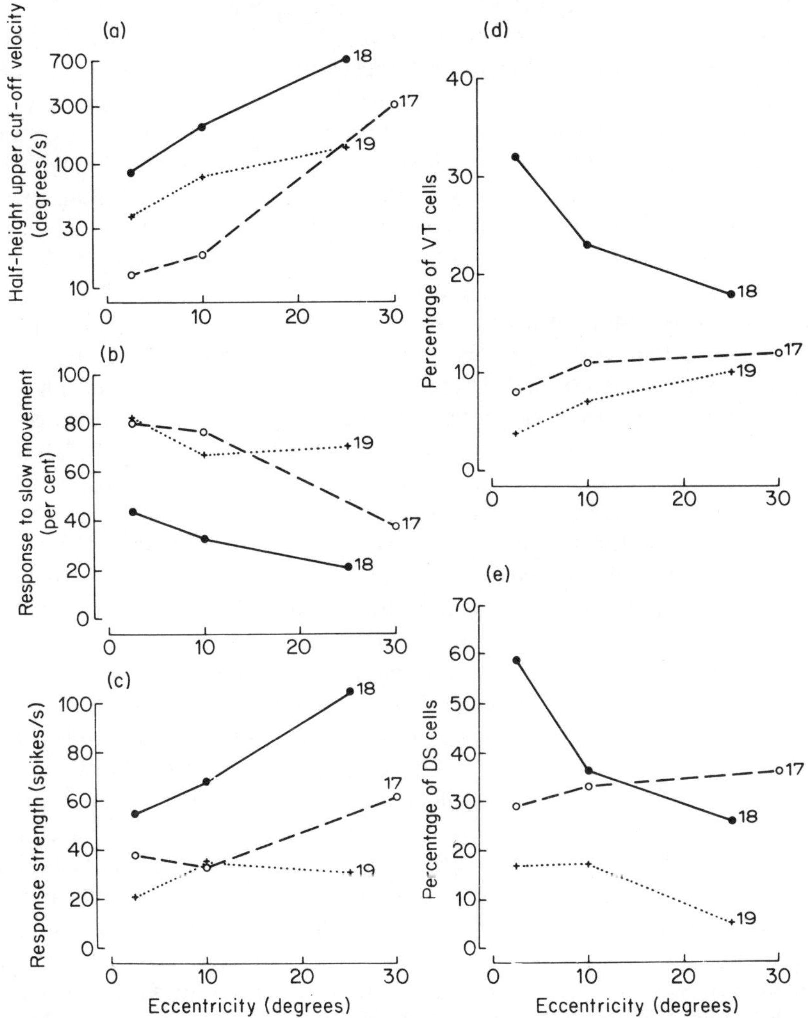

FIGURE 1 Changes in velocity sensitivity and direction selectivity as a function of eccentricity in areas 17, 18 and 19 of the cat: (a) half-height upper cut-off velocity; (b) response to slow movement; (c) response strength; (d) percentage of velocity-tuned (VT) cells; and (e) percentage of direction-selective (DS) cells. The sample sizes for the three eccentricity classes were: 153, 66 and 25 for eccentricity class 0–5°, 100, 66 and 42 for eccentricity class 5–15° and 49, 34 and 20 for eccentricity class over 15° in areas 17, 18 and 19 respectively.

movement' which is the response to the slowest velocity tested (0.3 or 0.5°/s) expressed as percentage of the maximum response (to the optimal velocity). The response to slow movement decreased with eccentricity in areas 17 and 18. The upper end of the range to which cells are sensitive is measured by the half-height upper cut-off velocity, i.e. the upper velocity at which the response falls to 50 per cent of the maximum response. This half-height upper cut-off velocity increased with eccentricity in all three areas, most clearly in area 17 and the least in area 19. In addition to this shift towards faster velocities with increasing eccentricity, the degree of selectivity decreased. Indeed, the proportion of velocity-tuned cells, defined as cells for which the width of velocity tuning (ratio of velocities corresponding to the intersection with the half-maximum response level) is 50 or less, decreased with eccentricity in area 18. In both other areas the proportions of these cells are small (Fig. 1d). The optimal velocity of velocity-tuned cells increased with eccentricity. More precisely, at larger eccentricities there were fewer cells tuned to slow velocities. With increasing eccentricity the degree of direction selectivity decreased. Direction-selective cells are defined as having an average response ratio of preferred over non-preferred directions of 3:1 or more (average over the 20 velocities tested). The proportion of direction-selective cells decreased with eccentricity in area 18, their proportions being lower in the two other areas (Fig. 1e).

It seems that the functional non-uniformities fall into three different types. A first type of physiological change seems to involve a decreased effectiveness of intracortical inhibition with increasing eccentricity. This is exemplified with the decrease in proportion of end-stopped and direction-selective cells (Table 2, Fig. 1e). Indeed, both end-stopping and direction selectivity depend on intracortical inhibition (Bishop, Kato and Orban, 1980; Emerson and Gerstein, 1977; Orban, Kato and Bishop, 1979; Sillito, 1977; Sillito and Versiani, 1977). This decreased effectiveness of intracortical inhibition could either be due to a decrease in intracortical inhibitory afferents, an increase in afferent excitatory input or decreased interaction between excitation and inhibition, e.g. because of temporal desynchronization. While there is little evidence for two of these changes (variations in intracortical inhibition or in excitatory–inhibitory interactions), an increase in afferent excitatory drive has been demonstrated anatomically in monkey striate cortex by the increase in the degree of convergence of the geniculocortical afferents with eccentricity (Hendrickson, Wilson and Ogren, 1978; Kaas, Lin and Casagrande, 1976; Myerson *et al.*, 1977; Tigges and Tigges, 1979; Tigges, Tigges and Perachio, 1977). In the monkey this increase in convergence of geniculocortical afferents with eccentricity is further supported by the steeper slope of the magnification–eccentricity relationship in area 17 as compared to the LGN. In the cat this is supported by the steeper slope of the RF size–eccentricity relationships in the cortical areas as compared to the LGN (Table 1) and by the increase in response strength with eccentricity (Fig. 1c). This increase with eccentricity of the strength in excitatory drive could decrease the efficiency of intracortical inhibition if the number of inhibitory inputs remains equal at all eccentricities as the uniformity of the intrinsic cortical connections suggests. In this way some functional non-uniformities would be the consequence of the intrinsic structural uniformity of the visual cortex. This view is supported by the observations made in area 18 of strobe-reared cats (Kennedy and Orban, 1983). The RF size did not change with eccentricity, indicating the same degree of excitatory subcorticocortical convergence at all eccentricities. In area 18 of those animals the proportion of direction selectivity and end-stopped cells was uniformly low. This can be taken as evidence that the inhibition became equally inefficient at all eccentricities.

The change of velocity sensitivity with eccentricity seems to be a second type of functional non-uniformity. The shift to faster velocities with increasing eccentricity could be due to the increase in RF dimensions. Indeed, for cortical cells there is correlation between RF width and upper cut-off velocity (Orban, Kennedy and Maes, 1981a). However, it should be noted that LGN cells, of which the RF is much smaller than that of cortical cells, have much higher upper cut-off velocities than cortical cells (Orban, Hoffmann and Duysens, 1981). It seems rather that, at least in area 17, the changes in velocity sensitivity are to a large extent due to the duration requirements of cells with central RFs. A large proportion of area 17 cells with central RFs are duration sensitive, i.e. require a long duration of visual stimulation in order to fire. The proportion of duration-sensitive cells decreases with eccentricity (Duysens, Orban and Cremieux, in preparation). It is unclear which mechanism underlies this duration threshold: a local in-field inhibition, a structural factor such as localization of excitatory and inhibitory synapses on the neuronal tree or membrane or transmitter release characteristics. Hence it is difficult to decide whether this functional uniformity requires a structural counterpart or not.

The third type of functional non-uniformity seems to require structural changes, as it involved properties linked to the columnar organization. The increase in binocularly driven cells with eccentricity in areas 17 and 18 (Table 2) could require that the ocular dominance columns of both eyes overlap more in the peripheral projections of the visual field than in the central ones. One should note, however, that in fact monocular cells receive excitatory input from one eye and inhibitory input from the other one (Kato, Bishop and Orban, 1981). Hence an alternative explanation of the decrease in the proportion of monocular cells with eccentricity could be the decreased effectiveness of intracortical inhibition. The changes in orientation preferences, at least in area 17, seem also to require some structural counterpart. In the projections of the central field some cell types prefer horizontal or vertical orientations (Leventhal and Hirsch, 1980; Orban and Kennedy, 1981, Orban

1984; Payne and Berman, 1983; Pettigrew, Nikara and Bishop, 1968) and with increasing eccentricity this organization seems to give way to a preference for radial orientations (Leventhal, 1983) which may reflect the bias of the retinal input (Levick and Thibos, 1982; Leventhal and Schall, 1983). It could well be that these meridional variations in orientation preference are stronger in the monkey than in the cat and affect not just S-cells but the overall population of area 17 with central RFs (Mansfield, 1974; Finlay, Schiller and Volman, 1976; Blakemore, Garey and Vital-Durand, 1981; Kennedy *et al.*, 1980). Finally, the functional change which may require the largest structural non-uniformity is the decrease in the proportion of non-oriented cells and of wavelength-selective cells in area 17 of the monkey (Poggio *et al.*, 1975; Zeki, 1983). According to Hubel and Livingstone (1981, 1982) the cytochrome oxidase patches in layers 2 and 3 correspond to agglomerations of non-oriented and colour-selective cells. Hence the changes in the proportion of non-oriented and wavelength-selective cells could imply a change in size or density of cytochrome oxidase patches. There is evidence that cytochrome oxidase patches may decrease in diameter with eccentricity (Hubel, personal communication). Hence an eccentricity-dependent cytochrome patch system would be superimposed on a more or less uniform system of hypercolumns.

As a conclusion one can say that at first glance the visual cortex seems to be structurally rather uniform and repetitive but that to consider the visual cortex as physiologically uniform is at least a partial view. Progress in physiological analyses has pointed towards marked non-uniformities, of which some may be a consequence of the intrinsic structural uniformity while others may depend on finer scale structural changes. Determination of these more subtle changes will require more refined anatomical techniques than those presently used.

REFERENCES

Albus, K. (1975a). Predominance of monocularly driven cells in the projection area of the central visual field in cat's striate cortex. *Brain Res.*, **89**, 341–347.

Albus, K. (1975b). A quantitative study of the projection area of the central and the paracentral visual field in area 17 of the cat. I. The precison of the topography. *Exp. Brain Res.*, **24**, 159–179.

Berman, N., Payne, B. R., Labar, D. R., and Murphy, E. H. (1982). Functional organization of neurons in cat striate cortex: variations in ocular dominance and receptive-field type with cortical laminae and location in visual field. *J. Neurophysiol.*, **48**, 1362–1377.

Bishop, P. O., Kato, H., and Orban, G. A. (1980). Direction-selective cells in complex family in cat striate cortex. *J. Neurophysiol.*, **43**, 1266–1283.

Blakemore, C. B., Garey, L. J., and Vital-Durand, F. (1981). Orientation preferences in the monkey's visual cortex. *J. Physiol. (Cond.)*, **319**, 78P.

Dow, B. M., Snyder, A. Z., Vautin, R. G., and Bauer, E. (1981. Magnification factor and receptive field size in foveal striate cortex of the monkey. *Exp. Brain Res.*, **44**, 213–228.

Duysens, J., Orban, G. A., van der Glas, H. W., and De Zegher, F. E. (1982a). Functional properties of area 19 as compared to area 17 of the cat. *Brain Res.*, **231**, 279-291.

Duysens, J., Orban, G. A., van der Glas, H. W., and Maes, H. (1982b). Receptive field structure of area 19 as compared to area 17 of the cat. *Brain Res.*, **231**, 293–308.

Emerson, R. C., and Gerstein, G. L. (1977). Simple striate neurons in the cat. II. Mechanisms underlying directional asymmetry and directional selectivity. *J. Neurophysiol.*, **40**, 136–155.

Finlay, B. L., Schiller, P. H., and Volman, S. F. (1976). Meridional differences in orientation sensitivity in monkey striate cortex. *Brain Res.*, **105**, 350–352.

Hendrickson, A. E., Wilson, J. R., and Ogren, M. P. (1978). The neuroanatomical organization of pathways between the dorsal lateral geniculate nucleus and visual cortex in Old World and New World Primates. *J. comp. Neurol.*, **182**, 123–136.

Henry, G. H., Bishop, P. O., Tupper, R. M., and Dreher, B. (1973). Orientation specificity and response variability of cells in the striate cortex. *Vision Res.*, **13**, 1771–1779.

Henry, G. H., Harvey, A. R., and Lund, J. S. (1979). The afferent connections and laminar distribution of cells in the cat striate cortex. *J. comp. Neurol.*, **187**, 725–744.

Hubel, D. H., and Livingston, M. S. (1981). Regions of poor orientation tuning coincide with patches of cytochrome oxidase staining in monkey striate cortex. *Soc. Neurosci. Abstr.*, **7**, 357.

Hubel, D. H., and Livingstone, M. S. (1982). Cytochrome oxidase blobs in monkey area 17: response properties and afferent connections. *Soc. Neurosci. Abstr.*, **8**, 706.

Hubel, D. H., and Wiesel, T. N. (1962). Receptive fields, binocular interaction and functional architecture in the cat's visual cortex. *J. Physiol. (Lond.)*, **160**, 106–154.

Hubel, D. H., and Wiesel, T. N. (1974). Uniformity of monkey striate cortex: a parallel relationship between field size, scatter, and magnification factor. *J. comp. Neurol.*, **158**, 295–305.

Hubel, D. H., and Wiesel, T. N. (1977). Functional architecture of macaque monkey visual cortex. *Proc. Roy. Soc. (Lond.) B*, **198**, 1–59.

Kaas, J. H., Lin, C.-S., and Casagrande, V. A. (1976). The relay of ipsilateral and contralateral retinal input from the lateral geniculate nucleus to striate cortex in the owl monkey: a transneuronal transport study. *Brain Res.*, **106**, 371–378.

Kato, H., Bishop, P. O., and Orban, G. A. (1978). Hypercomplex and the simple/complex cell classifications in cat striate cortex. *J. Neurophysiol.*, **41**, 1071–1095.

Kato, H., Bishop, P. O., and Orban, G. A. (1981). Binocular interaction on monocularly discharged lateral geniculate and striate neurons in the cat. *J. Neurophysiol.*, **46**, 932–951.

Kennedy, H., Martin, K., Orban, G. A., and Whitteridge, D. (1980). Neuronal properties in V1 and V2 of the baboon (*Papio ursinus*). *Exp. Brain Res.*, **41**, A20.

Kennedy, H., and Orban, G. A. (1983). Response properties of visual cortical neurons in cats reared in strobosopic illumination. *J. Neurophysiol.*, **49**, 486–704.

Leventhal, A. G. (1983). Systematic relationship between preferred orientation and receptive field position of neurons in cat striate cortex. *J. comp. Neurol.*, **220**, 476–483.

Leventhal, A. G., and Hirsch, H. V. B. (1980). Receptive-field properties of different classes of neurons in visual cortex of normal and dark-reared cats. *J. Neurophysiol.*, **43**, 1111–1132.

Leventhal, A. G., and Schall, J. D. (1983). Structural basis of orientation sensitivity of cat retinal ganglion cells. *J. comp. Neurol.*, **220**, 465–475.

Levick, W. R., and Thibos, L. N. (1982). Analysis of orientation bias in cat retina. *J. Physiol. (Lond.)*, **329**, 243–261.

Maes, H., and Orban, G. A. (1980). STIMUL: stimulus control and multihistogram analysis of single neuron recordings. *Med. & Biol. Eng. & Comput.*, **18**, 569–572.

Mansfield, R. J. W. (1974). Neural basis of orientation perception in primate vision. *Science*, **186**, 1133–1135.

Myerson, J., Manis, P., Miezin, F., and Allman, J. (1977). Magnification in striate cortex and retinal ganglion cell layer of owl monkey: a quantitative comparison. *Science*, **198**, 855–857.

Orban, G. A. (1984). *Neuronal Operations in the Visual Cortex*, Springer-Verlag, Berlin.

Orban, G. A., Hoffmann, K.-P., and Duysens, J. (1981). Influence of stimulus velocity on LGN neurons. *Soc. Neurosci. Abstr.*, **7**, 24.

Orban, G. A., Kato, H., and Bishop, P. O. (1979). Dimensions and properties of end-one inhibitory areas in receptive fields of hypercomplex cells in cat striate cortex. *J. Neurophysiol.*, **42**, 833–849.

Orban, G. A., and Kennedy, H. (1981). The influence of eccentricity on receptive field types and orientation selectivity in areas 17 and 18 of the cat. *Brain Res.*, **208**, 203–208.

Orban, G. A., Kennedy, H., and Maes, H. (1980). Functional changes across the 17–18 border in the cat. *Exp. Brain Res.*, **39**, 177–186.

Orban, G. A., Kennedy, H., and Maes, H. (1981a). Response to movement of neurons in areas 17 and 18 of the cat: velocity sensitivity. *J. Neurophysiol.*, **45**, 1043–1058.

Orban, G. A., Kennedy, H., and Maes, H. (1981b). Response to movement of neurons in areas 17 and 18 of the cat: direction selectivity. *J. Neurophysiol.*, **45**, 1059–1073.

Payne, B. R., and Berman, N. (1983). Functional organization of neurons in cat striate cortex: variations in preferred orientation and orientation selectivity with receptive-field type, ocular dominance, and location in visual-field map. *J. Neurophysiol.*, **49**, 1051–1072.

Pettigrew, J. D., Nikara, T., and Bishop, P. O. (1968). Responses to moving slits by single units in cat striate cortex. *Exp. Brain Res.*, **6**, 373–390.

Poggio, G. F., Baker, F. H., Mansfield, R. J. W., Sillito, A., and Grigg, P. (1975). Spatial and chromatic properties of neurons subserving foveal and parafoveal vision in rhesus monkey. *Brain Res.*, **100**, 25–29.

Sillito, A. M. (1977). Inhibitory processes underlying the directional specificity of simple, complex and hypercomplex cells in the cat's visual cortex. *J. Physiol. (Lond.)*, **271**, 699–720.

Sillito, A. M., and Versiani, V. (1977). The contribution of excitatory and inhibitory inputs to the length preference of hypercomplex cells in layers II and III of the cat's striate cortex. *J. Physiol. (Lond.)*, **273**, 775–790.

Tigges, J., Tigges, M., and Perachio, A. A. (1977). Complementary laminar terminations of afferents to area 17 originating in area 18 and in the lateral geniculate nucleus in squirrel monkey. *J. comp. Neurol.*, **176**, 87–100.

Tigges, J., and Tigges, M. (1979). Ocular dominance columns in the striate cortex of chimpanzee (*Pan troglodytes*). *Brain Res.*, **166**, 386–390.
Wilson, J. R., and Sherman, S. M. (1976). Receptive-field characteristics of neurons in cat striate cortex: changes with visual field eccentricity. *J. Neurophysiol.*, **39**, 512–533.
Zeki, S. (1983). The distribution of wavelength and orientation selective cells in different areas of monkey visual cortex. *Proc. Roy. Soc. Lond. B*, **217**, 449–470.

Models of the Visual Cortex
Edited by D. Rose and V. G. Dobson

CHAPTER 32

Are simple cells and complex cells distinctly different?

D.J. Tolhurst and A. F. Dean
The Physiological Laboratory, Cambridge, CB2 3EG, UK

Hubel and Wiesel (1962) reported, on the basis of qualitative observations, that two distinct types of cell could be found in the cat's striate cortex: simple cells and complex cells. This classification relies on apparently fundamental differences in receptive field organization. The simple-cell receptive field has two important defining features. First, it can be subdivided into discrete excitatory and inhibitory regions; within each region, the presentation of an appropriate visual stimulus elicits activity either at the stimulus onset or at the offset, *but not at both*. Second, there is spatial summation within each region and there is antagonism between excitatory and inhibitory regions. Thus, the greatest response is elicted by a visual stimulus whose width and length match those of the dominant receptive field region. If the stimulus is smaller, the response is less because less of that region is stimulated. If the stimulus is wider, the response will also be less because an antagonistic region is also stimulated. Subsequently, it has been shown that the stimulus preferences of simple cells can be predicted *quantitatively* from the receptive field map (e.g. Movshon, Thompson and Tolhurst, 1978), supporting the more stringent hypothesis that spatial summation in simple-cell receptive fields is linear.

Complex cells may be identified by either of two receptive field features. First, the field may not have separate excitatory and inhibitory regions; rather, there are roughly equal responses at the stimulus onset and offset in each retinal location. Such a cell may be regarded as a 'typical' complex cell. Second, there may be separate excitatory and inhibitory regions in the receptive field but the dimensions of these regions *do not allow a prediction of the width of the optimal visual stimulus*. Henry (1977) would call such cells A-cells.

A PROBLEMATICAL CELL

Figure 1 shows the quantitatively determined line weighting function of a cell recorded in the cat's striate cortex, taken from our own data. The histogram shows the relative frequency of generation of action potentials during the presentation of bright lines (open blocks) and dark lines (filled blocks) in different parts of the receptive field. The bright lines reveal excitatory regions; the dark lines reveal inhibitory regions. Above the histogram, the heavy bar shows the observed optimal width of the visual stimulus (0.57°), the width of one bright bar in a sinusoidal grating of optimal spatial frequency. Note that the optimal width is somewhat less than the width of the dominant, excitatory receptive field region (about 0.8°).

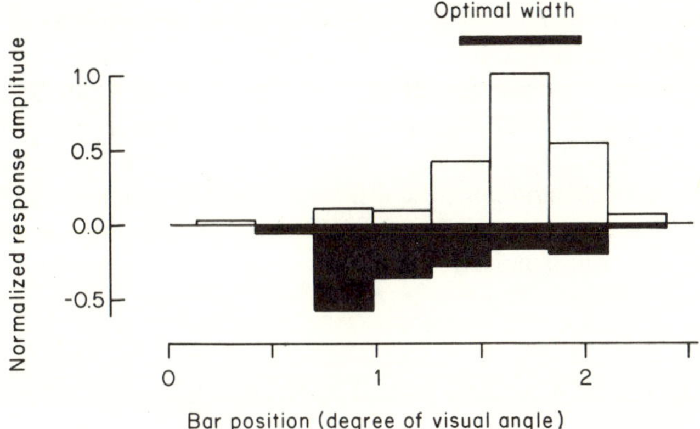

Figure 1 The line-weighting function of a cell recorded in cat striate cortex. The response amplitude is the average number of action potentials elicited by 60 presentations of each line, normalized to the response to the most effective stimulus. The lines were of optimal orientation and were 0.28 wide, which was also the distance between the centres of neighbouring lines. Full details of experimental procedures are given by Dean and Tolhurst (1983).

Is this cell simple or complex within the terms of Hubel and Wiesel's definitions? First, a simple cell should have *discrete* excitatory and inhibitory receptive field regions. The cell illustrated has a *predominantly* excitatory region (to the right) and a *predominantly* inhibitory region (to the left). However, these regions are not discrete: at all locations, the firing rate was increased (to different extents) by both bright and dark lines. Second, the optimal width of the stimulus for a simple cell should be the same as the width of the dominant receptive field region. For the illustrated neuron, the predicted and observed optimal widths are not the same, but the discrepancy is not great.

This cell seems to be transitional between 'typical' simple and complex classes. The receptive field regions partially overlap but are still distinguishable. The geometry of the optimal visual stimulus is not greatly different from that predicted. Whether one classes this cell as simple or complex depends largely upon how great a discrepancy between observation and prediction one is prepared to tolerate.

The existence of such cells invites a reexamination of the experimental basis for believing that simple and complex cells are distinct entities. It may be, instead, that they represent only the end-points of a continuum, with cells such as that depicted in Fig. 1 occupying a mid-point. The issue has important consequences for those who would ascribe separate ontogenies, morphologies or perceptual roles for 'simple' and 'complex' cells, and can be addressed only by examination of the behaviour of the *entire* sample of cells. DeValois, Albrecht and Thorell (1982) have pointed out that it is not adequate to illustrate the behaviour of only a few representative cells. The behaviour of all the cells must be summarized with some numerical index of the salient features of the receptive fields; such indices can be abstracted only from quantitative experiments. This kind of approach can be instructive. For instance, Rose (1977) measured the degree of end-inhibition in a population of cortical cells and concluded that it was continuously distributed; the distinctness of the hypercomplex class of receptive field (Hubel and Wiesel, 1965, 1968) was compromised.

A CONTINUOUS DISTRIBUTION OF CELL PROPERTIES

We have quantified the two relevant receptive field properties: the discreteness of the receptive field regions and the degree of predictability of spatial summation.

Consider that the amplitude of response has been measured for bright and dark lines in n non-overlapping but adjacent locations in the receptive field. At the ith location, the response to the bright line is $B(i)$ and to the dark line is $D(i)$; the latter is considered to be a negative number. Receptive field discreteness is defined as:

$$\frac{\sum_{i=1}^{n} |B(i) + D(i)|}{\sum_{i=1}^{n} |B(i)| + \sum_{i=1}^{n} |D(i)|}.$$

Discreteness can range from 0 to 1.0, the value expected for simple cells. The cell illustrated in Fig. 1 has a discreteness of 0.45.

The spatial summation ratio is a measure of the predictability of spatial summation. It is defined as the *predicted* optimal width of line divided by the *observed* optimal width. The former is the width of the dominant receptive field region estimated from a quantitative receptive field map; the latter is the width of one bar in a sinusoidal grating of optimal spatial frequency. For 'typical' complex cells, which do not have distinguishable regions, we use the width of the whole uniform receptive field. Simple cells should have a spatial summation ratio of 1. 0. The cell illustrated in Fig. 1 has an observed optimal width of 0.57° and a predicted optimum of 0.8°; the spatial summation ratio is, therefore, 1.4.

Figure 2a, b shows the distributions of receptive field discreteness and of the spatial summation ratio for all the cells in our sample. The filled blocks are for 'typical' complex cells, whose receptive fields could not be divided into regions which were predominantly excitatory or inhibitory. Discreteness is continuously distributed from about 0.1 to 1.0. Similarly, the spatial summation ratio is continuously distributed from about 0.8 to over 5. Neither measure alone shows sufficient discontinuity for it to be suitable as a means of confidently dividing cells into classes.

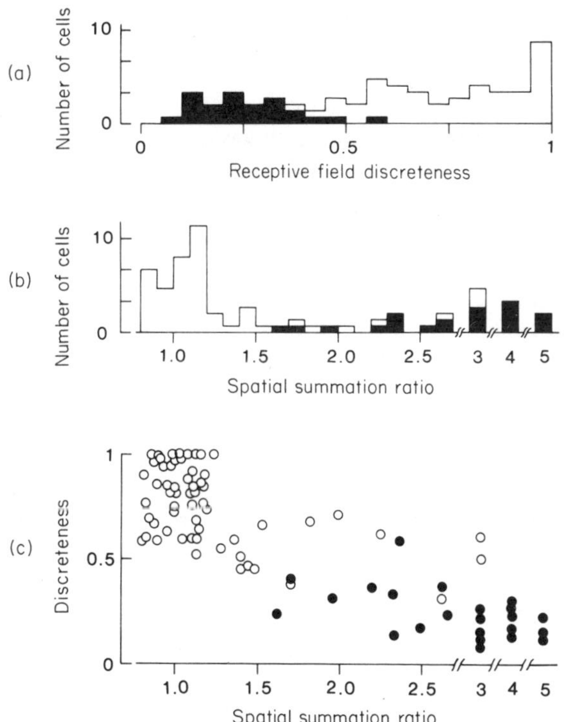

FIGURE 2 Distributions of and relation between the receptive field discreteness and spatial summation ratio for a sample of cells recorded in cat striate cortex. The filled histogram blocks and filled circles are for 'typical' complex cells, for which the width of the *whole* receptive field was used to calculate the spatial summation ratio.

TWO CLASSES OF CELL?

Although discreteness and spatial summation are inadequate *alone* as bases for classification, their association could be more useful. Figure 2c plots discreteness against the spatial summation ratio. The filled circles are for 'typical' complex cells. The two measures are not closely correlated, but nor are they completely dissociated. The data seem to fall into two loose clusters, which abut rather than merge. One cluster is of cells with a spatial summation ratio close to 1.0 and discreteness above 0.55. The other cluster is of cells with a spatial summation ratio above about 1.3 and discreteness usually below about 0. 7. The clustering of data in this fashion suggests that one can subdivide cortical cells into two groups using measures of receptive field discreteness and spatial summation. Since these two measures are the bases of Hubel and Wiesel's classification, one could call the two clusters simple (top left) and complex (bottom right).

There are no statistics which will allow us to decide whether the apparent clustering in Fig. 2c truly reflects the existence of two classes of cell or whether it is a chance sample from a single, continuously variable class. In either case, the distinction between classes does not seem to be great. Within the simple cell class and within the complex cell class, there is so much variety of behaviour that a simple cell, say, near the simple/complex border seems to show more affinity to some complex cells than it does to the bulk of simple cells. Thus, the classification still seems to be tenuous and, if simple cells and complex cells *are* distinct, this is not yet strikingly obvious.

It could be argued that simple and complex cells *are* distinct and that our failure to show this convincingly is because we have not examined the correct aspects of the cells' behaviour. However, we *have* examined the essence of Hubel and Wiesel's definition. Other quantitative measures of cell behaviour have been just as unconvincing as the bases for classification of cat cortical cells. Sherman, Watkins and Wilson (1976) examined the waveforms of the responses to moving lines; we (Dean and Tolhurst, 1983) have examined the responses to moving sinusoidal gratings; Hagerty *et al.* (1982) examined many aspects of cell behaviour including receptive field size, spontaneous activity and preferred velocity of movement. In no case is there a discontinuous distribution of cell behaviour; rather, the behaviour of simple cells merges with or abuts that of complex cells.

Hubel and Wiesel's classification had considerable appeal for many reasons. It seemed to rely on fundamental differences in receptive field organization; these differences seemed to imply that the cells processed their inputs in very different ways and that they might play very different roles in the analysis of visual information. The differences seemed to be so gross that they could be detected easily with qualitative tests. However, our experience suggests that a statistical analysis is necessary to allow a classification of *all* cells as either simple or complex. In the face of this conclusion, the classification loses most of its appeal and has little to recommend it over alternative classification schemes.

ACKNOWLEDGEMENT.

A.F.D. was supported by the Wellcome Trust.

REFERENCES

Dean, A. F., and Tolhurst, D. J. (1983). On the distinctness of simple and complex cells in the visual cortex of the cat. *J. Physiol.*, **344**, 305–325.

DeValois, R. L., Albrecht, D. G., and Thorell, L. G. (1982). Spatial frequency selectivity of cells in macaque visual cortex. *Vision Res.*, **22**, 545–589.

Hagerty, C. M., Lees, F. C., Tieman, S. B., and Hirsch, H. V. B. (1982). Principal components analysis of cells in cat visual cortex. *Brain Res.*, **251**, 45–53.

Henry, G. H. (1977). Receptive field classes of cells in the striate cortex of the cat. *Brain Res.*, **133**, 1–28.

Hubel, D. H., and Wiesel, T. N. (1962). Receptive fields, binocular interaction and functional architecture in the cat's visual cortex. *J. Physiol.*, **160**, 106–154.

Hubel, D. H., and Wiesel, T. N. (1965). Receptive fields and functional architecture in two nonstriate areas (18 and 19) of the cat. *J. Neurophysiol.*, **28**, 229–289.

Hubel, D. H., and Wiesel, T. N. (1968). Receptive fields and functional architecture of monkey striate cortex. *J. Physiol.*, **195**, 215–243.

Movshon, J. A., Thompson, I. D., and Tolhurst, D. J. (1978). Spatial summation in the receptive fields of simple cells in the cat's striate cortex. *J. Physiol.*, **283**, 53–77.

Rose, D. (1977). Response of single units in cat visual cortex to moving bars of light as a function of bar length. *J. Physiol.*, **271**, 1–23.

Sherman, S. M., Watkins, D. W., and Wilson, J. R. (1976). Further differences in receptive field properties of simple and complex cells in cat striate cortex. *Vision Res.*, **16**, 919–927.

Models of the Visual Cortex
Edited by D. Rose and V. G. Dobson

CHAPTER 33

Visual cortical processing: textural sensitivity and its implications for classical views

P. Hammond
Department of Communication and Neuroscience, University of Keele, Keele, Staffordshire ST5 5BG, UK

This chapter presents an overview of our investigations of visual cortical sensitivity to motion of visual texture and highlights some of the implications for classically held views of striate cortical function. Early studies were made in collaboration with MacKay (see Chapter 5 in this volume); others, notably B. Ahmed and A. T. Smith, have subsequently made significant contributions.

The classical approach, pioneered by Hubel and Wiesel (Hubel, 1982; Hubel and Wiesel, 1962, 1977), has been to characterize cortical neurones according to their responses to solitary oriented contours (moving or stationary light or dark bars or edges), presented against uniform backgrounds. Such an approach has led to a rigid classification of cells—simple, complex, hypercomplex—presumed to be linked serially in a hierarchical feature-extraction process, whose properties in the mature animal are unchanging. According to this scheme, higher order neurones build upon the properties of cells lower in the same chain; complex cells rely upon input from simple cells.

In parallel with our own investigations (Hammond, 1978, 1979a, 1979b, 1981; Hammond and MacKay, 1975, 1976, 1977, 1981; Hammond and Reck, 1980; Hammond and Smith, 1982, 1983), more recent work by others has rendered any strictly serial model of visual cortical processing untenable. For example, we now know that all cortical layers, except lamina II, receive direct afferent input (Stone and Dreher, 1973; Ferster and LeVay, 1978; Leventhal, 1979; Henry, Harvey and Lund, 1979; Bullier and Henry, 1979) and that all classes of cortical cells in all cortical laminae are potentially capable of receiving direct afferent input, in addition to input mediated by other cortical cells in the same or different layers (Bullier and Henry, 1979; Henry, Harvey and Lund, 1979). It has become abundantly clear that a significant element of parallel, in addition to serial (hierarchical), processing is involved and that any adequate description of cortical function requires knowledge of each cell's afferent input, lamina of origin, intracortical connectivity and efferent destination (see also Gilbert and Kelly, 1975; Gilbert, 1977; Sillito, 1977, 1979; Leventhal and Hirsch, 1978; Gilbert and Wiesel, 1979; Stone, Dreher and Leventhal, 1979; Lund *et al.*, 1979).

The assignation of a cell to one of only two categories, simple or complex, imposes stringent and unwarranted constraints on any concept of cortical processing. (For the striate cortex, there is now general agreement that hypercomplexity does not define a separate class of cell but is simply an additional restriction on length summation which both simple and complex cells may possess; see Dreher, 1972; Kato, Bishop and Orban, 1978.) The simple-cell category is well defined, although even then there are obvious laminar differences, e. g. in receptive field size, selectivity for orientation and eye dominance. The complex-cell category, on the other hand, is one of default—any cell which is not demonstrably simple—and embraces an enormous variety of cells with widely differing properties (Hubel and Wiesel, 1962; Hammond and MacKay, 1976, 1977; Gilbert, 1977; Leventhal and Hirsch, 1978; Henry, Harvey and Lund, 1979; Sillito, 1979).

The visual system routinely processes stimuli which are complex and are commonly perceived against backgrounds which are neither featureless nor stationary. We have therefore exploited a repertoire of randomly and regularly textured visual stimuli (Fig. 1), including pseudo-random static visual noise (composed of small, light or dark, elements retaining a fixed relationship to one another), variably scattered arrays of short line elements of common orientation, chequerboards and grating patterns. These stimuli are generated electronically as a 256 x 256 line raster at 50 frames/sec on a cathode ray tube display (average luminance 0.9 log cd/m^2, typically subtending 10 x 10°). They can be presented either alone or as backgrounds for foreground stimuli such as oriented bars. We have used the lightly anaesthetized adult cat, although our conclusions are likely to be valid for higher mammals.

In the visual cortex, only complex cells are responsive to motion of visual noise (Fig. 1a): simple cells are totally unresponsive (Hammond and MacKay, 1975, 1976, 1977). Both the major classes of cortical afferents—X- and Y-cells—are responsive to texture motion (Mason, 1976). The differential sensitivity of cortical cells to noise motion must therefore arise intracortically. Some properties of complex cells must be mediated by parallel pathways not involving simple cells, because simple cells are silent in situations where complex cells are driven and because, on the basis of synaptic vesicle shape,

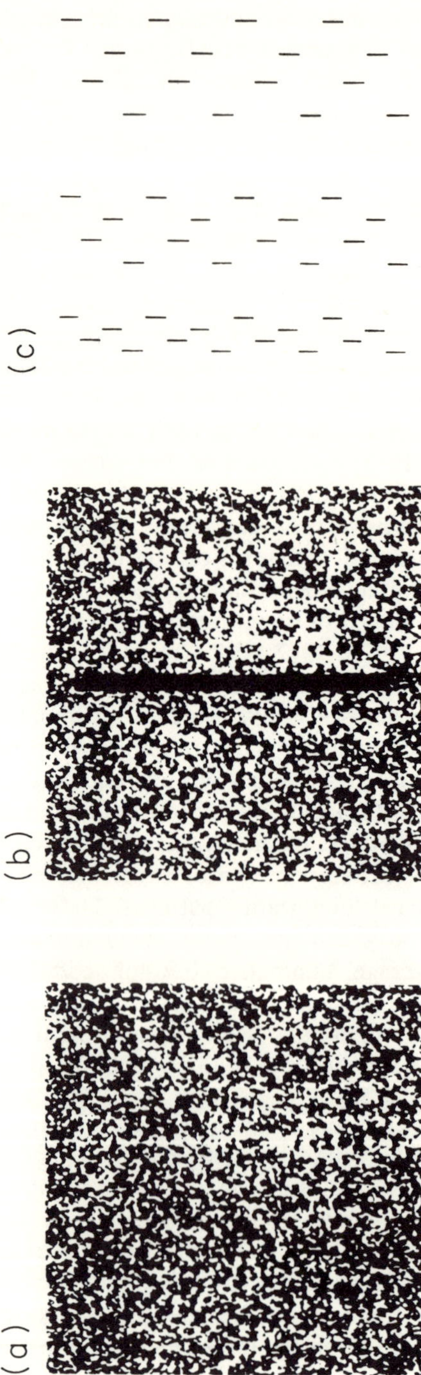

FIGURE 1 Examples of textured visual stimuli. Static visual noise, presented (a) alone or (b) as a background against which to present a foreground oriented bar. (c) A regularly textured array of short line elements of common orientation incorporating different degrees of scatter.

afferent input to the cortex is thought to be purely excitatory (Grey, 1959; Uchizono, 1965).

Responses of both simple and complex cells to moving bars of optimal orientation (Fig. 1b) can be powerfully modulated by simultaneous motion of textured backgrounds (Hammond and MacKay, 1981; Hammond and Smith, 1982, 1983). In simple cells this influence is inhomogeneous. Background motion confined to the receptive field centre exerts a suppressive influence on bar response. Background motion at either end of the receptive field typically exerts a facilitatory influence over an area which is greater than that of the simple cell's receptive field and which is comparable to that of the largest complex cell in the parent orientation column. This result runs counter to the classical picture and appears to be explicable only in terms of feedback from complex onto simple cells. These background influences on cortical cells seem unrelated to the 'periphery' or 'shift' effects observed subcortically (McIlwain, 1964; Fischer and Kruger, 1975). In complex cells, their area of influence is coextensive with the cells' mapped excitatory receptive fields, whereas the 'periphery' effect is spatially much wider ranging. Even very small contrasting elements, including specific features of moving textured backgrounds, may significantly bias a cell's response (see Chapter 5 by MacKay in this volume). The presence of stationary patterned backgrounds (Fig. 1b), however, has no influence on a cell's response to foreground objects. The directional tuning curves obtained from complex cells with a moving bar presented against either patterned or uniform stationary backgrounds of comparable mean luminance are identical.

Our results from complex cells argue against thinking of cortical cells as feature detectors. Thus the degree of responsiveness to motion of visual noise varies according to cell type and lamina of origin, ranging from cells which respond best to oriented stimuli to those responding preferentially or exclusively to noise (Hammond and MacKay, 1976, 1977; Hammond and Smith, 1982). Those complex cells most responsive to noise motion are found in two bands, sandwiching the principal afferent input layer IV, probably deep in layer III and in upper layer V (Hammond and Smith, 1982). They are typically large-field, special or standard, complex cells (Gilbert, 1977), which are usually direction selective and often show 'null' suppression of activity to movement in the non-preferred direction. Complex cells in superficial cortical laminae, especially those with small receptive fields, are characteristically less responsive to noise motion than those in deep laminae. More regularly textured arrays, including arrays of short line segments of common orientation (Fig. 1c), also reveal clear laminar differences. Simple cells respond preferentially to single lines of optimal orientation, rather than to such arrays, even though the total length of contour is the same. Complex cells in superficial laminae also respond preferentially to single lines or to close-knit arrays of line elements; those in deep laminae respond preferentially to arrays in which the scatter is wide. In summary, in response to a repertoire of different stimuli, each functional subclass of cells emphasizes certain features at the expense of others, rather than being selective for one. Orientation is neither the only nor the most significant attribute for many cells.

We have made extensive comparisons, monocularly and interocularly, of complex cells' directional tuning for the motion of oriented stimuli and visual noise. Our interest has been to establish the mechanisms underlying orientational and directional selectivity, the interactions between them and the likely afferent pathways involved. The classical view, based on responses to moving bars, is that a cell's preferred orientation and direction are similar for each eye. These properties, together with ocular dominance, are believed to be invariant in the adult and resistant to changes in bar width, length, contrast, contrast polarity and velocity of motion.

The directional tuning of complex cells for texture motion is strikingly different from that for oriented stimuli (Hammond, 1978, 1979a, 1979b). Preferred directions are dissimilar: in the direction optimal for a moving bar, the response to noise motion is weak; the best response to noise is elicited in directions disposed to either side of that preferred for a bar. Directional tuning for texture varies with the velocity of motion (Hammond and Reck, 1980; Hammond and Smith, 1983). Each cell's preferred velocity and velocity bandpass are higher for texture than for bar motion, and the above dissimilarities in tuning are most obvious at high velocities. Directional tuning is broader for noise than for bar motion, and the bias between preferred and opposite directions of motion is more marked (Hammond, 1978, 1979a, 1979b). A cell's directional tuning for oriented stimuli is the same for both eyes. For textured stimuli, this is frequently not the case: exceptionally, preferred directions through the two eyes may be opposite (Hammond, 1979a, 1979b, 1981). Ocular dominance for texture motion is also labile, even in the adult. Eye preference for texture alters over time and may differ from that for oriented stimuli (Hammond, 1979b, 1981).

All these results strongly imply that oriented and textured stimuli involve different pathways to the cortex, acting at discrete sites on the same complex cells and mediated by different cortical circuitry.

The detection and perception of foreground objects must be capable of withstanding motion and conformational changes to their backgrounds. Our most recent studies have therefore concentrated on interactions between oriented stimuli and textured backgrounds when *both* are in motion. We have compared directional tuning of complex cells, for simultaneous motion of bar and background in the same direction and at the same velocity, and for separate motion of bar and background (Hammond and Smith, 1983). The results indicate that an appropriately oriented foreground stimulus effectively gates out any response to the background. This implies that the input mediating a cell's orientation selectivity dominates and is more direct than that mediating directional (motion) selectivity.

When the object and background are in relative motion, however, so that they move simultaneously but in different directions and with different velocities, a different picture emerges. Even at the level of the striate cortex, the response of virtually all complex cells to an oriented stimulus is critically influenced by relative motion of a textured background. Whilst the response of most direction-selective complex cells to any relative motion configuration is broadly predictable from the separate responses to foreground or background alone, the same can not be said of bidirectional or partially direction-biased complex cells in the superficial cortical layers. A significant proportion of these are preferentially sensitive to relative motion between the foreground and background moving simultaneously in phase (in the same direction at different velocities); others respond preferentially to counterphase relative motion. This occurs regardless of the absolute direction of motion of the foreground stimulus and in a fashion not anticipated from the responses to the foreground or background alone (Hammond and Smith, 1982).

Our results with textural stimuli to date present a picture of striate cortical processing which, even beyond the period of maturational plasticity in early life, is far more dynamic and less stereotyped than the classical view. We believe that such approaches will continue to provide valuable fresh insight into connectivity and functional communication between cortical neurones.

REFERENCES

Bullier, J., and Henry, G. H. (1979). Laminar distribution of first-order neurones and afferent terminals in cat striate cortex. *J. Neurophysiol.*, **42**, 1271–1281.

Dreher, B. (1972). Hypercomplex cells in the cat's striate cortex. *Investig. Ophthalmol.*, **11**, 355–356.

Ferster, D., and LeVay, S. (1978). The axonal arborizations of lateral geniculate neurones in the striate cortex of the cat. *J. comp. Neurol.*, **182**, 923–944.

Fischer, B., and Kruger, J. (1975). Quantitative aspects of the shift effect in cat retinal ganglion cells. *Brain Res.*, **83**, 391–404.

Gilbert, C. D. (1977). Laminar differences in receptive field properties of cells in cat primary visual cortex. *J. Physiol. (Lond.)*, **268**, 391–421.

Gilbert, C. D. and Kelly, J. P. (1975). The projections of cells in different layers of the cat's visual cortex. *J. comp. Neurol.*, **163**, 81–106.

Gilbert, C. D., and Wiesel, T. N. (1979). Morphology and intracortical projections of functionally characterized neurones in the cat visual cortex. *Nature*, **280**, 120–125.

Grey, E. G. (1959). Axo-somatic and axo-dendritic synapses of the cerebral cortex: an electron microscopic study. *J. Anat. (Lond.)*, **93**, 420–433.

Hammond, P. (1978). Directional tuning of complex cells in area 17 of the feline visual cortex. *J. Physiol. (Lond.)*, **285**, 479–491.

Hammond, P. (1979a). Stimulus-dependence of ocular dominance and directional tuning of complex cells in area 17 of the feline visual cortex. *Exp. Brain Res.*, **35**, 583–589.

Hammond, P. (1979b). Lability of directional tuning and ocular dominance of complex cells in the cat's striate cortex. NATO Adv. Study Inst. Report A27, Plenum, pp. 163–174.
Hammond, P. (1981). Non-stationarity of ocular dominance in cat striate cortex. *Exp. Brain Res.*, **42**, 189–195.
Hammond, P., and MacKay, D. M. (1975). Differential responses of cat visual cortical cells to textured stimuli. *Exp. Brain Res.*, **22**, 427–430.
Hammond, P., and MacKay, D. M. (1976). Functional differences between cat visual cortical cells revealed by use of textured stimuli. *Exp. Brain Res.*, **1976**, Suppl. 1, 397–402.
Hammond, P., and MacKay, D. M. (1977). Differential responsiveness of simple and complex cells in cat striate cortex to visual texture. *Exp. Brain Res.*, **30**, 275–296.
Hammond, P., and MacKay, D. M. (1981). Modulatory influences of moving textured backgrounds on responsiveness of simple cells in feline striate cortex. *J. Physiol. (Lond.)*, **319**, 431–442.
Hammond, P., and Reck, J. (1980). Influence of velocity on directional tuning of complex cells in cat striate cortex for texture motion. *Neurosci. Lett.*, **19**, 309–314.
Hammond, P., and Smith A. T. (1982). On the sensitivity of complex cells in feline striate cortex to relative motion. *Exp. Brain Res.*, **47**, 457–460.
Hammond, P., and Smith, A. T. (1983). Directional tuning interactions between moving oriented and textured stimuli in complex cells of feline striate cortex. *J. Physiol. (Lond.)*, **342**, 35–49.
Henry, G. H., Harvey, A. R., and Lund, J. S. (1979). The afferent connections and laminar distribution of cells in the cat striate cortex. *J. comp. Neurol.*, **187**, 725–744.
Hubel, D. H. (1982). Exploration of the primary visual cortex, 1955–78. *Nature*, **299**, 515–524.
Hubel, D. H., and Wiesel, T. N. (1962). Receptive fields, binocular interaction and functional architecture in the cat's visual cortex. *J. Physiol.* (Lond.), **160**, 106–154.
Hubel, D. H., and Wiesel, T. N. (1977). The Ferrier Lecture: functional architecture of macaque monkey visual cortex. *Proc. Roy. Soc. Lond., B*, **198**, 1–59.
Kato, H., Bishop, P. O., and Orban, G. A. (1978). Hypercomplex and simple/complex cell classifications in cat striate cortex. *J. Neurophysiol.*, **41**, 1071–1095.
Leventhal, A. G. (1979). Evidence that the different classes of relay cells of the cat's lateral geniculate nucleus terminate in different layers of the striate cortex. *Exp. Brain Res.*, **37**, 349–372.
Leventhal, A. G., and Hirsch, H. V. B. (1978). Receptive field properties of neurones in different laminae of visual cortex of the cat. *J. Neurophysiol.*, **41**, 948–962.
Lund, J. S., Henry, G. H., MacQueen, C. L. and Harvey, A. R. (1979). Anatomical organization of the primary visual cortex (area 17) of the cat. A comparison with area 17 of the macaque monkey. *J. comp. Neurol.*, **184**, 599–618.
McIlwain, J. T. (1964). Receptive fields of optic tract axons and lateral geniculate cells: peripheral extent and barbiturate sensitivity. *J. Neurophysiol.*, **30**, 1–22.
Mason, R. (1976). Responses of cells in the dorsal lateral geniculate complex of the cat to textured visual stimuli. *Exp. Brain Res.*, **25**, 323–326.
Sillito, A. M. (1977). Inhibitory processes underlying the directional specificity of simple, complex and hypercomplex cells in the cat's visual cortex. *J. Physiol. (Lond.)*, **271**, 699–720.
Sillito, A. M. (1979). Inhibitory mechanisms influencing complex cell orientation selectivity and their modification at high resting discharge levels. *J. Physiol. (Lond.)*, **289**, 33–53.

Stone, J., and Dreher, B. (1973). Projection of X- and Y-cells of the cat's lateral geniculate nucleus to areas 17 and 18 of visual cortex. *J. Neurophysiol.*, **36**, 551–567.
Stone, J., Dreher, B., and Leventhal, A. G. (1979). Hierarchical and parallel mechanisms in the organization of visual cortex. *Brain Res. Rev.*, **1**, 345–394.
Uchizono, K. (1965). Characteristics of excitatory and inhibitory synapses in the central nervous system of the cat. *Nature*, **207**, 642–643.

Models of the Visual Cortex
Edited by D. Rose and V. G. Dobson

CHAPTER 34

Complex cells control simple cells

LAMBERTO MAFFEI
Istituto di Neurofisiologia del CNR, Via S. Zeno, 51, 56100-Pisa, Italy

Old as it might seem, the controversy on the order of processing of visual information in area 17 is not a quenched fire: the problem remains, even today, fresh and relevant. The original hypothesis proposed by Hubel and Wiesel (1962) twenty years ago was in favour of a serial model where simple cells receive the bulk of the input from the lateral geniculate nucleus and relay this information to complex cells. This theory is still in the battlefield, being either challenged or exploited by new works and models.

Subsequently, the alternative theory was proposed that simple and complex cells do not work in sequence but rather in parallel. The evidence in favour of this second model, which may well be coexisting with the first, is rather substantial and is still accumulating (Hoffman and Stone, 1971; Stone, 1972; Singer, Tretter and Cynader, 1975; Movshon, 1975; Hammond and MacKay, 1975, 1977, 1978). However, with the progression of research, it emerged that the concepts of a serial or parallel organization, useful though they have been in the past and might still be in explaining particular sets of results, risk being simple-minded when referred to the processing of visual information as a whole.

The concept of a hierarchy is neither necessary nor sufficient to explain all the properties of cortical cells in area 17 and neither is that of a purely parallel processing. What seems more likely is that simple and complex cells are bits of the same processor where the elements of a class interact with those of the other in the analysis of a given visual attribute. As a possible example of a cooperative network, we propose here, on the basis of recent results, that a class of complex cells might control and shape the function of simple cells and may even precede them in time in the visual information processing (Burr, Morrone and Maffei, 1981; Maffei, 1981; Morrone, Burr and Maffei, 1982).

Several years ago, Hammond and MacKay (1977, 1978) demonstrated that complex cells of area 17 respond to two-dimensional visual noise (random dot patterns) whereas simple cells do not. A random dot pattern (two-dimensional visual noise) can be considered as the sum of a large number of sinusoidal gratings of all spatial frequencies at all orientations, including those spatial frequencies at the cell's preferred orientation, i.e. that to which the simple cell is most sensitive. It is difficult to understand why a simple cell, which responds to a given set of spatial frequencies when presented in isolation, does not respond to the same set of spatial frequencies when these are embedded in a number of other spatial frequencies having the same and other orientations.

Recently Burr, Morrone and I (Burr, Morrone and Maffei, 1981; Morrone, Burr and Maffei 1982) repeated the experiments of Hammond and MacKay by using random dot patterns of large and known spectrum in order to be certain that the absence of response of simple cells was not due to a lack of certain spatial frequencies in the visual stimulus. In the majority of simple cells we have confirmed the findings by Hammond and MacKay. In addition we have found that simple cells do respond to one-dimensional noise. This stimulus has the same spatial frequencies and contrast as the two-dimensional visual noise, but only in one orientation. The fact that the one-dimensional noise is a subset of the two-dimensional noise excludes the possibility that simple cells fail to respond to the latter stimulus because it possesses insufficient energy. The only likely hypothesis to explain the silence of simple cells to two-dimensional noise is the coming into action of active inhibitory processes.

An experiment which supports this hypothesis, indicating the possible reasons why simple cells do not respond to two-dimensional noise is a recent one by Morrone, Burr and Maffei (1982). We have shown that gratings, drifting or alternated in phase at orientations outside the cell's orientation tuning curve, decrease the response of simple cells to other stimuli. The effect is most dramatic for the orientation orthogonal to the cell's preferred orientation. Since two-dimensional visual noise contains all possible orientations (as well as all possible spatial frequencies) it seems safe to conclude that simple cells do not respond to this stimulus because they are inhibited by distributions of energy in orientations outside the one preferred by the cell.

If inhibitory mechanisms are responsible for the fact that simple cells can respond to one-dimensional noise and cannot to two-dimensional noise, the problem arises concerning which type of cells provides these inhibitory mechanisms. We have found evidence suggesting that complex cells are responsible for the inhibition of simple cells. These conclusions are drawn from experiments which exploit the property of simple and complex cells that they respond in a different way to drifting and alternating gratings.

Simple cells respond to the single bars of a drifting grating with a modu lation of their discharge, whereas complex cells do not. Complex cells simply change their average rate of discharge as a function of the spatial frequency of the grating. This characteristic of simple and complex cells is illustrated in Fig. 1a, c. The response of simple and complex cells to phase-reversed sinusoidal gratings is also different. Simple cells modulate to the first and complex cells to the second harmonic of the stimulus (Fig. 1b, d).

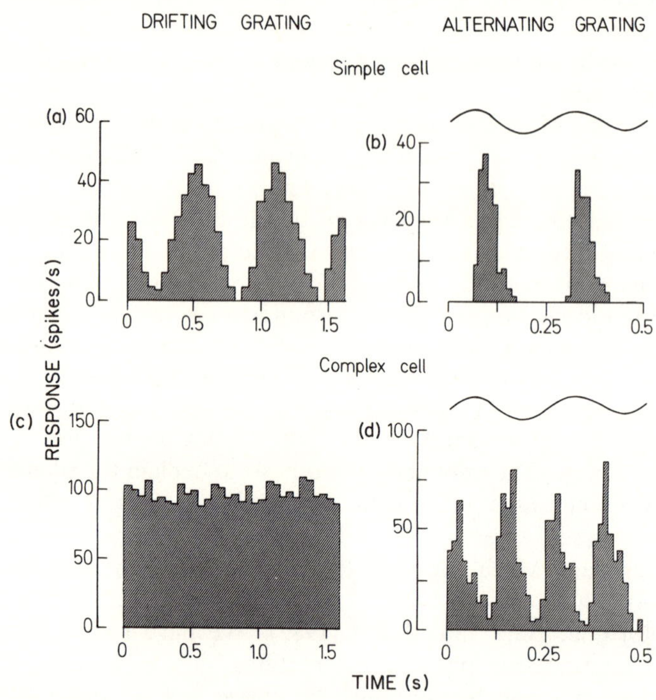

FIGURE 1 Response of a simple (a) and a complex (c) cell to a drifting sinusoidal grating. Spatial frequency 0.5 c.p.d., velocity 3.5°, contrast 20 per cent. Response of a simple (b) and a complex (d) cell to phase-reversed sinusoidal gratings. The sinusoids above the histograms indicate the temporal modulation of contrast of the grating. Spatial frequency 0.75 c.p.d., contrast 20 per cent. Note that the maxima of the responses in (b) correspond to minima in (d).

The main experiments which suggest that complex cells control simple cells are the following.

In the first, gratings alternated in phase outside the simple cell's orientation tuning curve decrease the responses of simple cells to other stimuli. Figure 2 illustrates an example of such findings. Figure 2a shows the response of a simple cell to a drifting one-dimensional noise. Figure 2b shows the response to the combination of one-dimensional noise and an alternating grating oriented orthogonally with respect to the cell's preferred orientation (and one-dimensional noise). The orthogonally oriented alternating grating not only depresses the mean firing rate of the cell but also causes the response to modulate at twice the temporal frequency of the stimulus. This inhibitory modulation of simple-cell activity has characteristics indicating that the modulating influences might come from complex cells.

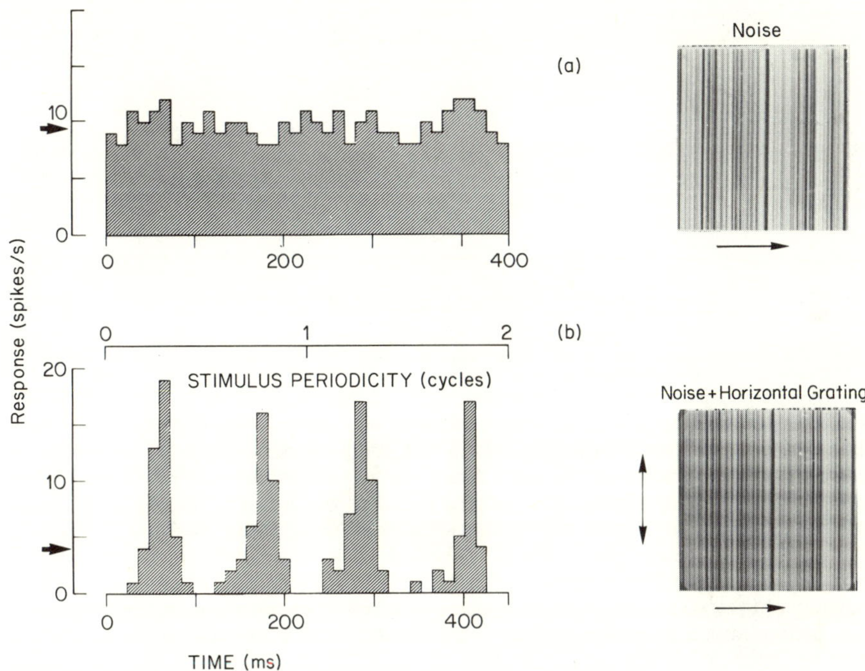

FIGURE 2 Responses of a simple cell to a vertical one-dimensional drifting noise superimposed either (a) on a homogeneous background or (b) on a horizontal sinusoidal grating alternated in phase. The stimuli are illustrated on the right part of the picture. The average of the responses have been computed for a time double that of the frequency of grating alternation. Average of 100 runs. Bin width 12.5 ms; arrows indicate the mean firing rates of the responses. Luminance 10 cd/m^2, drifting velocity of the visual noise 4/s, spatial frequency of the grating 1.0 c.p.d., contrast of the grating 20 per cent. Frequency of alternation of the grating 4 Hz. (Redrawn from results by Burr, Morrone and Maffei, 1981).

In the second, drifting gratings at orientations outside the simple cell's orientation tuning curve decrease uniformly the response of the simple cell to other stimuli. Figure 3 shows an example of this. The experiment was similar to that illustrated in the previous figure, but the grating, orthogonally oriented with respect to the cell's preferred orientation, was drifting. The response (Fig. 3b) is uniformly depressed and does not show any correlation

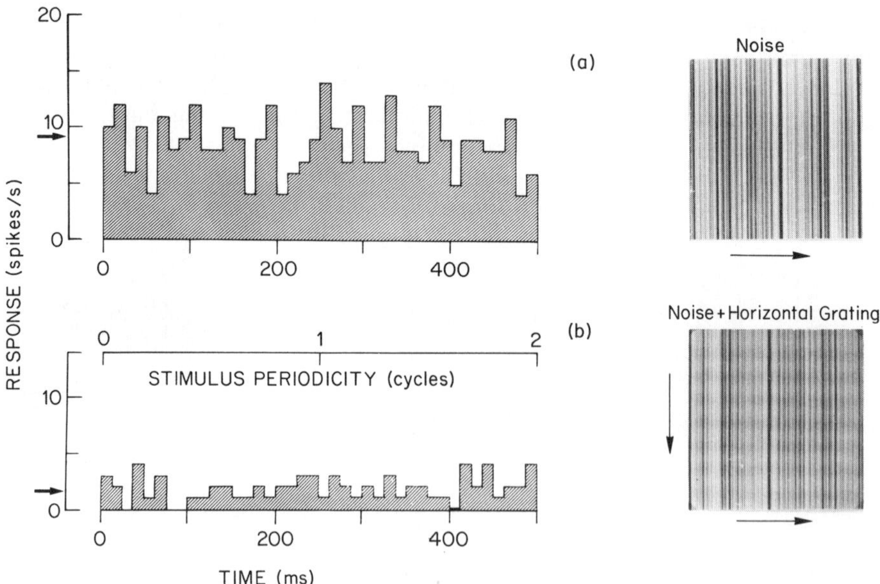

FIGURE 3 (a) Response of a simple cell to a one-dimensional drifting noise. Sums 100, bin width 2.5 ms, drifting velocity 5.5°/s, stimulus periodicity 3 s, luminance 10 cd/m². The arrow indicates the mean firing rate of the response. (b) Response of the same cell to the two-dimensional pattern reported on the right. The stimulus is the superposition of the vertical one-dimensional noise in (a) and a horizontal sinusoidal grating. The directions of movement of the two patterns are indicated by the arrows. Spatial frequency of the grating 1 c.p.d., contrast of the grating 20 per cent., drifting velocity of the two patterns 4°/s, luminance 10 cd/m². The poststimulus time histogram has been computed on time which is the double of the periodicity of the drifting grating. Average on 100 runs, bin width 12.5 ms. The black arrow indicates the mean firing rate of the response. (Redrawn from results by Burr, Morrone and Maffei, 1981).

with the passage of the single bars of the grating. This is expected if complex cells are responsible for this effect, since their response to a drifting grating is also not correlated with the passage of the bars of the grating and consists simply in a variation of their average discharge.

In summary, the results suggest that complex cells could be the controllers of the activity and function of simple cells through inhibitory circuits (Creutzfeldt, Innocenti and Brooks, 1974; Creutzfeldt, Kuhnt and Benevento, 1974; Singer, Tretter and Cynader, 1975; Hammond and MacKay, 1978). Complex cells exhibit a resting discharge which is rather high, particularly in one type of complex cell, as described by Palmer and Rosenquist (1974) and Gilbert (1977). This spontaneous discharge could generate a continuous inhibition of simple cells and possibly of other complex cells showing low spontaneous activity. The spontaneous tonic inhibition is most probably the cause of the fact that simple cells show little or no spontaneous discharge. Activation of complex cells (or a class of them) would modulate this tonic inhibition, and the modulation not surprisingly would exhibit complex-cell response characteristics.

The elaboration performed by complex cells which receive a substantial Y input through fast-conducting axons could even precede that of simple cells, despite the presence of an additional synapse. Indeed, Singer, Tretter and Cynader (1975) found that in response to electrical stimulation of the lateral geniculate body, the cortical inhibitory postsynaptic potentials often preceded the excitatory postsynaptic potentials, which frequently remained subthreshold.

In conclusion, the cortical scheme suggested by our results is one in which information is processed not by a hierarchical sequence (Hubel and Wiesel, 1962) but rather by an organized neural assembly in which the complex cells subserve the simple cells, shaping their selectivity and suppressing background noise.

REFERENCES

Burr, D. C., Morrone, C., and Maffei, L. (1981). Intra-cortical inhibition prevents simple cells from responding to textured patterns. *Exp. Brain Res.*, **43**, 455–458.

Creutzfeldt, O. D., Innocenti, G. M., and Brooks, D. (1974). Vertical organization in the visual cortex (area 17) in the cat. *Exp. Brain Res.*, **21**, 315–336.

Creutzfeldt, O. D., Kuhnt, U., and Benevento, L. A. (1974). An intracellular analysis of cortical neurones to moving stimuli: responses in a cooperative neuronal network. *Exp. Brain Res.*, **21**, 251–274.

Gilbert, C. D. (1977). Laminar differences in receptive field properties of cells in cat primary visual cortex. *J. Physiol. (Lond.)*, **268**, 391–421.

Hammond, P., and MacKay, D. M. (1975). Differential responses of cat visual cortical cells to textured stimuli. *Exp. Brain Res.*, **22**, 427–430.

Hammond, P., and MacKay, D. M. (1977). Differential responsiveness of simple and complex cells in cat striate cortex to visual texture. *Exp. Brain Res.*, **30**, 275–296.

Hammond, P., and MacKay, D. M. (1978). Modulation of simple cell activity in cat by moving textured backgrounds. *J. Physiol. (Lond.)*, **284**, 117P.

Hoffmann, K.-P., and Stone, J. (1971). Conduction velocity of afferents to cat visual cortex: a correlation with cortical receptive field properties. *Brain Res.*, **32**, 460–466.

Hubel, D. H., and Wiesel, T. N. (1962). Receptive fields, binocular interaction and functional architecture in the cat's visual cortex. *J. Physiol. (Lond.)*, **160**, 106–154.

Maffei, L. (1981). Processing of visual information in simple and complex cells: differences. In *Brain Mechanisms and Perceptual Awareness* (Eds. O. Pompeiano and C. Ajmone-Marsan), Raven Press, New York, pp. 37–51.

Morrone, C., Burr, D. C., and Maffei, L. (1982). Functional implications of cross-orientation inhibition of cortical visual cells I. Neurophysiological evidence. *Proc. Roy. Soc. Lond. B*, **216**, 335–354.

Movshon, J. A. (1975). The velocity tuning of single units in cat striate cortex. *J. Physiol., Lond.* **249**, 445–468.

Palmer, L. A., and Rosenquist, A. C. (1974). Visual receptive fields of single striate cortical units projecting to the superior colliculus in the cat. *Brain Res.*, **67**, 27–42.
Singer, W., Tretter F., and Cynader, M. (1975). Organization of cat striate cortex: a correlation of receptive field properties with afferent and efferent connections. *J. Neurophysiol.*, **38**, 1080–1098.
Stone, J. (1972). Morphology and physiology of the geniculocortical synapse in the cat: the question of parallel input to the striate cortex. *Invest. Ophthalmol.*, **11**, 338–346.

Models of the Visual Cortex
Edited by D. Rose and V. G. Dobson

CHAPTER 35

One, few, infinity: linear and nonlinear processing in the visual cortex

SHAUL HOCHSTEIN
Hebrew University of Jerusalem, Institute of Life Sciences, Jerusalem, Israel and Massachusetts Institute of Technology, Department of Psychology and Artificial Intelligence Laboratory, Cambridge, MA 02139, USA
HEDVA SPITZER
Hebrew University of Jerusalem, Institute of Life Sciences, Jerusalem, Israel and National Institutes of Mental Health, Bethesda, MD 20205, USA

INTRODUCTION

The function of the brain is information processing—nervous integration and decision making. It is becoming more and more apparent that this is not a matter of a voting process seeking a majority view (Minsky and Papert, 1972). Rather, the nervous system has at its disposal numerous nonlinear mechanisms applied as interactions amongst incoming signals in order to compute functional relationships and contingencies (see, for example, Marr and Poggio, 1977; Crick, Marr and Poggio, 1981; Marr, 1982; Poggio, 1982; and Koch and Poggio, Chapter 43 in this volume). The input signals to such a computation may be related to each other in various different ways, the simplest of which, perhaps, is that they represent information derived from the simultaneous reception of sensory data from neighboring locations. This 'spatial summation' may take on many forms, depending on the linear and nonlinear computations involved. We review here four types of spatial summation found in the visual system in the receptive fields of, respectively, X- and Y-cells of the retina and geniculate and simple and complex cells of the striate cortex. No attempt is made at an exhaustive review of the literature. This is, instead, a personal reflection on selected results and a presentation of a working hypothesis, a kind of opening position for a discussion of the kinds of mechanisms which may be responsible for the phenomena found.

LINEARITY CLASSIFICATIONS OF RETINAL AND GENICULATE CELLS

Nearly twenty years ago, Enroth-Cugell and Robson (1966) reported that cat retinal ganglion cells may be classified on the basis of their performing linear or nonlinear spatial summation respectively. They called these cells by the original and elucidating names, X and Y. Using counterphase gratings, they found that the responses of X-cells have a single peak during each period of stimulus modulation. Furthermore, the cells do not respond to the stimulus flicker at particular spatial phases, i. e. they have 'null' responses. On the other hand, Y-cells have frequency-doubled responses (two response peaks per stimulus modulation period) and at no spatial phase is the response 'null'.

Shapley and one of us (S.H.) studied X- and Y-cell receptive fields in the cat retina and lateral geniculate nucleus (Shapley and Hochstein, 1975; Hochstein and Shapley, 1976a, 1976b). It was found that X-cells have a center and a surround mechanism which integrate the illumination upon their respective receptive field subregions in a weighted linear fashion. The antagonistic interaction between these two regions is probably not simply linear subtraction. The well-known finding of a null position for counterphase sinusoidal grating stimulation is probably due to separate simultaneous nulling of the nearly concentric center and surround areas. Other center–surround interactions have recently been reviewed by Shapley and Enroth-Cugell (1984). Still, to a first approximation one may look upon the X-cell as modeled by a single difference of Gaussians (DOG) region. By a single region, we mean here a region over which responses are summed linearly.

In contrast, Y-cell receptive fields are made up of very many small scattered overlapping subunits, which are coextensive with the larger, linear, antagonistic center and surround regions (Fig. 1, top). Hochstein and Shapley (1976b) found that they could understand the response characteristics of Y-cells by assuming that the contributions of the small subunits are half-wave rectified before pooling. This implies that their excitatory elements are summed but that their inhibitory aspects are local and not shared. Thus a counterphase grating, in which the illumination on some areas is increased simultaneously to a decreasing illumination on other areas of the receptive field, leads to a response which is the sum of the *net local excitations unreduced by net distant inhibition*.

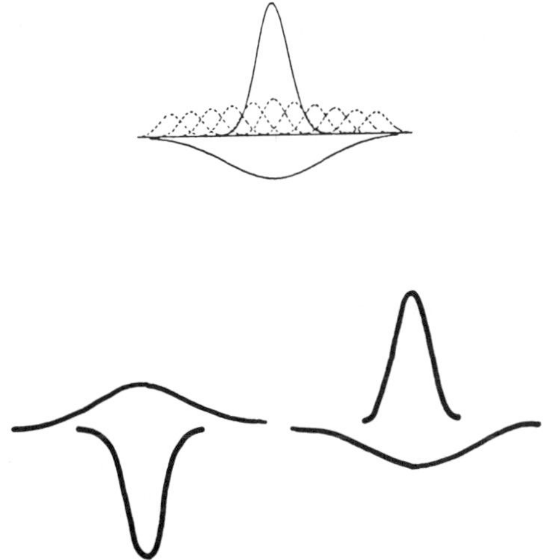

Figure 1 Receptive field models of a Y-cell and a cortical complex cell (top and bottom respectively). The Y-cell field is composed of a linear center and surround, forming together a single difference of Gaussians (DOG) region. Superposed upon this DOG there appear numerous scattered overlapping nonlinear subunits, whose responses are half-wave rectified before pooling (reproduced with permission from Hochstein and Shapley, 1976b). The complex cell receptive field is made up of a few (here two) subunits, each of which has a DOG profile. The responses of the subunits are again summed following half-wave rectification. Thus, the antagonistic effects of the surrounds are local and not pooled.

LINEARITY CLASSIFICATIONS OF STRIATE CORTEX CELLS

One aspect of the research of Hubel and Wiesel (1962, 1968) which is not always stressed is that the visual neurons which they classified on the basis of the 'complexity' of their receptive fields may be classified also on the basis of whether they perform spatial summation in a *linear* or a *nonlinear* fashion.

In their words, receptive 'fields were termed "simple" because:

- they were subdivided into distinct excitatory and inhibitory regions;
- there was summation within the separate excitatory and inhibitory regions;
- there was antagonism between excitatory and inhibitory regions; and
- it was possible to predict responses to stationary or moving spots of various shapes from a map of the excitatory and inhibitory areas.'

Thus, *simple* cell responses are *linear summations* of the excitatory and inhibitory effects of stimulation on the different areas of the receptive field.

On the other hand, complex cells were found to be nonlinear: 'The receptive fields of these cells were termed "complex", [because]:

- Unlike cells with simple fields, these responded to variously shaped stationary or moving forms in a way that could *not* be predicted from maps made with small circular spots.
- Often such maps could not be made, since small round spots were either ineffective or evoked only mixed (ON–OFF) responses throughout the receptive field.
- When separate ON and OFF regions could be discerned, the principles of summation and mutual antagonism, so helpful in interpreting simple fields, did not generally hold.'

Thus, complex cell responses are not linear summations of the excitatory and inhibitory effects of stimulation on the different areas of the receptive field. *Complex cell* 'properties cannot be derived in a single *logical* step from those of lateral geniculate cells (or, in the monkey, from the concentric cells of layer 4C)'. This additional logical step in complex cells may be the introduction of a nonlinearity in the receptive field spatial summation.

PARALLEL OR SERIAL PROCESSING?

Recognizing that cells of the retina, the lateral geniculate nucleus (LGN) and the visual cortex may all be classified by their spatial summation linearity, one is led to ask: Does the same type of mechanism which is responsible for the Y-cell nonlinearity also underlie the complex cell nonlinearity? Is the complex cell related to the Y-cell? Is the Y-cell nonlinearity the source of the complex cell nonlinearity? (See, for example, Movshon, Thompson and Tolhurst, 1978; also Lennie, 1980, and Stone, 1983, for reviews of this question.)

METHODS—CORTICAL RECORDING

In an attempt to answer these questions, we studied simple and complex cells of cat cortical areas 17 and adjacent 18 in the anesthetized paralyzed preparation (Spitzer and Hochstein, 1985a, 1985b, Hochstein and Spitzer, 1984). Visual stimuli were counterphase gratings displayed on an oscilloscope screen. These are standing sinusoidal gratings whose contrast is varied sinusoidally in time.

In order to satisfactorily characterize a unit one has to study its response dependence on both the stimulation grating spatial phase—the position of the grating in the visual field—and its spatial frequency—the number of grating cycles per unit visual angle.

In a linear filter system, a sinusoidal stimulus always leads to a sinusoidal response (of the same frequency), and one would expect both a sinusoidal response time course and a sinusoidal response amplitude dependence on the stimulus phase for stimulus gratings of all spatial frequencies. *No cortical cells were found to be linear in all these ways*. Cortical cells often have low maintained discharge rates so that net negative (inhibitory) inputs can not be expressed. These cells therefore have (temporally) half-wave rectified response forms, even though they display linear spatial summation (a sinusoidal response dependence on the stimulus phase at all spatial frequencies). In addition, for most stimulus parameters complex cortical cells have frequency-doubled (ON-OFF) responses, i. e. responses which have two peaks in the period of a single stimulus modulation cycle (full-wave rectification). The peaks are not necessarily identical, however, and their relative amplitudes were found to be a function of both spatial phase and spatial frequency for many of the cells studied.

RESULTS

The data indicate that there are some similarities and some differences between X-cells and simple cells, on the one hand, and between Y-cells and complex cells, on the other. Both the X-cell and the simple cell receptive fields perform linear spatial summation. Of course, simple cells are orientation and direction selective while LGN cells are not. Furthermore, the summed responses of the simple cell are half-wave rectified and contain large, even Fourier harmonics.

The complex cell receptive field has a spatial summation nonlinearity which may derive from a *mechanism* which is similar to that responsible for the Y-cell nonlinearity, namely *half-wave rectification* of the responses of subunits prior to pooling. However, the *number* and the *spatial arrangement* of the subunits of the complex cell are entirely different than those of the Y-cell. The simplest conclusion is thus that complex cell receptive field nonlinearities are not related to Y-cell nonlinearities.

The experimental evidence for these conclusions lies in the dependences of cortical cell responses on spatial phase and spatial frequency of the stimulation grating (Spitzer and Hochstein, 1985a, 1985b).

Cortical simple cells show some features which are the same as those of LGN X-cells, but they have some response characteristics which are very different, too. Simple cells seem to have a linear spatial summation mechanism so that for a contrast reversal grating of any spatial frequency a phase may be found which produces a null response. Simple cell responses also have a sinusoidal amplitude dependence on spatial phase at all spatial frequencies (Fig. 2, top row). In this sense, these cells appear to have a single receptive field linear summation region. Thus, the response of these cells to compound stimuli may be predicted on the basis of their responses to spots of light in different parts of their receptive field regions (the point spread function). Of course, simple cells are orientation selective, perhaps to a degree which is not explained in terms of summation and antagonism between excitatory and inhibitory parts of a single linear region. Thus, a second

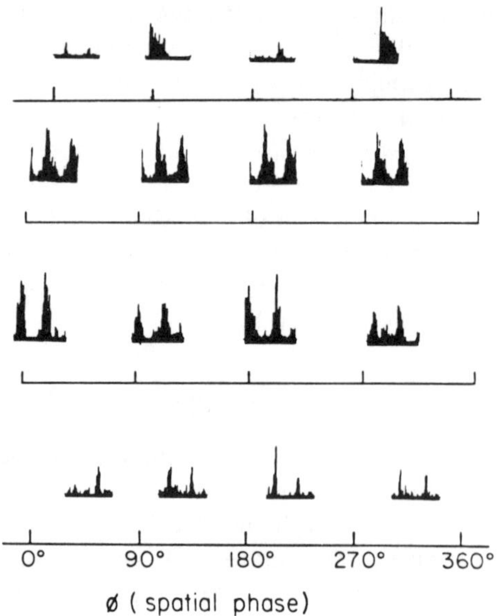

FIGURE 2 Responses from different cortical cell types. Responses to counterphase grating stimuli are shown for four different cortical cells. In each row PST histograms show the time course of the binned and averaged response firing rate of a single cortical neuron to stimuli of different spatial phases (about 90° apart). The top row shows responses of a simple cell, displaying single-peaked responses whose amplitude has a sinusoidal dependence on the spatial phase, indicating linear spatial summation. The second row shows frequency-doubled response forms with no spatial phase dependence. This is similar to the response characteristics of Y-cells and is rare in the cortex. The third row shows responses which are frequency doubled but whose amplitude has a distinct spatial phase dependence. The bottom row shows responses from another cell where the time-course of the response varied with stimulation spatial phase, being largely single peaked at some phases and largely frequency doubled at others. This variation with spatial phase (and with spatial frequency) would be expected from cells with receptive fields like that shown in Fig. 1 (bottom).

mechanism, probably a local inhibitory one, is in all likelihood present which does not allow the cell to produce responses to improperly oriented stimuli (see for example, Braitenberg, Chapter 50; Koch and Poggio, Chapter 43; Tieman and Hirsch, Chapter 45; and Sillito, Chapter 42, in this volume). A further difference between simple cells and X-cells is that simple cell responses are often half-wave rectified—a sign of (temporal) nonlinearity. This rectification follows spatial pooling, however, clearly differentiating these cells from complex cells.

If complex cell receptive fields derived their nonlinearity from Y-cells, one would expect the complex cell response to be independent of spatial phase, at least at high spatial frequencies. Spatial phase independence is found in

Y-cells of the retina and the geniculate and has its source in the multitude of scattered subunits of the Y-cell receptive field whose responses are dominant at high spatial frequencies. A few complex cells share this property (Fig. 2, second row).

Most complex cell responses, however, *did* show very definite spatial phase dependences. This was true even at high spatial frequencies, and despite the response being frequency doubled, i. e. having two response peaks in a single stimulation cycle (Fig. 2, third row).

Two additional characteristics of complex cell responses must be mentioned. Often the responses of a complex cell to counterphase gratings had two peaks of unequal amplitude (Fig. 2, bottom row). The relative amplitudes of the two peaks then varied as a function of spatial phase in a sinusoidal manner.

Finally, the response form itself was often a function of spatial frequency. Some cells had a frequency-doubled response with two equal peaks at some spatial frequencies, a response with two peaks of very different amplitudes with a spatial phase dependence at other spatial frequencies and a response which appeared with only a single peak at still other spatial frequencies. The variation between these different response shapes was not monotonic with spatial frequency but rather cyclic.

A MODEL

These characteristics appear puzzling at first but they may be understood quite readily in terms of a simple model (Spitzer and Hochstein, 1985b). The spatial frequency dependences may be understood if one assumes that the receptive field of the complex cell is made up of a *small number of subunits* or subunit regions, sometimes even as few as two (Fig. 1, bottom). In these cases, the relationship between the distances between the subunits and the spatial period of the stimulation grating determines the temporal character of the response and its spatial phase dependence. (Similar characteristics would result from models which build the receptive fields with clumps of subunits and/or regularly spaced subunits.)

For example, it is easy to see that if the distance between the subunits is exactly a grating cycle, then all the subunits will be stimulated identically, and the response will appear exactly as it might with a single linear subunit—namely with a single peak and a sinusoidal response dependence on phase. At other spatial frequencies, the distance between the subunits may correspond to half a grating cycle so that one subunit will see an increase in illumination while another sees an equal decrease in illumination. The nonlinear summation allows this to result in an ON-OFF type of response. This, too, will have a sinusoidal dependence on spatial phase. At intermediate spatial frequencies a mixed type of response will be seen, with two peaks of unequal amplitude and a complex dependence on spatial phase.

We analyzed this model quantitatively using half-wave rectification prior to pooling as the receptive field spatial summation nonlinearity and have found a good fit between the model and the responses of cortical complex cells. The receptive field subunits may each have the form of a DOG, as might be expected if they result from lateral geniculate nucleus X-cell input (Spitzer and Hochstein, 1985a). The subunits sum their excitatory elements while their inhibitory influences remain local.

CONCLUSION—COMPLEX CELLS VERSUS Y-CELLS

In conclusion, complex cells have the following properties:

1. A large number of scattered overlapping subunits.
2. *Linear* spatial summation *within* the subunits.
3. *Nonlinear* spatial summation *among* the subunits.
4. A nonlinearity which may derive from half-wave rectification before spatial pooling so that *inhibitory* influences remain local and only *net* excitatory effects are summed.
5. All subunits are of the same type—ON or OFF—so that responses to bar stimuli are ON or OFF, not ON-OFF.
6. The scatter of the subunits leads to a lack of spatial phase dependence of the response to counterphase grating stimuli.

These properties may be compared with those of Y-cell receptive fields:

1. A large number of scattered overlapping subunits.
2. *Linear* spatial summation *within* the subunits.
3. *Nonlinear* spatial summation *among* the subunits.
4. A nonlinearity which may derive from half-wave rectification before spatial pooling so that *inhibitory* influences remain local and only *net* excitatory effects are summed.
5. All subunits are of the same type—ON or OFF—so that responses to bar stimuli are ON or OFF, not ON-OFF.
6. The scatter of the subunits leads to a lack of spatial phase dependence of the response to counterphase grating stimuli.

Thus, complex cells have a lot in common with Y-cells, but not everything. They do not seem to derive their nonlinearity from Y-cells and though the underlying mechanism may be the same, the spatial arrangement and the number of subunits is entirely different.

As to the possible function of complex cells, let us say only that they may be signaling the simultaneous appearance of features at distinct locations within their receptive field areas. One example might be the appearance of a peak and a trough of illumination at neighboring positions forming a single edge. One type of complex cell may serve as a workable Marr zero-crossing filter (Marr, 1982; Hochstein and Spitzer, 1984). Other such coincidence functions may also be considered. In general, the number and the type of subunits in the complex cell receptive field determine their overall character. These may vary from cell to cell, forming a family of receptive field characters.

Further study, perhaps including recording from the alert behaving preparation, is required to determine what role these units actually play in visual information processing.

SUMMARY

Linear and nonlinear spatial summation in the receptive fields of four visual cell types were reviewed and compared. These are the X- and Y-cells of the retina and lateral geniculate nucleus, and the simple and complex cells of the striate visual cortex.

The responses to counterphase grating stimulation of X-cells and simple cells have sinusoidal response amplitude dependences on the stimulation grating spatial phase, and in particular they have 'null' positions. This indicates the presence of a linear spatial summation mechanism in the receptive fields of these neurons. In addition, the response-time histograms of these cells to counterphase gratings share the property of having a single peak per stimulus cycle (ON or OFF). Simple cell responses are generally more half-wave rectified than are X-cell responses, but this rectification follows the linear pooling of the effects from all regions of the receptive field.

Y-cell responses are spatial phase independent at high spatial frequencies where the time histograms have two peaks of equal amplitude. These characteristics may be modeled by a receptive field consisting of numerous scattered overlapping subunits whose responses are half-wave rectified before pooling. Complex cells, on the other hand, generally have distinct spatial phase dependences, even at high spatial frequencies, and even when the responses are frequency doubled. The responses of complex cells have a single peak or two equal or unequal peaks, as a function of spatial phase and frequency. These characteristics are modeled by a receptive field consisting of two (or a few) receptive field subunits whose responses are half-wave rectified before pooling. The relationship between the intersubunit distance and the stimulation grating period leads to the change of the response characteristics with spatial frequency.

Thus it is possible that a single nonlinearity—half-wave rectification—may appear *before* pooling of the responses of a *few* subunits in complex cells, *before* pooling of the responses of a *multitude* of subunits in Y-cells, *following* pooling of responses in simple cells, and perhaps *not at all* in X-cells.

ACKNOWLEDGMENT

We thank the Israel Academy of Arts and Sciences and the United States-Israel Binational Science Foundation (BSF) for supporting this study. S.H. thanks the Sloan Foundation for support at M.I.T.

REFERENCES

Crick, F. H. C., Marr, D. C., and Poggio, T. (1981). An information processing approach to understanding the visual cortex. In *The Cerebral Cortex* (Eds. F. O. Schmitt and F. G. Worden), MIT Press, Cambridge, Mass.

Enroth-Cugell, C., and Robson, J. (1966). The contrast sensitivity of retinal ganglion cells of the cat. *J. Physiol. (Lond.)*, **187**, 517–522.

Hochstein, S., and Shapley, R. M. (1976a). Quantitative analysis of retinal ganglion cell classifications. *J. Physiol. (Lond.)*, **262**, 237–264.

Hochstein, S., and Shapley, R. M. (1976b). Linear and nonlinear spatial subunits in Y cat retinal ganglion cells. *J. Physiol. (Lond.)*, **262**, 265–284.

Hochstein, S., and Spitzer, H. (1984). Zero-crossing detectors in primary visual cortex? *Biol. Cybernetics,* **51**, 195-199.

Hubel, D., and Wiesel, T. N. (1962). Receptive fields, binocular interactions and functional architecture in the cat's striate cortex. *J. Physiol. (Lond.)*, **160**, 106–154.

Hubel, D., and Wiesel, T. N. (1968). Receptive fields and functional architecture of monkey striate cortex. *J. Physiol. (Lond.)*, **195**, 215–243.

Lennie, P. (1980). Parallel visual pathways. *Vision Res.*, **20**, 561–594.

Marr, David (1982). *Vision*, Freeman, San Francisco.

Marr, D., and Poggio, T. (1977). From understanding computation to understanding neural circuitry. In *Neuronal Mechanisms in Visual Perception* (Eds. E. Poppel, R. Held and J. E. Dowling, Vol. 15, Neurosci. Res. Prog. Bull., pp. 470–488.

Minsky, M., and Papert, S. (1972). *Perceptrons, An Introduction to Computational Geometry, MIT Press, Cambridge, Mass.*

Movshon, J. A., Thompson, I. D., and Tolhurst, D. J. (1978). Receptive field organization of complex cells in the cat's striate cortex. *J. Physiol. (Lond.)*, **283**, 79–99.

Poggio, T. (1982). *Visual Algorithms*, AI Memo 683.

Shapley, R. M., and Enroth-Cugell, C. (1984). Visual adaptation and retinal gain controls. In *Progress in Retinal Research* (Eds. N. N. Osborne and G. J. Chader), Pergamon Press, New York.

Shapley, R. M., and Hochstein, S. (1975). Visual spatial summation in two classes of geniculate cells. *Nature (Lond.)*, **256**, 411–413.

Spitzer, H., and Hochstein, S. (1985a). Simple- and complex-cell response dependencies on stimulation parameters. *J. Neurophysiol.*, **53**, 1259-1280

Spitzer, H., and Hochstein, S. (1985b). A complex cell receptive-field model. *J. Neurophysiol.*, **53**, 1281-1301.

Stone, J. (1983). *Parallel Processing in the Visual System*, Plenum Press, New York.

Models of the Visual Cortex
Edited by D. Rose and V. G. Dobson

CHAPTER 36

Design duplication in streams of the striate cortex

G.H. Henry
Department of Physiology, Australian National University, PO Box 334, Canberra City, ACT 2601, Australia

PREAMBLE

The first, and possibly still the most significant, interpretation of neural sequences in the striate cortex was presented in 1962 by Hubel and Wiesel (Hubel and Wiesel, 1962) when it was proposed, for the cat, that simple, complex and hypercomplex cells form a hierarchical series. Since then X, Y and W streams have been identified in the afferent pathways (Enroth-Cugell and Robson, 1966; Cleland, Dubin and Levick, 1971; Stone and Fukuda, 1974) and it has also been suggested that retinal ganglion cells with on- and off-centre receptive fields contribute branches to the X and Y streams (Wässle, Peichl and Boycott 1981; Wässle, Boycott and Illing, 1981). The X, Y and W streams enter at different levels in the striate cortex (LeVay and Gilbert, 1976), although in most species (the mink may be an exception; see LeVay and McConnell, 1982) spatial segregation has not been observed in the on- and off-centre branches.

Over the years, there have been few modifications to the hierarchical classification, although one necessary adjustment requires that a preference for foreshortened stimuli, or hypercomplexity, should not be regarded as a class-defining characteristic but rather as a general property that may occur in cells that otherwise have simple or complex receptive fields (Bishop and Henry, 1972; Dreher, 1972).

In seeking a link between the hierarchical classification and streaming, it is no longer acceptable to suggest that simple and complex cells respectively act as exclusive representatives of the X and Y streams (Stone and Dreher, 1973), since it has now been shown that cells of each class may belong to either stream (Bullier and Henry, 1979a; Henry, Mustari and Bullier, 1983). Stream dependence, however, is to be observed in subclasses of the simple and complex categories (Henry, Mustari and Bullier, 1983; Mustari, Bullier and Henry, 1982) and it is through these that streaming should be followed into the striate cortex. Before taking this step, however, it should be noted that the on- and off-centre branches seem to converge onto single cortical cells and that on- and off-centre responses are to be observed in the receptive fields of both simple and complex cells. In a sizeable proportion of simple cells, however, on- and off-centre responses occur in isolation (Bullier, Mustari and Henry, 1982) and in these instances there is the potential for the on- and off-centre branches, like those of the X, Y, and W streams, to remain segregated as they pass through the cortex.

The stream-dependent subclasses, referred to above, do not display dramatic response differences and it has been necessary to take an alternative line to identify the stream to which the cell belongs. Approaches such as the measurement of differences in the velocity of conduction of signals initiated by electrical stimulation (Bullier and Henry, 1979a), the reversible blockade of one of the afferent streams (in the monkey) (Malpeli, Schiller and Colby, 1981) or branches (Schiller, 1982) and the cross-correlation of responses in striate and lateral geniculate nucleus (LGN) neurons (Tanaka, 1983) have all been used in stream identification. The general conclusion from these experiments is that streaming in the X and Y pathways persists into the striate cortex.

Experiments employing electrical stimulation (Bullier and Henry, 1979b; Henry, Mustari and Bullier, 1983) also indicate that both simple and complex cells of each stream distribute at different levels in lamina 4 to align with the point where the appropriate afferents terminate. Although a point of some controversy (Ferster and Lindström, 1983; Henry, Mustari and Bullier, 1983), it also now seems likely that complex cells of lamina 5 receive a direct input from the LGN and that this comes exclusively from cells in the Y stream (Henry, Mustari and Bullier, 1983; see also Schiller, Malpeli and Schein, 1979, for projection from lamina 5 to the superior colliculus in monkey). In lamina 6, cells belonging to different streams have also been identified (Bullier and Henry, 1979a), although as yet there have been no reports of spatial segregation in these streams.

STREAM DISTINCTIONS IN THE RESPONSES OF STRIATE NEURONS

It may be argued, for the X and Y pathways, that stream-induced differences, albeit at a quantitative level, may be extracted from the responses of both simple and complex cells. There is no indication, however, that any of these distinctions arise from cortical organization. The overriding impression is that cells, of a particular class have a consistent receptive field design and that stream-dependent differences are established before the cortex is reached.

Simple Cells

The receptive fields of simple cells are characterized by the orderliness of their internal construction. Created, as they appear to be, from short rows of on- and off-centre LGN units (Hubel and Wiesel, 1962; Rose, 1979; Bullier, Mustari and Henry, 1982), their receptive fields are made up of segregated on- and off-discharge regions that appear to be mutually antagonistic. The modulation of these discharges and the generation of a high degree of response specificity is achieved through cortical inhibition which lacks the specific character of antagonism in that it suppresses both on- and off-responses.

All the available evidence suggests that this design is common to simple cells of both streams and that their response differences simply reflect distinctions already apparent in their parent cells in the LGN. To be specific, size differences in the receptive field are carried through into the cortical cell (Mustari, Bullier and Henry 1982), resulting in larger discharge regions and a responsiveness to slightly faster stimuli in Y- than X-stream simple cells. Both also seem to suffer an almost identical reduction in the responsiveness to fast-moving stimuli, possessed by their parent cells in the LGN.

Less information is available on properties such as the level of persistence in the response and it is possible that the presence of a threshold in simple cells may render ineffective any sustained component in the parent cell. Stream differences in the level of response linearity also remain unassessed although there are two populations of simple cells—one with markedly non-linear and the other with close to linear responses (Movshon, Thompson and Tolhurst, 1978). These could be representatives of the Y and X streams respectively, but as yet the streams to these cells have not been confirmed by the alternative methods referred to above.

No stream differences, even at the quantitative level, have been reported for orientation specificity and direction selectivity in simple cells (Mustari, Bullier and Henry, 1982). In these features, as is the case with all cortically added properties, there is no sign of a disparate receptive field mechanism in the two streams.

Complex Cells

As with simple cells, the internal structure and organization of receptive fields appears to be similar in complex cells belonging to the two streams. Both appear to be constructed from a convergence of inputs that come either directly from the lateral geniculate nucleus (LGN) or through intermediate neurons in the cortex (Henry, Mustari and Bullier, 1983). In this merging of inputs there is an absence of spatial segregation of on- and off-regions and also of the mutual antagonism between these regions. Instead, composite on- and off-firing arises from all parts of the complex cell receptive field. Here, also, the stream-dependent distinctions established in the retina or the LGN appear to be responsible for most of the cortical differences. Thus, when compared to their Y-stream counterparts, complex cells of the X stream have smaller receptive fields, are not as responsive to faster moving stimuli and have a lower spontaneous activity and a more sustained response (Henry, Mustari and Bullier, 1983). It is possible that the sustained response and spontaneous activity may be seen as differentiating features in complex but not simple cells because the complex cell has a lower threshold (Henry, Goodwin and Bishop, 1978). In orientation specificity and direction selectivity the complex cell resembles the simple cell and there have been no reports of stream-dependent distinctions.

The identification of the W stream in the striate cortex has not been achieved with the same assurance as have the X and Y streams. In the few instances where a W-stream input seems likely the cells had a receptive field that was indistinguishable from that of complex cells in the X stream (Henry, Mustari and Bullier, 1983).

Cells with the Hypercomplex Property

Recently, there has been a growing doubt about the origin of the end-zone inhibition responsible for the hypercomplex property. Some, and possibly all, of the response reduction could come from the activation of the surrounds of LGN neurons or, alternatively, it could be a manifestation of cortical activity. Presumably, if mediated through interneurons in the cortex, the end-zone inhibition would be removed by the presence of bicuculline and would also display the stimulus selectivity of striate neurons.

On the first point, the hypercomplexity of some cells in lamina 3 has displayed a sensitivity to bicuculline (Sillito and Versiani, 1977), but the reduction in length preference was partial and there was no certainty about the relative incidence of these cells. With regard to stimulus selectivity, orientation dependence in the end-zone is difficult to assess without inducing changes in other stimulus parameters. In one well-controlled experiment, however, the end-zone inhibition was found to be weakly orientation specific (Orban, Kato and Bishop, 1979). Running counter to these results, direction selectivity in the end-zone, more readily assessed than orientation specificity, seldom appears to be present in regions of hypercomplexity.

Whatever the origin of end-zone inhibition it seems equally common in cells of both X and Y streams, although there is some indication, both in simple and complex cells, that the inhibition may be more intense in representations of the X- than the Y-stream (Mustari, Bullier and Henry,

1982; Henry, Mustari and Bullier, 1983). Once again, this tendency may reflect a distinction already present in the surrounds of parent cells in the LGN (Cleland, Levick and Sanderson, 1973).

SUMMING UP

At this stage we can only speculate on the functional reasons for streaming in the striate cortex. The on- and off-centre branches may play a variety of roles as they remain separate in some simple cells, unite in an apparently opponent organization, presumably to sharpen spatial discrimination, in other simple cells, and converge to produce the composite on/off-responses of complex cells.

It remains possible that simple and complex cells, whose differences in receptive field structure so impressed Hubel and Wiesel, may contribute to a hierarchical sequence, with the proviso that the complex cell has an additional direct input. By contrast, we must now be equally impressed with the similarity of receptive field design that occurs in striate neurons belonging to X and Y streams. All the distinctions in these cells appear to reflect differences already present in cells of the afferent pathway. This consistency in receptive field design points to the extraction of similar information in the two streams in the cortex.

If, as is often suggested, the X stream is responsible for high-resolution vision and the Y stream for motion detection, then these functional distinctions reside in the precortical responses. The cortical cells must all make similar abstractions and so take both form and motion information from the signal. In the striate cortex, the cells of the X stream may be principally concerned with high acuity vision but they are also influenced by image movement on the retina; similarly, although cells of the Y stream have a major interest in movement they are also modulated by the form of the image. Form and movement must be accurately correlated and it seems that this is achieved through design duplication in receptive fields of cells in both streams, rather than through elaborate cross-referencing between the two streams.

REFERENCES

Bishop, P. O., and Henry, G. H. (1972). Striate neurons: receptive field concepts. *Invest. Ophthalmol.*, **11**, 346–354.

Bullier, J., and Henry, G. H. (1979a). Neural path taken by afferent streams in striate cortex of the cat. *J. Neurophysiol.*, **42**, 1264–1270.

Bullier, J., and Henry, G. H. (1979b). Laminar distribution of first-order neurons and afferent terminals in cat striate cortex. *J. Neurophysiol.*, **42**, 1271–1281.

Bullier, J., Mustari, M. J., and Henry, G. H. (1982). Receptive field transformations between LGN neurons and S cells of cat striate cortex. *J. Neurophysiol.*, **47**, 417–438.

Cleland, B. G., Dubin, M. W., and Levick, W. R. (1971). Sustained and transient neurones in the cat's retina and lateral geniculate nucleus. *J. Physiol.*, **217**, 473–496.

Cleland, B. G., Levick, W. R., and Sanderson, K. J. (1973). Properties of sustained and transient ganglion cells in the cat retina, *J. Physiol.*, **228**, 649–680.

Dreher, B. (1972). Hypercomplex cells in the cat's striate cortex. *Invest. Ophthalmol.*, **11**, 355–356.

Enroth-Cugell, C., and Robson, J. G. (1966). The contrast sensitivity of retinal ganglion cells in the cat. *J. Physiol. (Lond.)*, **187**, 517–552.

Ferster, D., and Lindström, S. (1983). An intra-cellular analysis of geniculo-cortical connectivity in area 17 of the cat. *J. Physiol.*, **342**, 181–215.

Henry, G. H., Goodwin, A. W., and Bishop, P. O. (1978). Spatial summation of responses in the receptive fields of striate neurons. *Exp. Brain Res.*, **32**, 245–266.

Henry, G. H., Mustari, M. J., and Bullier, J. (1983). Different geniculate inputs to B and C cells of cat striate cortex. *Exp. Brain Res.*, **52**, 179–189.

Hubel, D. H., and Wiesel, T. N. (1962). Receptive fields, binocular interaction and functional architecture in cat's visual cortex. *J. Physiol.*, **160**, 105–154.

LeVay, S., and Gilbert, C. D. (1976). Laminar patterns of geniculo cortical projection in the cat. *Brain Res.*, **113**, 1–19.

LeVay, S., and McConnell, S. K. (1982). ON and OFF layers in the lateral geniculate nucleus of the mink. *Nature*, **300**, 350–351.

Malpeli, J., Schiller, P. H., and Colby, C. L. (1981). Response properties of single cells in monkey striate cortex during reversible inactivation of individual lateral geniculate laminae. *J. Neurophysiol.*, **46**, 1102–1119.

Movshon, J. A., Thompson, I. D., and Tolhurst, D. J. (1978). Spatial summation in the receptive fields of simple cells in the cat's striate cortex. *J. Physiol.*, **283**, 53–77.

Mustari, M. J., Bullier, J., and Henry, G. H. (1982). Comparison of the response properties of three types of monosynaptic S cell in cat striate cortex. *J. Neurophysiol.*, **47**, 439–454.

Orban, G. A., Kato, H., and Bishop, P. O. (1979). Dimensions and properties of end-zone inhibitory areas in receptive fields of hypercomplex cells in cat striate cortex. *J. Neurophysiol.*, **42**, 833–849.

Rose, D. (1979). Mechanisms underlying the receptive field properties of neurons in cat visual cortex. *Vision Res.*, **19**, 533–544.

Schiller, P. H. (1982). Central connections of the retinal ON and OFF pathways. *Nature*, **297**, 580–582.

Schiller, P. H., Malpeli, J. G., and Schein, S. J. (1979). Composition of geniculostriate input to superior colliculus of the rhesus monkey. *J. Neurophysiol.*, **42**, 1124–1133.

Sillito, A. M., and Versiani, V. (1977). The contribution of excitatory and inhibitory inputs to the length preference of hypercomplex cells in layers II and III of the cat's striate cortex. *J. Physiol.*, **273**, 775–790.

Stone, J., and Dreher, B. (1973). Projections of X-and Y-cells of the lateral geniculate nucleus to areas 17 and 18 of the visual cortex. *J. Neurophysiol.*, **36**, 551–567.

Stone, J., and Fukuda, Y. (1974). Properties of cat retinal ganglion cells: a comparison of W-cells with X- and Y-cells. *J. Neurophysiol.*, **37**, 722–748.

Tanaka, K. (1983). Cross-correlation analysis of geniculo-striate neuronal relationships in cats. *J. Neurophysiol.*, **49**, 1303–1318.

Wässle, H., Boycott, B. B., and Illing, R.-B. (1981). Morphology and mosaic of on- and off- beta cells in the cat retina and some functional considerations. *Proc. Roy. Soc. Lond. B*, **212**, 177–195.

Wässle, H., Peichl, L., and Boycott, B. B. (1981). Morphology and topography of on- and off- alpha cells in the cat retina. *Proc. Roy. Soc. Lond. B*, **212**, 157–175.

Models of the Visual Cortex
Edited by D. Rose and V. G. Dobson

CHAPTER 37

Receptive field organization of simple and complex cells

PAUL HEGGELUND
Neurobiological Laboratory, University of Trondheim, N-7055 Dragvoll, Norway

Despite the large variety of configurations of on- and off-zones which are encountered when the receptive field of single neurons in cat striate cortex are mapped with a stationary spot or slit, Hubel and Wiesel (1962) could divide the cells into the categories simple and complex. They defined simple cells as having spatially distinct on- and off-zones with spatial summation within each subregion and antagonism between the on- and off-regions. Cells lacking one or more of the defining characteristics of simple cells were defined as complex. The response properties of simple cells were explained by input from a row of lateral geniculate nucleus (LGN) cells with receptive fields centred along a common axis. Complex cells were presumed to be higher order cortical neurons with input from simple cells since they lacked the antagonistic subregions with spatial summation which are typical for retinal ganglion cells, LGN cells and cortical simple cells.

One difficult problem with receptive field analysis in the striate cortex is that the cells exhibit almost no spontaneous activity, and it is therefore usually impossible with extracellular recording techniques and a single stimulus to plot suppression zones in the receptive field. To avoid this problem I have combined extracellular recording with a dual-stimulus technique (static activation) where a stationary flashing *activation stimulus* (AS) produces activity against which both enhancement and suppression zones in the receptive field can be plotted with a stationary flashing *test stimulas* (TS) (Heggelund, 1981a, 1981b). These experiments have given evidence against the Hubel and Wiesel models, and led to alternative models for simple and complex cells.

In the static activation experiments the activation stimulus (usually an optimally oriented light slit) was placed in the position where the maximum response was elicited and the test stimulus was presented in a series of different receptive field positions to measure how the test stimulus enhanced or suppressed the response to the activation stimulus. In each test stimulus position a response histogram to a sequence of different stimulus conditions was recorded. From the response to the various stimulus conditions the effect of test stimulus both on and off on the on- and off-responses to the activation stimulus could be determined.

The class of simple cells were further subdivided into on- and off-dominant, depending on the type of response which was strongest (Heggelund, 1981a). Results from an on-dominant simple cell are shown in Fig. 1. Figure 1b, d shows how the on-response to an optimally oriented activation slit was modified by the test stimulus at a series of broadside distances from the activation slit. The test stimulus was a slit oriented parallel to the activation stimulus, and both slits had the same dimensions and luminance. The two response profiles show that the receptive field was subdivided into a central region and adjacent flanks which responded oppositely to the test stimulus. Corresponding results occur for off-dominant simple cells (Heggelund 1981a). For both subgroups of simple cells presentation of the two stimuli *in phase* (both on or both off simultaneously) produced enhancement in the central region and suppression on the flanks (Fig. 1b). The suppression was usually considerably stronger on one side. When the two slits were presented in *counterphase* (TS on–AS off or vice versa), strong suppression occurred in the central region and enhancement on that flank where the strongest suppression appeared by in-phase presentation (Fig. 1d). This means that the flank suppression is just as specifically linked to the light phase as the enhancement in the central region, and it occurs only by on in on-dominant simple cells and only by off in off-dominant simple cells.

Also the enhancement and suppression effects in the subregions of the response profiles show spatial summation (Heggelund, Krekling and Skottun, 1983). This is illustrated by Fig. 1c where a width summation curve over the strongest suppression flank is shown. The curve was made with a test slit of variable width which was presented in phase with the activation slit. The proximal edge of the slit was kept fixed on the border between the central region and the right flank. The curve shows that the suppression summated spatially just like the on- and off-responses of the simple cells.

The fact that the strengths of the on- and off-responses in simple cells are usually different, such that the cells could be classified as on- and off-dominant, indicates that these cells have excitatory input only from on- or only from off-centre LGN cells. This hypothesis is further supported by results from experiments where responses of LGN cells and simple cells to moving slits and gratings have been compared (Dreher and Sanderson 1973; Wilson

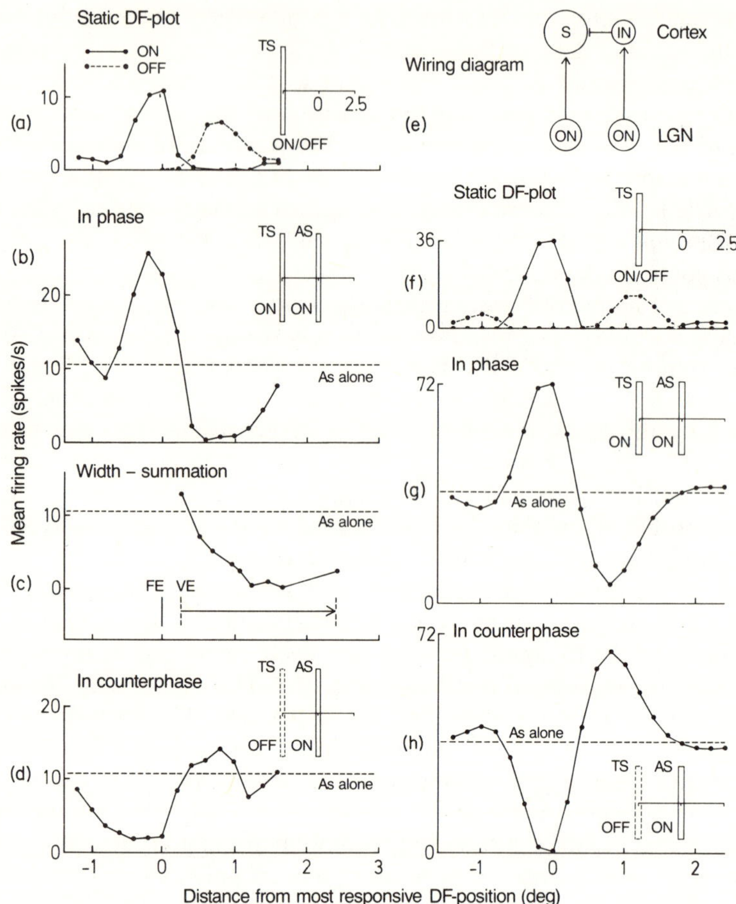

FIGURE 1 Results from an on-dominant simple cell. (a) Plot of the discharge field (DF) made with an optimally oriented light slit (0.17 x 3.4°) presented on and off in different broadside positions. (b) Static activation profile made with a test and an activation slit presented on simultaneously. The broadside position of the test slit varied. Both slits had the same dimensions (0.17 x 3.4) and luminance. The response to the activation stimulus (AS) alone is indicated by the broken horizontal line. Response above this value was termed enhancement, and response below, suppression. (c) Width-summation curve made with a test and an activation slit presented on simultaneously. The activation slit was in the same position as in (b). The test slit had variable width and one long edge was in a fixed edge position (FE) while the other variable edge (VE) was presented in a series of positions across the suppression flank. (d) Static activation profile made with the test slit presented off in various broadside positions while the activation slit was presented on. The time window of the on- and the off-periods was 500 ms and each data point was based on 25 repetitive measurements. (e) Schematic circuit showing the wiring diagram for the

and Sherman 1976; Lee, Elepfandt and Virsu, 1981) and from experiments with simultaneous recording from pairs consisting of a retinal ganglion cell and a cortical simple cell (Lee, Cleland and Creutzfeldt, 1977). The on-dominant simple cells therefore most probably get their excitatory input from on-centre LGN cells, and the off-dominant simple cells from off-centre LGN cells.

The maximum suppression which occurs on the flanks by in-phase presentation of the test and activation stimuli is strong (70 to 100 per cent), compared to the suppression found by comparable measurements in the receptive field periphery of LGN cells (10 to 25 per cent.; see Heggelund, 1981a). This clearly indicates that the suppression is due to intracortical inhibition, and not to the centre-surround antagonism in LGN as the model of Hubel and Wiesel presumes, and intracellular recordings have shown that inhibitory input to single cells in cat striate cortex actually produces sideband suppression (Creutzfeldt and Ito, 1968; Benevento, Creutzfeldt and Kuhnt, 1972).

The strong flank suppression in on-dominant simple cells occurs only by light on, and this indicates that on-dominant cells have inhibitory input from on-centre cells only. Mixed input from on- and off-centre LGN cells would have given flank suppression both by on and by off. In off-dominant simple cells the strong flank suppression occurs only by off, indicating that these cells have their inhibitory input from only off-centre LGN cells. This means that the simple cells have excitatory and inhibitory input from LGN cells of the same type, i.e. only from on- or off-centre cells. Hence, the stream of on- and off-information, which is kept separate at subcortical levels, is also kept separate in the class of simple cells in the striate cortex.

These properties are incorporated in the model for simple cells which is illustrated in Fig. 1. The model presumes that simple cells have excitatory and inhibitory input from only on- or off-centre LGN cells and that the excitatory and inhibitory fields are slightly displaced with respect to each other. The simplest possible wiring for an on-dominant simple cell is shown by the schematic circuit in Fig. 1e, where only input from two on-centre LGN cells is presumed. The wiring shows direct excitatory input, but inhibitory input via an interneuron in accordance with results from intracellular recordings which show that simple cells have monosynaptic excitatory and disynaptic inhibitory input from LGN (Ferster and Lindström, 1981). The model was used to calculate distributions of on- and off-responses to a single slit in positions across the receptive field (Fig. 1f), and static activation profiles made with in-phase (Fig. 1g) and in-counterphase (Fig. 1h) slit

---

simple cell model. ON indicates the LGN on-centre cell, IN the cortical interneuron, S the simple cell, ← excitatory synapse, – inhibitory synapse. (f) to (h) Simulated results based on the model. The values on the ordinates are in arbitrary units.

presentation. The simulated curves correspond well to the experimental results (Fig. 1a, b, d).

The slightly displaced excitatory and inhibitory fields explain orientation selectivity (Heggelund and Moors, 1983) and the elongation of the subregions of the receptive field along the optimal orientation. Contrary to the model of Hubel and Wiesel (1962) the present model also explains why these properties disappear when GABAergic intracortical inhibition is blocked (Tsumoto, Eckart and Creutzfeldt, 1979). The model explains in a simple manner how the circular receptive fields of the input to the striate cortex are combined to yield fields which consist of adjacent excitatory and inhibitory subregions, which is an ideal arrangement for detection of straight edges. An on-dominant simple cell, for example, would respond best when the light side of an edge covers the on activation zone at the same time as the darker side of the edge covers the adjacent on suppression zone. Off-dominant simple cells would respond to edges with the opposite luminance pattern. The model can furthermore account for the fact that the antagonism between the subregions in the cortex is stronger than in the LGN and for the lower spontaneous activity in the cortex. The large variety of on- and off-zone configurations amongst simple cells can also be accounted for.

Complex cells which seem to have a uniform receptive field when tested with a single slit turned out to have fields subdivided into antagonistic regions just like simple cells when tested with the dual stimulus technique (Heggelund, 1981b). Furthermore, the complex cells also showed spatial summation within the subregions just like simple cells. Figure 2a to d shows results from a complex cell which responded slightly better to on than to off (Fig. 1a). The response profiles made with the two parallel slits presented in phase showed enhancement in a central region and suppression on the flanks (Fig. 2b). The profiles made with in-counterphase presentation showed suppression in the central region and mainly enhancement on the flanks (Fig. 2d). Notice also that the complex cell had stronger suppression on one of the flanks by in-phase presentation (Fig. 2b), just like the simple cells.

The results showed that the receptive field organization of complex cells is strikingly similar to that of simple cells. Figure 2c also illustrates the spatial summation in the subregions of the complex fields. The upper curve shows summation across the central enhancement zone made by in-phase presentation of an activation slit of fixed size and a test slit of variable width. The lower response curve was made similarly and shows summation across the strongest suppression flank. The curve shows that the response decreased with test slit width, demonstrating spatial summation of the suppression. This also means that, with respect to summation, simple and complex cells are similar. The central receptive field region and the flanks were also in complex cells elongated along the optimal stimulus orientation, and the flank suppression was usually considerably stronger than surround suppression of LGN cells.

Since complex cells can be strongly activated by on and off in the central part of the receptive field it is reasonable to assume that these cells have mixed excitatory input from on- and off-centre LGN cells with overlapping receptive field centres. This would also explain the apparent contradiction between the uniform on/off receptive field and the lack of spatial summation over the whole discharge field (Hubel and Wiesel, 1962), since both the on- and off-responses have dual sources of input. The on-response in the central part of the receptive field is due to the receptive field centre of the excitatory on-centre LGN cells, but the on-response on the flanks comes from the surround of the excitatory off-centre LGN cells. Similarly, the off-response in the central region comes from activation of the centre of the excitatory off-centre LGN cells, but the off-response on the flanks is produced by off-activiation of the surround of the excitatory on-centre LGN cells. The spatial summation in complex cells is therefore limited to the subregions of the receptive field, as shown above, because a stimulus wider than a subregion would also stimulate antagonistic regions of the receptive field of the input fibres.

The model for complex cells illustrated in Fig. 2e presumes that both the excitatory and inhibitory input to these cells come from on- and off-centre LGN cells with overlapping receptive fields, but that the inhibitory field is slightly displaced with respect to the excitatory field. Two different versions of the model are shown in Fig. 2e. The diagram to the left presumes that complex cells receive input from pairs of on- and off-dominant simple cells (hierarchical version). The scheme to the right presumes that complex cells have direct excitatory input from LGN cells (parallel version).

Both the hierarchical and the parallel versions of the model can account for the data. Figure, 2f to h shows simulated results for complex cells based on the hierarchical version. On- and off-responses across the receptive field for a single light slit are shown in Fig. 2f. It shows a rather irregular profile, but such profiles are sometimes seen also in the experimental results. The simulations were based on input from only one on- and one off-centre LGN cell in the excitatory and the inhibitory field, and only a small increase in this number would give smoother profiles. In the simulated activation profiles (Fig. 2g, h) the characteristic division into a central region and flanks occurs with enhancement in the central region and suppression on the flanks by in-phase presentation, while the opposite pattern occurs by presentation in counterphase.

The basic receptive field arrangement in the model for complex cells is the same as in the model for simple cells, i.e. partly overlapping excitatory and inhibitory fields. This means that both cell types can be explained by this arrangement and that the basic difference between the two cell types seems to be their input, where simple cells have input from only on- and off-centre cells and complex cells from both on- and off-centre LGN cells. Thus, the

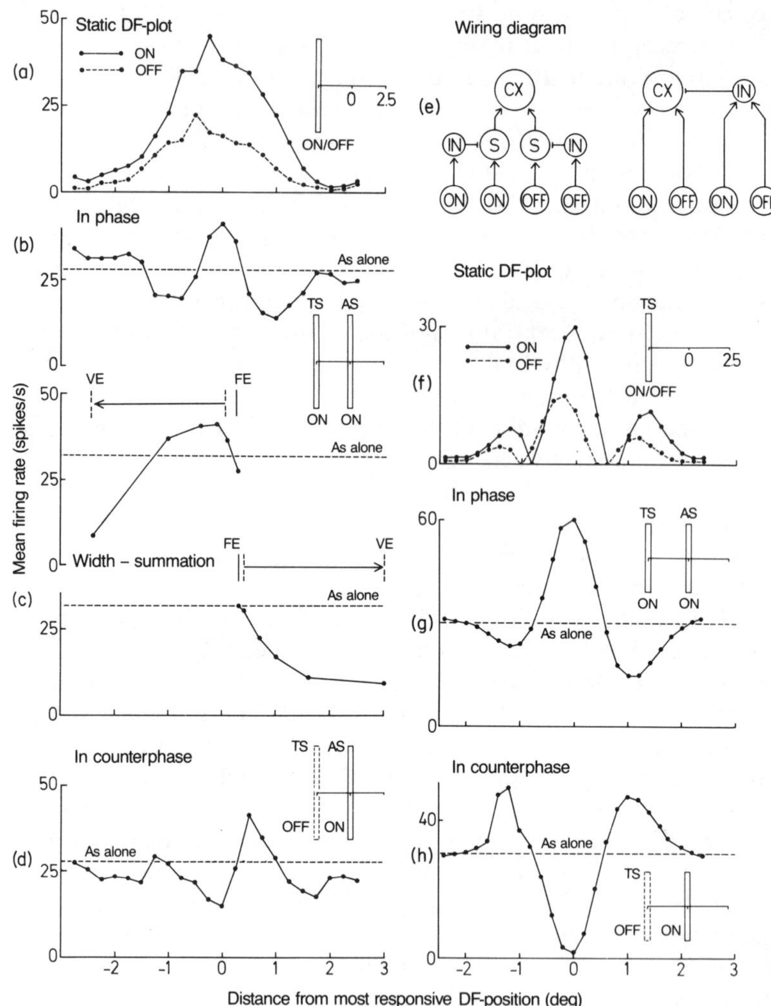

FIGURE 2 Complex cell. (a) Static discharge field plot for a complex cell from cat area 17 made with an optimally oriented light slit (0.2 x 3.9°). (b) Static activation profile made with test and activation slits (0.2 x 3.9°) presented on simultaneously. (c) Width response curves. The upper curve shows summation when the test slit was widened across the central region and into the left-side suppression flank. The position of the fixed edge (FE) and the variable test slit edge (VE) are indicated above the curve. The lower curve shows a corresponding summation across the right-side suppression flank. Both summation curves were made by in-phase presentation of the test and the activation slit. (d) Static activation profile made with presentation of the test slit off when the activation slit was presented on. Slit dimensions were the same as in (b). The window time for the on- and the off-periods was 500 ms and each data point was based on 40 repetitive measurements. (e) Schematic circuit showing the wiring diagram for the two versions of the complex cell model. The stream of on- and off-information is kept separated in the striate cortex in the class of simple cells, but is mixed in the class of complex cells.

REFERENCES

Benevento, L. A., Creutzfeldt, O. D., and Kuhnt, U. (1972). Significance of intracortical inhibition in the visual cortex. *Nature New Biol.*, **238**, 124–126.

Creutzfeldt, O. D., and Ito, M. (1968). Functional synaptic organization of primary visual cortex neurons in the cat. *Exp. Brain Res.*, **6**, 324–352.

Dreher, B., and Sanderson, K. J. (1973). Receptive field analysis. Responses to moving visual contours by single lateral geniculate neurons in the cat. *J. Physiol. (Lond.)*, **234**, 95–118.

Ferster, D., and Lindström, S. (1981). An intracellular study of geniculocortical connectivity in area 17 of the cat. *Soc. Neurosci. Abstr.*, **7**, 355.

Heggelund, P. (1981a). Receptive field organization of simple cells in cat striate cortex. *Exp. Brain Res.*, **42**, 89–98.

Heggelund, P. (1981b). Receptive field organization of complex cells in cat striate cortex. *Exp. Brain Res.*, **42**, 99–107.

Heggelund, P., Krekling, S., and Skottun, B. C. (1983). Spatial summation in the receptive field of simple cells in the cat striate cortex. *Exp. Brain Res.*, **52**, 87–98.

Heggelund, P., and Moors, J. (1983). Orientation selectivity and the spatial distribution of enhancement and suppression in receptive fields of cat striate cortex cells. *Exp. Brain Res.*, **52**, 235–247.

Hubel, D. H. and Wiesel, T. N. (1962). Receptive fields, binocular interaction, and functional architecture in the cat's visual cortex. *J. Physiol. (Lond.)*, **160**, 106–154.

Lee, B. B., Cleland, B. G., and Creutzfeldt, O. D. (1977). The retinal input to cells in area 17 of the cat's cortex. *Exp. Brain Res.*, **30**, 527–538.

Lee, B. B., Elepfandt, A., and Virsu, V. (1981). Phase of responses to sinusoidal gratings of simple cells in cat striate cortex. *J. Neurophysiol.*, **45**, 818–828.

Tsumoto, T., Eckart, W., and Creutzfeldt, O. D. (1979). Modification of orientation sensitivity of cat visual cortex neurons by removal of GABA-mediated inhibition. *Exp. Brain Res.*, **34**, 351–363.

Wilson, J. R., and Sherman, S. M. (1976). Receptive field characteristics of neurons in cat striate cortex: changes with visual field eccentricity. *J. Neurophysiol.*, **39**, 512–533.

---

right-side diagram illustrates a version which presumes direct LGN input (parallel version); the left-side diagram illustrates a version which presumes that the input to the complex cell comes from simple cell (hierarchical version). (f) to (h) Static plot of discharge field and static activation profiles simulated by the hierarchical version. To simulate the results in (a) the weight of the input from the off-centre cells was assumed to be only half of the weight of the input from the on-centre cells.

Models of the Visual Cortex
Edited by D. Rose and V. G. Dobson

CHAPTER 38

Neuronal circuitry in the cat visual cortex studied by cross-correlation analysis

KEISUKE TOYAMA
Department of Physiology, Kyoto Prefectural School of Medicine,
Kawaramachi-Hirokoji, Kamigyoku, Kyoto 602, Japan

INTUITIVE VERSUS EXPERIMENTAL APPROACHES FOR MODELLING OF VISUAL CORTICAL CIRCUITRY

A number of works has been devoted to investigation of the neuronal circuitry that explains the photic responsiveness of the cortical cells, since three basic response types (simple, complex and hypercomplex) were described in the cat striate cortex (Hubel and Wiesel, 1962, 1965). Two contrasting approaches have been made to this problem. The first approach was to construct a model that intuitively explains the differences in the receptive field organization between the three cell response types and the second was to construct a model according to the neuronal connectivities of these cells with the geniculate afferents, which were determined from their responses to electrical stimulation of the visual pathway. A serial model that assumes a sequential transfer of excitation from the geniculate to the simple, from the simple to the complex and from the complex to the hypercomplex cell was proposed by the first approach (Hubel and Wiesel, 1962, 1965), while parallel models that assume parallel transfer of geniculate excitation to the simple and the complex cells were constructed by the second approach (Stone and Dreher, 1973; Toyama, Maekawa and Takeda, 1973, 1977; Toyama *et al.*, 1977; Singer, Tretter and Cynader, 1975; Mitzdorf and Singer, 1978; Bullier and Henry, 1979a, 1979b; Tanaka, 1983).

CROSS-CORRELATION TECHNIQUE AS A TOOL TO INVESTIGATE NEURONAL NETWORKS.

Along the lines of the second approach we attempted to study interneuronal connectivities in the striate cortex, using a cross-correlation technique. This technique demonstrates an interaction between two simultaneously recorded cells during their responses to visual stimuli (Perkel, Gerstein and Moore, 1967; Kimura, Tanaka and Toyama, 1976; Toyama, Kimura and Tanaka, 1981a). Interneuronal connectivity can be identified according to the causal relationship of the interaction between the two cells. Although identification of neuronal connectivity in this manner is more indirect than the conventional method of identification based on the stimulus–response relationship, cross-correlation analysis provides several advantages over the conventional methods: (a) it reveals synaptic interaction between a selected pair of single neurons that may otherwise be concealed by the interactions amongst a number of cells, and therefore it is useful for elucidating the minute structure of neuronal networks; (b) it can be applied to responses evoked by natural stimuli and demonstrates interactions that underlie the photic responses of the two cells; (c) cross-correlation analysis avoids the complications introduced by electrical stimulation such as artefact potentials, spread of stimulus current to neighbouring structures and activation of axons of passage (Toyama, Kimura and Tanaka, 1981a).

RELATIONSHIP BETWEEN INTRACORTICAL CONNECTIONS AND RESPONSE TYPES

A pair of neurons was recorded simultaneously from the striate cortex of the cat, using two glass microelectrodes. Cross-correlation analysis of their impulse discharges revealed four basic patterns of interneuronal connectivities: (a) excitation of a pair of striate cells by common inputs from geniculate cells (common excitation), (b) excitation (intracortical excitation), and (c) inhibition (intracortical inhibition) of one cortical cell by another and (d) inhibition mutually exerted between a pair of cells. These were related to various cell response types. On the basis of an on- and off-response map, striate cells were classified into the following four response types: (a) the eon or eoff cell, with an exclusively on- or off-response area, (b) the simple cell, with discrete on and off areas, (c) the complex cell, with superimposed on and off areas and (d) the hypercomplex cell, with strong end-stop inhibition (Hubel and Weisel, 1962, 1965; Bishop, Kato and Orban, 1973, 1980; Toyama and Tokeda, 1974; Singer, Tretter and Cynader, 1975; Rose, 1977; Toyama, Maekawa and Takeda, 1977; Toyama *et al.,* 1977; Kato Bishop and Orban, 1978; Toyama, Kimura and Tanaka, 1981a, 1981b).

The first type of cell had only one kind of response area, either on or off, and it was differentiated from the simple cell category by the exclusive nature of the receptive field organization (Toyama and Takeda, 1974; Singer, Tretter and Cynader, 1975; Toyama, Maekawa and Takeda, 1977). Some workers (Emerson and Gerstein, 1977; Kato, Bishop and Orban, 1978) regarded these cells as a subtype of simple cell because they shared several response properties with simple cells: a relatively small response area, linear summation of photic response in the response area, sustained response to stationary stimuli and preference for slowly moving stimuli. However, these cells differed from simple cells in some aspects: (a) they exhibited more relaxed tuning for the orientation of light stimuli than simple cells, (b) they were most frequently encountered in layer IV, while the simple cells were distributed in layers III–V and VI and (c) cross-correlation analysis demonstrated that at least some of them inhibit simple cells but not vice versa (see below).

Common excitation occurred in all combinations except for those involving hypercomplex cells, indicating that common excitation is not strongly related to any specific neuronal response type. By contrast, intracortical excitation was more strongly related to response type, occurring in only two combinations: complex to complex cell and complex to hypercomplex cell. Intracortical inhibition was also response-type specific. It was observed in only three combinations: eon or eoff to simple cell, simple to complex cell and eon or eoff to complex cell. Mutual inhibition was found in only one combination, between eon and eoff cells.

THREE-DIMENSIONAL STRUCTURE OF INTRACORTICAL CONNECTIONS

The relationships between intracortical connections and cellular locations were studied by sampling a number of cells in a single track with a microelectrode, while continuously holding one reference cell with another electrode. Interneuronal connections between the sampled cells and the reference cell were determined by cross-correlation analysis.

Common excitation occurred in most neuronal pairs located in the middle layers (III–V) of the striate cortex with intercellular distances of less than several hundred micrometres, provided that the two cells shared the same ocular preference. Common excitation was found even in neuronal pairs with up to 40° differences in orientation preference. Cells in a cortical volume extending several hundred micrometres perpendicularly from the reference cell and spanning tangential distances of several orientation columns shared common inputs with the reference cell.

Intracortical excitation extended in a much more limited space in the striate cortex than common excitation. In all cases that revealed intracortical excitation, the source cell was located in the border zone between layers III and IV, and the target cells in layers II and III, roughly 200 to 400 μm superficial to the source cell. Intracortical excitation was strictly specific as to orientation and ocular preference. In all cases of intracortical excitation tested, paired cells shared the same ocular preference, and the orientation preference never differed by more than 15°. Therefore, intracortical excitation is confined tangentially to a single orientation and ocular dominance column and extends across a perpendicular distance of a few hundred micrometres.

A similar restricted spatial relationship between the source and target cells was found for intracortical inhibition. The source cells of intracortical inhibition were confined to the deeper half of layer IV, while the target cells frequently appeared in a cluster for a distance of a few hundred micrometres in layers III–V. Intracortical inhibition was found between paired cells with orientation preferences differing as much as 20 to 30°. Therefore, the target area of cortical inhibition extends perpendicularly for a few hundred micrometres in the middle layers of the striate cortex and tangentially at least a few orientation columns.

Mutual inhibition was found in one particular combination, eon versus eoff. All of these pairs were located in the deeper half of layer IV.

CROSS-CORRELATION STUDY OF GENICULATE INPUTS TO A SINGLE STRIATE CELL

The convergence of geniculate inputs to a single striate neuron was also studied by cross-correlation analysis (Tanaka, 1983). A number of geniculate cells with receptive fields overlapping that of the striate cell were sampled with a microelectrode, while a single striate cell was held continuously with the other microelectrode. The eon cells received a convergence of rather pure inputs from on-centre X-geniculate cells and the eoff cells from off-centre X-cells. Analysis of the contribution of geniculate excitation to the responses of these cells indicated that the convergence number is between 5 and 10 (Tanaka, 1983). Similarly, simple cells received a convergence of inputs from several X-geniculate cells with on- or off-centre fields. It was regularly found that the on-centre fields of the geniculate cells overlapped the on-response area of the simple cells and the off-centre fields overlapped the off-response area. Therefore, the on- and off-response areas of simple cells may be comprised of the on- and off-centre fields of geniculate cells respectively. The complex cells received mixed inputs from X- and Y-geniculate cells with on- or off-centre fields. The convergence number appears between 10 and 50 (Tanaka, 1983).

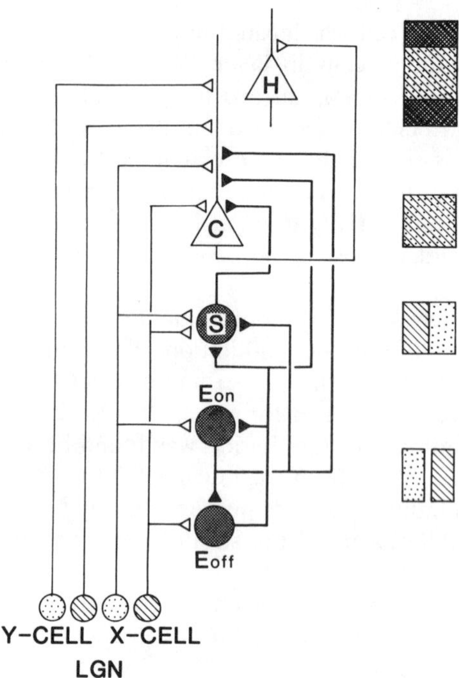

Figure 1 Models of striate cells. A diagram illustrating geniculate inputs to four different classes of striate cells. Cells and synapses or pathways presumed to be inhibitory are filled in black or indicated by thick solid lines. S, simple; C, complex; H, hypercomplex. Stippling and cross-hatching represent on- and off-areas respectively in the receptive fields of LGN cells (bottom left) and cortical cells (right).

MODELLING OF STRIATE CORTICAL CIRCUITRY

Figure 1 illustrates a model of striate cortical circuitry constructed on the basis of the experimental findings of intrinsic and extrinsic connectivity in the striate cortex. The eon cell receives a single line of inputs from X-geniculate cells with on-centre fields, and the eoff cell inputs from X-geniculate cells with off-centre fields. The eon and eoff cells are probably inhibitory cells exerting mutual inhibition. The simple cell receives double lines of input from X-geniculate cells with on-centre fields and off-centre fields, and receives inhibition from eon and eoff cells. The complex cell may receive mixed inputs from both X- and Y-geniculate cells with on- and off-centre fields. A population of simple cells is inhibitory, and they inhibit the complex cells. The hypercomplex cells may receive excitation from complex cells.

This model assumes parallel geniculate inputs to different classes of striate cells, pure X-inputs to the eon, eoff and simple cells, and mixed X- and Y-inputs to the complex cells. This is partially in agreement with Stone and Dreher's (1973) model which assumes parallel geniculate inputs: pure X-inputs to simple cells and pure Y-inputs to complex cells. However, it is contradictory to Hubel and Wiesel's (1962) model which assumes sequential transfer of geniculate inputs: from geniculate to simple cells and from simple to complex cells.

This model can at least qualitatively account for several major features of striate cell photic responsiveness. It explains response selectivity on the basis of inhibition exerted between striate cells. Reciprocal inhibition exerted between the eon and eoff cells might explain the directional selectivity of their responses to a moving stimulus (Toyama and Takeda, 1974), which in turn might be transmitted to the simple cells through inhibition. The inhibition by the eon and eoff cells might also account for the antagonism between the on- and off-response areas of simple cells (Hubel and Wiesel, 1962), which is probably the basis for their orientation and directional selectivities (Hubel and Wiesel, 1962; Henry, Bishop and Dreher, 1974; Henry, Dreher and Bishop, 1974; Goodwin, Henry and Bishop, 1975; Emerson and Gerstein, 1977). Inhibition of complex cells by the simple cells in neighbouring orientation columns may explain the orientation and directional selectivity of complex cells (Hubel and Wiesel, 1962, 1965; Goodwin and Henry, 1975; Bishop, Kato and Orban, 1980).

This view is consistent with the finding that the response selectivities of simple and complex cells were eliminated by iontophoretic ejection of bicuculline, which is a potent antagonist of GABA, a putative inhibitory transmitter in the visual cortex (Sillito, 1975, 1977, 1979). This model, however, provides no explanation for other important features of the responsiveness of striate cells such as end-stop inhibition in hypercomplex cells or binocular interaction occurring between the inputs from the two eyes. Finally, the present scheme does not perfectly reconcile with that suggested from the morphology of striate cells stained by intracellular injection of horseradish peroxidase (Gilbert and Wiesel, 1979, 1983). The former scheme does not contain some of the neuronal connections suggested in the latter scheme. The cause of the discrepancy is not quite clear at present. It may be that cross-correlation analysis is only able to demonstrate neuronal interaction converging from a limited number of cells with a relatively large contribution, and is unable to detect that converging from numerous cells with a small contribution, even though the total effect may be large. Further study will be necessary to solve these problems.

REFERENCES

Bishop, P. O., Coombs, J. S., and Henry, G. H. (1973). Receptive fields of simple cells in the cat striate cortex. *J. Physiol.*, **231**, 31–60.

Bishop, P. O., Kato, H., and Orban, G. A. (1980). Direction-selective cells in complex family in cat striate cortex. *J. Neurophysiol.*, **43**, 1266–1283.

Bullier, J., and Henry, G. H. (1979a). Ordinal position of neurons in cat striate cortex. *J. Neurophysiol.*, **42**, 1251–1263.
Bullier, J., and Henry, G. H. (1979b). Neural path taken by afferent streams in striate cortex of the cat. *J. Neurophysiol.*, **42**, 1264–1270.
Emerson, R. C., and Gerstein, G. L. (1977). Simple striate neurons in the cat. I. Comparison of responses to moving and stationary stimuli. *J. Neurophysiol.*, **40**, 119–135.
Gilbert, C. D., and Wiesel, T. N. (1979). Morphology and intracortical projections of functionally characterised neurons in the cat visual cortex. *Nature* (Lond.), **280**, 120–125.
Gilbert, C. D., and Wiesel, T. N. (1983). Clustered intrinsic connections in cat visual cortex. *J. Neurosci.* **3**, 1116–1133.
Goodwin, A. W., and Henry, G. H. (1975). Direction selectivity of complex cells in a comparison with simple cells. *J. Neurophysiol.*, **38**, 1524–1540.
Goodwin, A. W., Henry, G. H., and Bishop, P. O. (1975). Direction selectivity of simple striate cells: properties and mechanisms. *J. Neurophysiol.*, **38**, 1500–1523.
Henry, G. H., Bishop, P. O., and Dreher, B. (1974). Orientation, axis and direction as stimulus parameters for striate cells. *Vision Res.*, **14**, 767–777.
Henry, G. H., Dreher, B., and Bishop, P. O. (1974). Orientation specificity of cells in cat striate cortex. *J. Neurophysiol.*, **37**, 1394–1409.
Hubel, D. H., and Wiesel, T. N. (1962). Receptive fields, binocular interaction and functional architecture in the cat's visual cortex. *J,. Physiol. (Lond.)*, **160**, 106–154.
Hubel, D. H., and Wiesel, T. N. (1965). Receptive fields and functional architecture in two nonstriate visual areas (18 and 19) of the cat. *J. Neurophysiol.*, **28**, 229–289.
Kato, H., Bishop, P. O. and Orban, G. A. (1978). Hypercomplex and simple/complex cell classifications in cat striate cortex. *J. Neurophysiol.*, **41**, 1071–1095.
Kimura, M., Tanaka, K., and Toyama, K. (1976). Interneuronal connectivity between visual cortical neurons of the cat as studied by cross-correlation analysis of their impulse discharges. *Brain Res.*, **118**, 329–333.
Mitzdorf, U., and Singer, W. (1978). Prominent excitatory pathways in the cat visual cortex (areas 17 and 18): a current source density analysis of electrically evoked potentials. *Exp. Brain Res.*, **33**, 371–394.
Perkel, D. H., Gerstein, G. L., and Moore, G. P. (1967). Neuronal spike trains and stochastic point processes. II. Simultaneous spike trains. *Biophys. J.*, **7**, 419–440.
Rose, D. (1977). Responses of single units in cat visual cortex in moving bars of light as a function of bar length. *J. Physiol. (Lond.)*, **271**, 1–23.
Sillito, A. M. (1975). The contribution of inhibitory mechanisms to the receptive field properties of neurons in the striate cortex of the cat. *J. Physiol. (Lond.)*, **250**, 305–329.
Sillito, A. M. (1977). Inhibitory processes underlying the directional specificity of simple, complex and hypercomplex cells in the cat's visual cortex. *J. Physiol. (Lond.)*, **271**, 699–720.
Sillito, A. M. (1979). Inhibitory mechanisms influencing complex cell orientation selectivity and their modification at high resting discharge levels. *J. Physiol. (Lond.)*, **289**, 33–53.
Singer, W., Tretter., F. and Cynader, M. (1975). Organization of cat striate cortex: correlation of receptive-field properties with afferent and efferent connections. *J. Neurophysiol.*, **38**, 1080–1098.
Stone, J., and Dreher, B. (1973). Projection of X- and Y-cells of the cat's lateral geniculate nucleus to areas 17 and 18 of visual cortex. *J. Neurophysiol.*, **36**, 551–567.

Tanaka, K. (1983). Neuronal connectivity of a single striate cells with lateral geniculate cells studied by cross-correlation technique. *J. Neurophysiol.*, **49**, 1303–1319.
Toyama, K., Kimura, M., Shida, T., and Takeda, T. (1977). Convergence of retinal input onto visual cortical cells. II. A study of the cells disynaptically excited from the lateral geniculate body. *Brain Res.*, **137**, 221–231.
Toyama, K., Kimura, M., and Tanaka, K. (1981a). Cross-correlation analysis of interneuronal connectivity in cat visual cortex. *J. Neurophysiol.*, **46**, No. 2, 191–201.
Toyama, K., Kimura, M., and Tanaka, K. (1981b). Organization of cat visual cortex as investigated by cross-correlation technique. *J. Neurophysiol.*, **46**, No. 2, 202–214.
Toyama, K., Maekawa, K., and Takeda, T. (1973). An analysis of neuronal circuitry for two types of visual cortical neurons classified on the basis of their responses to photic stimuli. *Brain Res.*, **61**, 395–399.
Toyama, K., Maekawa, K., and Takeda, T. (1977). Convergence of retinal inputs onto visual cortical cells. I. A study of the cells monosynaptically excited from the lateral geniculate body. *Brain Res.*, **137**, 207–220.
Toyama, K., and Takeda, T. (1974). A unique class of cat's visual cortical cells that exhibit either ON or OFF excitation for stationary light slit and are responsive to moving patterns. *Brain Res.*, **73**, 350–355.

Models of the Visual Cortex
Edited by D. Rose and V. G. Dobson

CHAPTER 39

Retinocortical wiring of the simple cells of the visual cortex

DAN E. NIELSEN
Institute of Neurophysiology, The Panum Institute, Blegdamsvej 3 C, DK-2000 København N, Denmark

The behaviour and receptive field properties of the cells of the striate cortex have been studied for a long time and in great detail. But what are the mechanisms, the synaptic distributions, which are responsible for the observed properties? In order to study these mechanisms, I constructed a computer-implemented model of the retinocortical pathway, being of course well aware that such models cannot provide more than some possible answers to the above questions.

The model to be described in the following is very simple, yet it performs quite well in mimicking the behaviour of simple cells (S-cells). It contains three levels representing respectively: (a) receptor activity, (b) activity in centre/surround cells (CS-cells) and (c) activity in S-cells. Retinal ganglion cells, cells in the lateral geniculate body and cortical CS-cells were considered to be so much alike in behaviour (Nielsen, 1983) that they were treated together as one cell type.

The stimuli (spots, bars, edges and various periodic patterns) were generated by the computer as matrices of 84 × 42 real numbers. Each number represented the average light intensity in a square with side length 1.25′. As a linear relationship was assumed between the stimulus intensity and photoreceptor response (Nielsen, 1983), the stimulus matrix could also be considered as a matrix representing the average activity in groups of photoreceptors, each group covering 1.25′ of the visual field. Hence, this matrix was used directly as input to level two. Here each CS-cell was influenced by 12 × 12 points in the stimulus matrix (= 12 × 12 groups of photoreceptors) in such a way that the middle 4 × 4 values were summed, while the sum of the surrounding 128 values, multiplied by a weighting factor, were subtracted. To this difference was added a constant representing spontaneous activity, and remaining negative responses were set to zero. A matrix of the activity of CS-cells with on-centre/off-surround receptive fields was thus obtained. The purpose of the weighting factor was to adjust the strength of the inhibitory surround relative to the excitatory centre. (In biological systems dendrite density and/or the difference in excitability between the dendritic and somatic parts of the postsynaptic cell membrane may do the same job, although the exact balancing of excitation and inhibition assumed in the model is not likely to occur very often.) The number of cells in the CS-cell matrix depended on the degree of overlapping of the receptive fields of the individual cells. In all the simulations on which the present results are based maximal overlapping was chosen, because this was found to yield the best results. Maximal overlapping means that the centres of neighbouring cells were separated by 1.25′, so that the number of CS-cells came to 73 × 31.

Finally the activities of the CS-cells were read into overlapping simple cells, each S-cell receiving inputs from 13 × 13 CS-cells. The distribution of excitation and inhibition in the interior of the S-cell input fields could be varied in order to examine the effect of this variable on cell behaviour. The most extensively studied model S-cell had a central excitatory band flanked by two equally broad inhibitory ones (even symmetric cell). Other types had the excitatory band dispositioned laterally or contained only two subfields, one excitatory and one inhibitory (odd symmetric cells). The maps of input distributions thus corresponded to Hubel and Wiesel's early maps of receptive fields (Hubel and Wiesel, 1962). The wiring diagramme of the model and the input maps of two model S-cells are shown in Fig.1.

The aspects of cell function which could be studied by the model were: feature detection, spatial frequency filtering, orientation tuning, phase relationship between stimulus and response and uncertainty product. The stimulus situations were chosen so that the results from the model could be compared directly with existing electrophysiological data.

The model cells are poor feature detectors; they do respond when presented with isolated bars and edges, but if many bars and edges are present within a narrow region of space (i.e. the stimulus is a grating), the cell responses become correlated with one of the spatial frequencies contained in the stimulus. By using sinewave gratings as stimuli the optimal spatial frequency f_{opt} of the cell can be found. For the even symmetric model cells this procedure yielded the value 4.8 c. p. d. If a square-wave grating is so adjusted that the optimal frequency of the cell is equal to, say, the third harmonic of the grating, the peak responses of the model cells occur with the same periodicity as that of the third harmonic. This was shown to occur by Maffei *et al.* (1979) for living cells. The frequency filtering of the model cells, however, is not entirely linear, as frequencies outside of the frequency

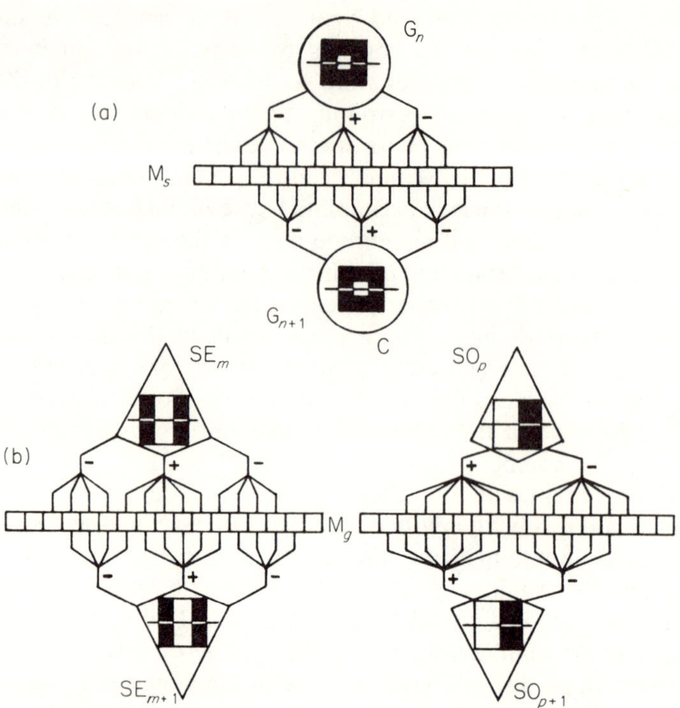

FIGURE 1 Schematic representation of the wiring diagramme used in the model. Cells are represented by circles (ganglion cells) and pyramids (simple cells). The figures inside the cells show input maps resulting from the connections whose distribution in one dimension is shown; the horizontal lines in the input maps show the sections through these input maps to which the connections shown correspond. The bar M_s represents a row of receptor groups, each group being equivalent to a 1.25′ x 1.25′ unit in the stimulus matrix. The two bars M_g are rows of ganglion cells where any pair of partitions in the bar represent the machinery shown in (a). The plus (+) and white part of the input map shows excitation, the minus (−) and black part of the input map shows inhibition. (a) The wiring of two nearest-neighbour ganglion cells (G_n, G_{n+1}). (b) The wiring of two nearest-neighbour even symmetric simple cells (SE_m, SE_{m+1}). (c) The wiring of two nearest-neighbour odd symmetric simple cells (SO_p, SO_{p+1}).

range of the cells (i.e. frequencies which do not elicit any response from the cell) are able to modify the cell responses. Thus, when presented with the optimal frequency plus one or more frequencies outside of the cell's frequency range the magnitude of the response is reduced relative to that of the response to the optimal frequency alone. Albrecht and DeValois (1981) have described this phenomenon in living cells. It is well known (Pollen and Ronner, 1981; Maffei *et al.*, (1979) that S-cells may respond in various phase relationships relative to the stimulus after elimination of delays due to transmission velocities. This is also the case in the model, where an even symmetric cell responds at a phase difference of 0°. But the more asymmetric the receptive field the more out of phase the response. Some important examples of the behaviour of the model cells are shown in Fig.2.

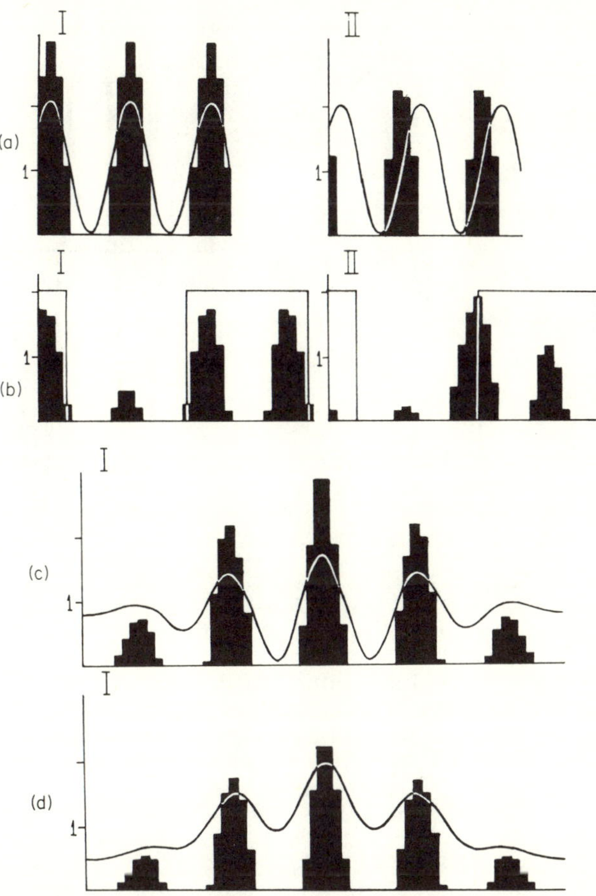

Figure 2 Histograms showing responses of model simple cells. I, even symmetric cell; II, asymmetric cell (left inhibitory flank seven times as broad as right one); continuous lines show stimulus profile; f_{opt}, optimal spatial frequency of the cell; f_n, *n*th harmonic in the stimulus; abscissae, position of cell centres relative to the stimuli; ordinates, response magnitude in arbitrary units. (a) Sinewave grating, $f_1 = f_{opt}$. (b) Square-wave grating, $f_3 = f_{opt}$. Note that the response peaks in (b) are in register with those in (a) and that f_3, is equal in phase (but not in amplitude) to f_1 in (a). (c) Grating composed of $1/4f_4 + 1/2f_5 + 1/4f_6$: $f_5 = f_{opt}$. (d) Grating composed of $1/1f_1 + 1/4f_4 + 1/2f_5 + 1/4f_6$; $f_5 = f_{opt}$. f_1 is outside of the frequency range, yet its presence decreases the response magnitude by approximately 1. 3. From Albrecht and DeValois (1981), the reduction factor in the same situation in living cells can be estimated to be about 1.4.

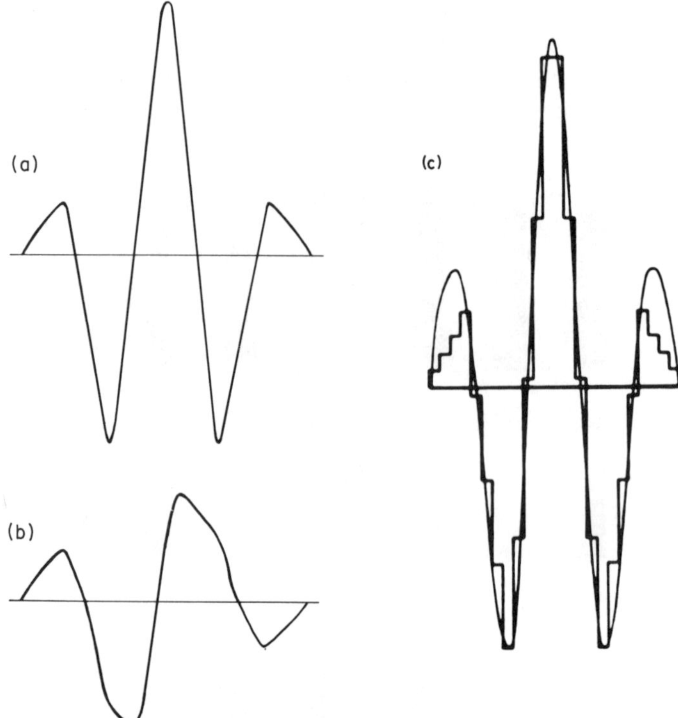

FIGURE 3 Profiles of the receptive fields of model simple cells for (a) an even symmetric call and (b) an odd symmetric cell. (c) A comparison of the receptive field profile (histogram) of an even symmetric model simple cell with the profile of a Gabor elementary signal (smooth curve) with the same σ_x and f_{opt}, as estimated for the cell. The Gabor signal has several (in principle an infinite number of) additional lateral fluctuations which are, however, not shown in the figure because of their minute amplitude.

The receptive field profiles of the model cells were examined by stimulating the cells with small spots. These profiles are shown in two dimensions in Fig.3a, b, while a three-dimensional representation of the distribution of excitation and inhibition can be found in Nielsen (1983, Fig.10). A comparison with, for example, Fig. 1 in Kulikowski, Marcelja and Bishop (1982) shows that these profiles closely resemble those of Gabor elementary signals. By assuming actual identity it was possible to determine σ_x (space constant along an axis perpendicular to the orientation of the excitatory subfields) and $\Delta x \Delta f$ (product of uncertainties in the space and spatial frequency domains) for the model cells. (A detailed description of the procedure is given in Nielsen, 1983). As σ_x and optimal frequency (f^{opt}) determine the profile of a Gabor elementary signal, the profile of such a

Gabor signal with the same characteristics as the model cells (i.e. $\sigma_x = 0.14°$; $f_{opt} = 4.8$ c.p.d.) could be calculated. Figure 3c shows a comparison between the model cell profile and the profile of the best-fitting Gabor signal. σ_x and $\Delta x \Delta f$ are shown in Table 1, which lists the available numerical data for the model cells.

TABLE 1 Numerical data for model cells

Physiological Properties	
Receptive field diameter	0.5°
Optimal Spatial frequency	4.8 c.p.d.
Spatial frequency bandwidth	1.2 octaves
Orientation bandwidth	± 20°
Mathematical Characteristics	
σ_x	0.14°
$\Delta x \Delta f$	0.56

As seen, the results obtained from the very simple model are in good agreement with what is known about the behaviour of simple cells, and it is concluded that the wiring scheme employed represents one anatomically and physiologically possible model of the neural mechanism endowing the simple cells of the striate cortex with the characteristics described above.

REFERENCES

Albrecht, D. G., and DeValois, R. L. (1981). Striate cortex responses to periodic patterns with and without the fundamental harmonics. *J. Physiol.*, **319**, 97–514.

Hubel, D. H., and Wiesel, T. N. (1962). Receptive fields, binocular interactions and functional architecture in the cat's visual cortex. *J. Physiol.*, **160**, 106–154.

Kulikowski, J. J., Marcelja, S., and Bishop, P. O. (1982). Theory of spatial position and spatial frequency relations in the receptive fields of simple cells in the visual cortex. *Biol. Cybernetics*, **43**, 187–198.

Maffei, L., Morrone, C., Pirchio, M., and Sandini, G. (1979). Responses of visual cortical cells to periodic and non-periodic stimuli. *J. Physiol.*, **296**, 27–47.

Nielsen, D. E. (1983). A functional model of the wiring of the simple cells of visual cortex. *Biol. Cybernetics*, **47**, 213–222.

Pollen, D. A., and Ronner, S. F. (1981). Phase relationships between adjacent simple cells in the visual cortex. *Science*, **212**, 1409–1411.

Models of the Visual Cortex
Edited by D. Rose and V. G. Dobson

CHAPTER 40

Retinal design of the striate cortex

AUDIE G. LEVENTHAL
Department of Anatomy, University of Utah School of Medicine, Salt Lake City, Utah 84132, USA

The organization of mammalian visual cortex has been studied intensively for over two decades. The first model of the visual cortex to gain widespread attention was proposed by Hubel and Wiesel in 1962. When this model was proposed, all retinal ganglion cells and lateral geniculate nucleus (LGN) neurons providing the input to the striate cortex were thought to have the circular, center-surround receptive fields originally described by Kuffler (1953); Hubel and Wiesel had to account for both the receptive field properties and functional architecture of the striate cortex in terms of a transformation of the input from cells having this type of receptive field. As a result, their model of the visual cortex as well as most subsequent models assume that intracortical mechanisms alone are responsible for the generation of the receptive field properties and functional architecture of cells in the striate cortex.

In recent years it has become clear that the afferents to the visual cortex are far more complex than early studies suggested. The striate cortex receives inputs via the LGN from a number of different classes of retinal ganglion cells which differ with respect to receptive field properties, morphology, axonal conduction velocity, retinal distribution and central projection. Even receptive field properties, such as orientation sensitivity, long thought to be exclusive features of cortical cells, are now known to be properties of retinal ganglion cells and LGN relay cells. Thus it has become necessary to revise models of the visual cortex in order to take the receptive field properties of retinal ganglion cells and LGN relay cells into account. It has already been suggested that many properties of cortical cells, such as receptive field size and velocity selectivity, are determined by the type of LGN afferent input received by different cortical cells (reviewed in Stone, Dreher and Leventhal, 1979). I now propose to add orientation sensitivity to the list of receptive field properties that originate in the retina.

In this chapter I put forth the thesis that orientation sensitivity, possibly the most distinctive feature of cortical cells, actually originates during development in the retina and that the orientation preferences of many striate cortical cells are *specified* during development by the orientation preferences of the LGN cells providing their afferent input; the function of the visual cortex is primarily to *amplify* and *process*, rather than *generate*, orientation sensitivity.

RETINAL ORIENTATION SENSITIVITY

Recent evidence suggests that neurons in peripheral parts of the visual pathways are orientation sensitive. Most ganglion cells in cat retina (Levick and Thibos, 1982) as well as most relay cells in the cat's dorsal lateral geniculate nucleus (dLGN) (Vidyasagar and Urbas, 1982) are sensitive to stimulus orientation. There is a systematic relationship between receptive field position and preferred orientation in the retina and LGN. Retinal ganglion cells respond best to stimuli oriented radially, i.e. oriented parallel to the line connecting their receptive fields to the area centralis projection (Levick and Thibos, 1982) (Fig. 1). A similar tendency has also been reported for the dLGN (Vidyasagar and Urbas, 1982).

A possible structural basis for the orientation sensitivity of cat retinal ganglion cells has been described (Leventhal and Schall, 1983). It appears that nearly all retinal ganglion cells have oriented dendritic fields and that dendritic field orientation is related in a systematic fashion to retinal position. This relationship appears to be a robust and intrinsic feature of the retina since it is found in normal cats as well as in Siamese cats and cats deprived of vision by lid-suture (Leventhal and Schall, 1983). In all regions of retina more than 0.5 mm from the area centralis the dendritic fields of retinal ganglion cells are oriented radially, i.e. like the spokes of a wheel having the area centralis at its hub (Fig. 1). This relationship is strongest for cells located close to the horizontal meridian (visual streak) of the retina. It appears that retinal ganglion cells are sensitive to stimulus orientation because they have oriented dendritic fields.

CORTICAL ORIENTATION SENSITIVITY

In area 17 (striate cortex) of cats and monkeys preferred orientation is also represented in a systematic fashion. Hubel and Wiesel (1963) first observed that cortical cells with similar preferred orientations are grouped into 'columns' or 'bands' which extend from the surface of the cortex to the white matter and that preferred orientation changes systematically from one 'column' to the next.

If the distribution of the orientation preferences of cells in the visual cortex depends upon the distribution of orientation-sensitive afferents, then the

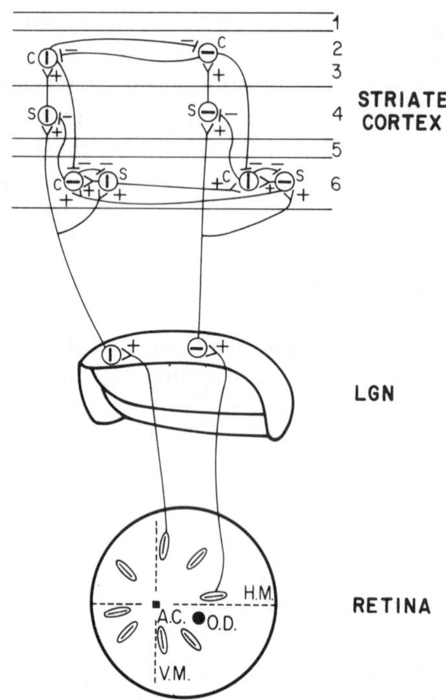

FIGURE 1 The generation of orientation sensitivity in cat striate cortex. The preferred orientation of each cell is indicated by the orientation of the line inside the cell body. Excitatory inputs are indicated by plus signs; inhibitory inputs are indicated by minus signs. Inhibitory inputs are shown to be direct only for ease of illustration. These inputs may be mediated by interneurons. Cortical cells having orientation preferences specified by excitatory, orientation-sensitive LGN afferents are marked S; cortical cells having orientation preferences specified by either excitatory or inhibitory inputs from other cortical cells are marked C. Some C-cells in layers 3, 5 and 6 are likely to receive direct LGN afferents. These connections are described in the text but are not shown in the figure. AC and OD refer to the area centralis and optic disk respectively; VM and HM refer to the vertical and horizontal meridians respectively. The specifics of the model are described in the text.

retinal relationship between the preferred orientation and receptive field position should be preserved in the visual cortex. Indeed, the distribution of the preferred orientations of cortical cells reflects the distribution of the preferred orientations of retinal ganglion cells; most cells in the striate cortex respond best to lines oriented radially, i.e. parallel to the line connecting their receptive fields to the area centralis projection. This relationship fails within the area centralis representation; the relationship is strongest for simple (Hubel and Wiesel, 1962) or S-type (Henry, 1977; Leventhal and Hirsch, 1978) cortical cells (many of which are first-order cortical cells), cells exhibiting the highest degree of orientation selectivity and cells in regions subserving the horizontal meridian (Leventhal, 1983).

The relationship between the preferred orientation and receptive field position is preserved even in the extrastriate cortex (Leventhal, Schall and Wallace 1984). As in area 17, outside of the representation of the area centralis most cells in areas 18 and 19 respond best to lines oriented radially.

Mechanisms mediating orientation selectivity in visual cortex

In the light of the foregoing studies I propose that cortical orientation sensitivity ultimately depends upon the orientation sensitivity of retinal ganglion cells. I hypothesize that the orientation preferences of first-order striate cortical cells are *specified* during development by orientation-sensitive LGN afferents whose preferred orientations are in turn specified by retinal ganglion cell afferents. Thus, the preferred orientations of S-cells in area 17 are strongly related to the receptive field position because most of these cells are first order; they have their orientation preferences specified by orientation-sensitive excitatory afferents they receive from the LGN (Fig. 1). However, since the orientation tuning of S-cells is much sharper than the tuning of retinal ganglion cells it is proposed that the orientation preferences of S-cells are *enhanced* by intracortical inhibition (Blakemore and Tobin, 1972; Sillito, 1975) and/or by excitatory convergence along the line of optimal orientation as proposed by Hubel and Wiesel in 1962 (Fig. 1). It must be stressed that once mature connectivity is achieved, the weak orientation sensitivity of cortical afferents may or may not be necessary for cortical orientation sensitivity. As just noted, the lion's share of orientation sensitivity in the adult cortex is likely to be a result of intracortical mechanisms.

The foregoing hypothesis accounts only for the orientation sensitivity of first-order cortical cells. We must now consider how higher order cortical cells come to be orientation sensitive during development. I suggest that there are two mechanisms, one for higher order cells in layers 2 and 3 and the other for higher order cells in layers 5 and 6. Since cells in layer 4 project heavily to layers 2 and 3, higher order cells in layers 2 and 3 are hypothesized to have their orientation preferences *specified* by *excitatory* inputs from first-order cortical cells in layer 4 (Fig. 1). The preferred orientations of higher order cells in layers 2 and 3 should reflect the preferences of cells in layer 4; cells in layers 2 and 3 should therefore show a strong tendency to prefer radially oriented stimuli. Since many striate cortical cells in layers 2 and 3 project to other cortical areas, cells in the extrastriate cortex should also tend to prefer radially oriented stimuli. Some complex (Hubel and Wiesel, 1962) or C-cells (Henry, 1977; Leventhal and Hirsch, 1978) in layers 2 and 3 have widely spread axonal arborizations which are periodically clustered (Gilbert and Wiesel, 1979, 1983). These cells are hypothesized to provide *inhibitory* connections (Matsubara and Cynader, 1983) required to *enhance* the orientation sensitivity of higher order cortical cells in layers 2 and 3 (Fig. 1).

afferents from the LGNd or from layer 4 S-cells. Collectively, these afferents respond to all orientations. I suggest that the orientation preferences of higher order cells in layers 5 and 6 are *specified* by intracortical *inhibitory* connections from highly selective higher order cells in layers 2 and 3 of the same column (Fig. 1). The preferred orientations of these cells should therefore be nearly orthogonal to the preferences of the cells in layers 2 and 3 that provide their inhibitory inputs; cells in layers 5 and 6 should not show a radial bias in the distribution of their preferred orientations. They should tend to prefer lines oriented more tangentially. Moreover, higher order cells in the deeper layers should be less selective than those in the upper layers since their orientation tuning curves should be wide and the complement of the narrow tuning curves of upper layer cells. Finally, cells in layers 5 and 6 project to the superior colliculus and LGN respectively. One function of these projections may be to modify the orientation sensitivity of cells in these subcortical structures. Cells in layers 5 and 6 also project heavily to layer 4; these cells may thus *enhance* the orientation sensitivity of first-order cells via inhibitory inputs (Fig. 1) (Gilbert and Wiesel, 1979, 1983).

I must emphasize that in the context of this model a higher order cell is one whose orientation preference is specified by an afferent from another cortical cell. These cells are higher order only with respect to preferred orientation; they may well have direct LGN afferents which contribute otherwise to their responses. In fact, cells in layers 5 and 6 are referred to as higher order even though it is proposed that they have excitatory afferents from the LGN or from large-field S-cells in layer 6 (Fig. 1).

There is some experimental support for the foregoing hypotheses. We have noticed that the relationship between the receptive field position and preferred orientation in area 17 is stronger in layers 2, 3 and 4 than in layers 5 and 6. This is related to our finding that the relationship between the preferred orientation and polar angle is strongest amongst S-cells in area 17. S-cells concentrate in layer 4, while C-cells having large receptive fields are common in the deeper layers (Gilbert, 1977; Leventhal and Hirsch, 1978). The finding that C-cells in layers 5 and 6 have larger receptive fields and are less sensitive to stimulus orientation then C cells in layers 2 and 3 is also consistent with the model (Leventhal and Hirsch, 1978). Moreover, it has been reported that marked shifts in preferred orientation often occur between layers 4 and 5 in both cat (Bauer, 1982) and monkey (Bauer, Dow and Vautin, 1980; Bauer *et al.*, (1983) striate cortex. Such shifts are expected if cells in layer 4 have their orientation preferences specified by their excitatory LGN afferents while many, though not necessarily all (Fig. 1), cells in layers 5 and 6 have their orientation preference specified by intracortical inhibitory inputs from upper layer cells of the same anatomical column. I suggest that one function of area 17 is to increase the number of orientation-sensitive cells that prefer lines oriented tangentially. Even though ganglion cells of all orientations are represented in all parts of the retina, tangentially oriented cells are rare and reasonable numbers of them seem necessary to allow the detection of lines of all orientations in all parts of the visual field.

Development of columnar organisation

The arrangement of cortical orientation 'columns' has been studied in detail (Hubel and Wiesel, 1963, 1977). Nevertheless, the mechanisms mediating the development of the 'columnar' organization in the striate cortex remain unclear. I suggest that the intrinsic framework within which the organized arrangement of cortical orientation 'columns' develops originates in the retina. Throughout the striate cortex a relationship between the preferred orientation and receptive field position is hypothesized to be guided by intrinsic mechanisms and to specify the orientation preferences of first-order cells in layer 4. These cells then regulate the formation of columns in the neonate by specifying the orientation preferences of cells above and below them; this process is hypothesized to depend upon early visual experience and to increase the number of cells preferring nonradially oriented stimuli.

There is some experimental support for these hypotheses. As already noted, there is a systematic relationship between the positions and orientations of ganglion cells in the retinae of visually deprived cats (Leventhal and Schall, 1983). This relationship, therefore, appears to be unaffected by early visual deprivation and so must be due to intrinsic mechanisms. All that is required for the development of a corresponding relationship in the visual cortex of deprived animals is that a topographic representation of the retina develop in the striate cortex in the absence of early visual experience.

There is also evidence for orientation clusters in layer 4 of visually inexperienced animals. It has been reported that many cells are not orientation sensitive in the striate cortex of visually deprived cats; those that are orientation sensitive are clustered together and have receptive field properties typical of layer 4 cells receiving direct LGN afferents (Leventhal and Hirsch, 1977, 1980). Thus, the rudimentary system of orientation 'clusters' in deprived striate cortex may well reflect the relationship between orientation and position in the retina.

Evidence also exists that early visual experience tends to increase the number of cells preferring lines having orientations which are not related to the receptive field position. Specifically, the relationship between the preferred orientation and receptive field position is strong for S-type and weak for C-type cells; it has been reported that nearly all orientation-sensitive cells in deprived striate cortex are S-type (Blakemore and Van Sluyters, 1975; Leventhal and Hirsch, 1978, 1980). Also supporting this idea is the finding that one group of cells in the striate cortex, most of which are S-type, is unaffected by early exposure to lines of one orientation. Other cortical cells, most of which are C-type and in the deeper layers, tend to respond best to lines of the same orientation as the animal viewed during early life (Hirsch *et al.*, 1983). Thus, the cells that show the weakest tendency to prefer radial stimuli are the most sensitive to early experience.

It must be noted that even though most ganglion cells prefer radially oriented stimuli, cells preferring all orientations are present in all parts of the retina. In the retina, ganglion cells having different preferred orientations seem to be intermingled while in the striate cortex, cells preferring different orientations are segregated. Thus, the retinal relationship between orientation and position cannot alone fully explain the columnar organisation of the striate cortex. Conceivably, early visual experience produces synchronous activity amongst retinal ganglion cells and LGN neurons that have the same innate orientation preference. As a result, cortical afferents having the same orientation preference are able to make selective connections with the same cells in the striate cortex; in this fashion synchronous activation could serve to guide the segregation of orientation-sensitive afferents into columns.

Implications for visual perception

If forced to use peripheral regions of retina, visual discrimination in humans is best for lines oriented radially, i.e. parallel to the visual field meridians (Rovamo *et al.*, 1982). I propose that the strong radial bias in the distribution of preferred orientations in regions of the retina and visual cortex subserving the peripheral visual field provides the basis of the visual system's preferential response to radially oriented contours.

If forced to use the central 10 to 15° of retina, humans are able to discriminate horizontal and vertical lines better than diagonal lines (see Appelle, 1972, for a review). In central retina there are more ganglion cells along the horizontal and vertical meridians than along the diagonal meridians (Stone, 1978). As a result there should be a preponderance of retinal ganglion cells preferring horizontal and vertical lines.

In the striate cortex magnification factors reflect ganglion cell density and are greatest in regions subserving the horizontal and vertical meridians (Tusa, Palmer and Rosenquist, 1978). As a result, there should be an over-representation of cells preferring horizontal and vertical stimuli in area 17. Such an over-representation has been reported in regions of the cortex subserving central vision in the cat and monkey (Mansfield and Ronner, 1977; Leventhal and Hirsch, 1978, 1980). Accordingly, I propose that the relationship between the preferred orientation and polar angle in the striate cortex is responsible for the visual system's preferential response to horizontal and vertical stimuli in central vision.

SUMMARY

Most retinal ganglion cells are orientation sensitive; a systematic relationship exists between the orientations and positions of retinal ganglion cells. This relationship is mirrored in the distribution of orientation-sensitive cells in the striate cortex. It is proposed that first-order striate cortical cells have their orientation preferences *specified* during development by orientation-sensitive afferents from the LGN. Intracortical inhibitory connections and/or excitatory convergence along the line of optimal orientation as originally suggested by Hubel and Wiesel (1962) *enhance* the orientation sensitivity of these cells. Higher order cortical cells are suggested to have their preferred orientations *specified* and *enhanced* by intracortical connections. The intracortical inputs which specify preferred orientation are primarily excitatory for cells in the upper layers and inhibitory for cells in the deeper layers. The systematic relationship between the orientations and positions of retinal ganglion cells is suggested to provide a starting point for the development of the organized arrangement of orientation-sensitive cells in the striate cortex.

The foregoing suggestions remain to be tested experimentally; we are still far from understanding the mechanisms underlying cortical orientation sensitivity and the development of the functional architecture of the striate cortex. Nevertheless, it would seem that we have an easier problem than we had twenty years ago when all of the properties of cortical cells had to be accounted for in terms of the transformation of inputs from LGN relay cells having unoriented receptive fields.

ACKNOWLEDGEMENTS

I am grateful to B. Dreher, H. V. B. Hirsch and J. D. Schall for helpful discussions and comments on the manuscript.

REFERENCES

Appelle, S. (1972). Perception and discrimination as a function of stimulus orientation: the 'oblique effect' in man and animals. *Psycholog. Bull.*, **78** 266–278

Bauer, R. (1982). A high probability of an orientation shift between layers 4 and 5 in central parts of the cat striate cortex. *Exp. Brain Res.*, **48**, 245–255.

Bauer, R., Dow, B. M., Snyder A. Z., and Vautin R., (1983). Orientation shift between upper and lower layers in monkey visual cortex. *Exp. Brain Res.*, **50**, 133–145.

Bauer, R., Dow, B. M., and Vautin, R. G. (1980). Laminar distribution of preferred orientations in foveal striate cortex of the monkey. *Exp. Brain Res.*, **41**, 54–60.

Blakemore, C., and Tobin, E. A. (1972). Lateral inhibition between orientation detectors in the cat's visual cortex. *Exp. Brain Res.*, **15**, 439–440.

Blakemore, C., and Van Sluyters, R. C. (1975). Innate and environmental factors in the development of the kitten's visual cortex. *J. Physiol. (Lond.)*, **248**, 663–716.

Gilbert, C. D. (1977). Laminar differences in receptive field properties of cells in cat primary visual cortex. *J. Physiol. (Lond.),* **368**, 391–421.
Gilbert, C. D., and Wiesel, T. N., (1979). Morphology and intracortical projections of functionally characterised neurons in the cat visual cortex. *Nature,* **289**, 120–125.
Gilbert, C. D., and Wiesel, T. N., (1983). Clustered intrinsic connections in cat visual cortex. *J. Neurosci.,* **3**, 1116–1133.
Henry, G. H. (1977). Receptive field classes of cells in the striate cortex of the cat. *Brain Res.,* **133**, 1–28.
Hirsch, H. V. B., Leventhal, A. G., McCall, M. A., and Tieman, D. G. (1983). Effects of exposure to lines of one or two orientations on different cell types in striate cortex of cat. *J. Physiol (Lond.),* **160**, 106–154.
Hubel, D. H., and Wiesel, T. N. (1962). Receptive fields, binocular interaction and functional architecture in the cat's visual cortex. *J. Physiol. (Lond.),* **160**, 106–154.
Hubel, D. H., and Wiesel, T. N. (1963). Shape and arrangement of columns in cat's striate cortex. *J. Physiol. (Lond.),* **165**, 559–568.
Hubel, D. H., and Wiesel, T. N. (1977). Ferrier Lecture. Functional architecture of macaque monkey visual cortex. Proc. Roy. Soc., B, **198**, 1–59.
Kuffler, S. W. (1953). Discharge patterns and functional organization of mammalian retina. *J. Neurophysiol.* **16**, 37–68.
Leventhal, A. G. (1983). Systematic relationship between preferred orientation and receptive field position of neurons in cat striate cortex, *J. comp. Neurol.* **200**, 476–483.
Leventhal, A. G., and Hirsch, H. V. B. (1977). Effects of early experience upon orientation sensitivity and binocularity of neurons in visual cortex of cats. *Proc. Natl. Acad. Sci.,* **74**, 1272–1276.
Leventhal, A. G., and Hirsch, H. V. B. (1978). Receptive-field properties of neurons in different laminae of visual cortex of the cat. *J. Neurophysiol.,* **41**, 948–962.
Leventhal, A. G., and Hirsch, H. V. B. (1980). Receptive-field properties of different classes of neurons in visual cortex of normal and dark-reared cats. *J. Neurophysiol.,* **43**, 1111–1132.
Leventhal, A. G., and Schall, J. D. (1983). Structural basis of orientation sensitivity of cat retinal ganglion cells. *J. comp. Neurol.,* **200**, 465–475.
Leventhal, A. G., Schall, J. D., and Wallace, W. (1984). Relationship between preferred orientation and receptive field position of neurons in cortical area 19 in cat. *J. comp. Neurol.* (in press).
Levick, W. R., and Thibos, L. N. (1982). Analysis of orientation bias in cat retina. *J. Physiol. (Lond.),* **329**, 243–261.
Mansfield, R. J. W., and Ronner, S. F. (1977). Orientation anisotropy in monkey visual cortex. *Brain Res.,* **149**, 229–234.
Matsubara, J. and Cynader, M. (1983). The role of orientation tuning on the specificity of local intracortical connections in cat visual cortex. *Soc. Neur. Abs.,* **9**, 475.
Rovamo, J., Virsu, V., Laurinen, P., and Hyvarinen, L. (1982). Resolution of gratings oriented along and across meridians in peripheral vision. *Invest. Ophthal. Vis. Sci.,* **23**, 666–670.
Sillito, A. M. (1975). The contribution of inhibitory mechanisms to the receptive field properties of neurons in the striate cortex of the cat. *J. Physiol. (Lond.),* **250**, 305–329.
Stone, J. (1978). The number and distribution of ganglion cells in the cat's retina. *J. comp. Neurol.,* **180**, 753–771.
Stone, J., Dreher, B., and Leventhal, A. G. (1979). Parallel and hierarchical mechanisms in the organization of visual cortex. *Brain Res. Rev.,* **1**, 345–394.

Tusa., R. J., Palmer, L. A. and Rosenquist, A. C. (1978). The retinotopic organization of area 17 (striate cortex) in the cat. *J. comp. Neurol.*, **177**, 213–236.

Vidyasagar, T. R., and Urbas, J. V. (1982). Orientation sensitivity of cat LGN neurons with and without inputs from visual cortical areas 17 and 18. *Exp. Brain Res.*, **46**, 157–169.

Models of the Visual Cortex
Edited by D. Rose and V. G. Dobson

CHAPTER 41

Geniculate orientation biases as Cartesian coordinates for cortical orientation detectors

T.R. VIDYASAGAR
Department of Neurobiology, Max-Planck-Institute for Biophysical Chemistry, Goettingen, Federal Republic of Germany

If spots of light are used as stimuli, lateral geniculate nucleus (LGN) neurons have approximately circular receptive fields (Hubel and Wiesel, 1961; Cleland, Dubin and Levick, 1971). In the striate cortex, however, Hubel and Wiesel demonstrated that a neuron there responds best to a line or an edge of a specific orientation and very poorly to diffuse flashes or circular spots of light (Hubel and Wiesel, 1962, 1968). When such elongated stimuli are used to study geniculate cell responses, some dependence on the orientation of the line is observed (Daniels, Norman and Pettigrew, 1977; Lee, Creutzfeldt and Elepfandt, 1979; Vidyasagar and Urbas, 1982). Receptive fields of cat retinal ganglion cells are also slightly elliptical (Hammond, 1974) and the responses show an orientation bias when tested with moving gratings (Levick and Thibos, 1980). A detailed study of the orientation biases in the LGN (Vidyasagar and Urbas, 1982) has revealed some aspects of the phenomenon that might have possible implications for the high orientation sensitivity of visual cortical cells.

In their pioneering work, Hubel and Wiesel suggested that the sensitivity of a visual cortical cell to the orientation of a line stimulus may be derived from a convergent excitatory input from a number of LGN cells that have their receptive fields arranged in a row in visual space (Hubel and Wiesel, 1962). Alternatively, an important role has been attributed to intracortical inhibition in achieving the orientation sensitivity (Creutzfeldt, Kuhnt and Benevento, 1974; Henry, Dreher and Bishop, 1974; Sillito, 1975). Intracellular recordings have shown the excitatory receptive fields of cortical cells to be circular (Creutzfeldt, Kuhnt and Benevento, 1974), and during microiontophoretic application of bicuculline, an antagonist of GABA-mediated inhibition, cortical cells lose much of their orientation sensitivity (Sillito 1975, 1979; Tsumoto, Eckhart and Creutzfeldt, 1979; Sillito *et al.*, 1980). However, in many of the above cases, iontophoretic bicuculline did not totally abolish the orientation sensitivity. This may be due to a residual excitatory convergence of the Hubel and Wiesel type, or to inadequate inactivation of GABA inhibition, or may reflect an orientation bias that may already be present in the geniculate input. In the light of the presence of appreciable orientation biases in the responses of LGN neurons, it is tempting to suggest that a simple, non-specific increase in the cortical firing threshold through intracortical inhibition, letting only the peaks of geniculate responses to get through, endows the cortical cell with a much higher orientation sensitivity than its geniculate input. But Sillito *et al.* (1980) found that after application of the more effective *N*-methylbicuculline, most of the cortical cells in their sample (9 out of 13) totally lost their orientation sensitivity. If this effect is indeed general, it excludes the importance of geniculate biases as an excitatory input. It is, however, possible that the LGN biases, conferred on inhibitory interneurons in the cortex, may be essential for the cortical orientation sensitivity.

The importance of LGN biases for the cortex is testable, if these biases can be selectively abolished when recording from a cortical cell. The orientation bias of an LGN neuron is in fact mostly due to a spatial asymmetry of the geniculate suppressive field. This can be seen in the responses of a cell to lines of different lengths moved in the optimum and non-optimum orientations. The end-inhibition in the optimum orientation is much less than in the non-optimum orientation. For short bars of optimum length, the bias is very low or absent (Vidyasagar and Urbas, 1982). The intrageniculate inhibition that contributes to the orientation bias seems to be mediated through GABAergic neurons, since local iontophoretic application of bicuculline abolishes or significantly reduces the orientation sensitivity of a geniculate cell (Vidyasagar, 1984; see Fig. 1a). While recording from a visual cortical cell, a significant broadening of the tuning would be predicted if bicuculline is iontophoresed in the topographically corresponding area of the geniculate in sufficient amounts to affect the orientation biases of a fairly large population of cells there. Experiments were done using a tungsten-in-glass electrode to record from the cortex and, simultaneously, a multibarrel electrode in the geniculate to abolish GABA inhibition in the topographically corresponding region. The geniculate electrode had four bicuculline-filled barrels glued to a central recording pipette with a lateral separation of 200 μm between the central pipette and each drug barrel. Preliminary results show that a cortical cell's orientation tuning can in fact be significantly broadened by a simple abolition of LGN orientation biases with bicuculline (see Fig. 1b), suggesting that the LGN biases do in some way contribute to the orientation sensitivity of striate cortical cells.

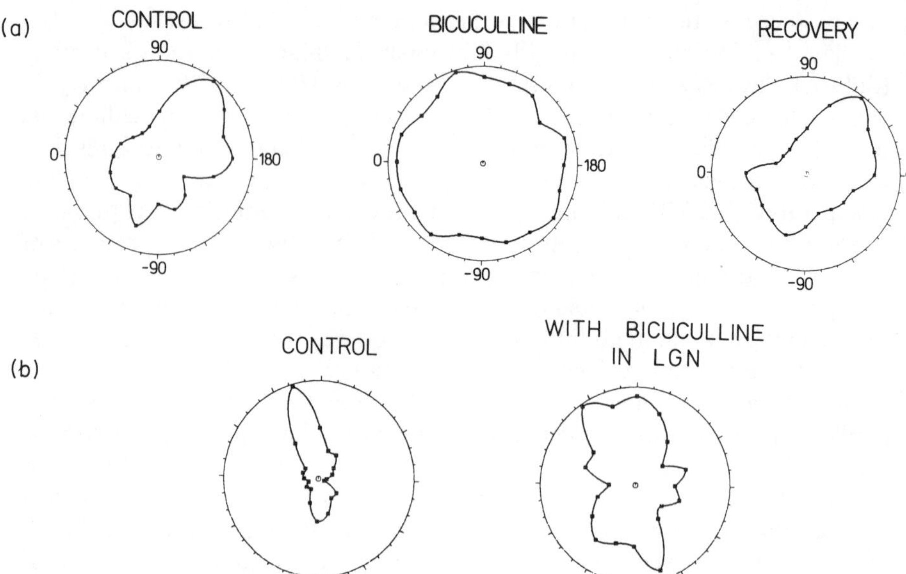

Figure 1 (a) Polar diagrams of normalized responses of a geniculate X-ON cell to a moving light bar (15° x 0.1°, 5°/s, contrast 6) at different orientations/directions are shown before, during and after local iontophoretic application of 50 nA of 0.01*M* bicuculline methiodide in normal saline. (b) Polar diagrams of normalized responses of an area 17 simple cell to a similar light bar moving at different orientations are shown before and during an iontophoretic application of bicuculline from four barrels in the topographically corresponding region of the ipsilateral dLGN. (Fig. 1a: from Vidyasagar, 1984, parts of Fig. 1. Reproduced by permission of Springer-Verlag.)

If we assume that LGN orientation biases are important, it would not be unreasonable to speculate a little on the implications of some of the known data on geniculate and cortical orientation sensitivities. The distribution of optimum orientations of geniculate neurons was not quite random, especially when the corticogeniculate feedback was removed by undercutting cortical areas 17 and 18 (Vidyasagar and Urbas, 1982). There was a high preponderance of cells with best orientations close to the vertical or horizontal over those tuned to the oblique orientation. Besides being a likely candidate for the neurophysiological origin of the psychophysically observed 'oblique effect' (Appelle, 1972; Campbell, Kulikowski and Levinson, 1968; Vandenbussche and Orban, 1981), this might also underlie the preferences seen for vertical and horizontal orientations in many single-unit studies on cat and monkey cortex (Pettigrew, Nikara and Bishop, 1968; Mansfield, 1974; Rose and Blakemore, 1974; Leventhal and Hirsch, 1977; Poggio and Fischer, 1977; Frégnac and Imbert, 1978; Blakemore, Garey and Vital-Durand, 1981; Orban and Kennedy, 1981). The orientational anisotropy could have a very basic function as well. The visual system may be using these two coordinates to code orientation information at the geniculate level. Since any orientation can be coded as the relative amounts of activity in two orthogonal sets of vectors, the visual cortical network may be designed to create the sensitivities for all possible orientations from the two pools of biased cells in the LGN. A similar scheme has already been proposed for the ontogenetic development of cortical orientation detectors (Leventhal and Hirsch, 1975; Frégnac and Imbert, 1978; von der Malsburg and Cowan, 1982). It is proposed here that this mechanism of generating oblique preferences in the visual cortex from vertical and horizontal preferences may be a dynamic one and not just an ontogenetic process. Deoxyglucose studies are also consistent with such a proposal, since presentation of vertical or horizontal gratings induces a pattern of labelling in the visual cortex different from that induced by oblique gratings (Silverman, Tootell and DeValois, 1980). In this scheme, the unambiguous distinction between the two obliques (45 and 135°) requires the use of an additional response property like directional selectivity. This break of symmetry can, however, be achieved by a self-organizing intracortical network that has a tendency to pool in columns not only cells preferring the same orientation (von der Malsburg, 1973) but also the same direction. There is some experimental evidence to support such grouping of preferred direction of movement in cortical columns (Payne, Berman and Murphy, 1980; Tolhurst, Dean and Thompson, 1981).

There are some advantages of coding a variable at the periphery in a limited number of very broadly tuned channels. This is very well exemplified in the domain of primate colour vision, where three different retinal pigments with broad and overlapping spectral sensitivity distributions form the basis for fine-colour discrimination across the whole visible spectrum. While a minimum of two pigments with broad-spectral sensitivities covering the entire spectrum are essential for any colour discrimination, too many pigments with fine tuning would reduce visual acuity and absolute sensitivity. The best scheme to preserve acuity, sensitivity and colour discriminative ability is to have a limited number of very broadly tuned photopigments. This is exactly what we find in primate colour vision. A similar strategy is probably adopted also in the orientation domain. Given the relatively small number of geniculate neurons (in the cat, three to five times the number of retinal ganglion cells), it would be impossible to code all orientations for any one visual field locus. Thus, if orientation information is somehow coded in the geniculate, the only possible way is to code it in a small number of broadly tuned overlapping channels with different peak sensitivities. The experimental findings discussed above are in good accordance with such a model. The orientation tuning of geniculate cells is broad enough to give good sensitivity and the optimum orientation of most neurons lies around either the vertical or horizontal orientation. The geniculocortical system probably depends on a few other mechanisms too, to improve the orientation sensitivity and discrimination. Specific intracortical inhibitory connections *per se* (i.e. without the geniculate biases acting through them) may confer a certain amount of orientational sensitivity on cortical cells. Further, the corticogeniculate projection may endow many uncommitted geniculate cells with oblique biases that may act as additional broadly tuned channels. The sharp tuning and the fairly uniform representation of all orientations can be achieved in the visual cortex with no loss of sensitivity or acuity since there are 300 to 400 striate cells for each ganglion cell.

Over twenty years ago, Hubel and Wiesel opened up a new field for investigation with their discovery of the functional organization of the visual cortex. The remarkable orientation sensitivity of cortical cells that they described may be the final outcome of many processes that include the orientation biases in the geniculocortical input and also many subtle intracortical and corticogeniculate connections.

REFERENCES

Appelle, S. (1972). Perception and discrimination as a function of stimulus orientation: the 'oblique effect' in man and animals. *Psychol. Bull.,* **78**, 266–278.

Blakemore, C. B., Garey, L. J., and Vital-Durand, F. (1981). Orientation preferences in the monkey's visual cortex. *J. Physiol. (Lond.),* **309**, 78P.

Campbell, F. W., Kulikowski, J. J., and Levinson, J. (1968). The effect of orientation on the visual resolution of gratings. *J. Physiol. (Lond.),* **187**, 427–436.

Cleland, B. G., Dubin, M. W., and Levick, W. R. (1971). Sustained and transient neurons in the cat's retina and lateral geniculate nucleus. *J. Physiol. (Lond.),* **217**, 473–496.

Creutzfeldt, O. D., Kuhnt, U., and Benevento, L. A. (1974). An intracellular analysis of visual cortical neurons to moving stimuli: responses in a co-operative neuronal network. *Exp. Brain Res.,* **21**, 251–274

Daniels, J. D., Norman, J. L., and Pettigrew, J. D. (1977). Biases for oriented moving bars in lateral geniculate nucleus neurons of normal and stripe-reared cats. *Exp. Brain Res.,* **29**, 155–172.

Frégnac, Y., and Imbert, M. (1978). Early development of visual cortical cells in normal and dark-reared kittens: relationship between orientation selectivity and ocular dominance. *J. Physiol. (Lond.),* **278**, 27–44

Hammond, P. (1974). Cat retinal ganglion cells: size and shape of receptive field centres. *J. Physiol. (Lond.),* **242**, 99–118.

Henry, G. H., Dreher, B., and Bishop, P. O. (1974). Orientation specificity of cells in cat striate cortex. *J. Neurophysiol.,* **37**, 1394–1409.

Hubel, D. H., and Wiesel, T. N. (1961). Integrative action in the cat's lateral geniculate body. *J. Physiol. (Lond.),* **155**, 385–398.

Hubel, D. H., and Wiesel, T. N. (1962). Receptive fields, binocular interaction and functional architecture in the cat's visual cortex. *J. Physiol. (Lond.),* **160**, 106–154.

Hubel, D. H., and Wiesel, T. N. (1968). Receptive fields and functional architecture of monkey striate cortex. *J. Physiol. (Lond.),* **195**, 215–243.

Lee, B. B., Creutzfeldt, O. D., and Elepfandt, A. (1979). The responses of magno-

and parvocellular cells of the monkey's lateral geniculate body to moving stimuli. *Exp. Brain Res.,* **35**, 547–557.
Leventhal, A. G., and Hirsch, H. V. B. (1975). Cortical effect of early selective exposure to diagonal lines. *Science,* **190**, 902–904.
Leventhal, A. G., and Hirsch, H. V. B. (1977). Effects of early experience upon orientation sensitivity and binocularity of neurons in visual cortex of cats. *Proc. Natl. Acad. Sci. USA.* **74**, 1272–1276.
Levick, W. R., and Thibos, L. N. (1980). Orientation bias of cat retinal ganglion cells. *Nature,* **268**, 389–390.
Mansfield, R. L. W. (1974). Neural basis of orientation perception in primate vision. *Science,* **186**, 1133–1135.
Orban, G. A., and Kennedy, H. (1981). The influence of eccentricity on receptive field types and orientation selectivity in areas 17 and 18 of the cat. *Brain Res.,* **208**, 203–208.
Payne, B. R., Berman, N., and Murphy, E. H. (1980). Organization of direction preferences in cat visual cortex. *Brain Res.,* **211**, 445–450.
Pettigrew, J. D., Nikara, T., and Bishop, P. O. (1968). Responses to moving slits by single units in cat striate cortex. *Exp. Brain Res.,* **6**, 373–390.
Poggio, G. F., and Fischer, B. (1977). Binocular interaction and depth sensitivity in striate and prestriate cortex of behaving rhesus monkey. *J. Neurophysiol.,* **40**, 1392–1405.
Rose, D., and Blakemore, C. (1974). An analysis of orientation selectivity in the cat's visual cortex. *Exp. Brain Res.,* **20**, 1–17.
Sillito, A. M. (1975). The contribution of inhibitory mechanisms to the receptive field properties of neurons in the striate cortex of the cat. *J. Physiol. (Lond.),* **250**, 305–329.
Sillito, A. M. (1979). Inhibitory mechanisms influencing complex cell orientation selectivity and their modification at high resting discharge levels. *J. Physiol. (Lond.),* **289**, 33–53.
Sillito, A. M., Kemp, J. A., Milson, J. A., and Berardi, N. (1980). A reevaluation of the mechanisms underlying simple cell orientation selectivity. *Brain Res.,* **194**, 517–520.
Silverman, M. S., Tootell, R. B., and DeValois, R. L. (1980). Deoxyglucose mapping of orientation and spatial frequency in cat visual cortex. *Invest. Ophthalmol. Vis. Sci. (ARVO Suppl.),* **19**, 225.
Tolhurst, D. J., Dean, A. F., and Thompson, I. D. (1981). Preferred direction of movement as an element in the organization of cat visual cortex. *Exp. Brain Res.,* **44**, 340–342.
Tsumoto, T., Eckhart, W., and Creutzfeldt, O. D. (1979). Modification of orientation sensitivity of cat visual cortex neurons by removal of GABA-mediated inhibition. *Exp. Brain Res.,* **34**, 351–363.
Vandenbussche, E., and Orban, G. A. (1981). Are there meridional differences in the orientation and grating acuity of the cat? *Neurosci. Lett. (Suppl.),* **7**, 146.
Vidyasagar, T. R. (1984). Contribution of inhibitory mechanisms to the orientation sensitivity of cat dLGN neurons. *Exp. Brain Res.,* **55**, 192–195.
Vidyasagar, T. R., and Urbas, J. V. (1982). Orientation sensitivity of cat LGN neurons with and without inputs from visual cortical areas 17 and 18. *Exp. Brain Res.,* **46**, 157–169.
von der Malsburg, C. (1973). Self-organization of orientation sensitive cells in the striate cortex. *Kybernetik,* **14**, 85–100.
von der Malsburg, C., and Cowan, J. D. (1982). Outline of a theory for the ontogenesis of iso-orientation domains in visual cortex. *Biolog. Cybernetics,* **45**, 49–56.

Models of the Visual Cortex
Edited by D. Rose and V. G. Dobson

CHAPTER 42

Inhibitory circuits and orientation selectivity in the visual cortex

ADAM MURDIN SILLITO
Department of Physiology, University College, PO Box 78, Cardiff, CF1 1XL, UK.

The extensive involvement of inhibitory circuits in the synaptic organization of the visual cortex was missed in the early models proposed by Hubel and Wiesel (1962) and as these in essence established the main approach to visual cortical function for the following two decades, there is still a tendency to underestimate the significance of inhibitory mechanisms. My purpose here is to discuss some of the key observations regarding the organization of inhibitory mechanisms that I believe should influence future models of visual cortical function. The approach I have adopted should be regarded as a provocation to thought rather than an attempt at a definitive account.

INHIBITORY SYNAPSES

It is clear on anatomical grounds that visual cortical cells are subject to a range of different types of inhibitory control. Using a pyramidal cell as an example it is possible to distinguish four or more types of inhibitory input. These all derive from cells that are broadly grouped in the non-spiny and partially spiny stellate categories, make the Gray's type II synapses which on morphological grounds are associated with inhibitory function (Gray, 1963; LeVay, 1973) and all appear to be GABAergic (Ribak, 1978; Sillito, 1974, 1975a; Peters *et al.*, 1982). Inhibitory inputs to the cell body and proximal dendrites derive from basket cells and short axon multipolar cells (Peters and Fairen, 1978; Peters and Regidor, 1981; Martin, Somogyi and Whitteridge, 1983). Further inputs to the distal parts of apical and basal dendrites arise from bitufted cells, basket cells and possibly certain other cell types (Peters and Regidor, 1981; Somogyi and Cowey, 1981; Martin, Somogyi and Whitteridge, 1983). A very specific type of axo-axonic input to the initial segment arises from the so-called chandelier cells (Somogyi, 1977). In addition to the fact that a given pyramidal cell may receive input from a range of different types of inhibitory interneuron, it should be noted that any given type of inhibitory input seems to derive from several input cells; thus, for example, a number of chandelier cells contribute to the axo-axonic synapses on the initial segment. As each of these inhibitory interneuron types have characteristic patterns of dendritic and axonal arborization, they thus must be judged to integrate rather different types of information and distribute their output over rather different spatial domains in the cortex. It follows from this that there are a range of distinct inhibitory circuits that must be presumed to govern different aspects of the visual response properties of visual cortical cells.

The various inhibitory inputs to a visual cortical cell are not only distinct in origin but in terms of the potential action they exert at the level of the recipient cell. Recent evidence for both hippocampal and visual cortical pyramidal cells (Andersen *et al.*, 1980; Kemp, 1984) show that GABAergic synapses on the cell body exert a different action to those on apical dendrites. The synapses on the cell body produce the expected hyperpolarization whilst those on apical dendrites cause a depolarization associated with an increased membrane conductance. Andersen *et al.* described the depolarizing effect as 'discriminative inhibition', because by shunting currents in the vicinity of the 'inhibitory input' it allows for a selective inhibitory influence of afferent excitatory inputs synapsing locally whilst even facilitating the effect of excitatory synapses at other locations in the dendritic tree. The chandelier cells also have to be regarded as having a potentially very distinctive influence. The initial segment of the axon is considered to be the site for initiation of the action potential (Spencer and Kandel, 1961) and hence the chandelier cell synapses to this zone must have the capability for a virtually preemptive control over the cell's activity. A potentially all or none influence of this type must provide the capability for an absolute control over the responses of the target cell population. It stands in contrast to the potentially more graded reaction that might be inferred to take place between the excitatory input and the inhibitory synapses on the soma and proximal dendrites. Whatever the synaptic basis, this type of graded interaction between inhibitory and excitatory inputs is known to occur because cells receive inhibitory inputs even when responding to an optimal stimulus and show increases in response magnitude when these inputs are blocked (Creutzfeldt, Kuhnt and Benevento, 1974; 1975a, 1975b). In summary, it seems that there are grounds for distinguishing three types of inhibitory effect at the level of the recipient cell: discriminative inhibition, an all or none preemptive inhibition and a graded interactive type.

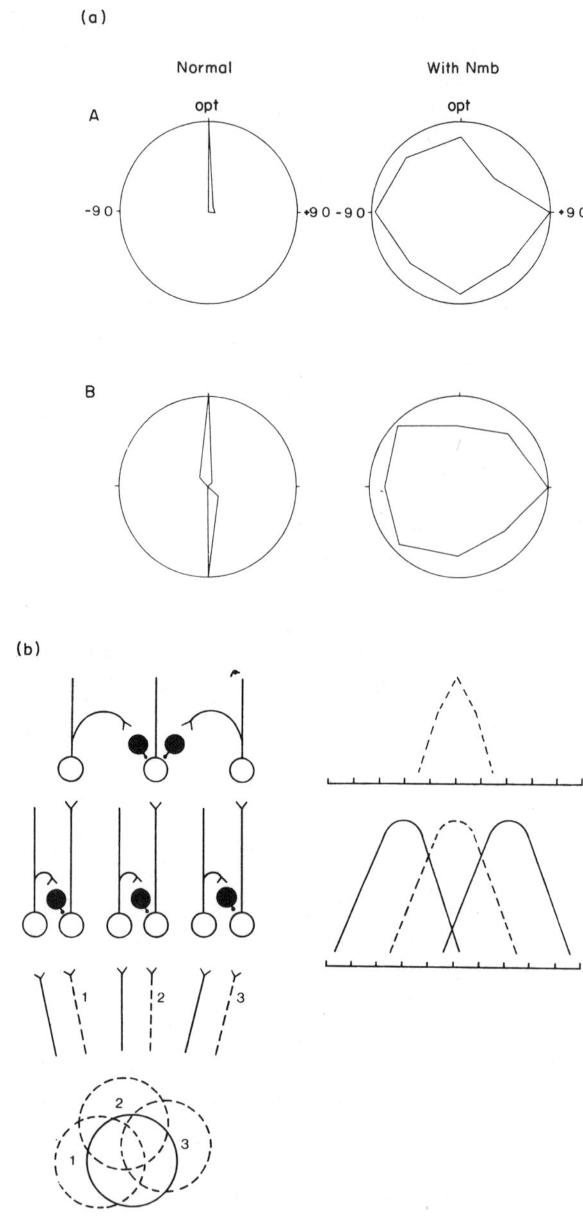

Figure 1 (a) Polar diagrams showing the orientation tuning of a simple cell (A) and a complex cell (B) before ('normal') and during iontophoretic application of the GABA antagonist *N*-methylbiculluline ('with Nmb'). Each point obtained from average for 25 trials expressed as peak frequency. Profiles are normalized to the maximum response in each situation. Testing orientations are as indicated. Nmb

What can we say of the functional contribution of these inhibitory processes to the observed response properties of visual cortical cells? From the experimental viewpoint, the role of inhibitory processes in the visual cortex is probably most directly demonstrated by data obtained from studies that have used iontophoretic application of the GABA antagonist bicuculline to produce a localized block of GABA-mediated inhibitory synapses. The evidence obtained in this way favours the not-altogether surprising view that virtually all aspects of the response selectivity exhibited by visual cortical cells are partially or wholly dependent on a structuring and control exerted by inhibitory circuits. Thus, for example, during a drug-induced localized blockade of intracortical inhibition, simple cells lose their spatially discrete 'on' and 'off' subdivisions, directional selectivity and orientation selectivity (Sillito, 1974, 1975a, 1975b, 1977; Tsumoto, Eckhart and Creutzfeldt, 1979; Sillito *et al.,* 1980. Complex cells are similarly affected. All show changes in orientation selectivity during bicuculline application, with some exhibiting a loss of the original selectivity and others retaining an orientation bias but showing a broadening of their orientation tuning curve (Sillito, 1979). The effects of bicuculline on simple and complex cell orientation selectivity are illustrated in Fig. 1a. Even ocular dominance, which is clearly originally established by the pattern of excitatory connections in layer IV, is influenced by a blockade of inhibitory synapses. Some 50 per cent of the cells showing a strong preference for one eye show a shift during bicuculline application, with some cells originally exclusively dominated by one eye showing a shift to group 4 (Sillito, Kemp and Patel, 1980.) These data and other observations made with bicuculline are discussed in the papers cited and are extensively reviewed elsewhere (Sillito, 1984). The primary conclusion that I would leave here is that there is a substantial breakdown in response selectivity (although not visual responsiveness) in the presence of the blockade of GABAergic inhibitory synapses. These conclusions do not stand in isolation; the intra-

---

ejection currents, 50 nA(A), 60 nA(B). (b) Concept diagram to illustrate some of the interactions that might contribute to orientation tuning. The receptive fields of the dLGN inputs are shown in the lower part of the diagram. Solid circle represents the 'receptive field' driving the first-stage inhibitory mechanism, dashed circles the excitatory fields of three cells receiving this inhibitory input. Orientation tuning curves on the lower right, derived from a quantitative simulation of the situation, show the selectivity that can be generated by overlapping partially shifted inhibitory and excitatory fields; the solid curves distinguish the orientation tuning of the cells providing the inhibitory drive to the second stage. The curve above shows the tuning profile after a 'second-stage' orientation domain filter. Horizontal bar under orientation tuning curves is marked in 45° steps. Solid neurons are inhibitory in function. The arrangement has been constructed for diagrammatic clarity. It is likely that in reality the input to the first-stage inhibitory interneurons would be monosynaptic from the dLGN and that the second stage would also be 'reflected' to the first stage, so that both processes operate in stage '1'. (See text for further discussion.)

cellular studies of Creutzfeldt's laboratory (e.g. Creutzfeldt, Kuhnt and Benevento, 1974) and a range of other observations (e.g. Ganz and Felder, 1984) all emphasize the importance of inhibitory synapses to the various aspects of the receptive field properties of visual cortical cells.

AN EVALUATION OF MECHANISMS UNDERLYING ORIENTATION SELECTIVITY

One of the primary issues deriving from the observations with bicuculline concerns the mechanisms underlying orientation selectivity. I will use a discussion of this as a basis for extrapolating several of the general issues that might influence thought on the relative contribution of inhibitory and excitatory mechanisms to visual cortical function. The general assumption has been, and still is represented as such in most neuroscience texts, that orientation selectivity is established at the level of the simple cell by the asymmetry in the spatial properties of the excitatory geniculate input to simple cells. It is important to state that there is no direct experimental evidence to support this view. In fact, the direct evidence suggests quite the converse: the excitatory discharge centres are basically round rather than elongated (Creutzfeldt, Kuhnt and Benevento, 1974; Cleland and Creutzfeldt, 1977), and the orientation selectivity is lost during blockade of inhibitory synapses (Sillito *et al.*, 1980). The observations that would have been taken to support an elongated excitatory discharge field would still apply if the response to the excitatory input were 'shaped' by the intracortical inhibitory network. Thus, as Heggelund (1981) indicated, there would be an apparent elongation of centre and flanks if the field comprised circular, overlapping, but acentric inhibitory and excitatory zones. Moreover, the inhibitory zones could restrict the plane of spatial summation that might derive from a large diffuse and subliminal intracortically mediated excitatory field, thus further enhancing the impression of spatial summation at the optimal orientation and the apparent 'length' of the simple cell field. The potential significance of the inhibitory mechanism is emphasized by recent observations of Albus and Wolf (1983), suggesting that the cells in layers IV and VI which seem to be the first to develop visual responses have inhibitory regions flanking the excitatory region.

A further factor that must influence an analysis of the mechanisms determining the responses of layer IV simple cells is that the majority of the excitatory inputs to layer IV cells derive from intracortical sources, with only 5 to 20 per cent. being of geniculate origin (Garey and Powell, 1971; Hornung and Garey, 1981). This raises the further possibility that during the blockade of intracortical inhibition, previously masked intracortical excitory inputs revealed by the enhanced excitability of the cell might 'swamp' spatial asymmetries in the geniculate input. However, the effects on orientation tuning are obtained at excitability levels below those causing response saturation, and hence unless the intracortical inputs are themselves biased to orientations away from the optimal, there is no reason to presume that their presence would eliminate orientation bias conferred by the geniculate input.

Other recent evidence indicates that there may be small but significant orientation biases in the retinal ganglion cell and dorsal lateral geniculate nucleus (dLGN) cell receptive fields (Levick and Thibos, 1982; Vidyasagar and Urbas, 1982). Whilst these biases are small in comparison with those seen in cortical cells, they could exert an influence on mechanisms establishing cortical orientation selectivity. Should these biases be detectable in simple cell responses during bicuculline application? They might be subtly masked by intracortical excitatory inputs gaining access to the cell during drug application, or they might be present or overlooked. Both the simple and complex cell illustrated in Fig. 1a, although showing a loss of the original selectivity, could still be argued to show residual orientation bias during drug application, but this bias is totally different to the original selectivity of the cell in question. It seems unlikely that the orientation selectivity of each simple cell is determined by biases in its excitatory input, but this matter is further discussed below.

Is there any evidence to favour the view that simple cells determine complex cell orientation selectivity? It would certainly seem plausible that the half of the complex cell population retaining some orientation bias during blockade of inhibitory synapses (Sillito, 1979) might receive an orientation biased excitatory input, although it is possible that the inhibitory blockade of these cells was not complete. Certainly the drug application procedure can never be assumed to block all inhibitory synapses; indeed, for example, it is highly unlikely that the iontophoretically applied bicuculline ever blocked inhibitory synapses reaching the more remote parts of apical dendrites. Nonetheless, were the input of these cells to be orientation biased, it would seem plausible that this derived from simple cells, although it is clear that simple cells in several orientation columns were sampled because the residual orientation selectivity of these cells was less than that of individual simple cells. Conversely, the orientation selectivity of those cells losing selectivity during drug application cannot be considered to be dependent on their excitatory input. They may be presumed to receive direct geniculate input (as many do: Singer, Tretter and Cynader, 1975; Toyama, Mackawa and Takeda, 1977; Martin and Whitteridge, 1981) and/or to sample intracortical excitatory connections from a large enough group of orientation columns for it to be essentialy non-orientation selective, or to do both.

The necessity for a serial relationship between simple and complex cells is questioned by a wide variety of evidence. Complex cells respond to moving noise, which simple cells do not (Hammond and MacKay, 1975); moreover, complex cells in laminae II, III and V retain apparently normal visual response properties in the face of a loss of responsiveness of cells in layers IV and VI, following cobalt chloride microinjection to the A laminae of the cat dLGN (Malpeli, 1983). This apparently paradoxical observation suggests that the responses of complex cells in layers II, III and V do not require input from the A laminae or, for example, simple cells in layers IV and VI. One must presume, although there are other possibilities, that the input from the C laminae is responsible for the visual responses of these cells and that their orientational and directional selectivities are generated by mechanisms which are independent of those operating in layers IV and VI. This latter view is supported by the fact that cells showing a loss of orientation selectivity during bicuculline application occur in all laminae (Sillito, 1979; Sillito *et al.*, 1980).

There is thus, from some viewpoints, a remarkable independence of the mechanisms generating the response selectivity of visual cortical cells in different laminae. Moreover, we know that the output connections from the different laminae are distinctive and that there is some laminar segregation of the inputs from X-, Y- and W-cells which further emphasizes the idea of a laminar segregation of function (LeVay and Gilbert, 1976; Ferster and LeVay, 1978). On the other hand, there is an equally notable degree of interconnectivity between the laminae and a powerful corticofugal feedback to the dLGN. Thus, taking the potential excitatory interconnections for the cat visual cortex, the A laminae of the dLGN project to laminae IV and VI in the cortex, lamina VI projects back to the dLGN and to lamina IV, lamina IV projects to laminae III and V, lamina III projects to V, and lamina V to III and VI (Lund *et al.*, 1979; Gilbert and Wiesel, 1979). This is an oversimplified account and ignores the organization of oblique and horizontal patterns of connectivity, but underlines the interdependence of the laminae, a view which is furthered by the axonal and dendritic arborizations of inhibitory interneurons such as the chandelier cells and double-bouquet cells (Somogyi, 1977; Somogyi *et al.*, 1981). These ideas do not easily fit together; it is difficult to see how the properties of layer V and III cells can be so independent of those in IV and VI. One possibility is that orientation tuning and directional selectivity only involve a small component of the available circuitry and that a more complex stimulus situation would be required to reveal the full significance of the various components of the synaptic operations. On another tack it is also plausible that the synaptic operations first establishing orientation selectivity take place in lamina III. The mosaic of cytochrome oxidase patches containing non-orientation specific cells in laminae II and III of the monkey visual cortex appear to have a special relationship to the columnar structure (Livingstone and Hubel, 1982), and given the association of cytochrome oxidase with GAD (Hendrickson, Hunt and Wu, 1981) are likely to contain a high density of inhibitory interneurons or terminals. Whilst these patches do not appear to be so well developed in the cat, their existence in the monkey does raise the possibility that the mechanisms establishing orientation selectivity might be initiated in layer III. It would be very useful to know whether orientation selectivity in layer IV survives a blockade of layer III. Given the existence of orientation columns which extend through the laminae and show a regularity in their sequence from one location to the next, it seems logical that some common process establishes the pattern and distributes its influence between laminae. This has always been inferred to take place in layer IV as a consequence of the organization of excitatory connections; it now seems likely that inhibitory processes are involved and the special role of layer IV is brought into question. I would argue that there is a logical case for using the inhibitory mechanism to maintain the coherence of orientation columns. Cells in the different cortical laminae gain access to a wide variety of excitatory inputs, both direct and intracortical, and in many cases individual cells sample a wide range of inputs. An example of the latter case is a lamina V pyramidal cell which may have basal dendrites sampling information from lamina V and an apical dendrite sampling information through laminae IV to I, with a fairly extensive arborization in lamina I. Were the excitatory connections to determine the selectivity, an extraordinary level of coherence would be required; conversely, a specific inhibitory input to the soma or initial segment could restrict the response to all the excitatory inputs to a particular part of the orientation domain (or any other parameter for that matter). Moreover, once established a common inhibitory process could regulate the response domain of a group of cells irrespective of their excitatory input. Given that orientation is a common parameter of the column whilst the excitatory input is not (in any other sense than location in visual space, that is), this would seem to be the most economical way of achieving the end result.

The mechanisms first establishing orientation selectivity do not have to be especially complex. Overlapping circular inhibitory and excitatory fields, requiring only a simple lateral inhibitory interaction in the spatial domain, would establish a broadly orientation tuned field. If this simple process generated a range of orientation biased channels for a given location in visual space, then a 'second'-stage lateral inhibitory interaction between these channels in the 'orientation' domain would create the observed levels of selectivity. The initial biases, which could be quite small, might of course derive from asymmetries in the geniculate input (e.g. Vidyasagar and Urbas, 1982), but such an assumption is not essential. It seems unlikely that orientation bias in the dLGN input would establish the pattern of cortical orientation selectivity via its direct excitatory drive to all individual cells contacted. This would raise the issue of coherence; the orientation bias of X, Y and W inputs to cells in a given orientation column would have to have a similar bias. Conversely, if the bias were fed into the inhibitory network in some way, it could provide the elemental asymmetry required to initiate the operation of a second-stage orientation domain filter and might be generalized to all cells in a column via this. It is relevant to note that Vidyasagar (1984) finds the retinal orientation bias to be translated into the geniculate via GABAergic inhibitory processes, which would be consistent with the view that any such bias would access the cortical mechanisms via inhibitory circuits.

Some of the ideas discussed here are illustrated in Fig. 1b, which assumes that the initial bias is established via partially shifted but overlapping inhibitory and excitatory zones. The connections from the dLGN input to the inhibitory interneurons is likely to be monosynaptic rather than disynaptic as illustrated, and it is envisaged that the second-stage inhibitory interaction would be reflected to the first stage, further enhancing selectivity at this level. The separation of the two stages is purely for diagrammatic simplicity. Two stages do seem to be necessary and bear comparison with some of the arguments put forward by Ganz and Felder (1984) for directional selectivity. For example, cooperativity between the elements of the second-stage interaction for different locations in visual space could provide a basis for the relationship between stimulus length and orientation tuning (Henry, Bishop and Dreher, 1974). There are some grounds, however, for believing that the mechanisms for directional and orientation selectivity are independent of each other. These include variations in the tuning observed for bar and texture motion (Hammond, 1981), differential effects of bicuculline on orientation and direction selectivity (e.g. Sillito, 1979), the nature of some of the observations of Ganz and Felder (1984) on the directional inhibition seen in cortical cells and the fact that nearby cells in an 'orientation column' can exhibit the same orientation selectivity but opposite directional selectivity. It seems that the range of inhibitory interneuron types discussed in the first section of this chapter may reflect the range of circuits that are required for orientation selectivity, directional selectivity, length preference, binocular inhibitory processes, spatial frequency tuning and the variety of other subtle interactions underlying the processing of the spatial components of the visual input. Furthermore, as suggested by Somogyi *et al.*, (1983), where cells receive inhibitory innervation from a number of the same type of cell, such as the basket cells, these may in fact actually subserve different functions. One of the difficulties in establishing clear functional correlations to the elements of the circuitry is that the degree of interconnectivity is such that all the elements have to be regarded as components of a large interactive network with parallel inputs and outputs.

ACKNOWLEDGEMENT

The support of the Wellcome Trust and the Medical Research Council is gratefully acknowledged.

REFERENCES

Albus, K., and Wolf, W. (1984). Early postnatal development of neuronal function in the kitten's visual cortex: a laminar analysis. *J. Physiol.*, **348**, 153–186.

Andersen, P., Dingledene, R., Gjerstad, L., Langmoen, I. A., and Laursen, A. M. (1980). Two different responses of hippocampal pyramidal cells to application of gamma-aminobutyric acid. *J. Physiol.*, **305**, 279–296.

Creutzfeldt, O. D., Kuhnt, U., and Benevento, L. A. (1974). An intracellular analysis of visual cortical neurons to moving stimuli: responses in a co-operative neuronal network. *Exp. Brain Res.*, **21**, 251–274.

Ferster, D., and LeVay, S. (1978). The axonal arborisation of lateral geniculate neurons in the striate cortex of the cat. *J. comp. Neurol.*, **182L**, 923–944.

Ganz, L., and Felder, R. (1984). Mechanism of directional selectivity in simple neurons of the cat's visual cortex analysed with stationary flash sequences, *J. Neurophysiol.*, **51**, 296–324.

Garey, L. J., and Powell, T. P. S. (1971). An experimental study of the termination of the lateral geniculo-cortical pathway in the cat and monkey. *Proc. Roy. Soc. Lond. B.*, **179**, 41–63.

Gilbert, D., and Wiesel, T. N. (1979). Morphology and intracortical projections of functionally characterised neurons in cat visual cortex. *Nature*, **280**, 120–125.

Gray, E. G. (1963). Electron microscopy of presynaptic organelles of the spinal cord. *J. Anat.*, **97**, 101–106.

Hammond, P. (1981). Simultaneous determination of directional tuning of complex cells in cat striate cortex for bar and texture motion. *Exp. Brain Res.*, **41**, 364–369.

Hammond, P., and MacKay, D. M. (1975). Differential responses of cat visual cortical cells to textured stimuli. *Exp. Brain Res.*, **22**, 427–430.

Heggelund, P. (1981). Receptive field organisation of simple cells in cat striate cortex. *Exp. Brain Res.*, **42**, 89–98.

Hendrickson, A. E., Hunt, S. P., and Wu, J. -Y. (1981). Immunocytochemical localisation of glutamic acid decarboxylase in monkey striate cortex. *Nature.*, **292**, 605–607.

Henry, G. H., Bishop, P. O., and Dreher, B. (1974). Orientation specificity of cells in cat striate cortex. *J. Neurophysiol.*, **37**, 1394–1409.

Hornung, J. P., and Garey, L. J. (1981). The thalamic projection to cat visual cortex: ultrastructure of neurons identified by Golgi impregnation on retrograde horseradish peroxidase transport. *Neurosci.*, **6**, 1053–1068.

Hubel, D. H., and Wiesel, T. N. (1962). Receptive fields, binocular interaction and functional architecture in the cat's visual cortex. *J. Physiol.*, **160**, 106–154.

Kemp, J. A. (1984). Intracellular recordings from rat visual cortical cells in vitro and the action of GABA. *J. Physiol.*, **349**, 13P.

Lee, B. B., Cleland, B. G., and Creutzfeldt, O. D. (1977). The retinal input to cells in area 17 of the cat's cortex. *Exp. Brain Res.*, **30**, 527–538.

LeVay, S. (1973). Synaptic patterns in the visual cortex of the cat and monkey. Electron microscopy of Golgi preparations. *J. comp. Neurol.*, **150**, 53–86.

LeVay, S., and Gilbert, C.D, (1976). Laminar patterns of geniculo-cortical projection in the cat. *Brain Res.*, **113**, 1–19.

Levick, W. R., and Thibos, L. N. (1982). Analysis of orientation bias in cat retina. *J. Physiol.* **329**, 243–261.

Livingstone, M. S., and Hubel, D. H. (1982). Thalamic inputs to cytochrome oxidase-rich regions in monkey visual cortex. *Proc. Natl. Acad. Sci. USA* **79**,6098–6101.

Lund, J. S., Henry, G. H., MacQueen, C. L., and Harvey, A. R. (1979). Anatomical organisation of the primary visual cortex (area 17) of the cat. A comparison with area 17 of the macaque monkey. *J. comp. Neurol.*, **184**, 559–576.

Malpeli, J. G. (1983). Activity of cells in area 17 of the cat in absence of input from layer A of lateral geniculate nucleus. *J. Neurophysiol.,* **49**, 595–620.

Martin, K. A. C., Somogyi, P., and Whitteridge, D. (1983). Physiological and morphological properties of identified basket cells in the cat's visual cortex. *Exp. Brain Res.,* **50**, 193–200.

Martin, K. A. C., and Whitteridge, D. (1981). Morphological identification of cells of the cat's visual cortex classified with respect to their afferent input and receptive field type. *J. Physiol.,* **320**, 14P.

Peters, A., and Fairen, A. (1978). Smooth and sparsely spined stellate cells in the visual cortex of the rat, a study using combined Golgi electron microscope techniques. *J. comp. Neurol., 181*, 129–172.

Peters, A., and Regidor, J., (1981). A reassessment of the forms of non-pyramidal neurons in area 17 of cat visual cortex. *J. comp. Neurol.,* **203**, 685–716.

Peters, A., Proskauer, C. C., and Ribak, C. E. (1982). Chandelier cells in rat visual cortex. *J. comp. Neurol.,* **206**, 397–416.

Ribak, C. E. (1978). Aspinous and sparsely-spinous stellate neurons in the visual cortex of rats contain glutamic acid decarboxylase. *J. Neurocytol.,* **7**, 461–478.

Schiller, P. H., Finlay, B. L., and Volman, S. F. (1976). Quantitative studies of single cell properties in monkey striate cortex, 1. Spatiotemporal organisation of receptive fields. *J. Neurophysiol.* **39**, 1288–1319.

Sillito, A. M. (1974). Modification of the receptive field properties of neurons in the visual cortex by bicuculline, a GABA antagonist. *J. Physiol.,* **239**, 36–37P.

Sillito, A. M. (1975a). The effectiveness of bicuculline as an antagonist of GABA and visually evoked inhibition in the cat's striate cortex. *J. Physiol.,* **250**, 287–304.

Sillito, A. M. (1975b). The contribution of inhibitory mechanisms to the receptive field properties of neurons in the striate cortex of the cat. *J. Physiol.,* **250**, 305–322.

Sillito, A. M. (1977). Inhibitory processes underlying the directional specificity of simple, complex and hypercomplex cells in the cat's visual cortex. *J. Physiol.,* **271**, 699–720.

Sillito, A. M. (1979). Inhibitory mechanisms influencing complex cell orientation selectivity and their modification at high resting discharge levels. *J. Physiol.,* **289**, 33–53.

Sillito, A. M. (1984). Functional considerations of the operation of GABAergic inhibitory processes in the visual cortex. In *The Cerebral Cortex* (eds. A. Peters and E. G. Jones), Vol. 2A, Plenum.

Sillito, A. M., Kemp, J. A., Milson, J. A., and Berardi, N. (1980). A re-evaluation of the mechanisms underlying simple cell orientation selectivity. *Brain Res.,* **194**, 517–520.

Sillito, A. M. Kemp, J. A., and Patel, H. (1980). Inhibitory interactions contributing to the ocular dominance of monocularly dominated cells in the normal cat visual cortex. *Exp. Brain Res.,* **41**, 1–10.

Singer, W., Tretter, F., and Cynader, M. (1975). Organisation of cat striate cortex: a correlation of receptive field properties with afferent and efferent connections. *J. Neurophysiol.,* **38**, 1080–1098.

Somogyi, P. (1977). A specific axo-axonal interneuron in the visual cortex of the rat. *Brain Res.,* **136**, 345–350.

Somogyi, P., and Cowey, A. (1981). Combined Golgi and electron microscopic study on the synapses formed by double bouquet cells in the visual cortex of the cat and monkey. *J. comp. Neurol.,* **195**, 547–566.

Somogyi, P., Cowey, A., Halasz, N., and Freund, T. F. (1981). Vertical organisation of neurons accumulating 3H-GABA in the visual cortex of the rhesus monkey. *Nature,* **294**, 761–763.

Somogyi, P., Kisvarday, Z. F., Martin, K. A. C., and Whitteridge, D. (1983). Synaptic connections of morphologically identified and physiologically characterized large basket cells in the striate cortex of cat. *Neuroscience,* **10**, 261–294.
Spencer, W. A., and Kandel, E. R. (1961). Electrophysiology of hippocampal neurons, III. Firing level and time constant. *J. Neurophysiol.,* **24**, 260–271.
Toyama, K., Maekawa, K., and Takeda, T. (1977). Convergence of retinal inputs onto visual cortical cells: I. A study of the cells monosynaptically excited from the lateral geniculate body. *Brain Res.,* **137**, 207–220.
Tsumoto, T., Eckhart, W., and Creutzfeldt, O. D. (1979). Modification of orientation sensitivity of cat visual cortex neurons by removal of GABA mediated inhibition. *Exp. Brain Res.,* **34**, 351–363.
Vidyasagar, T. R., and Urbas, J. V. (1982). Orientation sensitivity of cat dLGN neurons with and without inputs from visual cortical areas 17 and 18. *Exp. Brain Res.,* **46**, 157–169.
Vidyasagar, T. R. (1984). Contribution of inhibitory mechanisms to orientation sensitivity of cat dLGN neurones. *Exp. Brain Res.,* **55**, 192-195.

Models of the Visual Cortex
Edited by D. Rose and V. G. Dobson

CHAPTER 43

The synaptic veto mechanism: does it underlie direction and orientation selectivity in the visual cortex?

CHRISTOF KOCH and TOMASO POGGIO
Center for Biological Information Processing and Artificial Intelligence Laboratory, Massachusetts Institute of Technology, 545 Technology Square, Cambridge, MA 02139, USA.

THE BASIC PROBLEM

A major unsolved issue in the brain sciences is the question of what type of hardware is used by neurons to perform various computations. A closely related problem is what constitutes the basic unit of information processing in the central nervous system. Is it the neuron itself, or are the basic units of computation patches of membranes, single synaptic circuits, spines or dendrites (Bullock, 1959; Schmitt, Dev and Smith, 1976; Poggio and Torre, 1978; Koch, Poggio and Torre, 1982; Koch and Poggio, 1983; Koch, 1984; see also Swindale, 1983)? For silicon chips, the answer to both questions is simple: for metal oxide semiconductor technology the basic processing units are transistors, implementing NAND logic. Most of today's very large scale integrated (VLSI) circuits use this type of logic instead of AND, OR and NOT logics (Mead and Conway, 1980). One can of course synthesize all logical operations with NAND gates as basic building blocks. The early view of information processing in neurons was based on the threshold neuron model of McCulloch and Pitts (1943). In the most common version of this model, the neuron is considered the elementary processing unit that receives excitatory and inhibitory inputs leading to depolarizing (EPSP) and hyperpolarizing (IPSP) dendritic potentials. These are subsequently propagated to the soma where they contribute (in a linear fashion) to the somatic potential. If the somatic potential exceeds a given threshold, a spike is initiated and transmitted along the axon. Otherwise the neuron remains silent. The type of logic used by such cellular automata, linear addition or subtraction followed by a threshold operation, is the analog equivalent of binary OR or AND logic. If the threshold is low, either one of a multitude of inputs can elicit a spike, while a high threshold requires the conjunctive activation of several inputs. It is fairly straightforward to show that with an appropriate combination of these threshold-neurons one can reproduce every possible input–output behaviour (Kleene 1956; Palm 1982).

Evidence accumulated over the last forty years has shown that this is an overly simplified model of a nerve cell. The existence of spike-less neurons invalidates the view that all neurons use only all-or-none events, such as spikes, as their output signal. The discovery of dendrodendritic synapses, dendritic spikes, active voltage-dependent channels throughout the neuron, etc., indicates that the dendritic tree has a far greater role in information processing than previously thought. In this article we specifically consider one simple and attractive scheme to implement an analog computation of a binary AND-NOT gate in the dendritic tree (AND-NOT is different from NAND), contrasting it with the more usual AND type of logic requiring a threshold. Subsequently, we describe one instance where this veto operation may subserve neuronal function: direction selectivity in the visual cortex of cat and primates.

One has to add a note of caution here. Unlike silicon chips, the neuron does not function in a discrete binary manner. The fundamental biophysical processes underlying information processing, conductance and voltage changes, are smooth functions giving rise (with the exception of the spike) to graded, analog signals. Analyzing the possible elementary computations performed by a neuron in terms of binary gates is an oversimplified but suggestive way of representing these truly analog operations that emphasizes the similarities and differences with electronic circuits.

THE NEURONAL VETO OPERATION

An understanding of the biophysical basis for this mechanism lies in the observation that synaptic inputs are not current inputs, but rather conductance changes to specific ions. The resulting current is given by the product of the conductance change and the driving potential, i.e. the difference between the postsynaptic potential and the reversal potential of the synapse. Thus, synapses electrically close to each other will influence one another, leading to a somatic potential that is less than the sum of the individual contributions (Poggio and Torre, 1978; Koch, Poggio and Torre, 1982; Segev and Parnas, 1983). Probably the most common interaction of this type involves two synapses (or set of synapses), one excitatory and the other inhibitory. Activation of both synapses leads to an increase in their membrane conductance to specific ions. If the reversal potential of inhibition is very far from the resting potential of the cell then the interaction between the excitatory and the inhibitory synapse will be, to a good approximation, linear: the resulting somatic potential is the sum of the EPSP generated by the excitation and the IPSP generated by the inhibition.

The situation changes, however, if the inhibitory equilibrium potential is close to the resting potential. Activating this shunting inhibition by itself will not lead to a change in potential (silent inhibition); it may, however, have a very powerful effect in shunting a depolarization towards the resting state when a neighbouring excitatory synapse becomes activated (Torre and Poggio, 1978).

An extensive study of the precise conditions underlying the effectiveness and the specificity of this veto interaction was recently performed (Koch, Poggio and Torre, 1982, 1983). Golgi-stained cat retinal ganglion cells provided by Boycott and Wässle (1974) were modeled by a large number of cylinders of various lengths and diameters. The one-dimensional cable equation was solved for these highly branched structures by computer simulations (Koch and Poggio, 1985). Taking account of the nonlinear interaction between synaptic inputs, it was formally proven that in any dendritic tree the maximum veto effect is obtained when the shunting inhibition is situated on the direct path between the location of excitation and the soma (for an earlier observation of this effect, see Rall, 1964). The specificity for this 'on-the-path' condition can be quite strong for some dendritic trees, as demonstrated by computer simulations: inhibition located distal to excitation or on a nearby branch some 10 to 20 μm away from the branch point has a drastically reduced effect on excitation. This effect is relatively insensitive to large changes in various biophysical parameters, such as the membrane resistance and the intracellular resistance. It does, however, depend critically on the amplitude of the inhibitory conductance change involved. Our discussion of the veto mechanism is based on the assumption of passive membrane properties in the dendritic tree. Shunting inhibition, however, may have similar veto effects in the presence of dendritic spikes.

If one considers transient, time-varying synaptic inputs the specificity extends to the temporal domain. For inhibition to effectively veto excitation it must not only be on the direct path but its time course must substantially overlap with the time course of excitation. Inhibition is most effective if it lasts at least as long as excitation (Koch, Poggio and Torre, 1983; see also Segev and Parnas, 1983).

If the strength of inhibition exceeds a minimum value (typically between 10^{-8} and 10^{-7}nS) and the spatial and temporal conditions outlined above are fulfilled, inhibition can effectively veto all excitation lying more distally, implementing the analog of a logical AND-NOT gate. Such operations that exploit the branching geometry of the neuron could be distributed throughout the dendritic tree, suggesting the picture of a neuron resembling an analog

VLSI circuit with hundreds of elementary logical elements—the specific synaptic circuits (very specific arrangements of synapses have very recently been observed in an α retinal ganglion cell of the cat by Freed and Sterling, 1983). One pays for this increased complexity, however, by placing greater demands on the developmental mechanisms which guide the positioning of synapses onto dendrites (Swindale, 1983).

How important is the requirement of a shunting inhibition for our proposal? Although there is evidence for inhibitory synapses with a reversal potential close to the resting potential of the cell (see the next section), almost nothing is known about specific inhibitory channels in cortical cells. If the inhibitory reversal potential is far below the resting potential of the neuron, the nonlinear synaptic interaction loses much of its strength and specificity. Moreover, the influence of such an inhibition will spread throughout a large part of the dendritic tree, independent of the on-the-path condition (Koch, Poggio and Torre, 1982). In this case analog operations cannot be computed independently of one another: the dendritic tree appears to combine incoming signals in a linear fashion.

The specificity of the veto operation contrasts sharply with the lack of specificity of the AND mechanism based on a somatic threshold. Since this latter scheme only requires that the somatic potential be above a given threshold, any set of excitatory synapses can drive the cell, no matter where they are situated, as long as they act in conjunction, i.e. if they are all active within some temporal window. Another difference between these two types of mechanism is that the threshold scheme depends critically on the value of the threshold. Varying this threshold can change the whole input-output behaviour of the neuron, for instance from an AND-like logic into an OR-like logic. This point seems especially important in light of the observation that the threshold for eliciting spikes is known to change by a factor of 2 or 3 depending on previous spike activity (Raymond, 1979).

Unlike the Marr and Poggio (1977) approach, we think that the basic type of hardware used by the brain (or for that matter by any machine) to process information can influence decisively the type of computation performed (see Poggio, 1982). The AND type of threshold logic is well suited for performing a single operation on a large number of different inputs, regardless of their location. One of the most fundamental tasks for this scheme might be the integration of a large number of disparate synaptic inputs. A distributed AND-NOT type of logic is useful for performing complex computations that can be decomposed naturally into many simple independent operations in parallel throughout the dendritic tree (Abelson, 1978). The results of these computations are then combined at the soma. We propose that the synaptic veto mechanism is the biophysical basis for a variety of information-processing tasks in the brain. As one specific instance of such a task, we consider an operation carried out in the visual system of most animals: detecting the direction of motion.

The idea that a veto-like operation may play an important role for computations in the central nervous sytem is not new, though its specific synaptic nature and properties are. Barlow has stressed (e.g. Barlow, 1981) since his classical study of direction-sensitive ganglion cells (Barlow and Levick, 1965) that a veto-like operation is an important physiological mechanism for the retinal and cortical processes that underlie perception.

DIRECTION SELECTIVITY IN THE VISUAL CORTEX

Computing the direction of motion is one of the basic operations performed by the visual cortex in cat and primates (Hubel and Wiesel, 1959, 1962). Since the first model proposed by Hubel and Wiesel in 1962, many other models have been suggested (see Bishop, Coombs and Henry, 1971b; Goodwin, Henry and Bishop, 1975; Schiller, Finlay and Volman, 1976c; Marr and Ullman, 1981: Poggio, 1982). We consider an extension of a model by Torre and Poggio (1978) for direction selectivity in the retina to orientation and direction selectivity in the visual cortex (Poggio, 1982).

The basic idea is as follows. A single lateral geniculate on-center neuron (or a row of such neurons) excites a cortical cell in area 17 whenever a light-on stimulus falls within its center. A neighboring on-center geniculate neuron reduces the activity of the same cortical neuron by a delayed (or slower) shunting inhibition (for evidence pertaining to shunting inhibition in the central nervous system see Dingledine and Langmoen, 1980, or Gold and Martin, 1983). Because it is unlikely that the same cell type both excites and inhibits, we suppose that the second geniculate cell excites some interneuron, possibly in layer 4c, which in turn inhibits the direction-selective cell. This seems plausible in light of the fact that direction-selective cells in primate cortex first occur one synapse beyond layer 4c, e.g. in layer 4b (Dow, 1974). In fact, degeneration studies have shown that all terminals made by fibers from the lateral geniculate nucleus to area 17 make asymmetrical contacts commonly held to be excitatory (LeVay and Gilbert 1976; Peters *et al.*, 1979; Colonnier, 1981). A similar conclusion, i.e. that the cortical afferents of geniculate neurons only excite, has also been reached by Ferster and Lindström (1983) using intracellularly recorded latency measurements.

If the shunting inhibition is located on the direct path between soma and excitation, it will effectively veto excitation. In a model of a direction-selective simple cell a light-on edge moving in the preferred direction elicits EPSPs travelling unimpeded to the soma, followed by the later activation of the inhibitory synapses which does not influence the EPSPs significantly. Moving the edge in the opposite direction activates excitation and inhibition in such a way that the time-courses of the conductance changes overlap, preventing substantial EPSPs from reaching the soma (Fig. 1). If cells give a direction-selective response for very small movements of the stimulus in

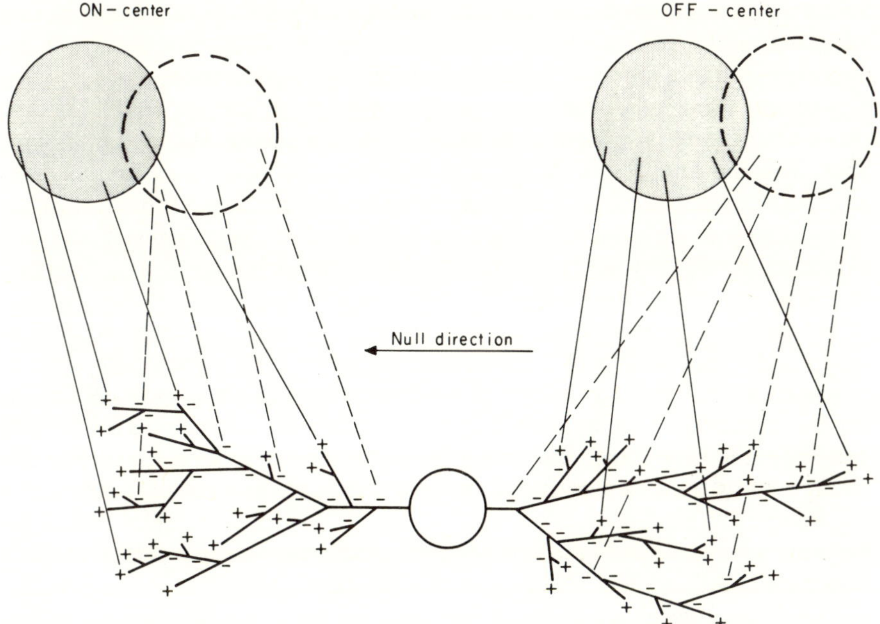

Figure 1 A mechanism for orientation and direction selectivity to motion in a simple cell in area 17. Two neighboring LGN neurons (either on- or off-center) excite (solid line) and inhibit via an interneuron (dashed line) the direction-selective neuron. Inhibition is slower and/or the inhibitory pathway contains an additional delay or a lowpass operation. Excitation is presumed to be more distal than the shunting inhibition. Because the on- and off-center systems innervate different parts of the simple cell dendritic tree, direction selectivity for light and dark edges are independent of each other. This simple cell will also be orientation selective, reacting preferentially to vertically orientated edges. The orientation tuning becomes sharper if the receptive field of the simple cell is built from a *row* of excitatory geniculate neurons facing a column of inhibitory neurons.

different parts of the receptive field (Goodwin, Henry and Bishop, 1975) one has to assume that a large number of geniculate neurons project to the cortical cell. In our scheme, pairs of geniculate neurons would be distributed throughout the receptive field, computing direction selectivity at the level of a dendrite of the cortical cell, independently of other geniculate cells.

Because inhibition has a key role in direction selectivity, we expect that if inhibition is abolished, e.g. by pharmacological means, the neuron will fire regardless of the direction of motion. In a series of experiments, Sillito and his group (Sillito, 1977; Sillito *et al.*, 1980) observed precisely such a phenomenon: direction selectivity of simple cells was severely reduced or abolished during application of bicuculline, an antagonist of GABA, a putative inhibitory neurotransmitter of the visual cortex. In a neuron with a nonzero spontaneous firing rate, inhibition should reduce this firing rate by blocking the EPSPs responsible for it. Accordingly, a stimulus moving in the nonpreferred direction should lead to a reduction in the spontaneous activity. Goodwin and Henry (Goodwin and Henry, 1975; Goodwin, Henry and Bishop, 1975; see also Orban, Kennedy and Maes, 1981) report that moving stimuli in the nonpreferred direction result in a reduced spontaneous firing rate for simple cells. These findings speak against models of direction selectivity based on an AND scheme, for which a stimulus moving in the null direction should only lead to an absence of excitation but never to a reduction in the spontaneous firing rate.

We would like at this point to clear up a possible source of confusion. The majority of studies in the visual cortex have used stimuli with optimal orientation and moving in the preferred direction. Under these stimulus conditions intracortical inhibition should be small; therefore these cells would not show the veto effect and instead behave linearly (with the exception of the rectifying spike-generating operation). Because the shunting inhibition is an inherently nonlinear process it is likely that cortical cells will show increasing evidence of nonlinearity for stimuli increasingly different from optimum (see Movshon, Thompson and Tolhurst, 1978)

The scheme described so far works only for light-on edges moving in one direction (one variant of the S1-cell described by Schiller, Finlay and Volman, 1976a). Adding a similar, but inverted, circuit constructed with geniculate off-center neurons endows a cell with direction selectivity for both light-on and light-off edges moving in the same direction—one of the most common types of direction-selective cells (Bishop, Coombs and Henry, 1971a; Goodwin, Henry and Bishop, 1975; the S2-cell of Schiller, Finlay and Volman, 1976a). These off-center neurons whose receptive fields possibly overlap with the fields of their on-center counterparts, map onto a different part of the dendritic field of the direction-selective cell, for instance onto a different branch in stellate cells or onto different parts of the basal dendrites of pyramidal cells (Poggio, 1982). This prediction, i.e. that direction selectivity for light-on and light-off edges results from independent convergence onto critical neurons, is supported by experiments done by Schiller (1982) in monkey and by Sherk and Horton (1984) in cat, using the pharmacological agent APB (DL-2-amino-4-phosphonobutyric acid). APB infusion of the retina, which reversibly blocks the on-pathway at the level of the retinal outer plexiform layer, eliminates the response of direction-selective cells to light edges while leaving the response to dark edges intact. This finding seems to jeopardize the AND mechanism needed for direction selectivity in models such as the one proposed by Marr and Ullman (1981).

The present scheme, unlike Marr and Ullman's proposal, requires no specific hypothesis about the relative contribution of the X and Y pathways to cortical direction selectivity. In fact, it may well be that both the parvo-

and the magnocellular pathways have similar but separate mechanisms for direction selectivity, in accordance with the finding of Malpeli, Schiller and Colby (1981).

Direction selectivity performs best within a given velocity range. Moving the stimulus too slowly or too quickly reduces or even abolishes direction selectivity, i. e. the response in one direction no longer differs substantially from the response to movement in the opposite direction. The range within which direction discrimination can be successfully performed is governed by the delay and the kinetics of the excitatory and inhibitory channels involved. From measurements of this velocity range (Goodwin, Henry and Bishop, 1975; Orban, Kennedy and Marr, 1981; Maunsell and Van Essen, 1983) it seems that cortical cells respond to a broad range of velocities. This finding suggests that the inhibitory conductance change lasts much longer than the corresponding excitatory conductance change. Thus, moving a stimulus in the preferred direction would give rise to a fast and relatively short-lasting EPSP following by a long-lasting shunting inhibition. Movement in the nonpreferred direction, even for very slowly moving stimuli, would activate the long-lasting inhibition, thereby shunting all EPSPs.

Different synaptic time-courses are one way of achieving an effective delay between excitation and inhibition. There are two other nonexclusive alternatives. It is widely held that inhibition in area 17 is never mediated directly by geniculate afferents, but requires an interneuron (see, for instance, Ferster and Lindström, 1983). As a result inhibition will always be somewhat delayed with respect to monosynaptic excitation. Second, an equivalent delay might be introduced by lowpass filtering in the inhibitory channel or by highpass filter (i.e. a derivate-like operation; see Koch, 1984) in the excitatory channel. Interestingly, Movshon, Thompson and Tolhurst (1978) inferred from their measurement of cat simple cells with odd-symmetric receptive fields a time lag between these two regions of about 25 ms (see also Maffei and Fiorentini, 1973).

The distribution of excitatory and inhibitory synapses onto the direction-selective cell is assumed to follow the on-the-path condition, i. e. excitatory synapses are expected to be localized on small distal dendrites and perhaps spines, while inhibitory synapses would be preferentially situated more proximally to the soma. Such a synaptic distribution is commonly observed in the cortex (for a review, see Colonnier, 1981). One intriguing possibility is that dendritic spines might be the specialized sites for the synaptic veto operation to take place. It has been repeatedly reported that 5 to 20 per cent. of all spines carry symmetrical and asymmetrical synaptic profiles on the same spine (see, for instance, Jones and Powell, 1969; Sloper and Powell, 1979). Such an arrangement can be used to perform a highly tuned temporal discrimination operation, essentially without influencing the rest of the neuron (Koch and Poggio, 1983). With a fast excitatory and a much slower inhibitory conductance change simultaneously occurring on the same spine, inhibition will effectively veto excitation if it sets in before the start of excitation (nonpreferred direction). Activating the inhibition some fraction of a millisecond after the start of excitation will not influence excitation to any significant degree (preferred direction).

A frequently raised question is whether the biophysical mechanisms for orientation and direction selectivity in the visual cortex are the same (see, for instance, Schiller, Finlay and Volman, 1976b). In the modeling of neurons which display both phenomena one has to start from the premise that a visual stimulus has to move in the right direction with the right orientation to elicit a maximal response from the cell. The synaptic veto mechanism may be used to detect oriented edges or zero-crossings within the receptive field of the cortical cell. As suggested earlier (Poggio, 1982), both operations could then be performed simultaneously using the same circuitry. The simple cell shown in Fig. 1 will respond (independently of contrast) as long as the stimulus is vertically orientated and moving from left to right. If one or more excitatory geniculate neurons facing a row of inhibitory geniculate neurons (inhibitory with respect to their indirect effect on the simple cell) converge onto the simple cell, the selectivity of the orientation tuning will be enhanced, (see Spitzer and Hochstein, 1985, for evidence about this kind of receptive field organization).

Another possibility is the combination of our synaptic veto mechanism with the scheme studied by Nielsen (1983) using linear addition and subtraction of geniculate on- or off-center neurons to model orientation selectivity. Orientation and direction selectivity could be performed within different, electrically independent parts of the dendritic tree, for instance in the apical tree and in the basal dendrites of a pyramidal cell as suggested by Schiller, Finlay and Volman, (1976b). EPSPs from both subunits would be required to elicit a spike. Because the detection of orientation and direction of motion should be independent of stimulus contrast, some contrast gain control is needed to avoid an activation of the cell by a high-contrast stimulus, independently of the direction of motion. In another scenario, both operations are computed throughout the neuron. The inhibition for direction selectivity would be of the shunting type, localized on proximal dendrites, while the linear summating excitation and inhibition underlying orientation selectivity (this inhibition should have a reversal potential much more negative than the resting potential of the cell) is preferrentially situated on distal dendrites. If the stimulus does not move in the preferred direction, the shunting inhibition will effectively block the output of the linear operation, independent of the stimulus orientation. This circuitry presupposes two different types of inhibitory synapses on the same cell, an arrangement known to occur in the LGN of the cat (Wilson, Friedlander and Sherman, 1984). Together with L. Mistler we are currently developing computer simulations of cortical cells to evaluate these different models.

ACKNOWLEDGEMENTS

We would like to thank E. Hildreth, S. Hochstein, J. Maunsell, L. Mistler and A. Yuille for much constructive criticism.

REFERENCES

Abelson, H. (1978). Towards a theory of local and global in computation. *Theoret. Comp. Sci.,* **6**, 41-67.

Barlow, H. B. (1981). Critical limiting factors in the design of the eye and visual cortex. *Proc. Roy. Soc. Lond., B,* **212**, 1–34.

Barlow, H. B., and Levick, W. R. (1965). The mechanism of directionally selective units in rabbit's retina. *J. Physiol.,* **178**, 477–504.

Bishop, P. O., Coombs, J. S., and Henry, G. H. (1971a). Responses to visual contours: spatio-temporal aspects of excitation in the receptive fields of simple striate neurons. *J. Physiol.,* **219**, 625–657.

Bishop, P. O., Coombs, J. S., and Henry, G. H. (1971b). Interaction effects of visual contours on the discharge frequency of simple striate neurons. *J. Physiol.,* **219**, 659–687.

Boycott, B. B., and Wässle, H. (1974). The morphological types of ganglion cells of the domestic cat's retina. *J. Physiol.,* **240**, 375–419.

Bullock, T. H. (1959). Neuron doctrine and electrophysiology. *Science,* **129**, 997–102.

Colonnier, M. (1981). The electron-microscopic analysis of the neuronal organization of the cerebral cortex. In *The Organization of the Cerebral Cortex* (Eds. F. O. Schmitt, F. G. Worden, G. Adelman and S. G. Dennis), MIT Press Cambridge, Mass., pp. 125–152.

Dingledine, R., and Langmoen, I. A. (1980). Conductance changes and inhibitory actions of hippocampal recurrent IPSPs. *Brain Res.,* **185**, 277–287.

Dow, B. M. (1974). Functional classes of cells and their laminar distribution in monkey visual cortex. *J. Neurophysiol.,* **37**, 927–946.

Ferster, D., and Lindström, S. (1983). An intracellular analysis of geniculo-cortical connectivity in area 17 of the cat. *J. Physiol.,***342**, 181–215.

Freed, N. A., and Sterling P. (1983). Spatial distribution of input from depolarizing cone bipolars to dendritic tree of on-center alpha ganglion cell. *Neurosci. Abst.,* **9**, 806.

Gold, M. R., and Martin, A. R. (1983). Inhibitory conductance changes at synapses in the lamprey brainstem. *Science,* **221**, 85–87.

Goodwin, A. W., and Henry, G. H. (1975). Direction selectivity of complex cells in a comparison with simple cells. *J. Neurophysiol.,* **38**, 1524–1540.

Goodwin, A. W., Henry, G. H., and Bishop, P. O. (1975). Direction selectivity of simple striate cells: properties and mechanism. *J. Neurophysiol.,* **38**, 1500–1524.

Hubel, D. H., and Wiesel, T. N. (1959). Receptive fields of single neurons in the cat's striate cortex. *J. Physiol.,* **148**, 574–591.

Hubel, D. H., and Wiesel, T. N. (1962). Receptive fields, binocular interaction and functional architecture in the cat's visual cortex. *J. Physiol.,* **160**, 106–154.

Jones, E. G., and Powell, T. P. S. (1969). Morphological variations in the dendritic spines of the neocortex. *J. Cell Sci.,* **5**, 509–529.

Kleene, S. C. (1956). Representation of events in nerve nets and finite automata. In *Automata Studies* (Eds. C. E. Shannon and J. McCarthy), Princeton University Press, Princeton, pp. 3–41.

Koch, C. (1984). Cable theory of neurons with active, linearized membranes. *Biolog. Cybernetics,* **50**, 15-33.

Koch, C. and Poggio, T. (1983). A theoretical analysis of electrical properties of spines. *Proc. Roy. Soc. Lond. B,* **218**, 455–477.
Koch, C., and Poggio, T. (1985). A simple algorithm for solving the cable equation in dendritic trees of arbitrary geometry. *J. Neurosci. Methods,* **12**, 303–315.
Koch, C., Poggio, T., and Torre, V. (1982). Retinal ganglion cells: a functional interpretation of dendritic morphology. *Phil. Trans. Roy. Soc. Lond. B,* **298**, 227–264.
Koch, C., Poggio, T., and Torre, V. (1983). Nonlinear interaction in a dendritic tree: localization, timing and role in information processing. *Proc. Natl. Acad. Sci. USA,* **80**, 2799–2802.
LeVay, S., and Gilbert, C. D. (1976). Laminar patterns of geniculocortical projection in the cat. *Brain Res.,* **113**, 1–19.
Maffei, L., and Fiorentini, A. (1973). The visual cortex as a spatial frequency analyser. *Vision Res.,* **13**, 1255–1267.
Malpeli, J. G., Schiller, P. H., and Colby, C. L. (1981). Response properties of single cells in monkey striate cortex during reversible inactivation of individual lateral geniculate laminae. *J. Neurophysiol.,* **46**, 1102–1119.
Marr, D., and Poggio, T. (1977). From understanding computation to understanding neural circuitry. *Neurosci. Res. Prog. Bull.,* **15**, 470–488.
Marr, D., and Ullman, S. (1981). Directional selectivity and its use in early visual processing. *Proc. Roy. Soc. Lond. B,* **211**, 151–180.
Maunsell, J. H. R., and Van Essen, D. C. (1983). Functional properties of neurons in middle temporal visual area of the macaque monkey. I. Selectivity for stimulus direction, speed and orientation. *J. Neurophysiol.,* **49**, 1127–1147.
Mead, C., and Conway, L. (1980). *Introduction to VLSI Systems,* Addison-Wesley, Reading, Mass.
McCulloch, W. S., and Pitts, W. (1943). A logical calculus of the ideas imminent in nervous activity. *Bull. Math. Biophys.,* **5**, 115–133.
Movshon, J. A., Thompson, I. D., and Tolhurst, D. J. (1978). Spatial summation in the receptive fields of simple cells in the cat's striate cortex. *J. Physiol.,* **283**, 53–77.
Nielsen, D. E. (1983). A functional model of the wiring of the simple cells of visual cortex. *Biolog. Cybernetics,* **47**, 213–222.
Orban, G. A., Kennedy, H., and Maes, H. (1981). Responses to movement of neurons in areas 17 and 18 of the cat: direction selectivity. *J. Neurophysiol.,* **45**, 1059–1073.
Palm, G. (1982). *Neural Assemblies,* Springer-Verlag, Berlin.
Peters, A., Proskauer, C. C., Feldman, M. L., and Kimerer, L. (1979). The projection of the lateral geniculate nucleus to area 17 of the rat cerebral cortex. V. Degenerating axon terminals synapsing with Golgi impregnated neurons. *J. Neurocytol.,* **8**, 331–357.
Poggio, T. (1982). Visual algorithms. In *Physical and Biological Processing of Images.* (Eds. O. J. Braddick and A. C. Sleigh), Springer-Verlag, Berlin, pp. 128–153; also published as Artificial Intelligence Laboratory Memo No. 683, Massachusetts Institute of Technology, Cambridge, Mass.
Poggio, T., and Torre, V. (1978). A new approach to synaptic interactions. In *Approaches to Complex Systems.* (Eds. R Heim, and G. Palm), Spring-Verlag, Berlin, pp. 89–115.
Rall, W. (1964). Theoretical significance of dendritic trees for neuronal input–output relations. In *Neural Theory and Modeling* (Ed. R. Reiss, Stanford University Press, Stanford, pp. 73–97.

Raymond, S. A. (1979). Effects of nerve impulses on threshold of frog sciatic nerve fibers. *J. Physiol.,* **290**, 273–303.
Schiller, P. H. (1982). Central connections of the retinal ON and OFF pathways. *Nature,* **297**, 580–583.
Schiller, P. H., Finlay, B. L., and Volman, S. F. (1976a). Quantitative studies of single-cell properties in monkey striate cortex. I. Spatiotemporal organization of receptive fields. *Neurophysiol.,* **39**, 1288–1319.
Schiller, P. H., Finlay, B. L., and Volman, S. F. (1976b). Quantitative studies of single-cell properties in monkey striate cortex. II. Orientation specificity and ocular dominance. *J. Neurophysiol.,* **39**, 1320–1333.
Schiller, P. H., Finlay, B. L., and Volman, S. F. (1976c). Quantitative studies of single-cell properties in monkey striate cortex. V. Multivariate statistical analyses and models. *J. Neurophysiol.,* **39**, 1362–1374.
Schmitt, F. O., Dev, P., and Smith, B. H. (1976). Electrotonic processing of information by brain cells. *Science,* **193**, 114–120.
Segev, I., and Parnas, I. (1983). Synaptic integration mechanisms. Theoretical and experimental investigation of temporal postsynaptic interactions between excitatory and inhibitory inputs. *Biophys. J.,* **41**, 41–50 (1983).
Sherk, H., and Horton, J. C. (1984). Receptive-field properties in the cat's area 17 in the absence of on-center geniculate input. *J. Neurosci.,* **4**, 374–380.
Sillito, A. M. (1977). Inhibitory processes underlying the directional specificity of simple, complex and hypercomplex cells in the cat's visual cortex. *J. Physiol.,* **271**, 699–720.
Sillito, A. M., Kemp, J. A., Milson, J. A., and Berardi, N. (1980). A re-evaluation of the mechanism underlying simple cell orientation selectivity. *Brain Res.,* **194**, 517–520.
Sloper, J. J., and Powell, T. P. S. (1979). An experimental electron microscopic study of afferent connections to the primate motor and somatic sensory cortices. *Phil. Trans. Roy. Soc. Lond. B,* **285**, 199–226.
Spitzer, H., and Hochstein, S. (1985). Simple and complex cell response depends on stimulation parameters. *J. Neurophysiol.* (in press).
Swindale, N. V. (1983). Anatomical logic of retinal nerve cells. *Nature,* **303**, 570–571.
Torre, V., and Poggio, T. (1978). A synaptic mechanism possibly underlying directional selectivity to motion. *Proc. Roy. Soc. Lond. B,* **202**, 409–416.
Wilson, J. R., Friedlander, M. J., and Sherman, S. M. (1984). Fine structural morphology of identified X- and Y-cells in the cat's lateral geniculate nucleus. *Proc. Roy. Soc. Lond. B,* **221**, 411–436.

Models of the Visual Cortex
Edited by D. Rose and V. G. Dobson

CHAPTER 44

A proposed mechanism and site for cortical directional selectivity

ROBERT C. EMERSON
Center for Visual Science, University of Rochester, Rochester, NY 14627, USA

MARK C. CITRON
Children's Hospital of Los Angeles, Neurology Research, PO Box 54700, Los Angeles, CA 90054, USA

DANIEL J. FELLEMAN* and JON H. KAAS
Department of Psychology, Vanderbilt University, 134 Wesley Hall, Nashville, TN 37240, USA

Objects moving in the visual world of an animal have great behavioural significance for the survival and success of that animal. Neural analysis of the direction of image movement appears to depend on directional selectivity (DS) a tendency for a visual neuron to respond to one direction of image movement through its receptive field (RF) but not to the opposite direction. In some animals, e. g. the rabbit, DS occurs as far peripherally in the nervous system as in the retina (Barlow and Levick, 1965). However, in the geniculocortical pathway of the cat and primate, DS does not occur with any regularity until the level of the visual cortex (Hubel and Wiesel, 1961, 1962, 1968).

Mechanisms underlying this useful RF property have been studied with great interest, beginning with work in the rabbit retina in 1963 by Barlow and Hill, and then in 1965 by Barlow and Levick. More recent literature on DS in the cortex of the cat includes work of Pettigrew, Nikara and Bishop (1968), Goodwin, Henry and Bishop (1975), Emerson and Gerstein (1977b) and Emerson and Coleman (1981), the latter two emphasizing the depen-

* Presently at the Department of Biology 216-76, California Institute of Technology, Pasadena, CA 91125, USA.

dence of mechanisms on nonlinear interactions. At present, we lack a detailed description of the temporal and spatial requirements of directional mechanisms, the exact functional nature of the apparent nonlinear transformation and the anatomical substrate.

Here we show for a DS cortical unit the distribution of nonlinear interactions around a single RF position, and a proposed membrane mechanism that can account for order detection. We also propose that dendritic spines provide the membrane site for the required nonlinear interaction.

METHODS AND RESULTS

Experiments involved single-unit measurements in area 17 of lightly anesthetized cats. Details of the physiological preparation and receptive field classification have been reported previously (Emerson and Gerstein, 1977a; Citron, Emerson and Ide, 1981; Emerson and Coleman, 1981). Spatial nonlinear interactions were measured with a new spatiotemporally random multiple-stripe stimulus that has been described recently (Citron and Emerson, 1983) and that was presented at the optimal orientation. Finally, we have illustrated the morphology of a DS spiny stellate cell by first studying its RF properties extracellularly and then injecting the cell iontophoretically with horseradish peroxidase (HRP) for subsequent morphological analysis (Adams, 1977).

Measurements of Spatial RF Interactions

At the top left of Fig. 1a is shown the responsivity for the RF of a striate complex cell, as represented by the response to onset and removal of a 512 ms bright bar at each position from 4 through 14. These values are the summed peak amplitudes of on- and off- responses shown for the same unit in Fig. 1B of Citron and Emerson (1983). Pairs of bright bars were then presented, as shown on the bottom left of Fig. 1a. An initial 16 ms bar presentation at position 6 through 12 was followed in 32 ms by a second 16 ms bar presentation at central position 9. After measuring the response to a single 16 ms bar presentation at each position (not shown), we calculated the pair response expected from a linear cell by adding the component responses, once they had been shifted a temporal interval equal to the stimulus separation. We then subtracted this superposition from the measured paired-bar response to assess the time-course of nonlinear suppression or facilitation caused by the second bar. This correction to the linear expectation is shown on the right of Fig. 1a, along with the time-course of the two stimulus presentations (below).

Note that, although successive bar presentations at central position 9 elicited only a weak negative nonlinear correction, the 10,9 sequence required a strong negative correction and the 8,9 sequence required a strong positive

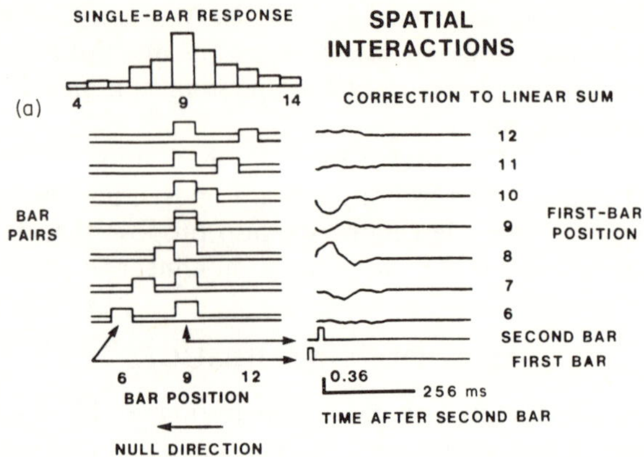

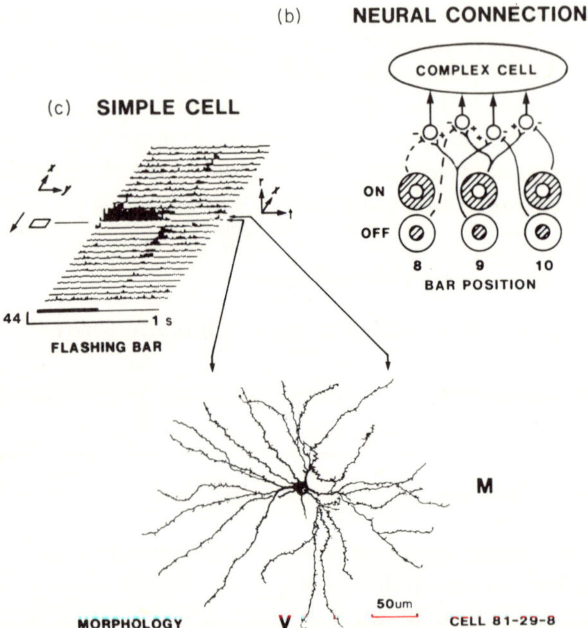

FIGURE 1 Physiology and anatomy of cortical directional selectivity (DS). (a) Spatial nonlinear interaction in the receptive field (RF) of a DS complex neuron. Retinal eccentricity was 12°, and bar widths and separations were multiples of 0.5°. Top left: averaged amplitude of response to onset and removal of a 512 ms bright bar at each of positions 4 through 14. Bottom left: seven combinations of two 16-ms bars used to test spatial RF interactions. Lower stimulus of each pair was presented 32 ms before the upper stimulus was presented at central position 9, as shown by stimulus markers below graphs in the right-hand portion of the figure. Right-hand graphs show the second-order nonlinear correction that would need to be added to the linear sum of responses to the two individual bars to obtain the response to the paired-bar

correction, in each case dimishing by about 150ms after the second bar was presented. The negative correction for the 10,9 sequence is evidence for the operation of a strong nonlinear suppressive interaction in the null direction of this RF. This suggestion agrees with the results of other physiological measurements of DS. Directional selectivity usually depends on suppression for the null direction rather than on facilitation for the preferred direction (e.g., Barlow and Levick, 1965; Goodwin, Henry and Bishop, 1975; Emerson and Coleman, 1981). We will return shortly to the apparent facilitation for the 8,9 sequence.

The rapid decline of either positive or negative interactions at distances greater than one bar position (i.e. more than 0.5°) from position 9 can be contrasted with the gradual decrease in response to a single bar shown in the top left of Fig. 1a. This restricted range of spatial interactions shows that DS depends on local spatial interactions. This value compares favorably with that of 0.63° found by Movshon, Thompson and Tolhurst (1978) for surround-type inhibition in nondirectional complex units, and of 0.25 to 0.75° for facilitation in DS units that were tested with pairs of bars whose onset was separated by more than 50 ms. The typical uniform distribution of directional selectivity across a cortical RF suggests that this global property depends on families of similar subunits that are distributed across the RF, as first suggested by Barlow and Levick in 1965 for ganglion cell RFs in the rabbit

---

stimulus. Amplitude of ordinate calibration (0.36 spikes/s), below, is small because of the short stimulus duration and low gain in a 'conditioned' experiment of this kind. (b) Neural connections proposed to account for nonlinear interactions of (a). Small circles below the complex cell represent points of nonlinear interaction between excitatory (+) and inhibitory (−) inputs. These inputs are specific to a position in the RF, as shown by numbers below, and come from an LGN unit with either an on-center (ON) or off-center (OFF) RF. Negative lateral interactions shown in solid lines between positions 10 and 9 account for the suppressive interaction shown at position 10 of (a). Dotted lines from position 8 are proposed to underlie the positive correction shown at position 8 of (a), because of the way the correction was calculated (see text). (c) Morphology of a DS simple cell. Retinal eccentricity was 2.5°. Top left: responses to a bright bar flashed for 512 ms at a sequence of positions across the RF. Positions were separated by 0.38°. Lowest histogram shows the stimulus-free (background) response, below that the time course of the stimulus at each position, and below that the responsivity (44 spikes/s) and time calibrations. The strong sustained on-response during presentations of the bar at central positions and the brief off-response following presentations of the bar at peripheral positions gives this RF the appearance of a typical ON-LGN unit, except for the strong DS designated by the arrow next to the hand-plotted RF at the far left. Below and to the right is shown the spiny stellate morphology of this cell injected iontophoretically with HRP. Arrows compare the broadside direction in the RF with the projected medial–lateral direction in the cortex. M, medial; V, ventral. Its appearance at the layer 3–layer 4 border suggests that it is probably transforming nondirectional LGN input into directional cortical output (see text).

retina. The interactions for the 8,9 sequence (shown here in Fig. 1a) are similar to those for the 9,10 sequence (shown in Fig. 3D of Citron and Emerson, 1983), which suggests that families of subunits do indeed operate in the complex cell shown here.

A Functional Model of Cortical Directional Selectivity

A cortical incarnation of the Barlow and Levick (1965) neural model that is simplified to explain only the nonlinear data of Fig. 1a is shown in Fig. 1b. The small circles below the complex cell are the sites of nonlinear interaction, the AND-NOT gates of Barlow and Levick (1965). They feed their outputs in some as yet unspecified way to sum, linearly, at the complex cell above. Inputs to the points of interaction, presumably taking origin from the lateral geniculate nucleus (LGN), come from below. They are numbered separately to designate three positions of interest in the data of Fig. 1a, and are segregated according to polarity, with schematic on-center (ON) LGN inputs (with bright RF centers) originating in a layer above and off-center (OFF) inputs (with shaded RF centers), below. Positive (excitatory) inputs are designated in the model by a Y-shaped terminal with an adjacent plus sign. For illustration purposes they are shown here originating only from position 9, because the second (excitatory) stimulus was always presented there. Inputs that can gate or veto the excitatory input to the circles, designated in the model by a flat terminal with an adjacent minus sign, arrive at the point of interaction after traveling laterally in the model a schematic distance corresponding to some lateral shift in the RF—here the distance between adjacent bars.

The model of Fig. 1b incorporates two kinds of suppressive lateral information transfer. Leftward transfers correspond to the null direction (designated by solid lines), while rightward transfers correspond to the preferred direction (designated by broken lines). For the null direction, the rightmost solid line shows the path over which ON inputs at RF position 10 can suppress the excitatory influence from ON inputs at position 9. The resulting interaction explains the negative correction at position 10 of Fig. 1a.

However, the actual stimulus sequence that was used to calculate the spatial corrections of Fig. 1a included more than the bright-bar pairs suggested by the stimulus markers below. To assess directional effects that would survive a reversal of stimulus polarity (e.g., bright to dark), we included as many dark-bar pair presentations as bright-bar pair presentations, at the same positions, with the same relative time sequence, and interleaved randomly with the bright bars. Responses to these dark-bar pairs were averaged into the correction, thereby cancelling the linear components, which automatically performed the subtraction of the linear superposition described above. The existence of suppressive effects at position 10 of Fig. 1a in an experiment that includes dark- as well as bright-bar pairs implies a probable contribution along the second solid lateral path shown from OFF inputs at position 10 onto transmission of OFF effects from position 9 in Fig. 1b. This connection is justified on the basis that complex cells always show the same preferred direction for bright and dark stimuli (Goodwin and Henry, 1975) and therefore depend on similar mechanisms for each polarity. The association of LGN ON RFs in the model with bright-bar response effects in Fig. 1a and of OFF RFs with dark-bar effects follows from the dominance of center responses in LGN RFs, even for bar stimuli (Citron, Emerson and Ide, 1981), and from the high threshold of most cortical units, which discriminates against surround effects.

Returning to the subject of facilitation, it is theoretically possible that a DS neuron may show nonlinear facilitation for ordered sequences in the preferred direction as well as nonlinear suppression for sequences in the null direction. However, the lack of an appropriate membrane mechanism (see below) and the paucity of physiological data supporting facilitation have forced us to consider alternative explanations for the positive correction shown in Fig. 1a. It is clear that this effect cannot be explained by a purely temporal nonlinearity, such as a threshold (Mancini, Madden and Emerson, 1982), because it is specific to the 8,9 sequence.

Because the corrections shown in Fig. 1a are of second order and therefore include the product of two luminance values, one at each stimulus site (Citron, Kroeker and McCann, 1981), they harbor an ambiguity in sign of the physiological interaction. We suggest that the alternative to facilitation between pairs of bright or dark bars is operating here, i.e. suppression between pairs that include one bright and one dark bar.

Neural connections that are compatible with this interpretation of the physiological data are shown in Fig. 1b as the broken, rightward-directed paths between positions 8 and 9. Suppressive lateral interactions must occur between OFF inputs at position 8 and ON inputs at position 9, and between ON inputs at position 8 and OFF inputs at position 9. Further evidence for this interpretation will be considered in the next section.

A Proposed Membrane Mechanism

Membrane mechanisms that might underlie the proposed suppressive interactions are a subject of considerable interest. In 1978 Torre and Poggio proposed a model to explain suppressive effects in directionally selective neurons of the rabbit retina. Nonlinear suppression occurs in their model because the suppressing, 'hyperpolarizing' input tends to clamp the membrane potential at its resting level, below the firing threshold, such that the normally excitatory, 'depolarizing' input cannot exert its influence. This mechanism, if it occurred in an isolated patch of membrane, would produce a multiplicative interaction, in the sense that the suppressing input would multiply the effect of the excitatory input by a number approaching zero. We avoid the term 'subtractive' because of its linear implication. The second-order correction, or second-order 'kernel' (Citron and Emerson, 1983), of Fig. 1a captures well the essence of such a multiplicative nonlinearity, because, as mentioned above, it depends heavily on the product of the stimulus luminances at the two positions of interest. Since this correction provides a consistent preferred direction for both positive and negative contributions, it provides indirect evidence for the operation of Torre and Poggio's mechanism in DS cortical units. Koch, Poggio and Torre (1982) and Koch and Poggio (Chapter 43 in this volume) provide convincing arguments for the utility of such a membrane mechanism in both the retina and the cortex.

Anatomical Studies of Suppressive Interactions

We have been correlating physiology with anatomy by using HRP-filled micropipettes to characterize extracellularly the RF properties of single units in the striate cortex of the cat. In successful cases, we inject these same cells with peroxidase for subsequent morphological analysis. An example of results is shown in Fig. 1c. At the top left is a series of poststimulus time histograms that shows the average response to a bright bar presented at each position in a closely spaced array for a duration shown by the heavy marker below. At the left is the position and shape of the region found responsive to a moving narrow bright bar, but only when the bar was moved in the downward left direction (note arrow). The ON center and OFF flanking regions, along with responses to moving edges (not shown), allow us to classify this unit not only as unambiguously simple, but also as having an RF very similar to that of an ON-LGN unit (Citron, Emerson and Ide, 1981). Subsequent intracellular staining and recovery of this cell, as shown in the lower portion of Fig. 1c, showed that it was nonpyramidal in morphology, and that its soma lay at the layer 3-layer 4 boundary where a large proportion of axons from the LGN terminate onto dendritic spines (LeVay and Gilbert, 1976). The rough appearance of dendrites classifies it as a spinous, spherical, multipolar neuron of Peters and Regidor (1981), and its appearance agrees with that of 'spiny stellates' found in layer 4 of Golgi preparations (Lund *et al.*, (1979) and in HRP injections of Gilbert and Wiesel (1979) and Lin, Friedlander and Sherman (1979).

A question that has puzzled us is why a cortical unit with an RF very similar to that of an LGN unit (Citron, Emerson and Ide, 1981)—or at most like two or three LGN RFs in a line—requires a 350 μm diameter dendritic tree with thousands of spines on which to form synaptic contacts from terminals of incoming LGN axons. The only RF 'complexity' (any property that can not be predicted with flashing spots) that we could find to correlate with the complexity of the dendritic tree in this cell was its strong DS for bright bars. Since there is no DS in the A laminae of the LGN, which project to layer 4 of the visual cortex (Ferster and LeVay, 1978; Lund *et al.*, 1979), and the cell of Fig. 1c represents the first layer of cortical processing, the cell must generate DS within its structure.

Although provisional at this preliminary stage of our physiological–morphological correlation, it seems to us an appropriate time to propose the dendritic spine as the primary site of nonlinear interaction underlying DS in the visual cortex. This proposed site conforms to the requirement of Torre and Poggio (1978) that the interaction should occur on a patch of membrane that is isolated from the rest of the neuron. Peters and Kaiserman-Abramof (1970) have shown that most visual cortical dendritic spines are of the thin (0.1 to 0.2 μm) type, with a slender stalk, averaging 1.1 μm in length, and a bulb at the end averaging 0.6 μm in diameter.

Our measurements of RF interactions are from a directionally selective complex cell and our morphological evidence is from a DS simple cell. However, recent evidence (e.g. Henry, Harvey and Lund, 1979; Toyama, Kimura and Tanaka, 1981) suggests that a large proportion of complex cells receive direct input from the LGN. Therefore, DS complex cells, whose morphology is strongly correlated with the pyramidal class (Gilbert and Wiesel, 1979; Lin, Friedlander and Sherman, 1979), could just as easily depend on dendritic spines of pyramidal cells as simple cells could depend on the demonstrated spines of the associated stellate or pyramidal classes. It seems to us prudent to examine the extent to which these first-order cortical neurons can generate nonlinear RF properties like DS within their structure (as in Fig. 1b) before being forced to conclude that complicated networks of cortical neurons are necessary. The latter approach does not answer the question of where and how DS of extraordinary spatial resolution can occur.

DISCUSSION AND CONCLUSIONS

Physiology

We have concluded here that directional selectivity depends on a single local suppressive mechanism, probably distributed in a family of subunits across the RF. Exceptional examples in the literature showing nonlinear facilitation (Emerson and Gerstein, 1977b; Movshon, Thompson and Tolhurst, 1978) may depend on threshold effects before the measured neuron, or on some other mechanism we have not considered here.

Our conclusion regarding suppressive interactions between ON and OFF inputs that project in a direction opposite from those involving pairs of ON inputs or pairs of OFF inputs was an unexpected finding, but one required by the paucity of demonstrated facilitation and the lack of an appropriate facilitative membrane mechanism. Although these suppressive interactions would not be expected to aid DS under normal circumstances, they at least do not hinder it under the normal circumstance where the image does not change its luminance drastically as it moves across the RF. These connections may represent a residue from an earlier developmental period, when inappropriate connections were pared from an initially random set of lateral interactions. Evidence that such a mechanism may operate, even in the human visual system, is provided by Anstis and Rogers (1975) and Anstis and Cavanagh (1983), who showed that reversal of stimulus contrast can reverse the direction of apparent (phi) movement. This result would follow from the opposite direction of transmission for lateral suppressive interactions along the dotted pathways of Fig. 1b, as compared with those along the solid pathways.

Hypothetical connections in Fig. 1b suggest a specific physiological test. Stepwise movement sequences covering only two RF positions have been shown to elicit directional selectivity in the rabbit's retina (Barlow and Levick, 1965) and in the cat's cortex (Emerson and Gerstein, 1977b). Therefore, reversal of the contrast of the second stimulus with respect to that of the first should reverse the preferred direction. Our most recent evidence from time-locked histograms (Emerson and Citron, 1984) suggests that in the present unit contrast reversal of either stimulus does reverse the direction of the preferred sequence. Furthermore, the effect depends on spatially specific suppression surrounded on all sides by facilitation. All cortical units show this facilitation, presumably because of their high thresholds (Mancini, Madden and Emerson, 1982).

Anatomy

Our proposal of the dendritic spine as a site for nonlinear interactions underlying DS, although somewhat speculative, is not a totally new one, since Torre and Poggio (1978) mention the spine as a possible site for nonlinear interaction. Before this, Diamond, Gray and Yasargil (1970) proposed a similar membrane function for the dendritic spine in the context of the fish spinal cord. Later Koch and Poggio (1983, and Chapter 43 in this volume) modeled the nonlinear advantages of synaptic interactions occurring on spines, but fell short of proposing the spine as *the* site for DS. By committing ourselves to a definite structural entity, we have perhaps made it easier to examine the consequences of such a site for DS.

One of those consequences is that the Torre and Poggio mechanism would require that dendritic spines engaging in this interaction have at least two synapses. The synapses would probably use different neurotransmitters and presumably exhibit distinguishable morphology because of purported differences observed between excitatory and inhibitory synapses (Peters, Palay and Webster, 1976; Colonnier, 1981). In a three-dimensional reconstruction of dendritic spines from the rat neostriatum, Wilson *et al.* (1983) have found that 8 per cent. of these spines exhibit a second synapse. One synapse is always excitatory (having asymmetrical membrane thickenings and round vesicles) and the other is always inhibitory (having symmetrical membrane thickenings and flat vesicles). There was never a case of two synapses of the same kind occurring on a dendritic spine, nor were single synapses on spines ever of the inhibitory type. A similar picture has been shown by Colonnier (1981) for the visual cortex of the cat, except that proportions of double synapses onto spines were not given. It remains to be seen whether the proportion of double-synaptic spines is higher in the visual cortex of the cat than in the rat neostriatum, and whether this feature is correlated with nonlinear RF interactions.

A second consequence of the spine hypothesis is that excitatory inputs to the cortex from the LGN must be inverted to provide the needed suppression to the spines, since inhibitory influences are thought not to be transmitted directly from the LGN to the cortex. The probable cellular sources are nonpyramidal cells (Peters and Regidor, 1981) with smooth or sparsely spined dendrites. To preserve the resolution inherent in the geniculate input without requiring a separate inhibitory interneuron for each interaction site, a single interneuron could be assigned to each LGN input and in turn distribute suppressive influences to subunits of all nearby cortical cells that require information about that point in visual space. The spine, with its isolated structural nature, seems to be the site of choice for the postsynaptic interaction. On the target neuron many sites can be accommodated without an increase in cell count and their effects can be summed linearly, as the sites are part of the dendritic structure of the neuron itself.

We have shown physiological evidence for multiplicative interactions being involved in DS, and invoked a specific membrane mechanism to account for them. We have also demonstrated an instance of dendritic spines in a neuron that was presumably generating DS responses from nondirectional inputs. Perhaps we have identified a practicable candidate for the elusive function of the dendritic spine (cf. Swindale, 1981; Gray, 1982). More importantly, if DS proves to depend solely on the proposed suppressive mechanism, we may have identified the fundamental logical gating function with which the brain performs orientation selectivity, size selectivity and other selective tasks leading eventually to pattern recognition.

ACKNOWLEDGEMENTS

We thank Drs. Alan Peters and Robert Chapman for helpful suggestions. This research was supported in part by Research Career Development Award

(RCDA) EY00081 and research grant EY04630 to R. C. E., Center Support Grant EY01319 to the Center for Visual Science, and RCDA EY00250 and research grant EY04711 to M.C.C.

REFERENCES

Adams, J. C. (1977). Technical considerations in the use of horseradish peroxidase as a neuronal marker. *Neuroscience,* **2**, 141.

Anstis, S. M., and Cavanagh, P. (1983). A minimum motion technique for judging equiluminance. In *Colour Vision,* (Eds.J.D. Mollon and L. T. Sharpe), Academic Press, London.

Anstis, S. M., and Rogers, B. J. (1975). Illusory reversal of visual depth and movement during changes of contrast. *Vision Res.,* **15**, 957–961.

Barlow, H. B., and Hill, R. M. (1963). Selective sensitivity to direction of movement in ganglion cells of the rabbit retina. *Science,* **139**, 412–414.

Barlow, H. B., and Levick, W. R. (1965). The mechanism of directionally selective units in rabbit's retina. *J. Physiol.,* **178**, 477–504.

Citron, M. C., and Emerson, R. C. (1983). White noise analysis of cortical directional selectivity in cat. *Brain Res.,* **279**, 271–277.

Citron, M. C., Emerson, R. C., and Ide, L. S. (1981). Spatial and temporal receptive-field analysis of the cat's geniculocortical pathway. *Vision Res.,* **21**, 385–396.

Citron, M. C., Kroeker, J. P., and McCann, G. D. (1981). Nonlinear interactions in ganglion cell receptive fields. *J. Neurophysiol.,* **46**, 1161–1176.

Colonnier, M. (1981). The electron-microscopic analysis of the neuronal organization of the cerebral cortex. In *The Organization of the Cerebral Cortex,* (Eds. F. O. Schmitt, F. G. Worden, G. Adelman and S. G. Dennis) MIT Press, Cambridge, Mass., pp. 125–152.

Diamond, J., Gray, E. G., and Yasargil, G. M. (1970). The function of the dendritic spine: an hypothesis. In *Excitatory Synaptic Mechanisms (Eds. P. Anderson and J. K. S. Jansen), Universitet Forlaget, Oslo, pp. 213–222.*

Emerson, R. C., and Citron, M. (1984). Suppressive interactions for the preferred direction in directionally selective cortical cells. *Invest. Ophthalmol. Vis. Sci., Suppl.,* **25**, 227.

Emerson, R. C., and Coleman, L. (1981). Does image movement have a special nature for neurons in the cat's striate cortex? *Invest. Ophthalmol. Vis. Sci.,* **20**, 766–783.

Emerson, R. C., and Gerstein, G. L. (1977a). Simple striate neurons in the cat. I. Comparison of responses to moving and stationary stimuli. *J. Neurophysiol.,* **40**, 119–135.

Emerson, R. C., and Gerstein, G. L. (1977b). Simple striate neurons in the cat. II. Mechanisms underlying directional asymmetry and directional selectivity. *J. Neurophysiol.,* **40**, 136–155.

Ferster, D., and LeVay, S. (1978). The axonal arborizations of lateral geniculate neurons in the striate cortex of the cat. *J. comp. Neurol.,* **182**, 923–944.

Gilbert, C. D., and Wiesel, T. N. (1979). Morphology and intracortical projections of functionally characterised neurons in the cat visual cortex. *Nature,* **280**, 120–125.

Goodwin, A. W., and Henry, G. H. (1975). Direction selectivity of complex cells in a comparison with simple cells. *J. Neurophysiol.,* **38**, 1524–1540

Goodwin, A. W., Henry, G. H., and Bishop, P. O. (1975). Direction selectivity of simple striate cells: properties and mechanism. *J. Neurophysiol.,* **38**, 1500–1523.

Gray, E. G. (1982). Rehabilitating the dendritic spine. *Trends in Neurosci.*, **5**, 5–6.
Henry, G. H., Harvey, A. R., and Lund, J. S. (1979). The afferent connections and laminar distribution of cells in the cat striate cortex. *J. comp. Neurol.*, **187**, 725–744.
Hubel, D. H., and Wiesel, T. N. (1961). Integrative action in the cat's lateral geniculate body. *J. Physiol.*, **155**, 385–398.
Hubel, D. H., and Wiesel, T. N. (1962). Receptive fields, binocular interaction and functional architecture in the cat's visual cortex. *J. Physiol. (Lond.)*, **160**, 106–154.
Hubel, D. H., and Wiesel, T. N. (1968). Receptive fields and functional architecture of monkey striate cortex. *J. Physiol. (Lond.)*, **195**, 215–243.
Koch, C., and Poggio, T. (1983). A theoretical analysis of electrical properties of spines. *Proc. Roy. Soc. Lond. B*, **218**, 455–477.
Koch, C., Poggio, T., and Torre, V. (1982). Retinal ganglion cells: a functional interpretation of dendritic morphology. *Phil. Trans. Roy. Soc. Lond. B.*, **298**, 227–264.
LeVay, S., and Gilbert, C. D. (1976). Laminar patterns of geniculocortical projection in the cat. *Brain Res.*, **113**, 1–19.
Lin, C.-S., Friedlander, M. J., and Sherman, S. M. (1979). Morphology of physiologically identified neurons in the visual cortex of the cat. *Brain Res.*, **172**, 344–348.
Lund, J. S., Henry, G. H., MacQueen, C. L., and Harvey, A. R. (1979). Anatomical organization of the primary cortex (area 17) of the cat: a comparison with area 17 of the macaque monkey. *J. comp. Neurol.*, **184**, 599–618.
Mancini, M., Madden, B. C., and Emerson, R. C. (1982). Temporal nonlinearities in simple cell responses. *Invest. Ophthalmol. Vis. Sci. Suppl.*, **22**, 118.
Movshon, J. A., Thompson, I. D., and Tolhurst, D. J. (1978). Receptive field organization of complex cells in the cat's striate cortex. *J. Physiol. (Lond.)*, **283**, 79–99.
Peters, A., and Kaiserman-Abramof, I. R. (1970). The small pyramidal neuron of the rat cerebral cortex. The perikaryon, dendrites and spines. *Amer. J. Anat.*, **127**, 321–356.
Peters, A., Palay, S. L., and Webster, H. de F. (1976). *The Fine Structure of the Nervous System: The Neurons and Supporting Cells*, W. B. Saunders, Philadelphia.
Peters, A., and Regidor, J. (1981). A reassessment of the forms of nonpyramidal neurons in area 17 of cat visual cortex. *J. comp. Neurol.*, **203**, 685–716.
Pettigrew, J. D., Nikara, T., and Bishop, P. O. (1968). Responses to moving slits by single units in cat striate cortex. *Exp. Brain Res.*, **6**, 373–390.
Swindale, N. V. (1981). Dendritic spines only connect. *Trends in Neurosci.*, **4**, 240–241.
Torre, V., and Poggio, T. (1978). A synaptic mechanism possibly underlying directional selectivity to motion. *Proc. Roy. Soc. Lond. B*, **202**, 409–416.
Toyama, K., Kimura, M., and Tanaka, K. (1981). Organization of cat visual cortex as investigated by cross-correlation technique. *J. Neurophysiol.*, **46**, 202–214.
Wilson, C. J., Groves, P. M., Kitai, S. T., and Linder, J. C. (1983). Three-dimensional structure of dendritic spines in the rat neostriatum. *J. Neurosci.*, **3**, 383–398.

Models of the Visual Cortex
Edited by D. Rose and V. G. Dobson

CHAPTER 45

Role of dendritic fields in orientation selectivity

Suzannah Bliss Tieman and Helmut V. B. Hirsch
Neurobiology Research Center, State University of New York at Albany, Albany, NY 12222, USA

A central assumption in neurobiology is that function follows form. Cells with different shapes are likely to have different functions and vice versa. For example, if the sensory input to a cell is organized topographically, then the structure of a cell's dendritic field may determine its receptive field (e.g. Kirk, Waldrop and Glantz, 1983; Bacon and Murphey, 1984). Thus, elongated dendritic fields have been suggested to generate elongated receptive fields and hence a preference for one orientation (Young, 1960; Colonnier, 1964; Leventhal and Schall, 1983). Colonnier (1964) observed that, in cat visual cortex, the dendritic fields of layer IV stellate cells are elongated in a plane tangential to the pial surface (i.e. in the plane of the retinotopic map) and that the long axis of these fields could be oriented at any angle. He postulated that these elongated dendritic fields produced the elongated receptive fields of the 'simple' cells described by Hubel and Wiesel (1962) (Fig. 1a). His model predicts that the preferred orientation of a simple cell matches the orientation within the cortical map of the long axis of its dendritic field. The changes we observed in the dendritic morphology of layer III pyramidal cells in cats reared viewing lines of only one orientation (stripe-reared cats; see Tieman and Hirsch, 1982) are consistent with a relationship between preferred orientation and dendritic field orientation for these cells.

In cats reared viewing only vertical lines, the dendritic fields of most layer III pyramidal cells are elongated perpendicular to the representation of the vertical meridian, and in cats reared viewing only horizontal lines, the dendritic fields of these cells are elongated parallel to the representation of the vertical meridian (Tieman and Hirsch, 1982). For these cells, dendritic field orientations are thus specifically related to the orientation of the lines presented during rearing. Although these results suggest that there is a relationship between dendritic field orientation and preferred orientation, it is possible that this relationship is an artifact of the stripe-rearing. We have described elsewhere (Tieman and Hirsch, 1982) two ways in which such an artifact might arise, but neither explanation accounts for the specificity of the results, i.e. that the dendritic fields in cats exposed to horizontal lines were in some sense perpendicular to the dendritic fields in cats exposed to vertical lines. Thus, because orthogonal stimuli resulted in orthogonal dendritic fields, we conclude that the elongated dendritic fields are related to the topography of the retinogeniculocortical pathway in stripe-reared cats, and, further, we suggest that the elongated dendritic field of a layer III pyramidal cell is related to the cell's preferred orientation in both normal and stripe-reared cats.

If, in referring to orientations within the cortical map, we use vertical to mean parallel to the representation of the vertical meridian and horizontal to mean perpendicular to the representation of the vertical meridian, then we observed vertical dendritic fields in cats exposed only to horizontal lines and horizontal dendritic fields in cats exposed only to vertical lines. Assuming that the preferred orientations of layer III pyramidal cells, like those of most cells in area 17 of stripe-reared cats (e.g. Blakemore and Cooper, 1970; Hirsch and Spinelli, 1970, 1971; Stryker *et al.*, 1978; Hirsch *et al.*, (1983) match the orientation of the lines presented during rearing, then, for these cells, the orientation of the dendritic field in the cortical map is perpendicular to the cell's preferred orientation. This relationship has recently been assessed directly by Martin and Whitteridge (1984). They recorded and filled 23 cells in visual cortex, and, when considering all layers combined, found no systematic relationship between preferred orientation and dendritic field orientation. However, for 5 of 6 layer III pyramidal cells, the difference between preferred orientation and dendritic field orientation was within 45° of that predicted by our data. Thus, their study of normal cats complements our own work in stripe reared cats, and confirms our suggestion that dendritic field orientaion tends to be perpendicular to preferred orientation for layer III pyramidal cells.

Given this relationship, orientation preferences of layer III pyramidal cells cannot be explained by the model that Colonnier (1964) proposed to account for orientation preferences of layer IV stellate cells. His model requires that preferred orientation match dendritic field orientation (Fig. 1a). In this chapter, we will consider three possible reasons why the dendritic field orientation of a layer III pyramidal cell may be orthogonal to the cell's preferred orientation. One of the major differences between these three explanations is in the nature of the input to the basal dendrites of the layer III pyramidal cells and its role in the genesis of orientation selectivity.

One explanation is that of Schiller, Finlay and Volman (1976), who proposed that the topography of both excitatory and inhibitory inputs

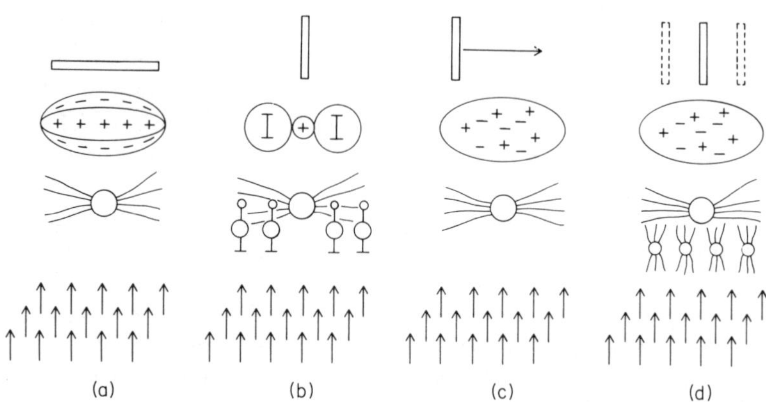

FIGURE 1 Four models relating the orientation of a cell's dendritic field to its preferred orientation. In each drawing, the arrows at the bottom represent the retinotopic array of incoming geniculocortical axons, the cell with horizontal dendrites represents a layer III pyramidal cell and the line at the top represents the cell's preferred orientation. Between the cell and the line is a schematic drawing of the cell's receptive field. (a) The cell's dendritic tree selects the appropriate input from the incoming array. This results in a horizontally elongated receptive field with separated on- and off-responses and a preference for horizontal lines (after Colonnier, 1964). (b) Geniculate axons end directly on the cell body and only indirectly (by way of inhibitory interneurons, shown as small, vertically oriented cells) on the cell's dendrites. This results in a receptive field with an excitatory center and inhibitory sidebands (I), and, hence, a preference for vertical lines (modified from Schiller, Finlay and Volman, 1976). (c) As in (a), the dendritic tree selects the appropriate inputs, but in this case it results in a preference for movement along the horizontal axis and, thus, to an apparent preference for vertical lines (after Tieman and Hirsch, 1982). (d) The geniculocortical axons influence the layer III pyramidal cell indirectly, through the layer IV stellate cells, which are shown as small, vertical cells. Illustrated here are four stellate cells that prefer vertical lines. These cells are scattered along a line that is horizontal in the plane of the map, and hence their receptive fields are also scattered along a horizontal line in the retina. The layer III pyramidal cell, which is one of Hubel and Wiesel's (1962) complex cells, optimizes the input from these stellate cells by having a dendritic field oriented horizontally in the plane of the map. Its receptive field is the sum of the receptive fields of the stellate cells and has mixed on- and off-responses. The pyramidal cell shares with the stellate cells a preference for vertical, and responds to a vertical line anywhere within its receptive field.

contributes to orientation selectivity (Fig. 1b). This is consistent with the data showing that intracortical inhibitory connections help to generate orientation selectivity (Sillito, 1975, 1979). Since their model requires that preferred orientation be perpendicular to dendritic field orientation, it is supported by our findings. However, their model also proposes that the inhibitory inputs end on the dendrites, while excitatory geniculocortical afferents end on or near the cell body. This is difficult to reconcile with electron microscopic data showing that presumed inhibitory synapses are usually located on the somata of pyramidal cells (Gray, 1959; Colonnier, 1968; Garey, 1971; LeVay,

1973) while the geniculocortical synapses are usually located on dendritic spines and shafts (Colonnier and Rossignol, 1969; Garey and Powell, 1971; LeVay and Gilbert, 1976; Tieman, 1984). Thus, as it stands, their model is only partly consistent with available data.

A second explanation is that the topography of excitatory inputs onto the basal dendrites of a layer III pyramidal cell determines a preference for one axis of movement, not a preference for one stimulus orientation. Most cells in area 17 respond best to moving stimuli and prefer movement in one or both directions perpendicular to the optimal stimulus orientation (Hubel and Wiesel, 1959, 1962; Henry, Bishop and Dreher, 1974). Such selectivity for a particular axis of movement can mimic orientation selectivity, and some cells that appear to be selective for stimulus orientation are, in fact, only axially selective (Henry, Bishop and Dreher, 1974; Henry, Dreher and Bishop, 1974; Pettigrew, 1974). If a dendritic field is elongated along the preferred axis of movement, this would permit greater spatiotemporal summation as the stimulus moves across the field (Fig. 1c), and would result in an orthogonal relationship between a cell's dendritic field orientation and its apparent preferred orientation.

A third explanation is that the primary excitatory input onto the layer III pyramidal cells is from other cortical cells (e.g. the layer IV stellate cells) that have the same preferred orientation. The shape of the basal dendritic field then may reflect the cortical distribution of these inputs. For example, the layer III pyramidal cells may correspond to 'complex' cells and receive inputs from several simple cells with the same preferred orientation and with receptive fields that are displaced relative to one another along a line perpendicular to their preferred orientations (Hubel and Wiesel, 1962). Since the retina projects topographically onto the visual cortex, these simple cells will themselves be displaced relative to one another along a line perpendicular (in the coordinates of the cortical map) to the cells' preferred orientations (Fig. 1d). Thus, a layer III pyramidal cell that preferred vertical stimuli would need a dendritic field that extended in a direction parallel to the projection of the horizontal meridian in order to select inputs from the appropriate simple cells. As a result, the layer III pyramidal cell would have a dendritic field that is perpendicular to its preferred orientation (vertical). Inputs to the layer III pyramidal cell from a set of simple cells preferring vertical lines and having receptive fields scattered along a horizontal line would give the cell a wide receptive field and a preference for stimulus widths smaller than the receptive field. In short, this third model is consistent with (a) the response properties of complex cells (Hubel and Wiesel, 1962), (b) physiological data that pyramidal cells are usually complex whereas stellate cells are usually simple (Kelly and Van Essen, 1974; Gilbert and Wiesel, 1979, 1983) and that complex cells predominate in layer III whereas simple cells predominate in layer IV (Hubel and Wiesel, 1962; Gilbert, 1977; Leventhal and Hirsch, 1978) and (c) anatomical data that cells in layer IV project onto cells in layer III (Gilbert and Wiesel, 1979; Lund *et al.*, (1979). It seems to us that this model offers the best explanation for the functional significance of the shape of the basal dendritic field of layer III pyramidal cells.

What about the layer IV stellate cells? Their dendritic fields are elongated in both normal and stripe-reared cats (Colonnier, 1964; Coleman *et al.*, 1981; Tieman and Hirsch, 1982). However, the distribution of their orientations is not affected by stripe-rearing (Tieman and Hirsch, 1982; but cf. Coleman *et al.*, 1981). We have suggested (Tieman and Hirsch, 1982) that the layer IV stellate cells correspond to the members of a physiologically characterized cell group whose members are orientation selective in normal (Leventhal and Hirsch, 1980) and stripe-reared cats (Hirsch *et al.*, 1983), but for which the distribution of preferred orientations is not affected by stripe-rearing (Hirsch *et al.*, 1983). Cells in this group have narrow receptive fields and do not respond to rapid stimulus motion, response properties associated in normal cats with direct excitatory input from X-type cells in the lateral geniculate nucleus (LGN) (Dreher, Leventhal and Hale, 1980). Because cells in this group are not affected by stripe-rearing, our data do not tell us whether dendritic field orientation and preferred orientation are parallel, orthogonal or unrelated.

It is unlikely that the shape of a stellate cell's dendritic field can be the sole determinant of its orientation selectivity. First, the fields are not sufficiently elongated to account for the narrow tuning curves of these cells. The dendritic fields of cortical cells appear to be no more elongated than those of retinal ganglion cells [compare Fig. 8 of Tieman and Hirsch, 1982, with Fig. 3b of Leventhal and Schall, 1983; our data for stellate cells in unpublished observations are very similar, yet cortical cells are much more finely tuned for orientation than are retinal ganglion cells (Hubel and Wiesel, 1959, 1962; compare the tuning curves of Rose and Blakemore, 1974, or Henry, Dreher and Bishop, 1974, with those of Levick and Thibos, 1982)].

Second, although the large tangential spread of the terminal arbors of individual geniculocortical axons (O'Leary, 1941; Sholl, 1967; Ferster and LeVay, 1978; Gilbert and Wiesel, 1979), apparently does not limit the resolution of the retinotopic map in the cortex (Hubel, Wiesel and LeVay, 1974; Hubel and Wiesel, 1977; Blasdel and Lund, 1983), it will decrease the effects of dendritic field elongation on receptive field elongation. Thus, if LGN terminal arbors are large, then cortical receptive fields will be larger and relatively less elongated than if LGN arbors are small (Fig. 2). The extent of the receptive field of a cortical cell is determined by the receptive fields of the most peripheral LGN arbors that contact the cell. If LGN arbors are large, then more can contact a given cortical cell, and their receptive field centers are more widely distributed. Thus, the receptive field will extend on all sides beyond that predicted by the dendritic field by an amount equal to the radius of the LGN arbors plus the radius of the receptive field of the LGN cell (Fig. 3). Consequently, the length-to-width ratio of the resulting receptive fields is closer to 1.0 than the length-to-width ratio of the dendritic field of the cortical cell. However, it is clear that those LGN cells whose arbors are peripheral to the dendritic field will contribute very little to the cortical cell's receptive field. Further, if cortical synapses are strengthened or weakened by activity (Hebb, 1949), then the inputs from these peripheral

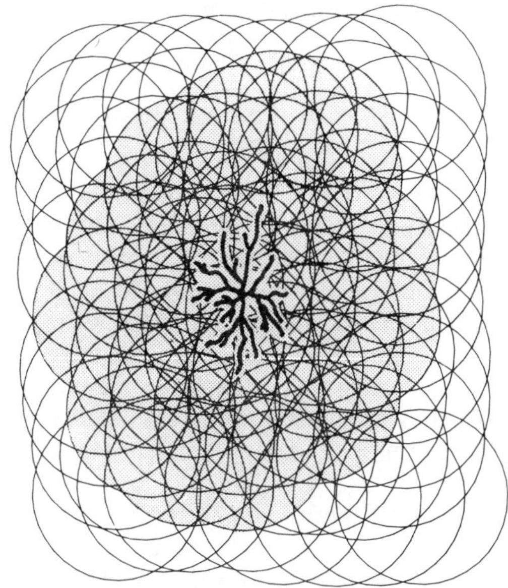

FIGURE 2 The effect of large afferent terminal arbors on the elongation of the receptive field that would be produced by an elongated dendritic tree. Each circle represents the terminal arbor of one LGN cell. These arbors are spaced, on the average, one half of a radius apart, and the centers of the arbors are assumed to form a retinotopic array. Every arbor that intersects the dendritic tree of the cortical cell is stippled. The stippled area thus gives an indication of the way in which extended terminal arbors affect the cortical cell's receptive field. Although the exact borders of the receptive field would depend on the receptive fields of the LGN axons (Fig. 3), it is clear that the cell's receptive field is less elongated than its dendritic field.

arbors may disappear (cf. Meyer, 1983; Schmidt and Edwards, 1983). The net effect of this will be to make the receptive field smaller and more elliptical. In short, the large LGN terminal arbors will reduce, but are unlikely to eliminate, the effect of elongation of the dendritic fields on elongation of receptive fields of layer IV stellate cells.

Several investigators have argued that the receptive fields of simple cells are round, not elongated, and that simple cells, therefore, do not receive input from LGN cells with receptive fields lined up in a row across the retina

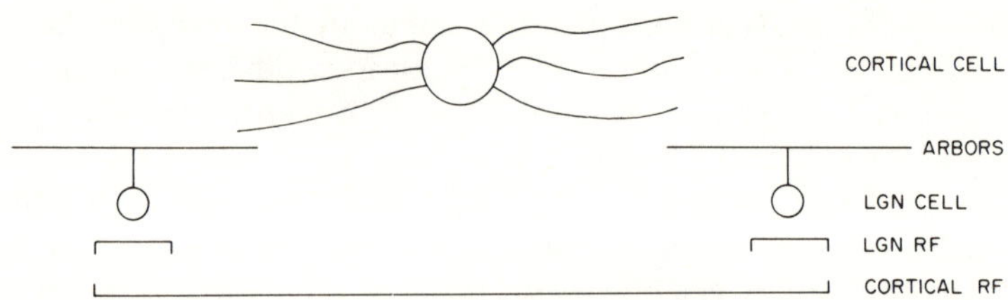

Figure 3 The genesis of a cortical cell's receptive field (RF). The visual field and the maps thereof in the LGN and visual cortex are shown in register in order to illustrate the manner in which the size of a cortical cell's receptive field is determined by the size of its dendritic field, the size of the most peripheral LGN arbors that contact the dendritic tree and the size of the receptive fields of the LGN cells that give rise to these arbors.

(Creutzfeldt, Kuhnt and Benevento, 1974; Tsumoto, Eckart and Creutzfeldt, 1979). If true, this clearly invalidates Colonnier's (1964) model. However, the length of a receptive field (i.e. the diameter of the receptive field along the line of an optimally oriented bar) is very difficult to measure (e.g. by some methods, negative lengths are possible). Those researchers who say that the receptive fields of simple cells are round tend to measure the dimensions of the receptive field using the width of the discharge peak in response histograms to stimuli moving along various axes (Creutzfeldt, Kuhnt and Benevento, 1974). While this is probably satisfactory for measuring the width of a cell's receptive field, it is not a good way to measure its length, since the cell will respond less reliably to stimuli moving along suboptimal axes and may not respond at all to stimuli moving along the length of the receptive field. Probably the best way to measure receptive field length is by length summation, and the fact that length summation is much greater than width summation (Hubel and Wiesel, 1962; Henry, Goodwin and Bishop, 1978) strongly suggests that receptive fields of simple cells are elongated. In summary, if we assume that the layer IV stellate cells are simple cells, then both their dendritic fields and their receptive fields appear to be elongated. However, although receptive field elongation may provide an orientation bias, additional factors are needed to explain the very limited range of stimulus orientations to which most simple cells respond.

One such factor could be the orientation bias that is already present in the afferents to area 17 (Vidyasagar and Urbas, 1982). Because LGN cells with receptive fields in a localized region of retina have similar preferred orientations, cells in the visual cortex that receive input from a small region of retina would be likely to have an overall bias for a particular orientation. The orientation of the dendritic field of the layer IV stellate cell could affect the width of this tuning. Given the radial distribution of preferred orientations in the LGN (Vidyasagar and Urbas, 1982) and retina (Levick and Thibos, 1982, and cf. Leventhal and Schall, 1983, and Leventhal, Chapter 40 in this volume), then the scatter of the preferred orientations of the afferents to a particular cell would be lowest when the cell's dendritic field is oriented radially. Thus, the elongation of the dendritic fields of layer IV stellate cells could help to sharpen the orientation bias provided by the LGN afferents. For cells with radially oriented dendritic fields, preferred orientation would match dendritic field orientation, and the resulting orientation selectivity would be greater than might be predicted on the basis of the stellate cell's dendritic field shape alone.

Inhibition also clearly plays a major role in determining orientation selectivity, a role which is ignored by all of the dendritic field models mentioned earlier except that of Schiller, Finlay and Volman (1976). Thus, for example, inhibitory sidebands contribute considerably to the orientation selectivity of simple cells (Henry, Dreher and Bishop, 1974) and the orientation selectivity of both simple and complex cells can be greatly reduced or even abolished by application of bicuculline or related drugs (Sillito, 1975, 1979; Tsumoto, Eckart and Creutzfeldt, 1979; Sillito, *et al.*, 1980). However, since the receptive field centers of simple cells show some orientation bias (Henry, Bishop and Dreher, 1974), and since it is quite difficult to abolish totally by pharmacological means the orientation selectivity of many simple and complex cells (Sillito, 1975, 1979; Tsumoto, Eckart and Creutzfeldt, 1979), we suggest that the stellate cell's elongated dendritic field provides it with an orientation bias that is enhanced by lateral inhibitory connections (cf. Blakemore and Tobin, 1972; Fries, Albus and Creutzfeldt, 1977; Hess, Negishi and Creutzfeldt, 1975; Nelson and Frost, 1978; Morrone, Burr and Maffei, 1982).

In summary, it is likely that dendritic field orientation is related to preferred orientation for at least some cells in area 17. The relationship between dendritic field orientation and preferred orientation need not be the same for all cells in area 17. For example, the mechanism responsible for the orientation selectivity of the layer III pyramidal cells may well differ from that of the layer IV stellate cells. For layer IV stellate cells, it is possible that the dendritic field orientation may determine the fundamental orientation preference, in much the way Colonnier (1964) has suggested. If most stellate cells can obtain an orientation bias in this fashion, then a set of lateral inhibitory connections amongst them can accentuate these biases and thus give rise to a population of cells that are very selective for stimulus orientation. The stripe-rearing data tell us nothing about the relation between dendritic field orientation and preferred orientation for layer IV stellate cells because they are little affected by environmental stimulation. However, the data of Martin and Whitteridge (1984) suggest that although Colonnier's (1964) model may hold for the cells of IVB, it does not hold for all layer IV cells.

Because both physiology and morphology of the layer III pyramidal cells are affected in a very specific fashion by early visual stimulation, we could deduce structure/function relationships of individual cells from the correspondences between physiological and morphological changes in populations of cells. Both our own stripe-rearing data and the data of Martin and Whitteridge (1984) suggest that, for layer III pyramidal cells, dendritic field orientation and preferred orientation are orthogonal. For these cells we think it likely that dendritic field shape is not directly related to the genesis of orientation selectivity, but rather functions to make it possible for the same preferred orientation to be expressed at different points within the cell's receptive field. In effect, dendritic field shape would function to generalize an orientation preference in much the way Hubel and Wiesel (1962) first suggested.

ACKNOWLEDGEMENTS

We thank David G. Tieman for helpful discussion and R. Loos and R. Speck for help in preparing the figures.

REFERENCES

Bacon, J. P., and Murphey, R. K. (1984). Receptive fields of the cricket giant interneurones are related to their dendritic structure. *J. Physiol. (Lond.)*, **352**, 601–623.

Blakemore, C. B., and Cooper, G. F. (1970). Development of the brain depends on the visual environment. *Nature*, **228**, 477–478.

Blakemore, C. B., and Tobin, E. A. (1972). Lateral inhibition between orientation detectors in the cat's visual cortex. *Exp. Brain Res.*, **15**, 439–440.

Blasdel, G. G., and Lund, J. S. (1983). Termination of afferent axons in macaque striate cortex. *J. Neurosci.*, **3**, 1389–1413.

Coleman, P. D., Flood, D. G., Whitehead, M. C., and Emerson, R. C. (1981). Spatial sampling by dendritic trees in visual cortex. *Brain Res.*, **214**, 1–21.

Colonnier, M. (1964). The tangential organization of the cortex. *J. Anat.*, **98**, 327–344.

Colonnier, M. (1968). Synaptic patterns on different cell types in the different laminae of visual cortex. An electron microscope study. *Brain Res.*, **9**, 268–287.

Colonnier, M., and Rossignol, S. (1969). Heterogeneity of the cerebral cortex. In *Basic Mechanisms of the Epilepsies* (Eds. H. H. Jasper, A. A. Ward and A. Pope), Little-Brown, Boston, pp. 29–44.

Creutzfeldt, O. D., Kuhnt, U., and Benevento, L. A. (1974). An intracellular analysis of visual cortical neurones to moving stimuli: responses in a co-operative neural network. *Exp. Brain Res.*, **21**, 251–274.

Dreher, B., Leventhal, A. G., and Hale, P. T. (1980). Geniculate input to cat visual cortex: a comparison of area 19 with areas 17 and 18. *J. Neurophysiol.*, **44**, 804–826.

Ferster, D., and LeVay, S. (1978). The axonal arborizations of lateral geniculate neurons in the striate cortex of the cat. *J. comp. Neurol.*, **182**, 923–944.

Fries, W., Albus, K., and Creutzfeldt, O. D. (1977). Effects of interacting patterns on single cell responses in cat's striate cortex. *Vision Res.*, **17**, 1001–1008.

Garey, L. J. (1971). A light and electron microscopic study of the visual cortex of the cat and monkey. *Proc. Roy. Soc., B*, **179**, 21–40.

Garey, L. J., and Powell, T. P. S. (1971). An experimental study of the termination of the lateral geniculo-cortical pathway in the cat and monkey. *Proc. Roy. Soc., B*, **179**, 41–63.

Gilbert, C. D. (1977). Laminar differences in receptive field properties in cat primary

visual cortex. *J. Physiol. (Lond.),* **268**, 391–421.
Gilbert, C. D., and Wiesel, T. N. (1979). Morphology and intracortical projections of functionally characterised neurons in the cat visual cortex. *Nature,* **280**, 120–125.
Gilbert, C. D., and Wiesel, T. N. (1983). Clustered intrinsic connections in cat visual cortex. *J. Neurosci.,* **3**, 1116–1123.
Gray, E. G. (1959) Axosomatic and axodendritic synapses of the cerebral cortex: an electron microscope study. *J. Anat. (Lond.),* **93**, 420–434.
Hebb, D. O. (1949). *The Organization of Behaviour,* Wiley, New York.
Henry, G. H., Bishop, P. O., and Dreher, B. (1974). Orientation, axis, and direction as stimulus parameters for striate cells. *Vision Res.,* **14**, 767–777.
Henry, G. H., Dreher, B., and Bishop, P. O. (1974). Orientation specificity of cells in the cat striate cortex. *J. Neurophysiol.,* **37**, 1394–1409.
Henry, G. H., Goodwin, A. W., and Bishop, P. O. (1978). Spatial summation of responses in receptive fields of single cells in cat striate cortex. *Exp. Brain Res.,* **32**, 245–266.
Hess, R., Negishi, K., and Creutzfeldt, O. D. (1975). The horizontal spread of intracortical inhibition in the visual cortex. *Exp. Brain Res.,* **22**, 415–419.
Hirsch, H. V. B., Leventhal, A. G., McCall, M. A., and Tieman, D. G. (1983). Effects of exposure to lines of one or two orientations on different cell types in striate cortex of cat. *J. Physiol. (Lond.),* **337**, 241–255.
Hirsch, H. V. B., and Spinelli, D. N. (1970). Visual experience modifies the distribution of horizontally and vertically oriented receptive fields. *Science,* **168**, 869–871.
Hirsch, H. V. B., and Spinelli, D. N. (1971). Modification of the distribution of receptive field orientation by selective visual exposure during development. *Exp. Brain Res.,* **12**, 509–527.
Hubel, D. H., and Wiesel, T. N. (1959). Receptive fields of single neurons in the cat's striate cortex. *J. Physiol. (Lond.),* **148**, 574–591.
Hubel, D. H., and Wiesel, T. N. (1962). Receptive fields, binocular interaction and functional architecture in the cat's visual cortex. *J. Physiol. (Lond.),* **160**, 106–154.
Hubel, D. H., and Wiesel, T. N. (1977). Functional architecture of macaque monkey visual cortex. *Proc. Roy. Soc. Lond., B,* **198**, 1–59.
Hubel, D. H., Wiesel, T. N., and LeVay, S. (1974). Visual field representation in layer IV C of monkey striate cortex. *4th Ann. Meet. Soc. Neurosci. Abstr.,* p. 264.
Kelly, J. P., and Van Essen, D. (1974). Cell structure and function in the visual cortex of the cat. *J. Physiol. (Lond.),* **238**, 515–547.
Kirk, M. D., Waldrop, B., and Glantz, R. M. (1983). A quantitative correlation of contour sensitivity with dendritic density in an identified visual neuron. *Brain Res.,* **274**, 231–237.
LeVay, S. (1973). Synaptic patterns in the visual cortex of the cat and monkey. Electron microscopy of Golgi preparations. *J. comp. Neurol.,* **150**, 53–86.
LeVay, S., and Gilbert, C. D. (1976). Laminar patterns of geniculocortical projection in the cat. *Brain Res.,* **113**, 1–19.
Leventhal, A. G., and Hirsch, H. V. B. (1978). Receptive field properties of neurons in different laminae of the visual cortex of the cat. *J. Neurophysiol.,* **41**, 948–962.
Leventhal, A. G., and Hirsch, H. V. B. (1980). Receptive field properties of different classes of neurons in the visual cortex of normal and dark-reared cats. *J. Neurophysiol.,* **43**, 1111–1132.
Leventhal, A. G., and Schall, J. D. (1983). Structural basis of orientation sensitivity of cat retinal ganglion cells. *J. comp. Neurol.,* **220**, 465–475.
Levick, W. R., and Thibos, L. N. (1982). Analysis of orientation bias in cat retina. *J. Physiol. (Lond.),* **329**, 243–261.

Lund, J. S., Henry, G. H., MacQueen, C. L., and Harvey, A. R. (1979). Anatomical organization of the primary visual cortex (area 17) of the cat. A comparison with area 17 of the macaque monkey. *J. comp. Neurol.,* **184**, 599–618.

Martin, K. A. C. and Whitteridge, D. (1984). The relationship of receptive field properties to the dendritic shape of neurones in the cat striate cortex. *J. Physiol. (Lond.)*, **356**, 291–302.

Meyer, R. L. (1983). Tetrodotoxin inhibits the formation of refined retinotopography in goldfish. *Dev. Brain Res.,* **6**, 293–298.

Morrone, M. C., Burr, D. C., and Maffei, L. (1982). Functional implications of cross-orientation inhibition of cortical visual cells. I. Neurophhhhysiological evidence. *Proc. Roy. Soc. Lond., B,* **216**, 335–354.

Nelson, J. I., and Frost, B. J. (1978). Orientation-selective inhibition from beyond the classic visual receptive field. *Brain Res.,* **139**, 359–365.

O'Leary, J. L. (1941). Structure of area striata of the cat. *J. comp. Neurol.,* **75**, 131–164.

Pettigrew, J. D. (1974). The effect of visual experience on the development of stimulus specificity by kitten cortical neurones. *J. Physiol. (Lond.),* **237**, 49–74.

Rose, D., and Blakemore, C. B. (1974). An analysis of orientation selectivity in the cat's visual cortex. *Exp. Brain Res.,* **20**, 1–17.

Schiller, P. H., Finlay, B. L., and Volman, S. F. (1976). Quantitative studies of single-cell properties in monkey striate cortex. V. Multivariate statistical analyses and models. *J. Neurophysiol.,* **39**, 1362–1374.

Schmidt J. T., and Edwards, D. L. (1983). Activity sharpens the map during the regeneration of the retinotectal projection in the goldfish, *Brain Res.,* **269**, 29–39.

Sholl, D. A. (1967). *The Organization of the Cerebral Cortex,* Hafner, New York.

Sillito, A. M. (1975). The contribution of inhibitory mechanisms to the receptive field properties of neurones in the striate cortex of the cat. *J. Physiol. (Lond.),* **250**, 305–329.

Sillito, A. M. (1979). Inhibitory mechanisms influencing complex cell orientation selectivity and their modification at high resting discharge levels. *J. Physiol. (Lond.),* **289**, 33–53.

Sillito, A. M., Kemp, J. A., Milson, J. A., and Berardi, N. (1980). A re-evaluation of the mechanisms underlying simple cell orientation selectivity. *Brain Res.,* **194**, 517–520.

Stryker, M. P., Sherk, H., Leventhal, A. G., and Hirsch, H. V. B. (1978). Physiological consequences for the cat's visual cortex of effectively restricting early visual experience with orientation. *J. Neurophysiol.,* **41**, 895–909.

Tieman, S. B. (1984). Effects of monocular deprivation on geniculocortical synapses in the cat. *J. comp. Neurol.,* **222**, 166–176.

Tieman, S. B., and Hirsch, H. V. B. (1982). Exposure to lines of only one orientation modifies dendritic morphology of cells in the visual cortex of the cat. *J. comp. Neurol.,* **211**, 353–362.

Tsumoto, T., Eckart, W., and Creutzfeldt, O. D. (1979). Modification of orientation sensitivity of cat visual cortex neurons by removal of GABA-mediated inhibition. *Exp. Brain Res.,* **34**, 351–363.

Vidyasagar, T. R., and Urbas, J. V. (1982). Orientation sensitivity of cat LGN neurons with and without inputs from visual cortical areas 17 and 18. *Exp. Brain Res.,* **46**, 157–169.

Young, J. Z. (1960). Regularities in the retina and optic lobe of octopus in relation to form discrimination. *Nature (Lond.),* **186**, 836–844.

Models of the Visual Cortex
Edited by D. Rose and V. G. Dobson

CHAPTER 46

Does the striate cortex contain a system of oriented axons?

G. J. Mitchison
MRC Laboratory of Molecular Biology, Hills Road, Cambridge, CB2 2QH, UK and Kenneth Craik Laboratory, Physiology Department, Cambridge, CB2 3EG, UK

In trying to understand the anatomy of nervous tissue, a powerful technique has been developed which makes use of the enzyme horseradish peroxidase (HRP). If HRP is injected into a neuron, the enzyme will be transported along its axons and dendrites, and will eventually come to be distributed through all of the cell's processes. The enzyme can be rendered visible, which reveals the structure of the cell in remarkable detail. In many ways, this technique has supplanted the Golgi technique, for the choice of the cell which is to be labelled lies with the experimenter rather than being left to the caprice of an unknown physicochemical process.

The injection of HRP into individual cells is not the only use which can be made of the enzyme. Recently, Rockland and Lund (1981, 1982) discovered that an intriguing pattern could be produced simply by injecting HRP with a comparatively coarse electrode (20 μm tip diameter) into the striate cortex of the tree shrew. The injection led to an initial distribution of HRP over an area of about 1 mm² of the cortex. When time had been allowed for the uptake and redistribution of the enzyme and the cortex was examined, it was found that the label was present not only in the original injection site but also in a pattern of stripes extending for several millimetres around this site. The strength of this technique is that a single injection appears to be telling us something important about the organization of the cortex. The disadvantage is that the interpretation of the patterns is by no means automatic. I shall give Rockland and Lund's own interpretation of their pattern and then discuss another interpretation which Francis Crick and I (Mitchison and Crick, 1982) have proposed.

Rockland and Lund's pattern consists, then, of a region of dense staining around the injection site, and beyond this a series of stripes whose intensity decreases with distance from the injection site. These stripes are seen only in the upper layers of the cortex; in the lower layers, a continuous pattern of label is seen. One curious feature of the pattern is its close resemblance to another pattern which occurs in the tree shrew striate cortex—the orientation stripes (Humphrey and Norton, 1980; Humphrey, Skeen and Norton, 1980). These are the loci of cells which share a common orientation preference (i.e. respond best to a light bar of a particular orientation). The orientation stripes can be made visible by injecting an animal with radioactively labelled deoxyglucose, a non-metabolizable analogue of glucose, and then exposing the animal to a visual stimulus of a fixed orientation. Subsequent autoradiography of the striate cortex shows which cells accumulated the label, and these are presumed to be those which responded to the stimulus. In the treeshrew, they are found to lie in stripes with a periodicity of about 1 mm (Humphrey, Skeen and Norton, 1980). Rockland and Lund's HRP pattern seems to have roughly the same periodicity and to run in the same overall direction as the orientation stripes.

To explain their pattern, Rockland and Lund suggested that the tree shrew striate cortex contains two systems, one consisting of cells with short axons, the other having cells with long axons. The latter system, they suggested, forms an array of stripes. When HRP is injected, the label would be carried farthest from the site by the long-axon cells and would then accumulate in locations corresponding to these stripes. Because of the resemblance of their pattern to the orientation stripes, they suggested that there might be some relationship between the underlying systems.

Let us consider what this relationship might be. The HRP is thought to be taken up by axons and other processes of cells in the injection site and transported throughout the cell (both retrogradely, i.e. back up an axon to the cell body, and orthogradely, i.e. from the cell body down its axon). Since processes from cells of all orientation preferences will be coursing through the injection site, there seems to be no compelling reason to suppose that the uptake will be restricted to cells of a particular orientation, even if the injection site is small. So it is unreasonable to suppose that the HRP stripes simply correspond to a subsystem of the orientation stripes consisting of cells with a particular orientation preference.

Suppose, however, that axonal fields are elongated, when viewed in a tangential section of the cortex (i.e. looking down on the cortical surface), and suppose that there is a systematic relationship between the direction of this elongation and the orientation preferences of cells. Then it turns out that one would indeed expect a set of stripes to be generated by the uptake and redistribution of HRP (Mitchison and Crick, 1982). To make this notion more precise, recollect that there is topographic mapping from the retina to the cortex. This allows us to translate a direction on the retina into a direction on the cortical surface. Our proposal is then that the axonal field of a cell is elongated in a direction corresponding, under this map from the retina to the cortex, to the preferred orientation. Another possibility is that the axonal field might be elongated at right angles to the cell's preferred orientation. As we shall see later, both versions of this rule can be given a natural interpretation in terms of the construction of receptive fields.

To see how the HRP patterns would be generated, suppose Fig. 1a represents part of a highly idealized orientation stripe system and suppose, for instance, that the direction which the stripes follow corresponds to the

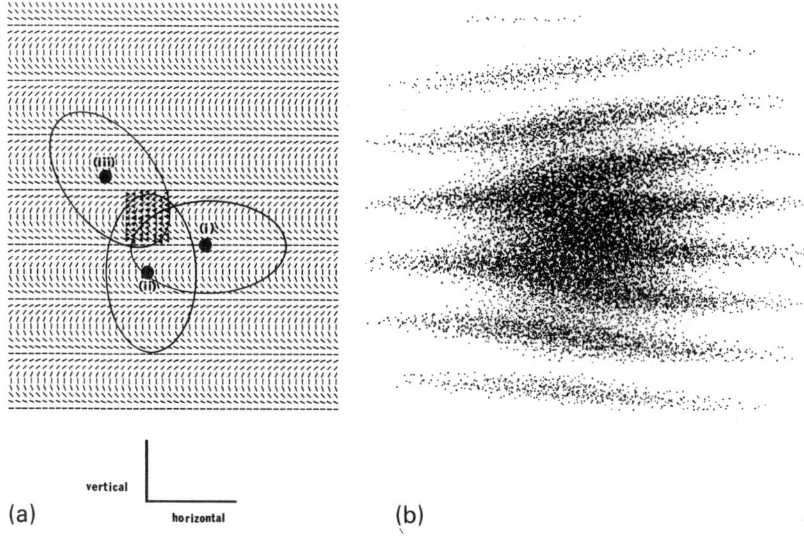

Figure 1 (a) A highly idealized depiction of orientation stripes. The directions of bars correspond to orientations in visual space. (The directions on the cortex corresponding to the vertical and horizontal in visual space are shown below.) The ellipses represent samples of axonal fields, shown elongated in the direction of the cells' orientation preference. Three cells are shown, with orientation preferences which are (i) horizontal, (ii) vertical and (iii) oblique. In another version of the rule, the elongation would be at right angles to that shown here. The injection site is marked by filled triangles. (b) The hypothetical HRP pattern generated by the rule that HRP is transported along axons which are elongated in the direction of the receptive field, as shown in (a). The uptake of HRP by a cell is assumed to follow a distribution which has an elliptical Gaussian shape. That is, the amount of HRP transported to a cell from a point (x, y), with coordinates along the principle axes of the ellipse, is proportional to exp $(-x^2/r^2-y^2/s^2)$. Here r and s are preassigned distances determining the dimensions of the uptake ellipse. The injection site is assumed to be a square whose side is equal to the period of the orientation stripes. The total concentration of HRP at a point is represented in the figure by the density of points. The pattern shown here was generated by taking $r=2s$, with s being the period of the orientation stripes (about 1 mm in the tree shrew).

horizontal direction in visual space. (This is approximately true for the tree shrew orientation stripes near to the area 17/18 border, for the stripes here run roughly at right angles to the border, corresponding to a horizontal direction in the region of visual space near to the vertical midline.) Let us consider the version of the rule where a cell's axonal field is elongated in a direction equivalent to the cell's orientation preference. If a stripe with cells of horizontal orientation preference passes through (or close to) the injection site, then all the cells in this stripe near to the injection site will be labelled (Fig. 1a (i)). Cells with a vertical orientation preference will be labelled only if they lie in stripes next to the injection site, in positions where their fields overlap the site (Fig. 1a (ii)). This will lead to patches of label on either side of the injection site. Oblique orientations (e.g. Fig. 1a (iii)) will fill in this pattern and join up the patches to make stripes.

A computer simulation (Fig. 1b) shows that these HRP stripes run roughly parallel to the orientation stripes, but show a phase shift varying with position relative to the injection site. The simulation assumed an injection site with a diameter equal to the period of the orientation stripes (about 1 mm in the tree shrew). Very similar results are obtained if the injection site is taken to be very small. This is a possibility if HRP uptake occurs predominantly from processes damaged by the injection needle, as has been proposed (Swindale 1982; Lund and Humphrey, 1982).

In our original paper (Mitchison and Crick, 1982) we formulated our rule in terms of connections: we required that a cell should make connections (axonal-dendritic synapses) with other cells of a similar orientation preference. As above, we required that these should occur within a region of cortex which was elongated in a direction related to the cell's orientation preference. The reason for this emphasis upon connections with cells of similar orientation tuning was that this gives a natural interpretation of the elongated axonal fields. Consider, for example, how a very elongated receptive field could be constructed. This could be achieved by stringing together several comparatively short receptive fields. Clearly, all the short fields must have the same orientation, and moreover the axons of the cells with these short fields must run in the direction of the long receptive field they contribute to. This is therefore an example of the version of our rule where we expect an axon to have the same orientation, under the map from the retina to the cortex, as the receptive field of its parent cell.

This idea was first proposed in a study carried out by Gilbert and Wiesel (1979), in which they determined the receptive field properties of cells in the cat striate cortex, then filled the cells (indivudually) with HRP. Amongst the cells they filled, there was one in layer 5 whose axon ran for a long distance in layer 6. As the cells in layer 6 are remarkable for their very long receptive fields (Gilbert, 1977), this suggested to them that the long axon was involved in creating such a field, in the fashion just described. They observed that the direction of the axon did appear to agree with the receptive field orientation.

More recently, Gilbert and Wiesel (1983) have given further examples of cells whose axonal fields are elongated in the direction of the receptive field. They have also found cells whose axons are elongated at right angles to the cell's orientation preference. They suggested that these cells might be involved in the generation of inhibitory sidebands (Bishop, Henry and Smith, 1971). This seems to us more reasonable than our original proposal that such connections were used to create complex receptive fields by joining simple receptive fields side by side. The sidebands often lie at some distance from the central part of the receptive field and therefore have dimensions more compatible with the spacing of the stripes in Rockland and Lund's patterns.

A striking feature of the HRP-filled cells (Gilbert and Wiesel, 1983) was that their axonal fields often showed clusters of connections. These clusters may, in some cases at least, correspond to orientation stripes. An axon which cuts across these stripes and only makes connections with cells of one orientation preference would be expected to show such clustering. This should be taken into account in simulating HRP patterns according to our rules, for the label is likely both to be taken up more readily and to accumulate more conspicuously in these clusters. We originally simulated an extreme version of this, where the uptake and deposition of HRP was assumed to be confined to particular orientation stripes. The only modification which needs to be made in using this rule is that a cell should take up HRP only from regions of the injection site where cells have the same orientation preference, rather than from every part of the site. The resulting patterns are very similar to those shown here (compare Fig. 1b with Fig. 2 in Mitchison and Crick, 1982).

Evidence from filling single cells therefore gives some support for the rules we have proposed, but this does not prove that the HRP patterns are generated in the way we suggest. Recently, Rockland, Lund and Humphrey (1982) have attempted to define the relationship between the HRP stripes and the orientation stripes by labelling with 2-deoxyglucose and HRP simultaneously. They reported that 'the two systems do not appear to be topographically identical, despite their similar overall geometry' (p.51). This finding is quite consistent with our hypothesis, since we predict that the two patterns should not be in register, but should show a relative phase shift depending upon the location relative to the injection site (Fig. 1b).

An experiment which might help to decide how the stripes are generated involves the injection of HRP at two sites. According to our hypothesis, the label from the two sites should overlap at a position along the line between the sites, but should occupy complementary positions at those points where the sites subtend a right angle (Mitchison and Crick 1982, Fig. 4). At the latter positions, therefore, we should predict a fairly uniform distribution of label. This would not be expected on Rockland and Lund's hypothesis.

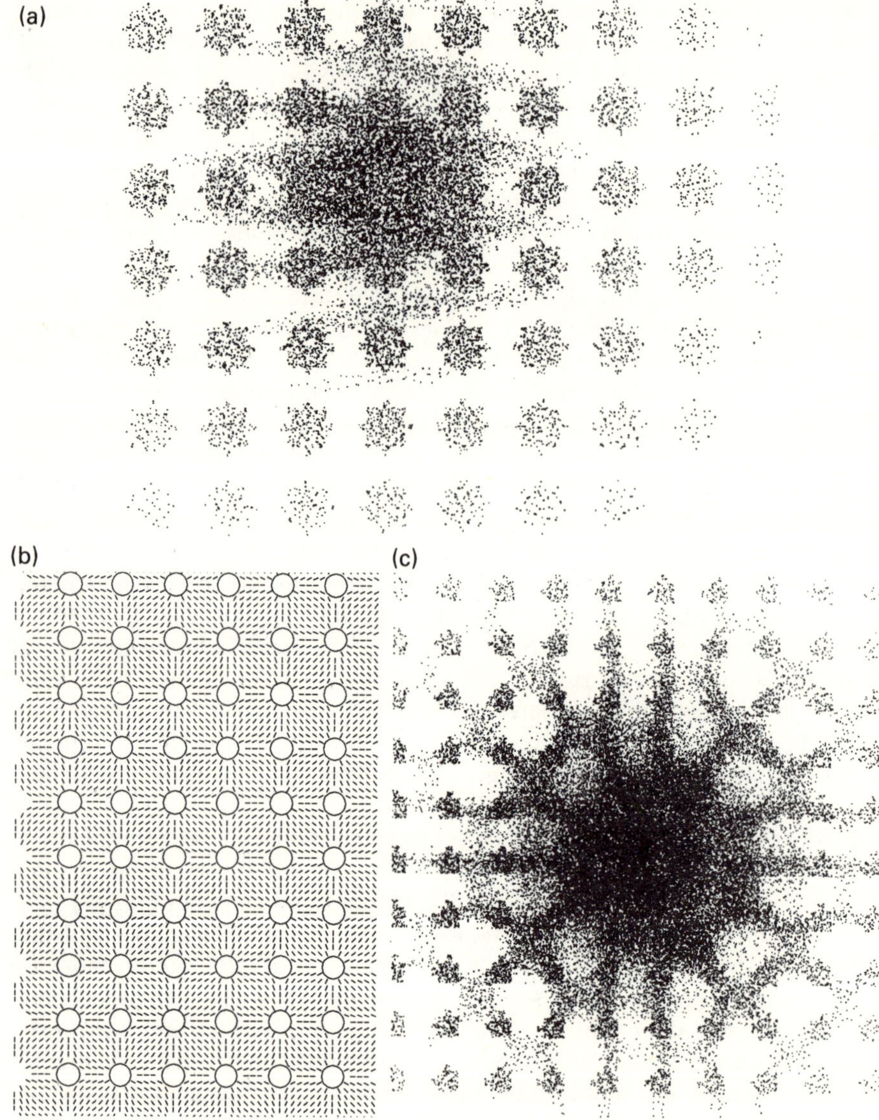

FIGURE 2 (a) An HRP pattern obtained by superposing a regular array of 'cytochrome oxidase' patches upon the kind of pattern shown in Fig. 1(b). This is highly artificial in many respects, not least in the regularity of the cytochrome oxidase array. This array has been given a slightly different period from the orientation stripes, and it has been assumed that HRP is transported isotropically between patches, falling off according to $\exp(-z^2/t^2)$, where t here is 3 periods of the orientation stripes. For movement of HRP along the oriented system, the same rule as in Fig. 1(b) has been used, with $r=2s$, but taking $s=0.8$ of the period of the orientation stripes. (b) A

Recently, Rockland and Lund (1982) have examined the HRP patterns in primates. In both the squirrel monkey and macaque they find that an injection of HRP in the striate cortex produces a 'radiating or lattice-like organization of HRP label, extending from the injection site . . .' (p.306). This pattern thins out, away from the injection site, into many small patches. It was discovered comparatively recently that, on staining with the enzyme cytochrome oxidase, the upper layers of the primate visual cortex show a pattern of patches (Humphrey and Hendrickson, 1980; Horton and Hubel, 1981). These patches, which are seen only in primates and not in the tree shrew (Horton and Hubel, 1981), consist of cells with relatively unoriented receptive fields (Hubel and Livingstone, 1981). Hubel and Livingstone (personal communication) have confirmed Rockland and Lund's findings for HRP injections in the primate cortex and have suggested that the HRP patches coincide with the cytochrome oxidase patches. One attractive possibility is therefore that Rockland and Lund's pattern in the primate results from the movement of HRP along a system of interconnections between cells in these cytochrome oxidase patches.

Although this hypothesis can explain the overall aspect of the pattern, it does not fully account for the 'lattice-like' character noted by Rockland and Lund, which seems to arise from comparatively weak labelling between the more evident patches. It is possible that the same rule for elongation of axonal fields that we have postulated for the tree shrew also applies here, restricted to the orientation-specific part of the striate cortex outside the cytochrome oxidase patches. If the range of this orientated system is less than the hypothetical cytochrome oxidase patch connections, then one might indeed expect something resembling a lattice, especially near to the injection site.

Unfortunately, the topography of the orientation system in primates is not well understood, and one can only examine various conjectures. Figure 2a shows what would be anticipated from combining HRP spread through an oriented system arranged in stripes with an isotropic spread between cytochrome oxidase patches. Figure 2b shows an orientation system arranged around centres, as proposed by Braitenberg and Braitenberg (1979), and

---

system of 'centres' (Braitenberg and Braitenberg, 1979) around which orientations are organized. This differs from Braitenberg and Braitenberg's exact proposals in having a square array, in which centres with the topological index +1 (where the orientations rotate in the same sense as a path taken round the centre) alternate with centres of index −1 (where the rotation is in the reverse sense). The circles are the centres and are assumed to coincide with the cytochrome oxidase patches. (c) The HRP pattern generated from this system, using the same rule for HRP spread between patches as in (a). For the oriented system, a Gaussian field has been assumed, with $r=2s$ and s taken to be 1.25 times the spacing between patches (i.e. the side of an individual square in the array which they form).

Fig. 2c shows the theoretical HRP pattern derived from it. As can be seen, this is more 'lattice-like' than the pattern generated from orientation stripes.

In conclusion, Rockland and Lund's HRP patterns may indeed reveal the presence of a cortical structure of the kind which they propose. Alternatively as Francis Crick and I suggest, the pattern of stripes in the tree shrew may arise naturally from a system of oriented axons, where this system forms an uninterrupted array of stripes. The story is less clear in the case of the primate. The structure of the primate cortex is known to be more complex and to have a substructure of cytochrome oxidase patches which are absent in the tree shrew. Moreover, the distribution of orientations in the primate cortex is not yet known. However, here also it may be unnecessary to posit any additional structure beyond the cytochrome oxidase pattern and a system of oriented axons.

ACKNOWLEDGEMENTS

I thank Drs. F. H. C. Crick and N. V. Swindale for helpful comments, and Dr. N. V. Swindale for a useful suggestion about the representation of computed patterns.

REFERENCES

Bishop, P. O., Henry, G. H., and Smith, C. J. (1971). Binocular interaction fields of single units in the cat striate cortex. *J. Physiol. (Lond.),* **216**, 39–68.

Braitenberg, V., and Braitenberg, C. (1979). Geometry of orientation columns in the visual cortex. *Biolog. Cybernetics,* **33**, 179–186.

Gilbert, C. D. (1977). Laminar differences in receptive field properties in cat primary visual cortex. *J. Physiol. (Lond.),* **268**, 391–421.

Gilbert, C. D., and Wiesel, T. N. (1979). Morphology and intracortical projections of functionally identified neurons in cat visual cortex. *Nature,* **280**, 120–125.

Gilbert, C. D., and Wiesel, T. N. (1983). Clustered intrinsic connections in cat visual cortex. *J. Neurosci.,* **3**, 1116–1133.

Horton, J. C., and Hubel, D. H. (1981). Regular patchy distribution of cytochrome-oxidase staining in primary visual cortex of macaque monkey. *Nature,* **292**, 762–764.

Hubel, D. H., and Livingstone, M. S. (1981). Regions of poor orientation tuning coincide with patches of cytochrome oxidase staining in monkey striate cortex. *Proc. Soc. Neurosci. (Abstr.),* **7**, 357.

Humphrey, A. L., and Hendrickson, A. E. (1980). Radial zones of high metabolic activity in squirrel monkey striate cortex. *Neurosci. Abstr.,* **6**, 315.

Humphrey, A. L., and Norton, T. T. (1980). Topographic organization of the orientation column system in the striate cortex of the tree shrew *(Tupaia glis)*. I. Microelectrode recording. *J. comp. Neurol.,* **192**, 531–548.

Humphrey, A. L., Skeen, L. C., and Norton, T. T. (1980). Topographic organization of the orientation column system in the striate cortex of the tree shrew *(Tupaia glis)*. II. Deoxyglucose mapping. *J. comp. Neurol.,* **192**, 549–566.

Mitchison, G. J., and Crick, F. H. C. (1982). Long axons within the striate cortex: their distribution, orientation, and patterns of connection. *Proc. Natl. Acad. Sci. USA,* **79**, 3661–3665.

Rockland, K. S., and Lund, J. S. (1981). Anatomical banding of intrinsic connections of tree shrew striate cortex. *Investig. Ophth. Vis. Sci. (ARVO Abstr.)* **22**, 176.

Rockland, K. S., and Lund, J. S. (1982). Widespread periodic intrinsic connections in tree shrew visual cortex (area 17). *Science,* **215**, 1532–1533.

Rockland, K. S., and Lund, J. S. (1983). Intrinsic laminar lattice connections in primate visual cortex. *J. comp. Neurol.,* **216**, 303–318.

Rockland, K. S., Lund, J. S., and Humphrey, A. L. (1982). Anatomical banding of intrinsic connections in striate cortex of tree shrews *(Tupaia glis). J. comp. Neurol.,* **209**, 41–58.

Swindale, N. V. (1982). An enlarged view of intracortical connectivity. *Nature,* **300**, 313–314.

Models of the Visual Cortex
Edited by D. Rose and V. G. Dobson

CHAPTER 47

Iso-orientation domains* and their relationship with cytochrome oxidase patches

N. V. SWINDALE†
Physiological Laboratory, Downing St., Cambridge CB2 3EG, UK

As an electrode moves laterally through the visual cortex the preferred orientation of the cells recorded changes systematically (Hubel and Wiesel, 1962, 1968, 1974). Graphs of the preferred orientation plotted against lateral distance shows sequences of change that are often linear over several cycles of orientation change (Fig. 1), with abrupt and apparently unpredictable reversals in slope and occasional large discontinuous changes of orientation up to 90° in size. A complete 180° cycle of orientation change occurs on average every 600 μm in the monkey (Hubel and Wiesel, 1974; Hubel, Wiesel and Stryker, 1978) and 1,200 μm in the cat (Albus, 1979; Singer, 1981), the periodicity being almost constant over the entire extent of the visual cortex. To this has to be added the recent discovery that in the upper layers of the monkey visual cortex there are, superimposed in some way upon the pattern of orientation domains, periodically spaced 'patches' of cortex in which orientation selectivity is poor or absent (Livingstone and Hubel, 1984). These patches also contain elevated levels of a variety of enzymes, most notably cytochrome oxidase (Hendrickson, Hunt and Wu, 1981; Horton, 1984; Horton and Hubel, 1981).

* Though the phrase orientation column is in general use, its meaning is not well defined. It is not even certain that orientation is represented in a strictly columnar fashion in the visual cortex (Bauer *et al.*, 1983; Bauer, Dow and Vautin, 1980; Kruger and Bach, 1982). Iso-orientation domains are defined for a given orientation range and consist of regions of cortex (whether columnar or not) containing cells whose orientation preferences lie within that range. If orientation is not columnar, my arguments could be generalized to three dimensions without affecting their main conclusions.

† Present address: Department of Physiology, Dalhousie University, Halifax, Nova Scotia B3H 4JI, Canada.

Despite agreement about the nature of the one-dimensional layout of orientations on the cortical surface (i.e. as seen on a single-electrode penetration) it is still not clear what the two-dimensional layout of orientations is and what relationship, if any, there is with the cytochrome oxidase patches. According to Hubel and Wiesel (1974), and more recently Livingstone and Hubel (1984), iso-orientation domains must be narrow flat parallel slabs of tissue, probably running at right angles to ocular dominance columns, and with no particular relationship to cytochrome oxidase patches. Braitenberg and Braitenberg (1979) have suggested that orientations are arranged radially in a full 360° cycle around periodically spaced centres (like a set of waggon wheels in a hexagonal or nearly hexagonal array). Each centre would be a cytochrome oxidase patch (Braitenberg, Chapter 50 in this volume). In this article I want to repeat suggestions for an alternative arrangement of the orientation domains, made as a result of considering how orientation selectivity might arise during development (Swindale, 1982). In this case, however, I want to base my arguments not on a possible mechanism of development but on a consideration of the features present in the one-dimensional layout of orientations, as seen by a recording electrode, and to generalize from this to the probable two-dimensional arrangement. The features of greatest interest (Fig. 1) are: (a) the discontinuities in the sequences; (b) the linear slope of orientation change graphed against distance and (c) the periodically spaced regions of low orientation selectivity, located in the cytochrome oxidase patches.

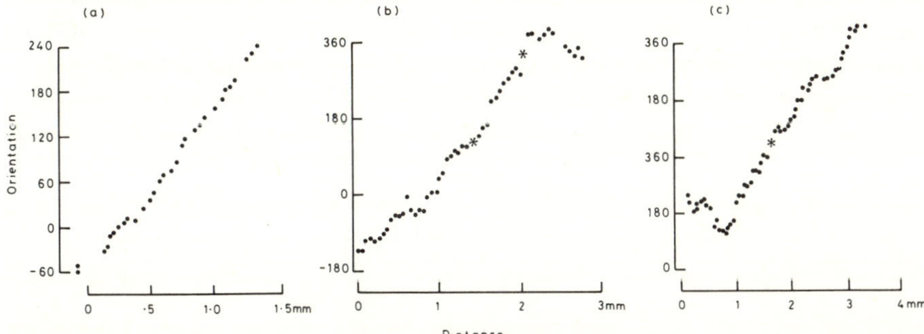

FIGURE 1 Graphs of preferred orientation versus distance along the recording electrode track: (a) is redrawn from Hubel and Wiesel (1974); (b) and (c) are redrawn from Livingstone and Hubel (1984) by permission of Society for Neuroscience, © 1984 Society for Neuroscience. The asterisks in (b) and (c) mark the location of non-oriented cells, located in cytochrome oxidase patches in the upper layers of the cortex.

DISCONTINUITIES

Though the existence of sudden changes of orientation of up to 90° (a larger change than this, e.g. one of 120°, is equivalent to a smaller jump of −60°)

may at first sight seem surprising, there are in fact very few ways of arranging a smoothly changing set of orientations on a surface without inevitably encountering discontinuities of this sort. The arguments that show this are worth giving, because they suggest an easy way of conceptualizing a smoothly varying layout of orientations: i.e. to represent orientation by a vector, $\mathbf{z}$, and to consider the two scalar components a and b of this vector. One problem that has to be faced in a representation of this sort is that while orientation is a number that is cyclic over the range 0 to 180°, the angle represented by a vector is cyclic over the range 0 to 360°. The simplest solution to this problem is to halve the angle represented by $\mathbf{z}$, i. e. to have $\theta = 0.5 \tan^{-1} (b/a)$, to obtain the orientation that would actually be measured by an experimenter. There is then a unique correspondence between the possible orientation values of $\mathbf{z}$ and those measured experimentally. It needs to be emphasized that the choice of this particular representation is mainly a matter of convenience; it does not affect the validity of conclusions based on its use, though it will (I hope) facilitate the forms of argument and calculation that I want to put forward.

If orientation changes smoothly and periodically, the two components of $\mathbf{z}$, a and b, must also change smoothly and periodically. The problem of representing orientations on the cortical surface is thus converted to one of considering how two periodically varying scalar functions are arranged. Next, consider that since all orientations are represented, there will be positions where a or b change sign and have a value of zero: if the functions vary periodically in two dimensions, the zeros will be two sets of periodically spaced lines which may be straight, curved or looped. The arrangement of these zeros tells one a good deal about the structure of the orientation domains, since they mark domain boundaries. (Note that a zero value for one function can code for two different orientations, depending on the sign of the other function at that point. Thus if $a=0$ and b is positive, the orientation represented is $\theta=45°$; if b is negative $\theta=135°$; conversely, when b is zero the orientations are 0 or 90° respectively.) If loops are absent (the zero values forming lines running between the boundaries of the visual cortex) and the zero values of a do not intersect or meet with those of b (Fig. 2a), then iso-orientation domains will be thin parallel unbranching strips (extending right across the cortex), as suggested by Hubel and Wiesel (1974) and Livingstone and Hubel (1984).

Alternatively, if loops are present, they could alternate in a quasi-concentric arrangement (resembling the contours on a map) without intersecting (Fig. 2b). Discontinuities would not arise as a natural consequence of either of these arrangements however; the fact that they seem to be fairly common suggests that neither of these configurations is generally the case. If the two sets of zero values meet, or intersect, as would be the case if the arrangement is at all different from either of the ones shown in Fig. 2a, b, then orientation

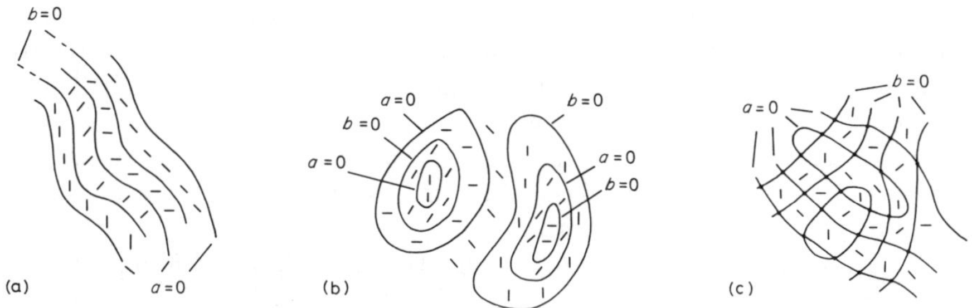

FIGURE 2 Three possible arrangements of the zero crossings of the functions *a* and *b* used to describe the organization of iso-orientation domains. (For the purpose of diagrammatic clarity, horizontal, vertical, left and right oblique orientations are shown on the figure; to do this the orientations actually represented by the values of *a* and *b* shown have been rotated by a constant 22.5°). (See text for a further description.)

jumps will be present at the intersections (Fig. 2c); this is clear from the fact that, since the zeros mark domain boundaries, different domains are forced to meet at the intersections; also, as mentioned above, if one function is zero, the angle represented depends only on the sign of the other function. If the sign changes, as it will at an intersection, the angle represented changes by 90°.

The existence of discontinuities in the orientation sequences thus suggest that if the cortical orientation map is resolved into two functions *a* and *b*, as defined above, their zero crossings will be found to intersect: *a* and *b* may in fact be found to be completely independent of each other. There would be two consequences of such an arrangement: discontinuities would be spaced with a periodicity roughly twice that of the iso-orientation domains and there should be little or no correlation between iso-orientation domains for orientations spaced 45° apart (a prediction testable by a double-label deoxyglucose experiment).

LINEAR ORIENTATION SEQUENCES

The existence of long linear sequences of orientation change spanning a 180° cycle or more (Fig. 1) implies that in these regions the functions *a* and *b* approximate closely to sinusoids that are 90° out of phase. This is easy to see, because the relationship $\theta = kx$, where x is distance and k is a constant, will be satisfied if $a = \sin(kx)$ and $b = \sin(kx + 90) = \cos(kx)$ (since $\theta = tan^{-1}[\sin(kx)/\cos(kx)] = kx$). The linear relationship conflicts with the Braitenbergs' model (and a variant of it: Dow and Bauer, 1984) because in their case orientation changes as a tangent, and not a linear, function of distance.

It is also interesting to note what happens if the phase relation between a and b is not 90°, as would often be the case if a and b varied independently. In this case the sequence wobbles about a sloping mean (Fig. 3b) with a periodicity equal to that of a and b. More than one obviously 'wobbly' sequence has been recorded (Figs. 1c and 3a, top) and analysis shows that the period of the wobble is about the same as the periodicity of the orientation domains (Fig. 3a, bottom).

The physiological data therefore suggests that both a and b approximate closely to sinusoids; in the language of Fourier analysis, this suggests that they are functions with a narrow bandwidth, centred on the basic periodicity of the domains. If the domains do not run in any overall preferred direction, as the evidence in the monkey suggests (Hubel *et al.*, 1978), then the energy in functions a and b may be dispersed over a wide range of orientations in

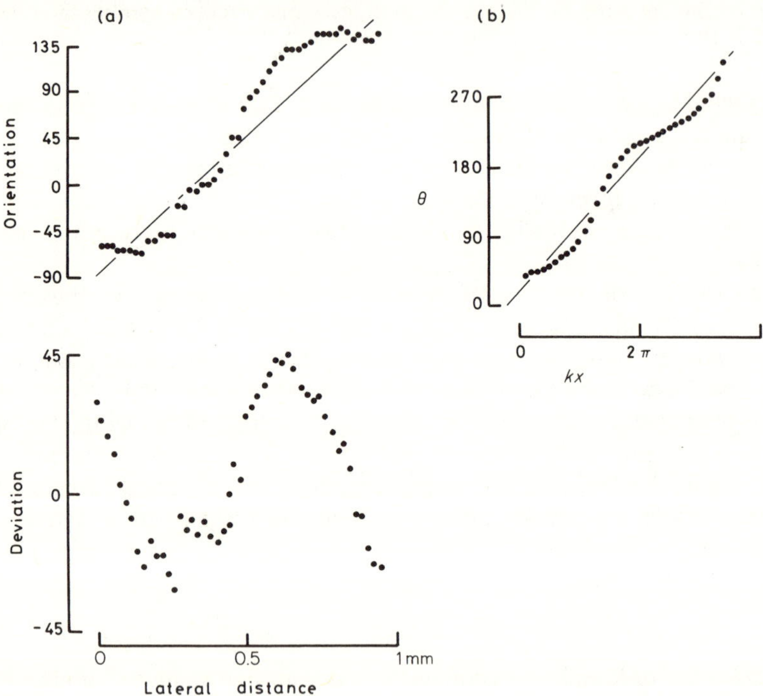

Figure 3 (a) Upper graph: an example of a 'wobbly' orientation sequence in the monkey visual cortex (taken from Hubel and Wiesel, 1974 by permission of Alan R. Liss, Inc.). Orientation change deviates approximately sinusoidally from a line of constant slope. Lower graph: the deviation in degrees of the data points from the sloping line drawn through them. The period of the deviation is about the same as the period of the orientation domains in the monkey. The latter part of the sequence shown in Fig. 1c behaves in the same way. (b) A graph of the function $\theta=0.5 \tan^{-1} [\sin(kx)/\sin(kx+\pi/4)]$; note the similarity in shape to the sequence actually observed in (a).

frequency space, or over a narrower range if the domains have a preferred direction, as in the cat (Singer, 1981). If *a* and *b* are uncorrelated (beyond having a similar peak frequency) the phase relations between them would be expected to be constant only over short distances (one or two periods); randomly occurring changes in the relative phase would account for the reversals in slope that are observed. Some forms of relationship between *a* and *b* need not be ruled out: e.g. there might be a tendency for their zero crossings to meet at right angles (as occurs in Fig. 2c). Alternatively, if the domains have an overall direction of elongation, as in the cat, the zero crossings might tend to avoid each other, giving rise to a more regular size and spacing of the domains and a smaller number of discontinuities.

If the zero crossings of *a* and *b* are curved rather than straight, it would mean that flat orientation sequences of any length would rarely be observed; iso-orientation lines could in principle have some degree of curvature almost everywhere, and at most might be approximately straight over a distance no more than about one domain period in some regions. The chance of an electrode travelling along an iso-orientation line would thus be very small: it would have to be both in the right place and moving in the right direction. The almost complete absence in the published experimental data of sequences that are flat over distances or more than a few hundred micrometres (of lateral movement through the cortex) is consistent with this, and argues against the idea that the orientation domains are flat slabs (Hubel and Wiesel, 1974; Livingstone and Hubel, 1984); if they were, it would not be difficult to orient a recording electrode parallel to them, and thereby prove the hypothesis.

REGIONS OF POOR ORIENTATION SELECTIVITY

The modulus of the vector $\mathbf{z}=(a^2+b^2)^{1/2}$ will be zero where the zero crossings of *a* and *b* coincide and small in the vicinity of this region. Since at such a point the orientation of **z** is undefined it is natural to suspect that the physiological counterpart might be a region where orientation selectivity is poor and that the modulus of **z** is related to the degree of orientation selectivity present at a particular cortical point. If the argument is correct and the zero crossings of *a* and *b* coincide about as frequently as one would expect if the two functions are uncorrelated, regions of poor orientation selectivity would be regularly spaced with a periodicity about twice that of the orientation domains. This corresponds well with the actual spacing of the cytochrome oxidase patches, which are about 250 to 300 μm apart on average (Horton and Hubel, 1981) compared with a spacing of about 600 μm for the orientation domains (Hubel and Wiesel, 1974; Hubel *et al.*, 1978). (The assignment of a physiological significance to the modulus of **z** in this way may seem somewhat *ad hoc*; after all, orientation selectivity is probably mechanistically rather complicated, and not likely to be adequately describable in such a simple way. The justification for the procedure, as in all theorizing, is that if it results in an economical description of phenomena, one should adopt it. In the physical world, and perhaps the biological one, quantities that have to be introduced in order to arrive at a simple description often turn out to have their counterpart in reality.)

If the postulated association between the modulus of $\mathbf{z}$ and orientation selectivity is correct, then discontinuities in the orientation sequences should be located in regions of poor orientation selectivity. It may be supposed that an orientation discontinuity could hardly be observed in a region in which orientation selectivity was absent. However, these regions are probably rather small and located within the central region of each cytochrome oxidase patch (Livingstone and Hubel, 1984). Outside this central region, and above and below the patches, orientation selectivity may be sufficiently marked to define a discontinuity. (This means that functions a and b must differ in some way between layers or that the relationship between the modulus of z and orientation selectivity is different in the upper and lower cortical layers.)

Experimentally, one should look to see if orientation selectivity does decrease in the neighbourhood of a discontinuity or whether the discontinuities are located above or below cytochrome oxidase patches. Evidence for a correlation already exists. For example, Hubel and Wiesel (1974) noted the existence of a group of colour-selective cells close to one discontinuity: such cells are known to be located within cytochrome oxidase patches (Livingstone and Hubel, 1984) and will probably be present above and below them since their distribution is columnar (Michael, 1981). One of the sequences illustrated in Livingstone and Hubel (1984), and reproduced here (Fig. 1b), shows an orientation jump located exactly above a region of poor orientation selectivity in a cytochrome oxidase patch. In other cases, however, the orientation sequences seem to continue undisturbed through such regions (Fig. 1b, c), and on this observation principally the authors base their conclusion that the patches and the orientation domains are independent. However, a surprisingly small distortion of a sequence can result in a discontinuity. Thus although the orientations on either side of the non-orientated region in Fig. 1c appear to be part of the same linear sequence, they in fact differ by nearly 90°, though only 150 μm apart. Study of the computer model used to simulate the arrangement of orientation domains also shows (Fig. 4a) that the discontinuities often exist only as very local distortions of otherwise nearly linear sequences. Thus there may be only a small region within each patch where a discontinuity is evident. The evidence that any of the discontinuities lie within the cytochrome patches conflicts with the Braitenbergs' model, since they would have them occurring outside the patches.

Deoxyglucose experiments should show that orientation domains coalesce with the cytochrome oxidase pathces, irrespective of the orientation used as a stimulus. Although the interpretation of deoxyglucose autoradiographs in the monkey is problematic (reviwed by Swindale, 1981), the evidence has been interpreted by two separate groups (Horton and Hubel, 1980, 1981; Tootel *et al.*, 1982) to suggest that orientation domains, independent of the orientation used to stimulate them, are confluent with the cytochrome oxidase patches.

Figure 4b summarizes the conclusions of the model: a computer was used to generate two independently varying functions with a narrow spatial frequency bandwidth and energy in all directions. The density of dots is inversely proportional to the modulus and is intended to simulate the appearance of the cytochrome oxidase patches. Two sets of iso-orientation domains are superimposed on this pattern, one set centred on a vertical orientation range, the other centred on a horizontal one. The domains meet in the patches and their curved boundaries run between them.

CONCLUSIONS

I suggest that iso-orientation domains in the visual cortex can be described by a vector whose two components are smoothly periodic with a narrow spatial frequency bandwidth and probably (though not necessarily) spatially uncorrelated with one another. The orientation of the vector describes the preferred orientation of cells at that point, and its magnitude is proportional to the degree of orientation selectivity. This simple model accounts for several apparently unrelated phenomena: the existence of discontinuities in the orientation sequences; the linear relationship over short distances between preferred orientation and distance; the occurrence of 'wobbly' sequences where there is an approximately sinusoidal deviation from linearity; and the existence of periodically spaced regions where orientation selectivity is poor. If the organization of the visual cortex can indeed be described in such a simple way, it surely will have significance for theories of how information about the visual world is represented within the cortex and how it is encoded and transmitted to other parts of the brain.

REFERENCES

Albus, K. (1979). C-deoxyglucose mapping of orientation subunits in the cat's visual cortical area. *Exp. Brain Res.*, **37**, 609–613.

Bauer, R., Dow., B. M., Snyder, A. Z., and Vautin, R. S. (1983). Orientation shift between upper and lower layers in the monkey visual cortex. *Exp. Brain Res.*, **50**, 133–145.

Bauer, R., Dow., B. M., and Vautin, R. G. (1980). Laminar distribution of preferred orientations in foveal striate cortex of monkey. *Exp. Brain Res.*, **41**, 54–60.

Braitenberg, B., and Braitenberg, C. (1979). Geometry of orientation columns in the visual cortex. *Biolog. Cybernetics,* **33**, 179–186.

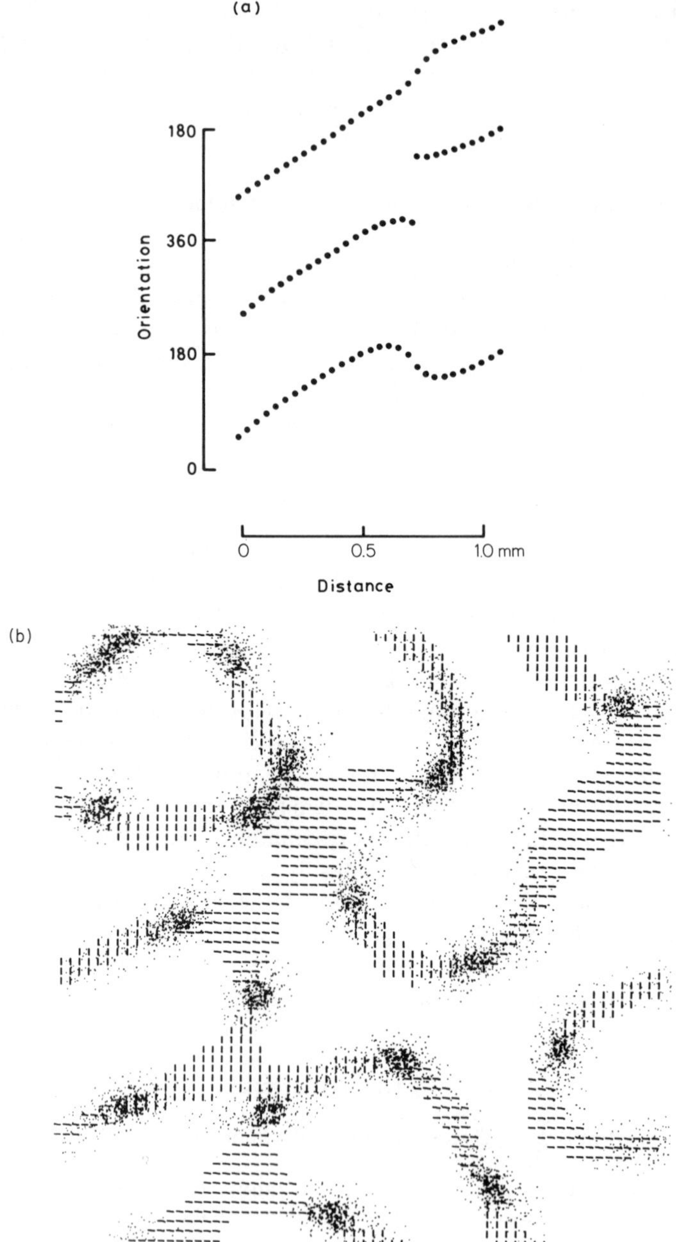

FIGURE 4 (a) Three adjacent orientation sequences generated by the model, spaced vertically by 180° for clarity. The sequences are parallel to one another, and spaced a distance equal to about one tenth the period of the domains (one and a half dot

Dow, B. M., and Bauer, R. (1984). Retinotopy and orientation columns in the monkey: a new model. *Biolog. Cybernetics,* **49**, 189–200.
Hendrickson, A. E., Hunt, S. P., and Wu, J.-Y. (1981). Immunocytochemical localisation of glutamic acid decarboxylase in monkey striate cortex. *Nature,* **292**, 605–607.
Horton, J. C. (1984). Cytochrome oxidase patches: a new cytoarchitectonic feature of monkey visual cortex. *Phil. Trans. Roy. Soc. Lond., B,* **304**, 199–253.
Horton, J. C., and Hubel, D. H. (1980). Cytochrome stain preferentially labels intersection of ocular dominance and vertical orientation columns in macaque striate cortex. *Soc. Neurosci. Abstr.,* **6**, 315.
Horton, J. C., and Hubel, D. H. (1981). Regular patchy distribution of cytochrome oxidase staining im primary visual cortex of macaque monkey. *Nature,* **292**, 762–764.
Hubel, D. H., and Wiesel, T. N. (1962). Receptive fields, binocular interaction, and functional architecture in the cat's visual cortex. *J. Physiol. (Lond.),* **160**, 106–154.
Hubel, D. H., and Wiesel, T. N. (1968). Receptive fields and functional architecture of monkey striate cortex. *J. Physiol. (Lond.),* **195**, 215–243.
Hubel, D. H., and Wiesel, T. N. (1974). Sequence regularity and geometry of orientation columns in the monkey striate cortex. *J. comp. Neurol.,* **158**, 267–294.
Hubel, D. H., Wiesel, T. N., and Stryker, M. P. (1978). Anatomical demonstration of orientation columns in macaque monkey. *J. comp. Neurol.,* **177**, 361–380.
Krüger, J., and Bach, M. (1982). Independent systems of orientation columns in upper and lower layers of monkey visual cortex. *Neurosci. Lett.,* **33**, 225–230.
Livingstone, M. S., and Hubel, D. H. (1984). Anatomy and physiology of a colour system in the primate visual cortex. *J. Neurosci.,* **4**, 309–356.
Michael, C. R. (1981). Columnar organization of colour cells in the monkey's striate cortex. *J. Neurophysiol.,* **46**, 587–604.
Singer, W. (1981). Topographic organization of orientation columns in the cat visual cortex. *Exp. Brain Res.,* **44**, 431–436.
Swindale, N. V. (1981). Patches in monkey visual cortex. *Nature,* **293**, 509.
Swindale, N. V. (1982). A model for the formation of orientation columns. *Proc. Roy. Soc. Lond., B.,* **215**, 211–230.
Tootell, R. B. H., Silverman, M. S., Switkes, E., and DeValois, R. L. (1982). *Soc. Neurosci. Abstr.,* **8**, 707.

---

spacings: equivalent to 60 μm in the monkey). The middle sequence passes through a discontinuity; to either side of it the sequences are smooth. In the uppermost sequence only a slight steepening of the curve is evident in the region where the discontinuity is going to develop. In the middle sequence, the curve jumps upwards by about 90°; the reversal in the lowest sequence takes the curve down by 90°, but continuity on the other side of the jump is preserved because the operations of adding 90° and subtracting 90° leave the orientation unchanged. (b) Computer simulation of a pattern of iso-orientation domains showing their postulated relationship with regions of low orientation selectivity. The density of dots is inversely proportional to the degree of orientation selectivity, and two sets of iso-orientation domains are shown: vertical lines indicate orientations within 15° of vertical and horizontal lines those within 15° of horizontal.

Models of the Visual Cortex
Edited by D. Rose and V. G. Dobson

CHAPTER 48

A proposed mechanism for the origin and development of iso-orientation columns

J.D. Cowan
Mathematics Dept., The University of Chicago, Chicago, Ill 60637, USA.
and
C. von der Malsburg
Max-Planck-Institut fur biophysikalische Chemie, Gottingen-Nikolausberg D-3400, Federal Republic of Germany.

INTRODUCTION

One of the most striking of Hubel and Wiesel's many discoveries is that most visual cortical cells are *orientation selective* (Hubel and Wiesel, 1962). A further striking discovery is that orientation selective cells are arranged in *iso-orientation columns* (Hubel and Wiesel, 1963, 1974a; Hubel, Wiesel and Stryker, 1977, 1978). These observations are not yet understood in terms of a blueprint or wiring diagram, detailing the circuits which produce such properties.

ORIENTATION SELECTIVITY

An obvious arrangement for orientation selectivity was first proposed by Hubel and Wiesel (1962), in which afferent fibers connect to a cortical cell to produce an elongated receptive field. Colonnier (1964) refined this arrangement in terms of elongated cortical dendritic fields. However, given the scatter in visual field positions of afferent fibers (Hubel and Wiesel, 1974b; Albus, 1975; Horton, Greenwood and Hubel, 1979; Williams and Rakic, 1984), such an arrangement seems unworkable. Indeed, the evidence does not support it. Creutzfeldt, Kuhnt and Benevento (1974) found that only a small proportion of cortical cells have elongated receptive fields, and Lund (1981) found no marked asymmetry in cortical dendritic fields. It would seem that the circuitry which produces orientation selectivity is mainly *intra-cortical*. Creutzfeldt, Kuhnt and Benevento (1974) and Finette, Harth and Csermely (1978) both suggested this, and in fact Sillito (1979, 1980) has shown that orientation selectivity is abolished if intra-cortical *inhibition* is blocked.

There are a number of model circuits which produce elongated receptive fields after a training process (Bienenstock, Cooper and Munro, 1982; Cooper, Liberman and Oja, 1979; Legendy, 1978; Nass and Cooper, 1975; von der Malsburg, 1973). However, orientation selectivity is already present in immature animals (Sherk and Stryker, 1976; Stryker and Sherk, 1975; Wiesel and Hubel, 1974).

ISO-ORIENTATION COLUMNS

Iso-orientation columns have been found in both cats and primates (Hubel and Wiesel, 1963, 1974a). These are slab-like aggregates of cortical cells, all of which have the same orientation preference. The slabs extend, more or less, through the cortical laminae, are 25 to 50 μm wide and can be several millimetres long (Hubel, Wiesel and Stryker, 1977, 1978). Thus a single iso-orientation column can comprise as many as 40 to 50 thousand cells. The receptive fields of these cells are not homogeneous and vary considerably in size and field position (Hubel and Wiesel, 1974a).

There has been considerable debate concerning the exact nature of iso-orientation columns. Braitenberg and Braitenberg (1979) proposed that organizing *centers* exist, in a doubly periodic array, and that iso-orientation domains are arranged like the spokes of a wheel, around such centers. Cytochrome oxidase staining (Wong-Riley, 1979) does indeed reveal a regular pattern of cortical patches or blobs (Horton and Hubel, 1981), but it now appears that they comprise cells tuned for *colour*, not for orientation (Livingstone and Hubel, 1984). It seems that iso-orientation columns are indeed slabs arranged in a periodic *stripe* pattern, as revealed by deoxyglucose staining (Hubel, Wiesel and Stryker, 1978). Horseradish peroxidase (HRP) stain (Rockland and Lund, 1982) reveals a periodic pattern of widespread intrinsic connections in cortical layers II and III of tree shrews and other primates. Mitchison and Crick (1982) interpret these observations to indicate that the HRP label is carried by a system of oriented axons which link cells with similar orientation preferences in a narrow cortical strip, the direction of which is related to such preferences. This is consistent with observations by Gilbert and Wiesel (1981) of widespread intrinsic connections of cortical layer V cells in cats, but not with the connections of cat supragranular cells.

INTERACTIONS BETWEEN ISO-ORIENTATION COLUMNS

The other striking aspect of orientation selectivity is of course that orientation preference changes in a piecewise *continuous* fashion across the cortical surface—i.e. in going from column to column (Hubel and Wiesel, 1974a). This bears on the nature of the interactions between iso-orientation columns. It seems reasonable that cells with the same orientation preference should facilitate each other, and therefore that the intrinsic connections described above should be *excitatory*. Enough is known about interactions produced by differing orientations (Blakemore and Tobin, 1972) to suggest that *inhibition* operates between columns. Thus the rules are: excitation between cells with similar orientation preferences, and inhibition between cells with differing preferences. The model introduced by von der Malsburg (1973) has nearest neighbor excitatory and next-nearest neighbor inhibitory interactions. Training sequences provided by external patterns produce a distribution of orientation preferences consistent with the above rules. A more recent simulation by Swindale (1982) embodies the rules from the start and shows rapid convergence to the iso-orientation columnar distributions found experimentally. Iso-orientation stripes, however, are obtained only if anisotropic interactions between neighboring orientations are incorporated.

A NEW MECHANISM FOR THE ORIGIN OF ISO-ORIENTATION COLUMNS

Both models are unsatisfactory in certain respects—von der Malsburg's because the formation of orientation preferences and columns is prenatal (Sherk and Stryker, 1976) and Swindale's because it does not address the problem of the ontogenesis of orientation selectivity. We describe here a model which addresses such problems (von der Malsburg and Cowan, 1982). The basic idea is as follows:

1. We suppose the cortex to be interconnected in such a way that, during its early development, it spontaneously generates periodic cellular activity patterns (see Fig. 1). It has been shown that the adult cortex could support such patterns, under the influence of hallucinogenic drugs (Ermentrout and Cowan, 1979; Cowan, 1982). We assume that the immature cortex can also support such patterns, and that prior to eye opening the cortex is active in the form of a random sequence of periodic (standing) wave patterns of all phases.
2. We assume that amongst the afferents to the visual cortex, there is a small subset of *orientation-inducing* fibers, possibly W-cell fibers of retinal origin (Stone and Hoffmann, 1972; Cleland and Levick, 1974), carrying only *vertical* (V) and *horizontal* (H) preferences. We suppose the synapses

between such fibers and cortical cells to be *modifiable*, so that spontaneous random activation of the V-fibers becomes associated with one of the cortical wave patterns and activation of the H-fibers with patterns of the opposite phase.

3. We suppose that following eye opening there are two further stages of development:

(a) during the critical period immediately following eye opening (Blakemore and Mitchell, 1973) and

(b) after the critical period is over.

In stage (a) we propose that patterned visual stimulation modulates the development of afferent connections and receptive fields, along the lines proposed by von der Malsburg (1973), mainly to generate sensitivity to *oblique* orientations. Some intrinsic cortical reorganization may also be expected to occur at this time, to reinforce the emergence of iso-orientation columns, in a fashion consistent with Swindale's simulation (1982). The process is essentially a filling-in of oblique orientations, between H- and V-fibers. Thus, if a retinal location is stimulated with an oblique orientation, say 45°, local H- and V-fibers will be stimulated, more or less symmetrically. As a result there will be two possible positions for the resulting cortical pattern, at phase angles of $\pm\pi/2$ relative to the phases of H and V patterns (see Fig. 2), i.e. if retinal stimulus orientation *rotates* with a particular sense, the corresponding cortical pattern can *translate* in either of two directions. This symmetry is supposedly broken spontaneously by random fluctuations of cortical activity, prior to eye opening, or by irregularities in the distribution of cortical afferents, during the initial critical period.

In stage (b), we propose that individual cells can adapt in such a way as to sharpen their orientation preferences, to generate elongated receptive fields and perhaps direction selectivity. The mechanisms proposed by von der Malsburg (1973), Nass and Cooper (1975), Perez, Glass and Schlaer (1975), Legendy (1978), Cooper, Liberman and Oja (1979) and Bienenstock, Cooper and Munro (1982) may operate at this stage.

DISCUSSION

The mechanism outlined in this paper for the development of iso-orientation columns is two-fold: intracortical interactions spontaneously generate stripe-like aggregates of coupled cells, following which peripheral afferents carrying orientation preferences become associated with such aggregates. The mechanism is essentially self-organizing and cooperative. As such, it has the property that very weak afferent modulations are sufficient to bias the process by which orientation preferences become associated with systems of stripe-like aggregates. Thus only a small fraction of cortical afferents need be

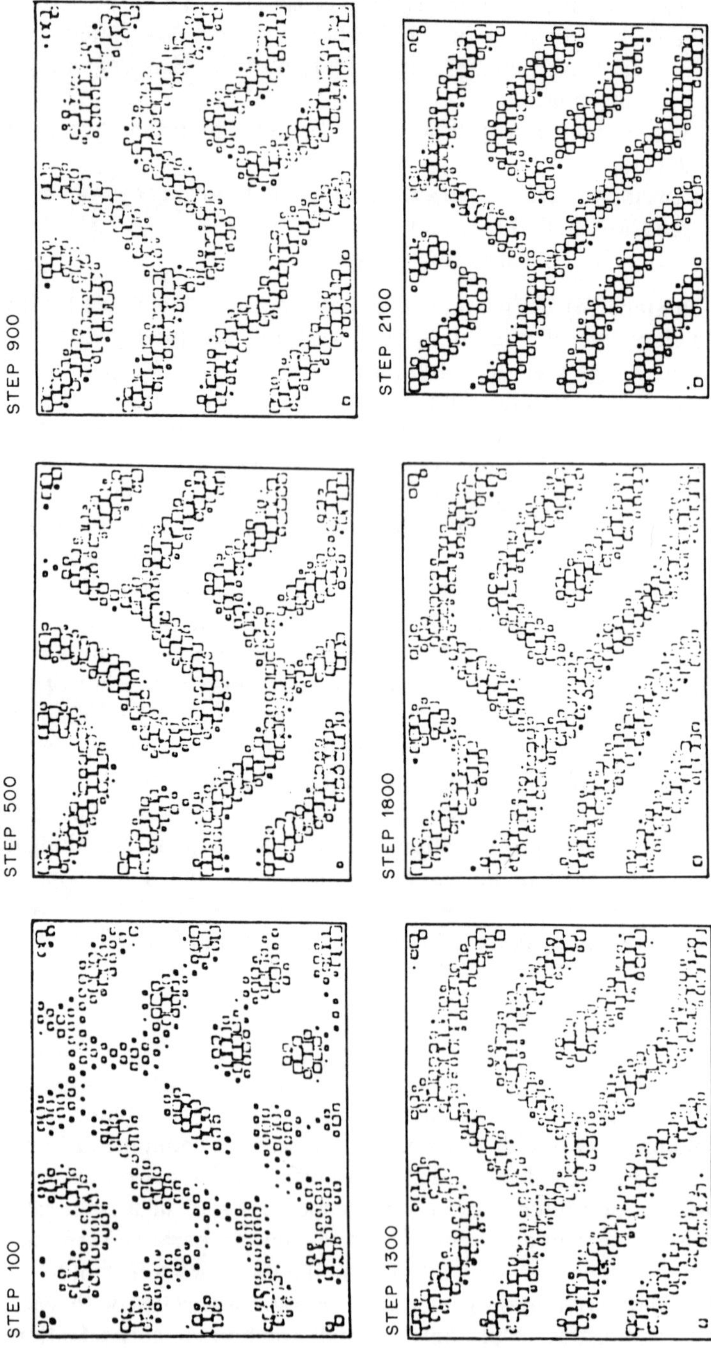
STEP 100
STEP 500
STEP 900
STEP 1300
STEP 1800
STEP 2100

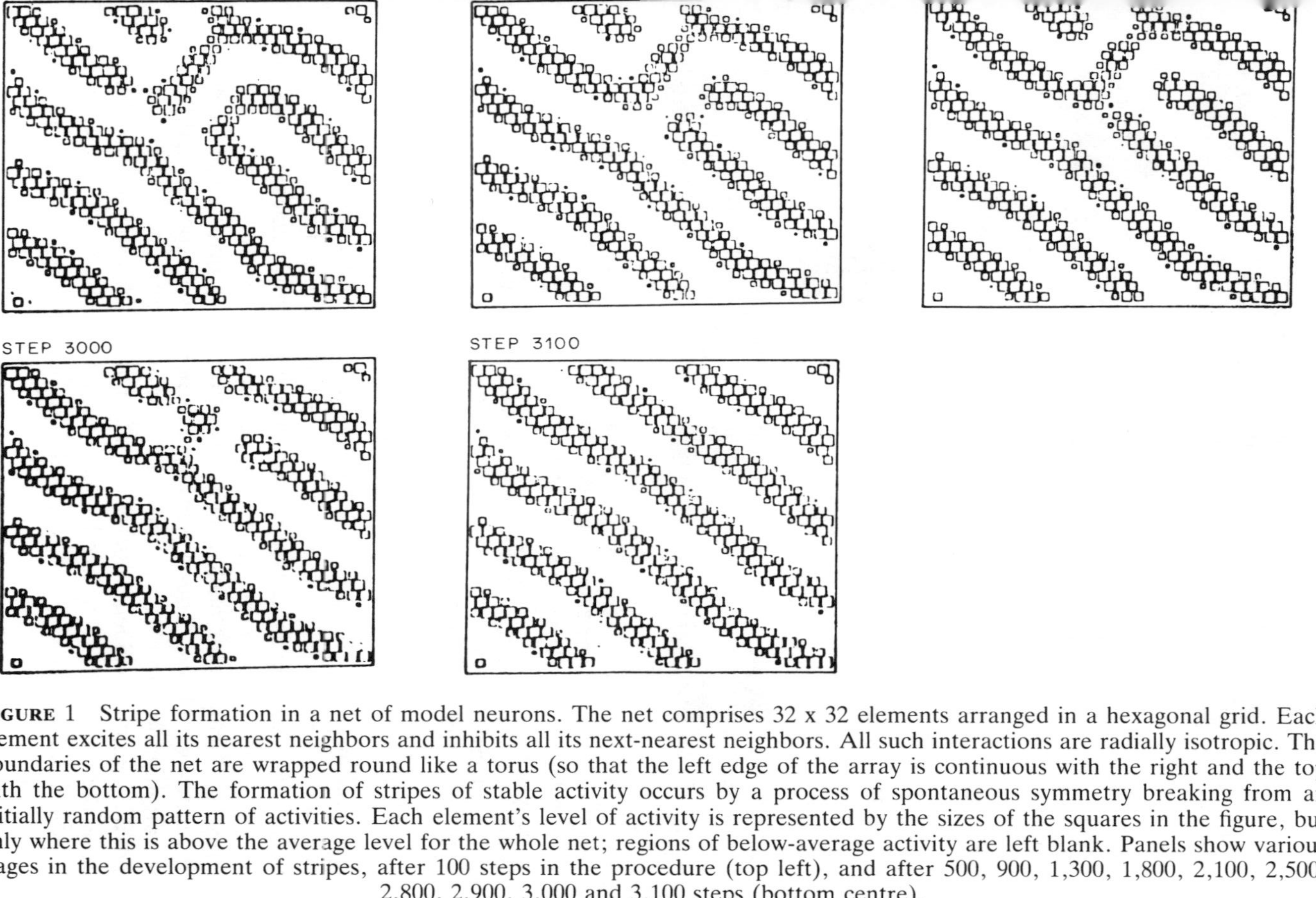

Figure 1 Stripe formation in a net of model neurons. The net comprises 32 x 32 elements arranged in a hexagonal grid. Each element excites all its nearest neighbors and inhibits all its next-nearest neighbors. All such interactions are radially isotropic. The boundaries of the net are wrapped round like a torus (so that the left edge of the array is continuous with the right and the top with the bottom). The formation of stripes of stable activity occurs by a process of spontaneous symmetry breaking from an initially random pattern of activities. Each element's level of activity is represented by the sizes of the squares in the figure, but only where this is above the average level for the whole net; regions of below-average activity are left blank. Panels show various stages in the development of stripes, after 100 steps in the procedure (top left), and after 500, 900, 1,300, 1,800, 2,100, 2,500, 2,800, 2,900, 3,000 and 3,100 steps (bottom centre).

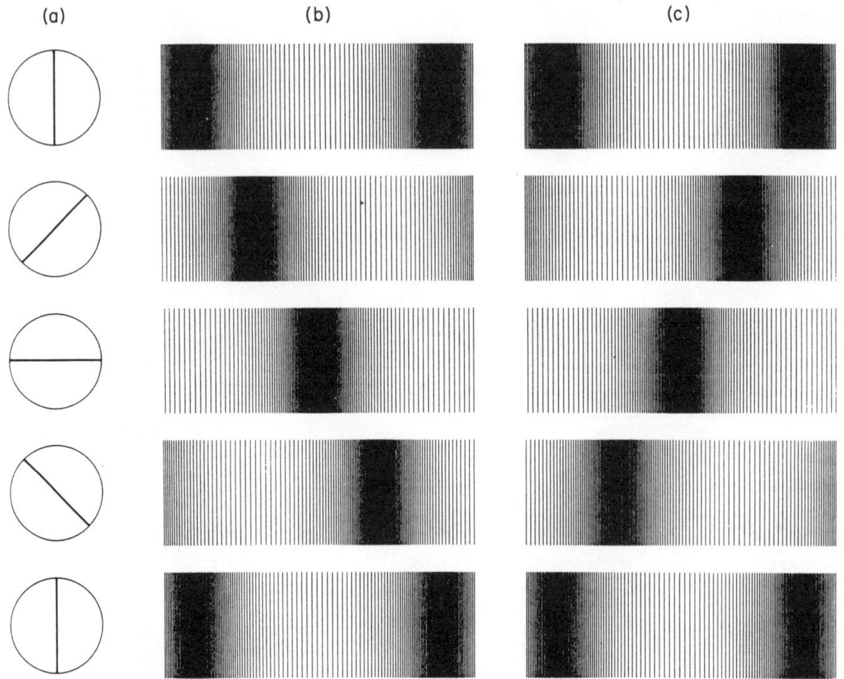

Figure 2 Coupling between cortical stripe activity patterns and oriented stimuli. (a) Local retinal stimuli-oriented light bars. (b) Levels of cortical activity in a two-dimensional strip of cortical tissue. The response to vertical stimuli (rows 1 and 5) and to horizontal stimuli (row 3) is fixed in the previsual stage by orientation-inducing fibers. In the early visual stage, responses to obliquely oriented stimuli are interpolated, giving a continuous map from orientations to cortical patterns; (b) shows one such map and (c) the other. (From von der Malsburg and Cowan, 1982, Fig. 2. Reproduced by permission of Springer-Verlag.)

orientation inducing. It may therefore be very difficult to find such fibers experimentally, especially since their cells of origin are unknown.

Such cells may be in the retina, lateral geniculate nucleus (LGN) or superior colliculus, or even in oculomotor nuclei. This latter possibility is interesting. Oculomotor nuclear cells respond only to oriented movements of the entire visual field. Thus activity in optic nerve fibers conveying differing orientation preferences would tend to be negatively correlated. On the other hand, if orientation-inducing fibers originate in retinotopically organized nets, such as the retina or LGN, then for each cortical hemisphere the fibers must be monocular (probably contralateral). It is hard to imagine how the necessary negative correlation of activity could be produced between cells in differing retinae or LGN layers.

There is some experimental evidence to support the postulated mechanism. Amongst the W-cells of the retina, there is at least one subset involved in transmitting orientation preferences to the superior colliculus (Hoffmann, 1973). Conceivably this is relayed to the cortex. There is also a direct W-cell pathway to the LGN, and thence to the cortex, some of the fibers of which could transmit the requisite preferences (Cleland *et al.*, 1975, 1976; Dreher and Sefton, 1979; LeVay and Ferster, 1977; Wilson, Rowe and Stone, 1976; Wilson and Stone, 1975). Levick and Thibos (1980) have found appreciable orientation selectivity in 70 per cent. of all cat retinal ganglion cells, while Daniels, Norman and Pettigrew (1977) and Vidyasagar and Urbas (1982) and Lee, Creutzfeldt and Elephandt (1979) have found orientation preferences in cat and primate LGN cells respectively. Interestingly, Vidyasagar and Urbas found that cooling the cortex abolished all the oblique LGN orientation preferences, but not the horizontal and vertical ones. Similarly, Leventhal and Hirsch (1975, 1977) found that oblique orientation preferences are much more easily obliterated by deprivation, compared with horizontal and vertical ones. There is also a greater representation of horizontal and vertical preferences, particularly in immature cats (Frégnac and Imbert, 1978; Frégnac *et al.*, 1981). In such animals there is an initial preponderance of monocular, contralaterally driven cells tuned to horizontal or vertical orientations. All these observations are consistent with our postulate that horizontal and vertical orientation preferences are subcortically induced, and that oblique preferences result from a subsequent postnatal visually organized process.

ACKNOWLEDGEMENTS

Supported in part by NATO grant 1791, and by grants from the System Development Foundation, Palo Alto, CA., and the Brain Research Foundation, University of Chicago.

REFERENCES

Albus, K. (1975). A quantitative study of the projection area of the central and the paracentral visual field in area 17 of the cat. 1. The precision of the topography. *Exp. Brain Res.*, **24**, 159–180.

Bienenstock, E. L., Cooper, L. N., and Munro, P. W. (1982). Theory for the development of neuron selectivity: orientation specificity and binocular interaction in visual cortex. *J. Neurosci.*, **2**, 32–48.

Blakemore, C., and Mitchell, D. E. (1973). Environmental modification of the visual cortex and the neural basis of learning and memory. *Nature,* **241**, 467–468.

Blakemore, C., and Tobin, E. A. (1972). Lateral inhibition between orientation detectors in the cat's visual cortex. *Exp. Brain Res.,* **15**, 439–440.

Braitenberg, V., and Braitenberg, C. (1979). Geometry of orientation columns in the visual cortex. *Biolog. Cybernetics,* **33**, 179–186.

Cleland, B. G., and Levick, W. R. (1974). Properties of rarely encountered types of ganglion cells in the cat's retina and an overall classification. *J. Physiol. (Lond.),* **240**, 457–492.

Cleland, B. G., Levick, W. R., Morstyn, R., and Wagner, H. G. (1976). Lateral geniculate relay of slowly conducting retinal afferents to cat visual cortex. *J. Physiol. (Lond.),* **255**, 299–320.

Cleland, B. G., Morstyn, R., Wagner, H. G., and Levick, W. R. (1975). Long latency retinal input to lateral geniculate neurons of the cat. *Brain Res.,* **91**, 306–310.

Colonnier, M. (1964). The tangential organization of the visual cortex. *J. Anat.,* **98**, 327–344.

Cooper, L. N., Liberman, F., and Oja, E. (1979). A theory for the acquisition and loss of neuron specificity in visual cortex. *Biolog. Cybernetic,* **33**, 9–28.

Cowan, J. D. (1982). Spontaneous symmetry breaking in large scale nervous activity. *Int. J. Quant. Chem.,* **XXII**, 1059–1082.

Creutzfeldt, O. D., Kuhnt, U., and Benevento, L. A. (1974). An intracellular analysis of visual cortical neurons to moving stimuli. *Exp. Brain Res.,* **21**, 251–274.

Daniels, J. D., Norman, J. L., and Pettigrew, J. D. (1977). Biases for oriented moving bars in lateral geniculate nucleus neurons of normal and stripe-reared cats. *Exp. Brain Res.,* **29**, 155–172.

Dreher, B., and Sefton, A. J. (1979). Properties of neurons in the cat's dorsal lateral geniculate nucleus: a comparison between medial intralaminar and laminated parts of the nucleus. *J. comp. Neurol.,* **183**, 47–64.

Ermentrout, G. B., and Cowan, J. D. (1979). A mathematical theory of visual hallucination patterns. *Biolog. Cybernetics,* **34**, 137–150.

Finette, S., Harth, E., and Csermely, T. J. (1978). Anisotropic connectivity and cooperative phenomena as a basis for orientation sensitivity in the visual cortex. *Biolog. Cybernetics,* **30**, 231–240.

Frégnac, Y., and Imbert, M. (1978). Early development of visual cortical cells in normal and dark-reared kittens: relationship between orientation selectivity and ocular dominance. *J. Physiol.,* **278**, 27–44.

Frégnac, Y., Trotter, Y., Bienenstock, E., Buisseret, P., Gary-Bobo, E., and Imbert, M. (1981). Effect of unilateral enucleation on the development of orientation sensitivity in the primary visual cortex of normal and dark-reared kittens. *Exp. Brain Res.,* **42**, 453–466.

Gilbert, C. D., and Wiesel, T. N. (1981). Laminar specialization and intracortical connections in cat primary visual cortex. In *The Organization of the Cerebral Cortex* (Eds. F. O. Schmitt *et al.*), MIT Press, Cambridge, Mass., pp.163–191.

Hoffman, K. P. (1973). Conduction velocity in pathways from retina to superior colliculus in the cat: a correlation with receptive field properties. *J. Neurophysiol.,* **36**, 409–424.

Horton, J. C., Greenwood, M. M., and Hubel, D. H. (1979). Non-retinotopic arrangement of fibres in cat optic nerve. *Nature,* **282**, 720–722.

Horton, J. C., and Hubel, D. H. (1981). Regular patchy distribution of cytochrome oxidase staining in primary visual cortex of macaque monkey. *Nature,* **292**, 762–764.

Hubel, D. H., and Wiesel, T. N. (1962). Receptive fields, binocular interaction, and functional architecture in the cat's visual cortex. *J. Physiol. (Lond.),* **160**, 106–154.

Hubel, D. H., and Wiesel, T. N. (1963). Shape and arrangement of columns in cat's striate cortex. *J. Physiol. (Lond.),* **165**, 559–568.

Hubel, D. H., and Wiesel, T. N. (1974a). Sequence regularity and geometry of orientation columns in the monkey striate cortex. *J. comp. Neurol.,* **158**, 267–294.

Hubel, D. H., and Wiesel, T. N. (1974b). Uniformity of monkey striate cortex: a parallel relationship between field size, scatter, and magnification factor. *J. comp. Neurol.,* **158**, 295–306.

Hubel, D. H., Wiesel, T. N., and Stryker, M. P. (1977). Orientation columns in monkey visual cortex demonstrated by the 2-deoxyglucose autoradiographic technique. *Nature.*, **269**, 328–330.
Hubel, D. H., Wisel, T. N., and Stryker, M. P. (1978). Anatomical demonstration of orientation columns in macaque monkey. *J. comp. Neurol.*, **177**, 361–380.
Lee, B. B., Creutzfeldt, O. D., and Elepfandt, A. (1979). The responses of magno- and parvo-cellular cells of the monkey's lateral geniculate body to moving stimuli. *J. comp. Neurol.*, **35**, 547–577.
Legendy, C. R. (1978). Cortical columns and the tendency of neighboring neurons to act similarly. *Brain Res.*, **158**, 89–105.
LeVay, S., and Ferster, D. (1977). Relay cell classes in the lateral geniculate nucleus of the cat and the effects of visual deprivation. *J. comp. Neurol.*, **172**, 563–584.
Leventhal, A. G., and Hirsch, H. V. B. (1975). Cortical effects of early selective exposure to diagonal lines. *Science*, **190**, 902–904.
Leventhal, A. G., and Hirsch, H. V. B. (1977). Effects of early experience upon orientation sensitivity and binocularity of neurons in visual cortex of cats. *Proc. Natl. Acad. Sci. USA*, **74**, 1272–1276.
Levick, W. R., and Thibos, L. N. (1980). Orientation bias of cat retinal ganglion cells. *Nature*, **268**, 389–390.
Livingstone, M. S., and Hubel., D. H. (1984). Anatomy and physiology of a color system in the primate visual cortex. *J. Neurosci.*, **4**, No. 1, 309-340.
Lund, J. S. (1981). Intrinsic organization of the primate visual cortex, area 17, as seen in Golgi preparations. In *The Organization of the Cerebral Cortex* (Eds. F. O. Schmitt *et al.*), MIT Press, Cambridge, Mass, pp.125–152.
Mitchison, G., and Crick, F. (1982). Long axons within the striate cortex: their distribution, orientation, and patterns of connection. *Proc. Natl. Acad. Sci. USA*, **79**, 3661–3665.
Nass, M. M., and Cooper, L. N. (1975). A theory for the development of feature detecting cells in visual cortex. *Biolog. Cybernetics*, **19**, 1–18.
Perez, R., Glass, L., and Schlaer, R. (1975). Development of specificity in the cat visual cortex. *J. Math. Biol.*, **1**, 275–288.
Rockland, K. S., and Lund, J. S. (1982). Widespread periodic intrinsic connections in the tree shrew visual cortex. *Science*, **215**, 1532–1534.
Sherk, H., and Stryker, M. P. (1976). Quantitative study of cortical organization in visually inexperienced kitten. *J. Neurophysiol.*, **39**, No. 1, 63–70.
Sillito, A. M. (1979). Inhibitory mechanisms influencing complex cell orientation selectivity and their modification at high resting discharge levels. *J. Physiol. (Lond.)*, **289**, 33–53.
Sillito, A. M. (1980). Orientation selectivity and the spatial organization of the afferent input to the striate cortex. *Exp. Brain Res.*, **41**, A9.
Stone, J., and Hoffmann, K. P. (1972). Very slow conducting ganglion cells in the cat's retina: a major new functional type? *Brain Res.*, **43**, 610–616.
Stryker, M. P., and Sherk, H. (1975). Modification of cortical orientation selectivity in the cat by restricting visual experience: a reexamination. *Science*, **190**, 904–905.
Swindale, N. V. (1982). A model for the formation of orientation columns. *Proc. Roy. Soc. Lond. B*, **215**, 211–230.
Vidyasagar, T. R., and Urbas, J. V. (1982). Orientation sensitivity of cat LGN neurons with and without inputs from visual cortical areas 17 and 18. *Exp. Brain Res.*, **46**, 157–169.
von der Malsburg, C. (1973). Self-organization of orientation sensitive cells in the striate cortex. *Kybernetik*, **14**, 85–100.

von der Malsburg, C., and Cowan, J. D. (1982). Outline of a theory for the ontogenesis of iso-orientation domains in visual cortex. *Biolog. Cybernetics,* **45**, 49–56.
Wiesel, T. N., and Hubel, D. H. (1974). Ordered arrangement of orientation columns in monkeys lacking visual experience. *J. comp. Neurol.,* **158**, 307–318.
Williams, R. W., and Rakic, P. (1984). Axon-axon order is not preserved in the embryonic primate optic nerve. An electron microscope analysis from serial sections. *Invest. Ophthalmol. Suppl.* **4**.
Wilson, P. D., Rowe, M. H., and Stone, J. (1976). Properties of relay cells in cat's lateral geniculate nucleus: a comparison of W-cells with X- and Y-cells. *J. Neurophysiol.,* **39**, 1193–1209.
Wilson, P. D., and Stone, J. (1975). Evidence of W-cell input to the cat's visual cortex via the C laminae of the lateral geniculate nucleus. *Brain Res.,* **92**, 472–478.
Wong-Riley, M. (1979). Changes in the visual system of monocularly sutured or enucleated cats demonstrable with cytochrome oxidase histochemistry. *Brain Res.,* **171**, 11–28.

Models of the Visual Cortex
Edited by D. Rose and V. G. Dobson

Chapter 49

Mathematical link between the width of cortical columns and the range of intracortical inhibition

CHARLES R. LEGÉNDY*
Department of Neuroscience, Albert Einstein College of Medicine, 1300 Morris Park Avenue, Bronx, NY 10461, USA

INTRODUCTION

A little over ten years ago, von der Malsburg (1973) discovered, in computer simulation, that a straightforward set of assumptions leads to the creation of a column-like arrangement of preferred trigger feature orientations in a network of idealized striate cortex neurons. Subsequently I have succeeded in formulating the corresponding problem for mathematical treatment and solving it analytically (Legéndy, 1978). Obtaining an analytic solution in this case has the advantage over computer simulation that it yields formulae, indicating what depends on what and how, and that it gives us a hint as to what is the simplest set of assumptions under which a certain result appears. The computer only gives us numbers and graphs, without comment.

In this chapter the resulting model is described without the mathematics, its properties listed and the experimental evidence discussed.

SKETCH OF THE MODEL

The model describes cortical columns in terms of a tendency of neighboring neurons to act similarly, and describes this tendency, termed *lateral assimilation*, as arising under the influence of synaptic inputs from the thalamus, an assumed plastic modifiability of thalamocortical synapses, and short-range excitatory interactions between cells. The fact that such a tendency can be described in a mathematically sound manner and without coming into conflict with the available data is considered to be an indication that a genetically

* Present address: ITT Avionics Division, 390 Washington Avenue, Nutley, N.J. 07110, USA.

determined column pattern is not the only possible explanation of the related physiological and anatomical observations.

It is possible to sketch the model, illustrating what it does and why, in a nonmathematical way, by the following schematic picture. Let us consider two neurons in the striate cortex, A and B, both receiving excitatory synaptic input from the thalamus, with B receiving also an excitatory synaptic input from A. Let us assume that the thalamocortical synapses are modifiable in accordance with Hebb's (1949) rule, meaning that excitatory synaptic contact becomes more effective if the presynaptic and postsynaptic neuron fire together often.

Let us consider a simple and frequent retinal image, keeping in mind the example of an oriented bar or edge, and designate the set of thalamic neurons responding to it by S. Finally, let us assume that S excites A strongly enough to make it respond, and excites B somewhat more weakly—not enough to make it respond without help from the excitatory inputs from A to B, but enough to make it respond with that help. (In the more detailed description A is a *set* of neurons and their combined help is accordingly stronger.) Because of the Hebbian synapse modification rule, the excitatory contacts from S to B are strengthened every time B fires in response to S. Every time this happens, in other words, the cell B becomes more responsive to the elementary retinal image in question; the response properties of A and B become more similar. The reason they do, it can be said, is the excitatory contact from A to B.

CONSEQUENCES AND PREDICTIONS OF THE MODEL

The formal argument is more general, but the foregoing simple argument is enough to show the main mechanisms and make the main consequences plausible. In particular:

1. Since neighboring cells are more likely to be excitatorily interconnected than cells far apart, neighboring cells will have a greater chance of acquiring similar response properties, resulting in a tendency of neighboring neurons to act similarly (*lateral assimilation*).
2. After many repetitions of the elementary step just sketched, the cortico-cortical excitation becomes no longer necessary for making the neuron B respond. The response similarity becomes anchored in the pattern of thalmocortical contacts.
3. The expected range of response similarity is about the same as the range of strongest intracortical excitation.
4. By the same token, responses tend towards greatest dissimilarity in the approximate range of strongest intracortical inhibition. In terms of cortical columns, the distance of greatest dissimilarity of response is the same

as the width of a column. For orientation columns, it thus makes sense to redefine the word 'column' to refer to a region of cortex which spans 90°; the earlier concept of orientation steps of 10° or less (Hubel, Wiesel and LeVay, 1975) should more properly be referred to as 'minicolumns'. The distance of greatest dissimilarity of response is then across one column width, both for orientation columns (a change of orientation of 90°) and for ocular dominance columns (from the midline of a left-eye column to the midline of a right-eye column). This, then, is the mathematically created link mentioned in the title between the width of cortical columns and the range of intracortical inhibition.

5. If the range of intracortical excitation is greater along the cortical vertical than along the cortical horizontal, as may be caused by electrotonic interaction amongst pyramidal cells in the immature cortex, then lateral assimilation will result in a vertically elongated pattern of response similarity; in other words, in a column-like pattern.
6. Lateral assimilation does not require visually caused inputs, only inputs; therefore spontaneous inputs from the thalamus may under certain conditions set up a columnar pattern within the cortex, explaining reported columnar organization in animals never exposed to patterned visual stimulation.

DISCUSSION

The most readily testable prediction of the model is that the distance of strongest intracortical inhibition should approximate the distance of greatest dissimilarity of neuron response. It has been found by Hubel, Wiesel and LeVay (1975) that the horizontal distance in the visual cortex at which the preferred orientation changes by 180° is roughly the same as the combined width of a left-eye and right-eye ocular dominance column. The model requires them both to agree with the distance of strongest intracortical inhibition. By scanning the available data, it is found that the range at which cortical cells are most frequently found to inhibit each other is between 100 and 400 μm (Asanuma and Rosen, 1973; Curtis and Felix, 1971; Hess, Negishi and Creutzfeldt, 1975), in approximate agreement with the widths of single columns (250 μm).

The model, whose results are in probabilistic terms, does not describe columns or lateral assimilation as a strict identity, for instance, of preferred orientations along vertical penetrations through the visual cortex, as appears in the early Hubel and Wiesel papers. In fact, Albus (1975) and Lee *et al.* (1977) find a substantial (roughly 10°) scatter about the smoothly varying means.

The 'column-like' nature of striate orientation columns, in other words their integrity from top to bottom of the cortical sheet, has recently come under fire, when a gross discontinuity has been discovered in the preferred orientations at the transition between the deeper and superficial layers in the central striate cortex of the cat (Bauer, 1982), as well as in the foveal striate cortex of the monkey (Bauer, Dow and Vautin, 1980). The discontinuity is not surprising in view of the present model, at least not in the monkey, where a layer of cells without orientation selectivity (Hubel and Wiesel, 1968) has been described as the region where continuity was disrupted, the columns being continuous above and below the layer (Bauer, Dow and Vautin, 1980). If lateral assimilation is carried by the orientation responsiveness of the assimilating cells, then disruption of orientation responsiveness will disrupt the assimilation.

The present model uses Hebb's (1949) synapse modification hypothesis in deriving lateral assimilation. It is the first success of this oft-quoted hypothesis in explaining a known feature of the nervous system. The only attempt to my knowledge to test the Hebbian rule was in *Aplysia* (Wurtz, Castellucci and Nusrala, 1967) and produced a negative result.

In estimating the ontogenetic time of lateral assimilation, it is useful to recall some of the data available on the establishment of thalamocortical connections in the visual system of the cat. Geniculate fibers begin to synapse in area 17 before birth, in area 18 in the third week and in area 19 in the second week after birth (Anker and Cragg, 1974). The early geniculostriate synapses are in the cortical layers I and II; only after about two weeks does the massive projection to layer IV appear (Laemle, Benhamida and Purpura, 1972). The total number of synapses on the average visual cortical neuron reaches its maximum, estimated to be about 12,000 synapses per neuron, around the fortieth day of life, and subsequently drops to the adult level of about 9,000 synapses per neuron (Cragg, 1975). At the end of the third week about 67 per cent. of the optic radiation fibers are myelinated (Moore, Kalil and Richards, 1976). The first evoked potentials in the visual cortex appear on the second day of life, but these are all surface-negative and their latency is as much as ten times the adult value. Adult latency is reached between the sixth week and the third month. The positive wave, believed to be caused by corticocortical inhibitory contacts (Creutzfeldt, Lux and Nacimiento, 1964), appears between the second and fifth week (Ellingson and Wilcott, 1960). Unit responses appear about the same time as evoked potentials (Huttenlocher, 1967). The eyes open at 3 to 15 days; the optic media clear at about 4 weeks after birth (Vital-Durand and Jeannerod, 1974).

The brains of kittens raised without any visual experience at all ('naive' kittens) have been reported to contain strip-like regions of alternating eye preference (Hubel and Wiesel, 1974; Pettigrew, 1974), which survive binocular deprivation. A continuous variation of preferred orientations in near-horizontal penetrations has been observed in naive kittens 1 to 3 weeks old (Blakemore and Van Sluyters, 1975), and the rudiments of column-like groupings of cells with like directional preference in vertical penetrations has been reported in naive kittens about 3 weeks old (Sherk and Stryker, 1976).

Since lateral assimilation evidently arises very early in life, it may be tentatively concluded that the Hebbian synapse modification rule is in effect in certain synapses, at this very early time; although not necessarily later.

ACKNOWLEDGEMENT

This work was supported by the National Eye Institute, under grant no. 5 RO1 EY03072.

REFERENCES

Albus, K. (1975). A quantitative study of the projection area of the central and paracentral visual field in area 17 of the cat. II. The spatial organisation of the orientation domain. *Exp. Brain Res.*, **24**, 181–202.

Anker, R. L., and Cragg, B. G. (1974). Development of the extrinsic connections of the visual cortex of the cat. *J. comp. Neurol.* **154**, 29–42.

Asanuma, H., and Rosen, I. (1973). Spread of mono- and polysynaptic connections within cat's motor cortex. *Exp. Brain Res.*, **16**, 507–520.

Bauer, R. (1982). A high probability of an orientation shift between layers 4 and 5 in central parts of the cat striate cortex. *Exp. Brain Res.*, **48**, 245–255.

Bauer, R., Dow, B. M., and Vautin, R. (1980). Laminar distribution of preferred orientations in foveal striate cortex of the monkey. *Exp. Brain Res.*, **41**, 54–60.

Blakemore, C., and Van Sluyters, C. (1975). Innate and environmental factors in the development of the kitten's visual cortex. *J. Physiol.* (*Lond.*), **248**, 663–716.

Cragg, B. G. (1975). The development of synapses in the visual system of the cat. *J. comp. Neurol.*, **160**, 147–166.

Creutzfeldt, O. D., Lux, H. D., and Nacimiento, A. C. (1964). Intracellulaere Reizung corticaler Nervenzellen. *Pfluegers Arch.*, **281**, 129–151.

Curtis, D. R., and Felix, D. (1971). The effect of bicuculline upon synaptic inhibition in the cerebral and cerebellar cortices of the cat. *Brain Res.*, **34**, 301–321.

Ellingson, R. J., and Wilcott, R. C. (1960). Development of evoked response in visual and auditory cortices of kittens. *J. Neurophysiol.*, **23**, 363–375.

Hebb, D. O. (1949). *The Organization of Behavior*, Wiley, New York.

Hess, R., Negishi, K., and Creutzfeldt, O. D. (1975). The horizontal spread of intracortical inhibition in the visual cortex. *Exp. Brain. Res.*, **22**, 415–419.

Hubel, D. H., and Wiesel, T. N. (1968). Receptive fields and functional architecture of monkey striate cortex. *J. Physiol.*, **195**, 215–243.

Hubel, D. H., and Wiesel, T. N. (1974). Ordered arrangement of orientation columns in monkeys lacking visual experience. *J. comp. Neurol.*, **158**, 307–318.

Hubel, D. H., Wiesel, T. N., and LeVay, S. (1975). Functional architecture of area 17 in normal and monocularly deprived macaque monkeys. *Cold Spring Harbor Symp. Proc.*, **40**, 581–589.

Huttenlocher, P. R. (1967). Development of cortical neuronal activity in neonatal cat. *Exp. Neurol.*, **17**, 247–262.

Laemle, L., Benhamida, C., and Purpura, D. P. (1972). Laminar distribution of geniculocortical afferents in visual cortex of the postnatal kitten. *Brain Res.*, **41**, 25–37.

Lee, B. B., Albus, K., Heggelund, P., Hulme, M. J., and Creutzfeldt, O. D. (1977). The depth distribution of optimal stimulus orientations for neurons in cat area 17. *Exp. Brain Res.*, **27**, 301–314.

Legéndy, C. R. (1978). Cortical columns and the tendency of neighboring neurons to act similarly. *Brain Res.*, **158**, 89–105.

Moore, C. L., Kalil, R., and Richards, W. (1976). Development of myelination in optic tract of the cat. *J. comp. Neurol.*, **165**, 125–136.

Pettigrew, J. D. (1974). The effect of visual experience on the development of stimulus specificity by kitten cortical neurons. *J. Physiol.* (*London.*), **237**, 49–74.

Sherk, H., and Stryker, M. P. (1976). Quantitative study of cortical orientation selectivity in visually inexperienced kitten. *J. Neurophysiol.*, **39**, 63–70.

Vital-Durand, F., and Jeannerod, M. (1974). Role of visual experience in the development of optokinetic response in kittens. *Exp. Brain Res.*, **20**, 297–302.

von der Malsburg, C. (1973). Self-organization of orientation sensitive cells in the striate cortex. *Kybernetik*, **14**, 85–100.

Wurtz, R. H., Castellucci, V. F., and Nusrala, J. M. (1967). Synaptic plasticity: the effect of the action potential in the postsynaptic neurol. *Exp. Neurol.*, **18**, 350–368.

Models of the Visual Cortex
Edited by D. Rose and V. G. Dobson

CHAPTER 50

An isotropic network which implicitly defines orientation columns: Discussion of an hypothesis

V. Braitenberg
Max-Planck-Institute for Biological Cybernetics, Spemannstrasse 38, D-7400 Tübingen, Federal Republic of Germany

The following facts of visual cortical physiology constitute a puzzle when put in terms of neuronal modelling:

1. There is an orderly projection of points in the visual field onto a cortical map. The grain of this map is that of the separation of geniculocortical input fibres, of the order of 0.1 mm or better (Hubel, Wiesel and LeVay, 1974). However, most neurons of the cortex have receptive fields comprising points in the visual field whose cortical representations may be well over a millimetre apart (Hubel and Wiesel, 1974b; Dow *et al.*, 1981).
2. Within these fields, elongated visual details influence the neuronal activity. The optimal stimulus is often much thinner than the field is wide (Hubel and Wiesel, 1968). The exact position of the stimulus within the field is often not critical, but the orientation almost always is.
3. The orientation preferences of neurons picked up by an electrode proceeding tangentially through the cortex change in regular sequences (Hubel and Wiesel, 1974a). When the preferred orientations are drawn onto the cortical map, they show a tendency to swirl around centres situated 0.4 to 0.5 mm from each other (Braitenberg and Braitenberg, 1979).
4. The positions of receptive fields within the visual field scatter irregularly for neurons situated in any narrow region of cortex. The scatter is about as much as the individual field is large (Hubel and Wiesel, 1974b). The scatter of orientation preference for the same neurons is less imposing.

5. Perceptual experiments in humans (Wilson, 1983) suggest that information about the periodicity of visual patterns is processed through a small number of discrete channels, having characteristic space frequencies of 16, 8, 4, 2.8 c. p. d. and lower (in the central visual field).

These facts are well explained by the following model, based on the assumption that special inhibitory neurons, or a greater number of inhibitory neurons, are located in the cortex at positions about 0.4 to 0.5 mm apart, perhaps corresponding to the 'cytochrome oxydase blobs' discovered in monkey cortex by Horton and Hubel (1980) and Humphrey and Hendrikson (1980). This assumption is suggested by the finding (Carroll and Wong-Riley, 1982) that the blobs are due to the staining of mitochondria mainly within neurons that have the characteristics of spineless stellate cells. Figure 1 shows the principle by which the dendritic field of a pyramidal cell (P) between two dendritic fields of inhibitory cells (I) acts as a detector for elongated stimuli (dashed outline) at a certain orientation. The pyramidal cell will be excited maximally when the stimulus does not fall on the dendritic trees of the neighbouring inhibitory cells (Fig. 1, bottom) and will respond less or not at all to other orientations (Fig. 1, top). We may say that each pyramidal cell has a receptive field which is that of its own dendritic tree minus the area occupied by dendritic trees of inhibitory cells under whose influence it is. Depending on the range of inhibition and on the relative position of pyramidal and inhibitory cells, these fields ('microfields') will be biconcave or crescent shaped.

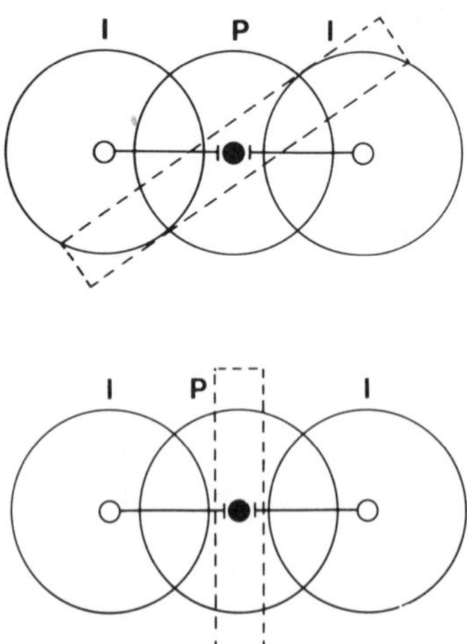

FIGURE 1 See text for details.

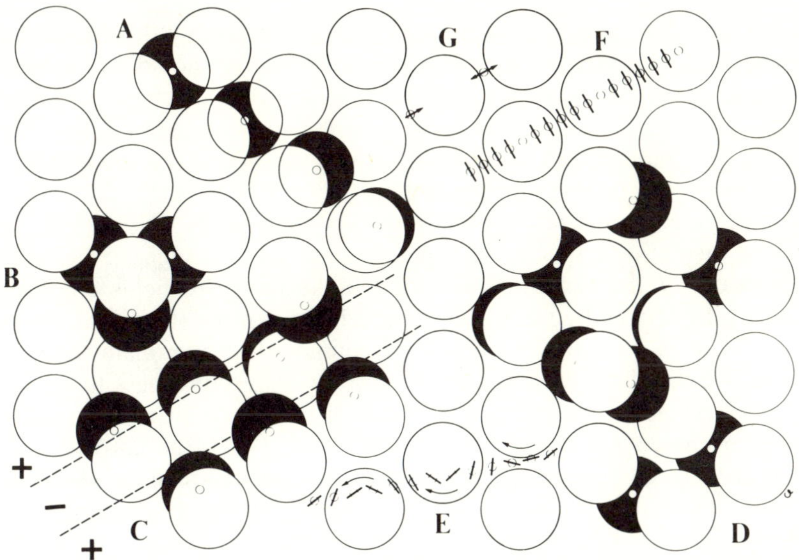

Figure 2 See text for details.

This is shown in Fig. 2, which represents a schematic view of the visual cortex from the top. The circles arranged in a hexagonal array are the dendritic fields of inhibitory neurons. We assume their centres to be at a distance of 0.4 mm from each other. Other circles (Fig. 2A) represent dendritic fields of some pyramidal cells. Their biconcave or crescent-shaped receptive fields are shown in black, the positions of the corresponding perikarya being indicated by small circles. The microfields of individual pyramidal cells have their long axes oriented along circles surrounding the inhibitory centres (Fig. 2B). The system of microfields pertaining to one centre may be called a hypercolumn. A probe traversing one hypercolumn would encounter a sequence of orientations (Fig. 2E) much as described in Hubel and Wiesel (1974a).

The microfields are very small, only a few minutes of arc in visual field coordinates (with the magnification factor of the fovea). In order to account for the much larger 'simple' and 'complex' receptive fields, we assume that groups of pyramidal cells located in corresponding positions of different hypercolumns (and therefore responding to the same orientation) are tied together into Hebbian cell assemblies by early learning, due to their often becoming active within a short time when contours sweep the visual field. The optimal stimulus would still be a narrow slit with the orientation imposed by the microfields, but it would be irrelevant which of the individual components are excited first to ignite the whole assembly. A further consequence is the scatter of the fields, since several pyramidal cells of one assembly share the same receptive field and nearby pyramidal cells may belong to different assemblies, the outline of the assembly determining the receptive field. The sizes of the fields and their scatter should be limited by the length of the fibres subserving the associative connections within the assembly. The assumption that these are the axon collaterals and basal dendrites of pyramidal cells is compatible with the size of receptive fields, of the order of 1 mm in cortical coordinates.

I propose that the discrete spatial frequency channels of Wilson (1983) are related to the periodic spacing of hypercolumns (an idea suggested to me by Jack Cowan). If the excitatory portions of the receptive fields are always located in the spaces between the hypercolumn centres (the black regions on Fig. 2), large receptive fields will be composed of microfields one, two or more hypercolumn distances apart. They will respond best to periodic functions having maxima spaced at these distances. With a magnification factor of 6 mm/degree (Daniel and Whitteridge, 1961), the cortical hypercolumn distance of 0.4 mm corresponds to 4 minutes of arc in the visual field and multiples of that distance to 8', 12' and 16'. The corresponding spatial frequencies are 15, 7.5, 5 and 3.75 c. p. d. perhaps compatible with the first four of Wilson's values.

The present model leads to testable consequences. Fields with well-separated excitatory (+) and inhibitory (−) subfields (simple cells) should have orientations corresponding to the three axes of the array (Fig. 2C), assuming it to be the hexagonal kind. Complex fields in the same region could have the other orientations (Fig. 2D). Also, an electrode proceeding tangentially through the cortex along a row of hypercolumn centres (in the direction of one of the axes of the hexagonal array; (Fig. 2F) should encounter neurons with the same orientation preference, or with no orientation preference, in the following sequence: no orientation (hypercolumn centre), orientation and direction selectivity (margin of the hypercolumn), orientation selectivity but no direction selectivity (between hypercolumns), orientation selectivity and opposite direction selectivity (margin of the adjoining hypercolumn), no orientation selectivity, etc., under the assumption that directional selectivity in a line detector is due to the asymmetrical action of inhibition (Fig. 2G), as proposed by Barlow and Levick (1965) for retinal neurons.

Confirmation of the model may also come from an analysis of receptive fields of single neurons with an accuracy exceeding that of the first pioneering accounts. The report by Schiller, Finlay and Volman (1976a, 1976b, 1976c) may well serve as a model for such a high resolution analysis. It already provides evidence for a microstructure of receptive fields which is reminiscent of the microfields postulated here.

A criticism which has been advanced against the idea of a centric arrangement of orientation preferences is based on some records (Hubel and Wiesel, 1974a) of continually turning orientation preference over long stretches along a straight electrode path. This is only an apparent argument against the scheme which I propose. Actually, according to the model of Fig. 2, a penetration which stays parallel to one of the rows of the hexagonal array may produce a rotation of orientation preference of almost 180° within each hypercolumn, the rotation in adjoining hypercolumns continuing so smoothly as to be indistinguishable, within the experimental error, from that predicted by Hubel and Wiesel's scheme of 'orientation slabs'.

To my mind the main virtue of the model is that it does not require any special oriented systems of fibres in the cortex. The inhibitory contacts which it postulates are quite in accordance with the known morphology of axonal ramifications in stellate cells, namely short range and diffuse. The excitatory contacts which are necessary to tie pyramidal cells together into cell assemblies are also there in the system of axon collaterals and basal dendrites, one of the main constituents of the cortical neuropil. In order to appreciate the advantage of such simple and homogeneous wiring, the reader is invited to consider alternative schemes which one obtains by assuming special connections to every cortical neuron from all the input elements which contribute to its receptive field. Not only is it practically impossible to draw such a scheme for more than two or three neighbouring neurons, but it also becomes immediately apparent that this would put an impossible burden on the developmental mechanisms which are known to produce 'orientation columns' before experience sets in to mould them. The orderly distribution of inhibitory neurons amongst a majority of randomly connected pyramidal cells, which I propose, can do with much less genetic specification.

ACKNOWLEDGEMENT

I would like to express my gratitude to Jack Cowan for suggesting a connection between my model and Wilson's frequency channels.

REFERENCES

Barlow, H. B., and Levick, R. W. (1965). The mechanism of directionally selective units in rabbit's retina. *J. Physiol.* (*Lond.*), **178**, 477–504.

Braitenberg, V., and Braitenberg, C. (1979). Geometry of orientation columns in the visual cortex. *Biolog. Cybernetics*, **33**, 179–186.

Carroll, E., and Wong-Riley, M. (1982). Light and EM analysis of cytochrome oxidase-rich zones in the striate cortex of squirrel monkeys. *12th Ann. Meeting Soc. Neurosci. (Minneapolis), Abstr.*, **8**, Part 2, 706.

Daniel, P. M., and Whitteridge, D. (1961). The representation of the visual field on the cerebral cortex in monkeys. *J. Physiol.*, **159**, 203–221.

Dow, B. M., Snyder, A. Z., Vautin, R. G., and Bauer, R. (1981). Magnification factor and receptive field size in foveal striate cortex of monkey. *Exp. Brain Res.*, **44**, 213-228.

Horton, J. C., and Hubel, D. H. (1980). Cytochrome stain preferentially labels intersection of ocular dominance and vertical orientation columns in macaque striate cortex. *10th Ann. Meeting Soc. Neurosci. (Cincinnati, Ohio), Abstr.*, p. 315.

Hubel, D. H., and Wiesel, T. N. (1968). Receptive fields and functional architecture of monkey striate cortex. *J. Physiol. (Lond.)*, **195**, 215–243.

Hubel, D. H., and Wiesel, T. N. (1974a). Sequence regularity and geometry of orientation columns in the monkey striate cortex. *J. comp. Neurol.*, **158**, 267–294.

Hubel, D. H., and Wiesel, T. N. (1974b). Uniformity of monkey striate cortex: a parallel relationship between field size, scatter, and magnification factor. *J. comp. Neurol.*, **158**, 295–306.

Hubel, D. H., Wiesel, T. N., and LeVay, S. (1974). Visual-field representation in layer IVC of monkey striate cortex. *4th Ann. Meeting Soc. Neurosci. (St. Louis), Abstr.*, p. 264.

Humphrey, A. L., and Hendrikson, A. E. (1980). *Neurosci. Abstr.*, **6**, 315.

Schiller, P. H., Finlay, B. L. and Volman, S. F. (1976a). Quantitative studies of single-cell properties in monkey striate cortex. I. Spatiotemporal organization of receptive fields. *J. Neurophysiol.*, **39**, 1288–1319.

Schiller, P. H., Finlay, B. L., and Volman, S. F. (1976b). Quantitative studies of single-cell properties in monkey striate cortex. II. Orientation specificity and ocular dominance. *J. Neurophysiol.*, **39**, 1320–1333.

Schiller, P. H., Finlay, B. L., and Volman, S. F. (1976c). Quantitative studies of single-cell properties in monkey striate cortex. III. Spatial frequency. *J. Neurophysiol*, **39**, 1334–1351.

Wilson, H. R. (1983). Psychophysical evidence fur spatial channels. In *Physical and Biological Processing of Images*. (Eds. O. J. Braddick and A. C. Sleigh), Springer-Verlag, Berlin, Heidelberg and New York.

Models of the Visual Cortex
Edited by D. Rose and V. G. Dobson

CHAPTER 51

A microelectrode study of the spatial arrangement of iso-orientation bands in the cat's striate cortex

K. Albus
Abt. für Neurobiologie, Max-Planck-Institut für Biophysikalische Chemie, Am Faßberg, 3400 Göttingen, Federal Republic of Germany

In the cat stimulation of the visual field with a light pattern of a particular orientation activates an assembly of neurons in area 17 which has been named the orientation subunit (Albus, 1975). In sections perpendicular to the cortical surface the orientation subunits often appear as columns whereas in sections parallel to the cortical surface they appear as bands (iso-orientation bands) (Albus, 1975, 1979; Hubel and Wiesel, 1962, 1963, 1974; Schoppmann and Stryker, 1981; Singer, 1981). Single-unit recordings (Albus, 1975) have suggested that iso-orientation bands representing one particular orientation form a rather complex network within the visual cortex. The basic pattern of this network has been identified as a set of 2 to 4 iso-orientation bands which run, either straight or curving, parallel to each other over some cortical distance. Individual bands may either end abruptly or fuse with adjacent bands. Less frequently than parallel bands, circular arrangements or small isolated patches are seen. No particular direction of the bands was identified in early investigations using physiological methods. Later, however, a predominance of directions orthogonal to the vertical meridian was suggested for some parts of the striate cortex (Albus, 1979).

These suggestions about the spatial arrangement of the iso-orientation bands have received only partial support from recent experimental and theoretical studies. In one of these studies (Singer, 1981), iso-orientation bands were reconstructed from ^{14}C-deoxyglucose autoradiographs and it was found that iso-orientation bands run strictly perpendicular to the cortical representation of the vertical meridian at the 17/18 border and that they are more or less continuous throughout the mediolateral extent of area 17. Fusions between adjacent bands were extremely rare in these reconstructions and circular formations were virtually nonexistent. In the following assessment, this type of organization will be referred to as the continuous band system.

An arrangement quite different from the continuous band system was proposed by Braitenberg and Braitenberg (1979). These authors suggested that orientations are arranged radially around centres. They presented some evidence for this on the basis of data from Hubel and Wiesel (1974) and Albus (1975) by showing that their theoretical plots fit quite well the plots of orientation versus cortical distance (orientation–distance plots) seen in those two studies.

In order to test the models proposed by Braitenberg and Braitenberg (1979) and Singer (1981), I have reinvestigated the orientation representation by making long microelectrode penetrations through the striate cortex. The experiments were performed on anaesthetized and paralysed cats. With one exception (C18a), the penetrations were confined to parts of the striate cortex located in the medial bank of the lateral gyrus and the suprasplenial gyrus, where eccentricities between 5 and 20° are represented. In C18a, central parts of the visual field (eccentricity between 0 and 2°) located in the lateral bank of the postlateral gyrus were explored. The penetrations were made in the anterior to posterior or posteromedial directions. The course of the penetration with respect to the direction of the lower vertical meridian at the 17/18 border, and accordingly the direction of the penetration in the visual field, varied from animal to animal. The preferred orientation and orientation selectivity (Albus, 1975) of single units or cell clusters were determined at regular intervals of 100 μm. Since the penetrations were approximately tangential, the distances between recording points approximate cortical distances as measured parallel to the cortical surface. For a detailed description of the experimental methods see Albus (1975).

The results of eight long penetrations through area 17 of six animals are shown in Fig. 1. As in earlier experiments (Hubel and Wiesel, 1974; Albus, 1975), a systematic and highly ordered spatial arrangement of orientation-sensitive cells is revealed. A close inspection of Fig. 1 indicates that neither the continuous band system of Singer (1981) nor the centric organization model of Braitenberg and Braitenberg (1979) can account for the shape of these orientation versus distance plots. A continuous band system would predict that a microelectrode penetration forming an acute angle with the vertical meridian representation at the 17/18 border should show regular sequential changes in optimal orientation with high drift rates over long cortical distances. By contrast, penetrations forming an obtuse or perpendicular angle with the vertical meridian should reveal either sequential changes in optimal orientation, with low drift rates, or no changes in preferred orientation. As can be seen from Fig. 1, these predictions are not fulfilled. Sequential changes with high drift rates also occur in penetrations running nearly perpendicular to the vertical meridian (second half of C2, C3, C4), whereas sections with constant preferred orientation or with low drift rates are also recorded in penetrations forming an acute angle with the vertical meridian (C18a, C18b, H6).

Other findings are inconsistent with the continuous band system of Singer (1981). For example, regular sequential changes in orientation generally persist only for cortical distances of less than 3 mm. Also, abrupt shifts in preferred orientation interrupting periods of regular orientation representation occur relatively frequently in some penetrations (C18a, H6, C2 and C4). In fact, the results shown in Fig. 1 exclude the posibilities that iso-orientation bands always run orthogonal to the vertical meridian and that they are continuous through the lateromedial extent of the striate cortex. The results do not, however, preclude the possibility of parallel arrangements of iso-orientation bands. These arrangements seem to be restricted to cortical volumes of less than 3 to 4 mm across and their direction may assume various angles with the 17/18 border.

The disagreement between the reconstruction in this study and that of Singer could possibly be attributed to interindividual differences in the spatial organization of the orientation domain. One would need to know also whether or not the restriction of the visual environment in the postnatal period, as done in three of the four cases in Singer's experiments, changes the 'normal' pattern of iso-orientation bands.

Braitenberg and Braitenberg (1979) proposed that orientations are arranged radially around centres and on the basis of their model formulated some rules on how orientation–distance plots derived from tangential penetrations should appear. These are summarized as follows:

Rule 1. An orientation-distance plot must consist of sigmoid segments, spanning no more than the distance between two neighbouring centres in the cortex and no more than 180 in the orientation domain.
Rule 2. Two consecutive ascending (descending) pieces in the orientation-distance plots must be separated by an abrupt shift in the orientation domain.
Rule 3. All the ascending (descending) segments of one plot must have their points of maximum gradient at one and the same orientation, namely the orientation at right angles to the direction of the penetration in the visual field. Where there are abrupt shifts in the orientation domain, they must occur at orientations orthogonal to those at the gradient maxima of the orientation–distance plots.

The plots shown in Fig. 1 do not strictly conform to the rules given in Braitenberg and Braitenberg (1979). In relation to rule 1, the distance between two neighbouring centres was calculated according to Braitenberg and Braitenberg (1979) matching the penetrations of Fig. 1 with centre maps (see Fig. 2). Distances varied between 0.3 and 1.2 mm (mean ± 1/SD = 0.9

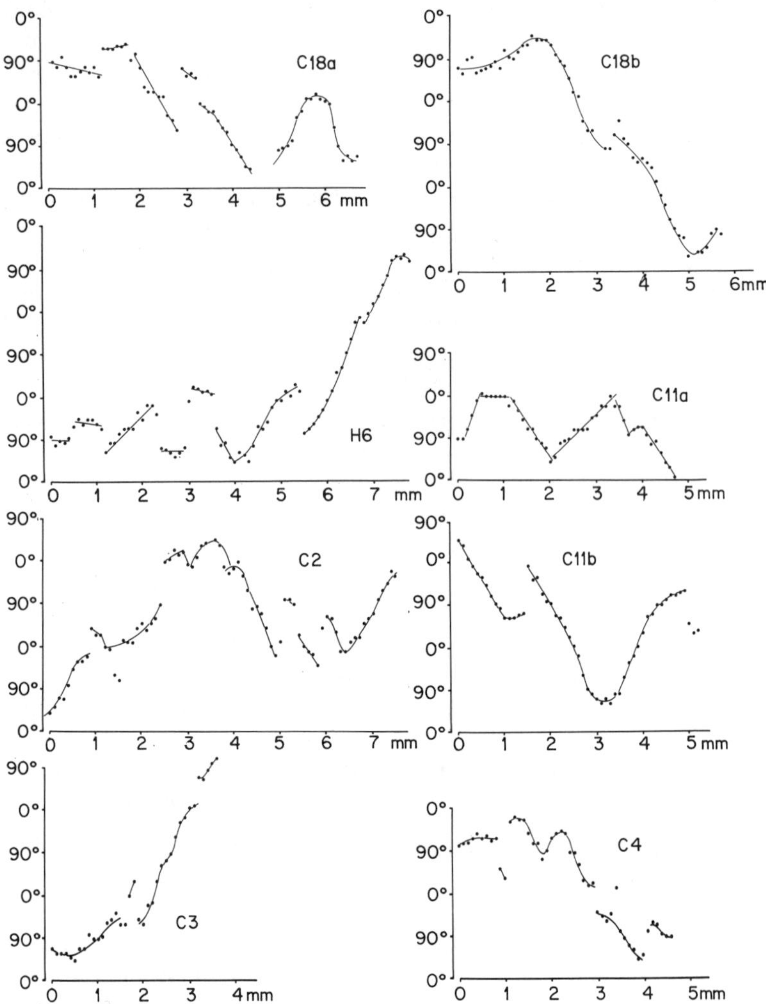

FIGURE 1 Spatial arrangement of orientation-sensitive cells in area 17 of the cat. The preferred orientation of single neurons or cell clusters (ordinate) is plotted as a function of cortical distance (abscissa). The plots were fitted by hand with curves or with straight lines. In each of the cases C18 and C11 two penetrations were performed. In the cortex penetrations started on the lateral gyrus, at its medial edge; in the visual field, the first receptive field was always recorded close to or at the vertical meridian, in the lower hemi-field. Subsequently, recorded fields approached the area centralis, or more peripheral parts of the horizontal meridian. The angle between the direction of the vertical meridian and the direction of the penetration in the visual field varied from animal to animal. It was between 10 and 20° in C18b, about 30° in H6, between 30 and 40° in C18a, between 40 and 50° in C11a and C11b, about 70° in C3 and 80° in C4. In C2 the angle changed slowly from 50 during the first 2 or 3 mm to about

± 0.3 mm) which is similar to values found by Braitenberg and Braitenberg (1979). In four out of eight penetrations (C18b, H6, C11b, C2), the ascending (descending) segments span cortical distances longer than 1.2 mm (the distances ranging from 1.3 mm in C2 to 2.2 mm in H6). In five penetrations, the sequential changes in preferred orientation span more than 180° (C18b, H6 (360° !!), C11b, C3 and C2). In the remaining cases (C4, C18a, C11a), the sequential changes in preferred orientation span less than 180°.

Referring to rule 2, it can be seen that, in the orientation-distance plots, shifts in preferred orientation often separate consecutive ascending or descending portions of the curves. It should be remembered, however, that most of these segments span more than 180° and/or more than 1.2 mm cortical distance (see above), thus violating rule 1. Abrupt shifts in preferred orientation also separate sections with constant or nearly constant orientation (C18a, H6, C4). This is difficult to reconcile with a centric organization.

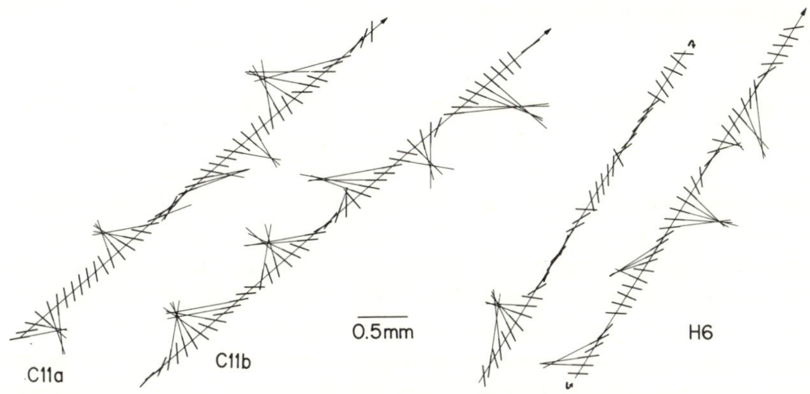

FIGURE 2 Reconstruction of common centres according to Braitenberg and Braitenberg (1979) for the orientation–distance plots C11a, C11b and H6 (see Fig. 1). The preferred orientations are given by small bars drawn along the path of the electrode. The direction of the latter, as shown in this figure, is identical with the actual direction of the penetration through the visual field. In both C11a and C11b, 60 per cent. of all orientations recorded were matched by centres.

Rule 3 is applicable only for penetrations and/or parts of penetrations having a constant direction in the visual field. This condition was fulfilled in all of the penetrations shown in Fig. 1 the only exception being C2. In only two cases (C18b, C11a) did the orientations at the maximum gradients of the orientation–distance plots and the orientations indicating the positions of abrupt shifts remain approximately constant. In the remaining penetrations,

---

70 in the second half of the penetration. Thus C18b runs nearly parallel to the vertical meridian and C3 and C4 nearly parallel to the horizontal meridian, through the visual field.

the orientations at the maximum gradients and/or at the positions of orientation shift varied considerably (C18a, H6, C11b).

The findings presented in Fig. 1 thus do not support the hupothesis of a centric organization of iso-orientation bands. Accordingly, the matching of the penetrations to a map of centres is not particularly successful. Only in two cases (C11a and C11b, Fig. 2) were more than 50 per cent. of all the orientations recorded matched by centres. In the remaining penetrations, this proportion was much less than 40 per cent. (e.g. 32 per cent. in H6, Fig. 2).

The result of this study does not exclude the possibility that locally iso-orientation bands are arranged around centres as proposed by Braitenberg and Braitenberg (1979). In such regions, the shape of the orientation–distance plots conforms to the rules set down on the basis of a centric organization, and the plots are matched quite well by centre maps (e.g. second half of C18a, second half of C4).

In conclusion, long tangential penetrations through the striate cortex suggest that neither the centric model (Braitenberg and Braitenberg, 1979) nor continuous band organization (Singer, 1981) can alone account for the organization of the orientation domain. We have recently attempted to reconstruct the spatial distribution of iso-orientation bands over broader regions of cat's striate cortex (Albus and Sieber, 1984), using the ^{14}C-deoxyglucose technique. The results of this study suggest that the iso-orientation bands form a complex network which closely resemble the reconstructions based on physiological experiments (Albus, 1975). The organizational rules of the network cannot be explained in a straightforward manner. It would appear, however, that both the centric as well as the continuous band system are in operation and local interactions between both types of organization generate the iso-orientation network in the striate cortex of the cat.

ACKNOWLEDGEMENT

It is a pleasure to thank D. Tigwell for improving the English version of this paper.

REFERENCES

Albus, K. (1975). A quantitative study of the projection area of the central and the paracentral visual field in area 17 of the cat: II. The spatial organization of the orientation domain. *Exp. Brain Res.*, **24**, 181–202.

Albus, K. (1979). ^{14}C-deoxyglucose mapping of orientation subunits in the cats viaual cortical areas. *Exp. Brain Res.*, **37**, 609–613.

Albus, K., and Sieber, B. (1984). On the spatial arrangement of iso-orientation bands in the cats visual cortical areas 17 and 18: a ^{14}C-deoxyglucose study. *Exp. Brain Res.*, **56** 384–388.

Braitenberg, V., and Braitenberg, C. (1979). Geometry of orientation columns in the visual cortex. *Biolog. Cybernetics*. **33**, 179–186.
Hubel, D. H., and Wiesel, T. N. (1962). Receptive fields, binocular interaction and functional architecture in the cat's visual cortex. *J. Physiol. (Lond.)*, **160**, 106–154.
Hubel, D. H., and Wiesel, T. N. (1963). Shape and arrangement of columns in the cat's striate cortex. *J. Physiol. (Lond.)*, **165**, 559–568.
Hubel, D. H., and Wiesel, T. N. (1974). Sequence regularity and geometry of orientation columns in the monkey striate cortex. *J. comp. Neurol.*, **158**, 267–294.
Schopmann, A., and Stryker, M. P. (1981). Physiological evidence that the 2-deoxyglucose method reveals orientation columns in cat visual cortex. *Nature*, **293**, 574–576.
Singer, W. (1981). Topographic organization of orientation columns in the cat visual cortex. A deoxyglucose study. *Exp. Brain Res.*, **44**, 431–436.

Models of the Visual Cortex
Edited by D. Rose and V. G. Dobson

CHAPTER 52

Neuronal composition and circuitry of rat visual cortex

ALAN PETERS
Department of Anatomy, 80 East Concord Street, Boston, MA 02118, USA

The present chapter will be confined to a consideration of the neuronal composition of rat area 17, the synaptic relations and possible neurotransmitters used by the neurons and the sites of termination of the geniculocortical afferents.

In the rat, the primary visual cortex, area 17, is on the dorsal surface of the occipital pole where it is surrounded medially by area 18 (or 18b) and laterally by area 18a (Krieg, 1946; Schober and Winkelmann, 1975) and the anatomical extent of area 17 coincides with that portion of the cortex which Montero, Rojas and Torrealba (1973) have shown to be the primary visual cortex on the basis of physiological recordings. As in other species there is a precisely arranged retinotopic organization in area 17 (also see Adams and Forrester, 1968) and a proportionally larger expanse of the cortex is devoted to central than to peripheral vision.

NEURONAL COMPOSITION

For purposes of description, the neurons visualized in Golgi preparations of the rat visual cortex can be considered as being either pyramidal or nonpyramidal in form. The separation of neurons into morphological types is primarily based upon the distribution of the dendrites and although dendrites are often ignored in consideration of the functions of neurons, the form of the dendritic tree along with the location of the neuron determines its synaptic input and hence its physiological response properties. Thus, the pyramidal cells with their long apical dendrites can sample input from a number of layers of cortex, while a small multipolar cell may have its input confined to afferents in only one layer.

Pyramidal Cells

The pyramidal cells in rat area 17 have been described by Parnavelas, Lieberman and Webster (1977) and by Werner *et al.* (1979). Typical pyramidal cells have oval- or conical-shaped cell bodies and an apical dendrite which ascends into layer I. These neurons also have a basal skirt of dendrites, and all of the dendrites have numerous spines protruding from them. These protrusions receive most of the axon terminals which synapse with the pyramidal cells and the terminals form asymmetric synapses with the spines. Why these spines are present is not yet known but a number of hypotheses have been put forward to account for their presence. One is that their function may be solely to reach out and contact axons passing through the neuropil (Peters and Kiaserman-Abramof, 1970). Another is that the electrical resistance of the stalk might reduce the 'weight' of the spine synapse (Rall and Rinzel, 1971), and Crick (1982) has recently suggested reasons for considering that spines might be involved in memory.

Many of the pyramidal cells have axons which project out of area 17, but so far little is known about the connections that the axons of pyramidal cells make within the cortex itself. It is known, however, that the axon terminals of pyramidal cells form asymmetric synapses (see Peters, Palay and Webster, 1976), the majority of which involve dendritic spines and the remainder, dendritic shafts (see Parnavelas *et al.*, 1977; Somogyi, 1978). Thus it is likely that the pyramidal cells are excitatory neurons, but their neurotransmitter has still to be defined. Pyramidal cells dominate the cortex and we have calculated that in layers II to VIa they account for some 90 per cent. of the neuronal population (Peters and Kara, unpublished data).

As to arrangements of pyramidal cells, it is interesting to note that the apical dendrites of layer V pyramidal cells are organized into groups which have been termed 'clusters' (Feldman and Peters, 1974; Winkelmann, Brauer and Berger, 1975). These clusters have a center-to-center spacing of between 30 and 40 μm. As the clusters of layer V apical dendrites ascend through the cortex the apical dendrites of some layer III pyramidal cells join them, so that eventually the apical dendrites are organized into quite large groups. Similar clusters are also present in the cat and monkey, but although their significance is not known they offer the only real morphological sign of a definite neuronal arrangement in the cortex, and it may be that each cluster of neurons functions as a discrete unit.

Nonpyramidal Cells

The nonpyramidal cells of rat area 17 are those neurons which lack apical dendrites; they can be broadly classified on the basis of the forms of their dendritic trees, the abundance of spines borne by their dendrites (Feldman and Peters, 1976) and the forms of their axonal plexuses. In contrast to the pyramidal cells which have only symmetrical axosomatic synapses, the cell bodies of nonpyramidal cells have both symmetrical and asymmetrical axosomatic synapses.

Multipolar and Bitufted Neurons with Smooth and Sparsely spinous Dendrites.

Multipolar and bitufted neurons with dendrites possessing few, if any, spines are considered together since they appear to form one group in which the multipolar cells, with dendrites radiating from a number of sites on the cell body, are at one extreme and the bitufted cells, with dendrites mainly arising from the upper and lower poles of the vertically elongate cell body, are at the other end of the morphological spectrum. These neurons have either myelinated or unmyelinated axons. The distribution of the myelinated axons is unknown, because they do not impregnate in Golgi preparations (Peters and Proskauer, 1980a), but when the axon is unmyelinated it is seen to form an extensive local plexus which largely overlaps some or all of the dendritic tree in its distribution (Parnavelas *et al.*, 1977; Peters and Fairén, 1978; Peters and Proskauer, 1980b). The terminals of the axons of these neurons form symmetric synapses and they synapse with the cell bodies of both pyramidal and nonpyramidal cells, as well as with the dendritic shafts of such neurons and with axon initial segments. This distribution of axon terminals is apparently identical with that of axon terminals which form symmetric synapses and give a positive reaction with an antibody to glutamic acid decarboxylase (GAD), the enzyme which synthesizes GABA. Ribak (1978) has also shown that some of the cell bodies of smooth and sparsely spinous multipolar cells are GAD-positive. Further, both Chronwall and Wolff (1980) and Somogyi *et al.* (1981) have demonstrated that such nonpyramidal cells accumulate [^{3}H]GABA. Consequently, there is strong evidence to support the concept that these neurons are inhibitory in function. Interestingly, these same types of neuron also give a positive reaction to an antibody to somatostatin, and McDonald *et al.* (1982a) state that as many as 2 to 3 per cent. of the neurons in rat area 17 label with this antibody.

As to the proportion of multipolar and bitufted neurons in area 17, we have found that they account for about 6 per cent. of all neuron profiles in layers II to VIa. The greatest number is in layer II/III, in which they make up about 8 per cent. of the neuronal population.

Chandelier Cells.

In addition to the smooth and sparsely spinous neurons with local axonal plexuses, there are other multipolar and bitufted cells in layer II/III with quite different and distinct axonal plexuses. These are the chandelier cells, whose axons form vertically oriented strings largely confined to layer II/III in distribution. There the strings of axonal boutons synapse quite specifically with the axon initial segments of pyramidal cells (Somogyi, 1977; Peters, Proskauer and Ribak, 1982; Somogyi, Freund and Cowey, 1982), leading Somogyi to conclude that a better name for these neurons might be axo-axonic interneurons. As shown by Peters, Proskauer and Ribak (1982), their axonal boutons are GAD-positive, so that, like the cells with local axonal plexuses, the chandelier cells appear to be GABA-ergic inhibitory neurons. It has also been shown that strings of boutons from several chandelier cells can converge upon a single layer II/III pyramidal cell and since a chandelier cell can form numerous strings of boutons, each one probably contributes to the inhibition of perhaps a hundred or more supragranular pyramidal cells. In area 17, however, many of the supragranular pyramidal cells lack chandelier terminals on their axon initial segments and, indeed, chandelier cells are probably most common not in area 17 but at the 17/18a border and to a lesser extent at the 17/18 border (Peters, Proskauer and Ribak, 1982).

Bipolar Cells.

The other common type of nonpyramidal cell in layers II to VIa is the bipolar cell, characterized, as the name suggests, by having two primary dendrites which extend vertically, one leaving the upper pole of the cell body and the other the lower pole to form a long, slim dendritic tree (Feldman and Peters, 1978). Typically the axons of bipolar cells arise from one of the primary dendrites at some distance from the cell body and the axon also extends vertically. In Golgi–electron microscopic preparations, Peters and Kimerer (1981) have found that the axons of bipolar cells form asymmetric synapses. Most of the postsynaptic elements are dendritic spines, but they also include dendritic shafts. Parnavelas *et al.* (1977), on the other hand, suggest that bipolar cells form symmetric synapses. Consequently, the synaptic connections of these neurons should be investigated further.

In an electron microscopic survey of rat area 17 we have found the cell bodies of bipolar cells to account for about 4 per cent. of the total number of neuronal profiles in layers II to VIa. Again, the greatest number of them occurs in layer II/III, in which they account for 7 per cent. of all neuronal cell bodies. The other layer containing significant numbers of bipolar cells is layer V, in which they form about 4 per cent. of the population, while in layers IV and VIa only 1 to 2 per cent. of neurons are bipolar.

Bipolar cells have recently gained some prominence because of the frequency with which they appear when antibodies to some peptides are used on area 17 of the rat visual cortex. Thus, antibody to vasoactive intestinal polypeptide (VIP) reacts with bipolar cells (e.g. Fuxe *et al*, 1977; Loren *et al.*, 1979; Emson and Tindwall, 1979; Sims *et al.*, 1980). The reactive bipolar cells are present throughout layers II to VIa, but they are most common in layers II/III, and McDonald *et al.* (1982b) estimate that VIP-positive neurons, the great majority of which are bipolar cells, make up about 3 per cent. of the neuronal population of rat area 17. Antibodies to cholecystokinin (CCK) also react with bipolar cells, amongst other nonpyramidal cell types (Peters, Miller and Kimerer, 1983), and again the bipolar cells are most prominent in layer II/III. On the other hand, the bipolar cells which react with antibody to avian pancreactic polypeptide (APP) are scattered throughout the cortex (McDonald *et al.*, 1982c), as are bipolar cells which are visualized by using an antibody to choline acetyltransferase (Houser *et al.*, 1983). If each of these antibodies reacts with a different subpopulation of bipolar cells in area 17, then there must be a significant number of these neurons in this cortex, as our analysis of the frequency of bipolar cell bodies suggests.

Neurons of Layers I and VIb.

These layers appear to contain only nonpyramidal cells and most of them are quite different from the nonpyramidal cells encountered in layers II to VIa. At present, however, there is little information about these neurons although Bradford, Parnavelas and Lieberman (1977) suggest that layer I contains at least four different types of neurons. In layer VIb, however, most of the neurons appear to be horizontally elongated cells with dendrites which arise as tufts from the ends of the cell body (see Feldman and Peters, 1978).

Comments.

Before leaving the topic of the neuronal composition of area 17 of the rat, it should be pointed out that although our estimate of the proportion of pyramidal cells present (90 per cent.) is in reasonable agreement with that obtained by Werner *et al.* (1982) (93 per cent.) a quite different conclusion has been reached by Winfield, Gatter and Powell (1980). They estimate the pyramidal cells to account for only 62 to 72 per cent. of neurons, the remainder of the neuronal population being formed by large stellate cells (3 to 7 per cent.) and small stellate cells (23 to 33 per cent.) and they find similar values for the cat and monkey visual cortex. At present the reason for the discrepancy is not apparent, unless Winfield, Gatter and Powell (1980) consider the neurons of layer IV in the rat cortex to be small stellate cells and not small pyramidal cells as we believe them to be.

Multipolar and bitufted neurons with local axonal plexuses and chandelier cells are the main types of inhibitory neurons present in rat area 17. Both the cat (e.g. Lund *et al.*, 1979; Peters and Regidor, 1981) and monkey (e.g. Lund, 1973; Somogyi *et al.*, 1982) have similar types of neurons, but in addition these animals possess other neurons with axons forming symmetric synapses. These are the basket cells (Somogyi *et al.*, 1983) and double-bouquet cells (Somogyi and Cowey, 1981) which seem to be absent from rat area 17. A further difference is that few, if any, spiny stellate cells of the type present in layer IV of the cat and primate cortex have been encountered in the rat. Instead, the prevalent cell type in layer IV of the rat is a small pyramidal cell.

THALAMOCORTICAL INPUT

Area 17 receives thalamic afferents from the dorsal lateral geniculate nucleus (dLGN) which is not obviously laminated in the rat. The primary site of termination of these afferents is layer IV and lower layer III, with secondary sites in lower layer I and layer VI (see Ribak and Peters, 1975). If lesions are made in the dLGN it is seen that the degenerating thalamic axon terminals form asymmetric synapses, indicating, as the physiology suggests, that the geniculocortical input is excitatory (e.g. Toyama *et al.*, 1974). In layer IV about 83 per cent. of the degenerating terminals synapse with dendritic spines, 15 per cent. with dendritic shafts and 2 per cent. with neuronal perikarya (Peters and Feldman, 1976; also see Schober and Winkelmann, 1977). Using the combined Golgi–electron microscope technique it is found that the neuronal elements receiving these thalamic afferents include the spines of apical dendrites of layer V pyramidal cells; spines of basal and proximal dendrites of layer III pyramids; spines, dendritic shafts and perikarya of multipolar cells (Peters *et al.*, 1979); and the dendrites of bipolar cells (Peters and Kimerer, 1981). These relationships are shown diagramatically in Fig. 1. Thus, no specific neuronal element in layer IV and lower layer III receives the geniculocortical afferents. Instead, it seems that all neuronal elements in those layers which are capable of forming asymmetric synapses can receive these afferents (Peters, 1979). Further, it can be concluded that since the majority of thalamic afferents synapse with spines, almost all of which must be derived from the population of pyramidal cells, the pyramids must be the major recipients of the thalamic afferents. It is not known, however, whether some neuronal types receive more thalamic afferents than others, although this seems to be the case in the mouse barrel field (see White and Hersch, 1981).

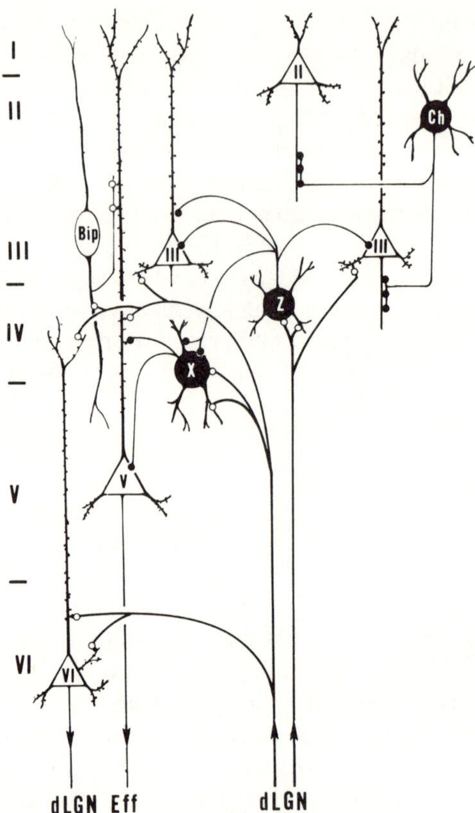

FIGURE 1 Diagram to illustrate some of the known synaptic relationships of neurons in area 17 of the rat visual cortex. Pyramidal cells are shown as having triangular-shaped cell bodies and are numbered according to the layers in which they are contained. The diagram also includes two multipolar neurons with local axonal plexuses (X and Z), a layer III bipolar cell (Bip) and a chandelier cell (Ch). Asymmetric (excitatory) synapses are shown as open circles and symmetric (inhibitory) synapses by blackened circles. The connections shown in this diagram are considered in the text.

Less is known about the termination of the geniculocortical afferents in layers I and VI, although in layer I most of the terminals synapse with spines. The same is true in layer VI (Peters and Saldanha, 1976) where the spines are probably derived from the layer VIa pyramidal cells, which project back to the dLGN to produce a monosynaptic feedback loop (see Fig. 1). These layer VIa pyramidal neurons also probably receive a second thalamic input in layer IV as they do in the mouse SmI cortex (see Hersch and White, 1981).

Presumably, the sites of termination of the geniculocortical afferents must to some extent determine the receptive field properties of neurons, and in a recent study of functionally identified neurons in the rat visual cortex Parnavelas, Burne and Lin (1983) have found that layer IV contains the largest percentage of simple cells. These are both pyramidal and nonpyramidal in form. However, Parnavelas, Burne and Lin (1983) encountered simple, complex and hypercomplex cells in all layers and, with the exception of layer IV, these different functional types of neurons appear to be evenly distributed through the cortex. This is unlike the distribution of receptive field types in the cat and monkey cortex. The rat also differs in that functional columns of neurons have not yet been recognized in its visual cortex.

NEURONAL CONNECTIONS

Some of the synaptic connections in cat area 17 described in the preceding pages are shown diagrammatically in Fig. 1. The thalamic afferents to layer IV and lower layer III can be expected to have an excitatory effect on layer VI, layer V and lower layer III pyramids, as well as those in layer IV. However, this same input would also excite smooth and sparsely spinous multipolar and bitufted cells (e.g. cells X and Z). The effect of this would be to produce a disynaptic inhibition of the pyramidal cells contained within the axonal fields of these GABA-ergic neurons. Moreover, since these inhibitory neurons synapse with each other, they can also produce disinhibition, so that, for example, cell Z could inhibit cell X and so prevent it from inhibiting the layer V pyramidal cell with which it is shown to synapse in the diagram. Continued excitation of pyramidal cells might also be enhanced by the action of the bipolar cells (Bip), for, as shown, the axons of some of these neurons form asymmetric synapses with dendritic spines of pyramids. Since the bipolar cells have long and slender axonal plexuses, they might excite a small group of pyramidal cells, such as those contained within one or several adjacent clusters.

The chandelier cells (Ch) are not contained within the field of termination of the main thalamic input. The inputs which might excite these neurons are not yet known, but clearly their role in the cortex is to produce inhibition of large numbers of supragranular pyramidal cells. Since the chandelier cells are most common at the 17/18a border where callosal connections exist, and where midline vision is represented, both they and the pyramidal cells around them might be excited by the callosal input. The chandelier cells could then produce a disynaptic inhibition of the pyramidal cells to prevent them from continued firing. Indeed, the basic principle of organization of neurons in the rat visual cortex seems to be that the neurons are interconnected in such a way that excitation is followed by inhibition.

CONCLUSIONS

1. Pyramidal cells are the most common neuronal cell type in rat area 17 and some of these cells are arranged into clusters.
2. Compared to the cat and monkey, the variety of nonpyramidal cells is limited.
3. The lateral geniculate nucleus shows no lamination and unlike the cat and monkey visual cortex there is no evidence for geniculocortical afferents terminating in sublaminae in layer IV.
4. Although rat area 17 contains neurons with receptive field properties similar to ones present in the cat and monkey visual cortex, there is no evidence for a columnar organization in the rat.
5. The geniculocortical afferents form asymmetric synapses with all neuronal elements in layer IV capable of forming this type of synapse.
6. The intrinsic connections of neurons in rat area 17 seem to be organized in such a way that excitation produced by geniculocortical and other afferents would be followed by inhibition.

ACKNOWLEDGEMENT

Supported by US Public Health Service Research Grant NB-07016 from the National Institute of Neurological and Communicative Disorders and Stroke.

REFERENCES

Adams, A. D., and Forrester, J. M. (1968). The projection of the rats visual field on the visual cortex. *J. Exp. Physiol.*, **53**, 327–336.

Bradford, R., Parnavelas, J. G., and Lieberman, A. R. (1977). Neurons in layer I of the developing occipital cortex of the rat. *J. comp. Neurol.*, **176**, 121–132.

Chronwall, B., and Wolff, J. R. (1980). Prenatal and postnatal development of GABA-accumulating cells in the occipital neocortex of rat. *J. comp. Neurol.*, **190**, 187–208.

Crick, F. (1982). Do dendritic spines twitch? *Trends in Neurosci.*, **5**, 44–46.

Emson, P. C., and Lindvall, O. (1979). Distribution of putative neurotransmitters in the neocortex. *Neurosci.*, **4**, 1–30.

Feldman, M. L., and Peters, A. (1974). A study of barrels and pyramidal dendritic clusters in the cerebral cortex. *Brain Res.*, **77**, 55–76.

Feldman, M. L., and Peters, A. (1978). The forms of nonpyramidal neurons in the visual cortex of the rat. *J. comp. Neurol.*, **179**, 761–794.

Fuxe, K., Hokfelt, T., Said, S. Z., and Mutt, V. (1977). VIP and the nervous system: immunohistochemical evidence for localization in central and peripheral nerves, particularly intracortical neurons of cerebral cortex. *Neurosci. Lett.*, **5**, 241–246.

Hersch, S. M., and White, E. L. (1981). Quantification of synapses formed with apical dendrites of Golgi-impregnated pyramidal cells: variability in thalamocortical inputs, but consistency in the ratios of asymmetrical to symmetrical synapses. *Neurosci.*, **6**, 1043–1051.

Houser, C. R., Crawford, G. D., Barber, R. P., Salvaterra, P. M., and Vaughn, J. E. (1983). Organization and morphological characteristics of cholinergic neurons: an immunocytochemical study with a monoclonal antibody to choline acetyltransferase. *Brain Res.*, **266**, 97–119.

Krieg, W. J. S. (1946). Connections of the cerebral cortex. 1. Albino rat. A. topography of the cortical areas. *J. comp. Neurol.*, **84**, 221–275.

Loren, I., Emson, P. C., Fahrenkrug, J., Bjorklund, A., Alumets, J., Hakanson, R., and Sundler, F. (1979). Distribution of vasoactive intestinal polypeptide in the rat and mouse brain. *Neurosci.*, **4**, 1953–1976.

Lund, J. S. (1973). Organization of neurons in the visual cortex, area 17, of the monkey (*Macaca mulatta*). *J. comp. Neurol.*, **147**, 455–496.

Lund, J. S., Henry, G. H., MacQueen, C. L., and Harvey, A. R. (1979). Anatomical organization of the primary visual cortex (area 17) of the cat. A comparison with area 17 of the macaque monkey. *J. comp. Neurol.*, **184**, 599–618.

McDonald, J. K., Parnavelas, J. G., Karamanlidis, A. N., Brecha, N., and Koenig, J. I. (1982a). The morphology and distribution of peptide-containing neurons in the adult and developing visual cortex of the rat. I. Somatostatin. *J. Neurocytol.*, **11**, 809–824.

McDonald, J. K., Parnavelas, J. G., Karamanlidis, A. N., and Brecha, N. (1982b). The morphology and distribution of peptide-containing neurons in the adult and developing visual cortex of the rat. II. Vasoactive intestinal polypeptide. *J. Neurocytol.*, **11**, 825–837.

McDonald, J. K., Parnavelas, J. G., Karamanlidis, A. N., and Brecha, N. (1982c). The morphology and distribution of peptide-containing neurons in the adult and developing visual cortex of the rat. IV. Avian pancreatic polypeptide. *J. Neurocytol.*, **11**, 985–995.

Montero, V. M., Rojas, A., and Torrealba, F. (1973). Retinotopic organization of striate and prestriate visual cortex in the albino rat. *Brain Res.*, **53**, 197–201.

Parnavelas, J. G., Burne, R. A., and Lin, C. S. (1983). Distribution and morphology of functionally identified neurons in the visual cortex of the rat. *Brain Res.*, **261**, 21–29.

Parnavelas, J. G., Lieberman, A. R., and Webster, K. E. (1977). Organization of neurons in the visual cortex, area 17, of the rat. *J. Anat.*, **124**, 305–322.

Parnavelas, J. G., Sullivan, K., Lieberman, A. R., and Webster, K. E. (1977). Neurons and their synaptic organization in the visual cortex of the rat. Electron microscopy of Golgi preparations. *Cell Tissue Res.*, **183**, 499–517.

Peters, A. (1979). Thalamic input to the cerebral cortex. *Trends in Neurosci.*, **2**, 183–185.

Peters, A., and Fairén, A. (1978). Smooth and sparsely-spined stellate cells in the visual cortex of the rat: a study using a combined Golgi-electron microscope technique. *J. comp. Neurol.*, **181**, 129–172.

Peters, A., and Feldman, M. L. (1976), The projection of the lateral geniculate nucleus to area 17 of the rat cerebral cortex. I. General description. *J. Neurocytol.*, **5**, 63–84.

Peters, A., and Kaiserman-Abramof, I. R. (1970). The small pyramidal neuron of the rat cerebral cortex. The perikaryon, dendrites and spines. *Amer. J. Anat.*, **127**, 321–356.

Peters, A., and Kimerer, L. M. (1981). Bipolar neurons in rat visual cortex: a combined Golgi-electron microscope study. *J. Neurocytol.*, **10**, 921–946.

Peters, A., Miller, M., and Kimerer. (1983). Cholecystokinin-like immunoreactive neurons in rat cerebral cortex. *Neurosci.*, **8**, 431–448.

Peters, A., Palay, S. L., and Webster, deF.H. (1976). *The Fine Structure of the Nervous System: The Neurons and Supporting Cells*, Saunders, Philadelphia.

Peters, A., and Proskauer, C. C. (1980a). Smooth and sparsely-spined cells with myelinated axons in rat visual cortex. *Neurosci.*, **5**, 2079–2092.

Peters, A., and Proskauer, C. C. (1980b). Synaptic relationships between a multipolar stellate cell and a pyramidal neuron in the rat visual cortex. A combined Golgi-electron microscope study. *J. Neurocytol.*, **9**, 163–183.

Peters, A., Proskauer, C. C., Feldman, M. L., and Kimerer, L. (1979). The projection of the lateral geniculate nucleus to area 17 of the rat cerebral cortex. V. Degenerating axon terminals synapsing with Golgi impregnated neurons. *J. Neurocytol.*, **8**, 331–357.

Peters, A., Proskauer, C. C., and Ribak, C. E. (1982). Chandelier cells in rat visual cortex. *J. comp. Neurol.*, **206**, 397–416.

Peters, A., and Regidor, J., (1981). A reassessment of the forms of nonpyramidal neurons in area 17 of cat visual cortex. *J. comp. Neurol.*, **203**, 685–716.

Peters, A., and Saldanha, J. (1976). The projection of the lateral geniculate nucleus to area 17 of the rat cerebral cortex. III. Layer VI. *Brain Res.*, **105**, 533–537.

Rall, W., and Rinzel, J. (1971). Dendritic spine function and synaptic attenuation calculations. *Soc. Neurosci. Abstr.*, p. 64.

Ribak, C. E. (1978). Aspinous and sparsely spinous stellate neurons in the visual cortex of rats contain flutamic acid decarboxylase. *J. Neurocytol.*, **7**, 461–478.

Ribak, C. E., and Peters, A. (1975). An autoradiographic study of the projections from the lateral geniculate body of the rat. *Brain Res.*, **92**, 341–368.

Schober, W., and Winkelmann, E. (1975). Der visuelle Kortex der Ratte. Cytoarchitektonic und Stereotaktische Parameter. *Z. mikrosk.-anat. Forsch.*, **89**, 431–446.

Schober, W., and Winkelmann, E. (1977). Die geniculo-kortikale Projection bei Albinoratten. *J. fur Hirnforschung.*, **18**, 1–20.

Sims, K. B., Hoffman, D. L., Said, S. I., and Zimmerman, E. A. (1980). Vasoactive intestinal polypeptide (VIP) in mouse and rat brain: an immunocytological study. *Brain Res.*, **186**, 165–183.

Somogyi, P. (1977). A specific 'axo-axonal' interneuron in the visual cortex of the rat. *Brain Res.*, **136**, 345–350.

Somogyi, P. (1978). The study of Golgi stained cells and of experimental degeneration under the electron microscope: a direct method for the identification in the visual cortex of three successive links in a neuron chain. *Neurosci.*, **3**, 167–180.

Somogyi, P., and Cowey, A. (1981). Combined Golgi and electron microscopic study on the synapses formed by double bouquet cells in the visual cortex of the cat and monkey. *J. comp. Neurol.*, **195**, 547–566.

Somogyi, P., Freund, T. F., and Cowey, A. (1982). The axo-axonic interneuron in the cerebral cortex of the rat, cat and monkey. *Neurosci.*, **7**, 2577–2607.

Somogyi, P., Freund, T. F., Halasz, N., and Kisvarday, Z. F. (1981). Selectivity of neuronal [^{3}H] GABA accumulation in the visual cortex as revealed by Golgi staining of the labelled neurons. *Brain Res.*, **225**, 431–436.

Somogyi, P., Kisvarday, Z. F., Martin, K. A. C., and Whitteridge, D. (1983). Synaptic connections of morphologically identified and physiologically characterized large basket cells in the striate cortex of cat. *Neurosci.*, **10**, 261–294.

Toyama, K., Matsunami, K., Ohno, T., and Tokashiki, S. (1974). An intracellular study of neuronal organization in the visual cortex. *Exp. Brain Res.*, **21**, 45–66.

Werner, L., Hedlich, A., Winkelmann, E., and Brauer, K. (1979). Versuch einer Identifizierung von Nervenzellen der visuellen Kortex der Ratte nach Nissl—und Golgi-Kopsch-Darstellung. *J. Hirnforsch.*, **20**, 121–139.

Werner, L., Wilke, A., Blodner, R., Winkelmann, E., and Brauer, K. (1982). Topographical distribution of neuronal types in the albino rats area 17. A qualitative and quantitative Nissl study. *Z. mikrosk.-anat. Forsch.*, **96**, 433–453.
White, E. L., and Hersch, S. M. (1981). Thalamocortical synapses of pyramidal cells which project from SmI to MsI cortex in the mouse. *J. comp. Neurol.*, **198**, 167–181.
Winfield, D. A., Gatter, K. C., and Powell, T. P. S. (1980). An electron microscopic study of the types and proportions of neurons in the cortex of the motor and visual areas of the cat and rat. *Brain*, **103**, 245–258.
Winkelmann, E., Brauer, N., and Berger, U. (1975). Zur columnaren Organisation von Pyramidenzellen in visuellen Cortex der Albinoratte. *Z. mikroskop.-anat. Forsch.*, **89**, 239–256.

Models of the Visual Cortex
Edited by D. Rose and V. G. Dobson

CHAPTER 53

Local excitatory circuits in area 17 of the cat

KEVAN A. C. MARTIN and PETER SOMOGYI
Department of Experimental Psychology, South Parks Road, Oxford, OX1 3UD, UK

INTRODUCTION

Our understanding of the anatomy of local neuronal circuits in the brain is derived almost solely on extrapolations made from studies of single cells that have been stained by the Golgi process (e.g. Ramón y Cajal, 1911). Although it is possible to obtain data about the synaptic relationships between neurons contributing to long axon tracts and their targets, there has been no standard method that can be applied to the study of the synaptology of local circuits. Direct examination of the intracortical circuitry has only been possible through fortuitous Golgi staining (e.g. Peters and Proskauer, 1980) or in situations where the postsynaptic target is a specialized and easily indentifiable structure (Somogyi, 1977, 1979). The purely anatomical approach has the further obvious disadvantage that it provides no direct information about the functional relationships.

Indirect physiological methods (Benevento, Creutzfeldt and Kuhnt, 1972; Bullier and Henry, 1979a, 1979b, 1979c; Ferster and Lindström, 1983; Hoffman and Stone, 1971; Hubel and Wiesel, 1962; Mitzdorf and Singer, 1977; Singer, Tretter and Cynader, 1975; Toyama, Maekawa and Takeda, 1977; Toyama *et al.*, 1974; Toyama, Kimura and Tanaka, 1981) have also been used to analyse the cortical circuitry, and these provide the basis for current models of cortical organization. The major drawback of these methods is that, being indirect, they have produced conflicting hypotheses of cortical mechanisms. It has now become necessary to use a more direct approach to the problem. Direct analysis of structure/function relationships has become possible through intracellular recording and marking techniques

(Gilbert and Wiesel, 1979; Kelly and Van Essen, 1974; Lin, Friedlander and Sherman, 1979). In order to analyse the cortical circuitry we have extended the technique by combining electrical stimulation, receptive field mapping and intracellular injection of horseradish peroxidase (HRP; see Martin and Whitteridge, 1981, 1982, 1984), and in this way have traced out some of the main paths of the spiny cells (the putative excitatory cells) in cat area 17.

THE RELAY OF EXCITATION THROUGH AREA 17

The first stage

Using these combined techniques it has been possible to study directly three major correlations that have been suggested in models of area 17:

1. Simple cells get direct input from the lateral geniculate nucleus (LGN); complex cells get indirect LGN input via simple cells (Hubel and Wiesel, 1962).
2. Simple cells get direct input from LGN X-cells; complex cells get direct input from LGN Y-cells (Hoffman and Stone, 1971).
3. Simple cells are stellate cells; complex cells are pyramidal cells (Kelly and Van Essen, 1974).

To give a short answer, our studies suggest that there is no clear separation of neuron populations, although the above statements may express a trend. The results of our investigations are summarized in schematic form in Fig. 1 and the paragraphs below describe and discuss the issues in more detail.

It is striking to note (Fig. 1) that cells driven monosynaptically either by X or Y LGN afferents are found in all cortical layers except layers 1 and 2. As we have commented (Martin, 1984; Martin and Whitteridge, 1982, 1984), the finding that there is a population of cells in layer 4A and 3 that receive direct input from a slow-conducting (X-like) input from the LGN is unexpected, given the data of Ferster and LeVay (1978) and Gilbert and Wiesel (1979) who described X-cells of the LGN terminating exclusively in layer 4B and upper layer 6. We have since followed up this issue by injecting HRP into the axons of physiologically characterized LGN afferents. Amongst the axons recovered were some X afferents whose terminal arbor occupied the entire width of laminae 4 and 6 (Martin, 1984), thus providing an anatomical substrate for our earlier observation.

Ultrastructural examination of these same HRP-filled afferents (Freund *et al.*, in preparation) has shown that the X afferents have fewer postsynaptic targets per bouton than the Y afferents, in agreement with the suggestions made by Winfield and Powell (1983). The Y afferent arborizations are generally much larger than those of the X afferents and contain more boutons

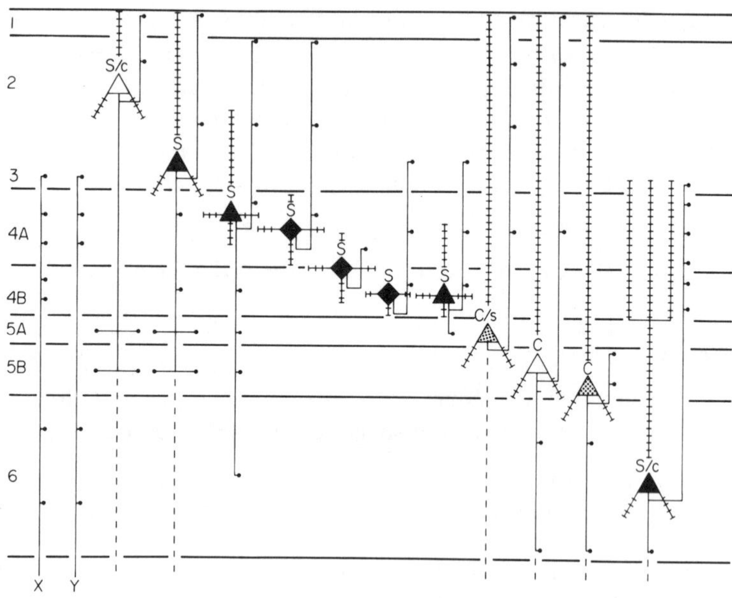

FIGURE 1 A schematic summary of the main classes of spiny neurons found after intracellular recordings and injections of HRP. Cortical layers are marked 1 to 6. Position of the cell body is indicated by triangles for pyramidal cells and diamonds for stellate cells. Filled symbols: cells receiving monosynaptic input from the afferents of the lateral geniculate nucleus (LGN), labelled according to their physiological type, X and Y. Unfilled symbols: cells activated indirectly (polysynaptically) by LGN afferents. Stippled symbols indicate that most cells received a polysynaptic input, but a reasonable minority were driven monosynaptically by LGN afferents. Dendrites are indicated by hatched lines, axons by lines with the clubs, indicating the layers in which most boutons were found. The triangles with the stellate cell dendritic configuration indicate the 'star' pyramidal cells of layer 4. 'S' and 'C': receptive field (RF) type of the majority of cells in that layer. 's' and 'c': the RF of a reasonable minority of cells in that layer. C-type RFs have overlapping ON and OFF areas. S-type RFs have separate ON and/or OFF subfields. Dashed lines: some cells had axons that left area 17.

(Ferster and LeVay, 1978; Gilbert and Wiesel, 1979; and Freund, *et al.*, in preparation). Thus, in addition to the amplification of the Y system which occurs in the LGN (Friedlander *et al.*, 1981), there may be a further amplification of the Y system in area 17.

The upper portion of layer 4 thus contains a mixture of cells that can be driven monosynaptically by either the X or Y afferents, but not both. Although the X and Y afferents are selective for particular cells in that individual cells are not driven by both afferent types, they are not selective for particular cell types: all the three main morphological classes (pyramidal cells, spiny stellate cells and cells with smooth dendrites) can get input from either pathway. Anatomical confirmation of this finding has been possible through a combination of intracellular HRP, electron microscopy and Golgi staining (Freund *et al.*, in preparation). The vast majority of cells in layer 4 have simple or S-type receptive fields (RFs). Thus, with regard to point 2 above, there is no simple correlation between the receptive field type and the type of afferent input, at least in layer 4. Nor, with regard to point 3, is there a strict correlation of the morphology and RF type, although all spiny stellate cells have S-type RFs, as suggested by Kelly and Van Essen (1974).

Intracellular injections (Gilbert and Wiesel, 1983; Martin and Whitteridge, 1982, 1984) have also solved the problem of how X-afferent input is relayed out of layer 4B to the superficial layers (1, 2 and 3) of the cortex. Previously, Golgi studies had shown that the spiny stellate cells of layer 4B cells have local collaterals within layer 4 but do not project directly to the superficial layers (Lund *et al.*, 1979). However, some layer 4 spiny stellate and star pyramidal cells do project directly to layers 2 and 3 (Gilbert, 1983; Gilbert and Wiesel, 1983; Martin and Whitteridge, 1982, 1984). Compared to the projections of the cells in layer 4A that receive direct X-like input, the layer 4B cells injected so far have a much smaller collateral system. Thus the main X-input to superficial layers is probably provided by the layer 3 and 4A cells. The nature of the difference, if any, between the 4A and 4B X-like input remains to be determined.

Cells whose somas lie outside the main termination zone of the LGN afferents may also receive direct input from the LGN fibres. This applies particularly to the pyramidal cells of layer 3 whose basal dendrites descend into layer 4, but also to the pyramidal cells of layer 5, some of whose apical dendrites receive direct LGN input (Hornung and Garey, 1981; Freund *et al.*, in preparation). However, most layer 5 cells, despite having an apical dendrite passing through layer 4, do not appear to get a strong monosynaptic activation (Martin and Whitteridge, 1984). Some layer 5 cells that receive direct LGN input have basal dendrites that descend into layer 6 and could presumably get their input from the LGN collaterals in layer 6 (Martin and Whitteridge, 1984). The majority of cells in layer 5 have complex or C-type RFs and are driven indirectly by the LGN afferents, but it should be emphasized that, with regard to point 1 above, some layer 5 cells with typical complex RFs do receive direct LGN input. Those that receive direct LGN input tend to be activated by Y-fibres, as suggested by Hoffman and Stone (1971).

The layer 6 pyramidal cells have characteristic apical dendrites that branch widely in layer 4. These cells may receive their main monosynaptic activation via LGN input to their apical dendrites, because the collateral projection of the LGN afferents in layer 6 is weak (Bullier and Henry, 1979c; Ferster and LeVay, 1978; Freund *et al.*, in preparation; Martin, 1984). Synapses made on the distal portion of the apical dendrite are not electrically distant from the soma (probably less than one space constant; see Jack, Noble and Tsien,

1975) and therefore will be quite effective in producing a potential change at the soma.

The synthesis of receptive fields

The cells that are intracellularly filled with HRP show far more extensive collateral systems than have been seen in Golgi-stained material and their axonal projections provide important data as to the interlaminar projections of the different cell types. A major problem in our interpretation of how RFs are synthesized lies in the lack of data on the postsynaptic targets of the spiny neurons. In view of the data from the hippocampus (e.g. Lorente de Nó, 1934) that different afferent pathways arrive at different positions on the apical dendrites of pyramidal cells, it would be rash to suppose that the neocortex operates on a completely different principle, although it is often assumed that the main postsynaptic targets of an intracortical projection are cells whose somas lie in the layer to which the axon projects (Ferster and Lindström, 1983; Gilbert and Wiesel, 1979; Mitzdorf and Singer, 1977; Toyama *et al.*, 1974). This problem of interpretation lies not so much at the first synapse between the LGN afferents and the cortical cells, since anatomical methods have told us something about the targets of the afferents, but at the second synapse. In dealing with this problem the combination of methods outlined above has been most useful.

The major output from the cells receiving direct LGN input in layer 4 is to the superficial layers (see Fig. 1). The potential dendritic targets of this projection are the cells in layers 2 and 3, and the apical dendrites of the layer 5 pyramidal cells. The latency of response of the layer 2 and 3 pyramidal cells that receive indirect activation suggests that they are the main targets of the output of layer 4. However, cells in layers 2 and 3 remain responsive even when layer 4 cells have been inactivated (Malpeli, 1983) and this parallel drive is probably provided by the cells in layer 5 that receive direct LGN input, although these latter cells seem few in number (Martin and Whitteridge, 1984).

In the models of Hubel and Wiesel (1962) and Gilbert and Wiesel (1979, 1981), the layer 4 to layer 3 projection involves a transformation from simple RFs to complex RFs. However, the actual structure of the RFs of the layer 2 and 3 cells may be quite similar to the RFs of cells in layer 4 (Kato, Bishop and Orban, 1978; Henry, Harvey and Lund, 1979; Bullier and Henry, 1979a; Duysens *et al.*, 1982; Martin and Whitteridge, 1984). The minority of cells in layers 2 and 3 whose complex or C-type RFs make them quite distinct from the RFs of layer 4 cells could receive their input either from the simple cells in layer 4 or the complex cells in layer 5 (Fig. 1). Thus there could be a serial progression, from simple to complex, as originally suggested (Hubel and Wiesel, 1962), or a parallel, simple-to-simple, complex-to-complex, arrangement.

About 70 per cent. of the pyramidal cells of the superficial layers that were filled with HRP had an axon which leaves area 17. Most of these same cells also participated in local circuits through collateral projections in the superficial layers and the descending collateral network in layer 5 (Fig. 1). This descending collateral system is of interest, not only because of the link it provides between superficial and deep layers (layers 5 and 6) but because layer 5 contains a majority of complex cells (Gilbert, 1977; Hubel and Wiesel, 1962) that may receive their main input from simple cells in layers 2 and 3, thus providing a serial transformation. As yet we do not know the postsynaptic targets of the collateral projection of the superficial pyramidal cells to the deep layers (Fig. 1), but the principal candidates are the pyramidal cells in layers 5 and 6.

There are several possible sites at which the hypothesis of a simple-to-complex cell progression may be reversed. One such site is the projection of the layer 5 complex cells to the simple cells of the superficial layers. Anatomically this is a rich projection though, apart from our study (Martin and Whitteridge, 1982, 1984), it has only been seen in Golgi-stained material (Lorente de Nó, 1949; Lund *et al.*, 1979) and its functional significance is unknown. A second site at which the simple-to-complex rule may be reversed is that of the projection of the layer 5 complex cells to layer 6, which contains a majority of cells with simple RFs (Hubel and Wiesel, 1962). It should be noted that no layer has an exclusive and pure population of cells of one RF type. Thus, until the RFs of both pre- and postsynaptic cells are determined, virtually any model involving serial transformation or parallel processing remains possible.

If complex cells do indeed make synapses on cells with simple RFs, why is this input not reflected in the RF properties of the simple cell? There is in fact some physiological evidence for a complex-to-simple link. Simple cells can be facilitated (Hammond and McKay, 1981) or suppressed (Burr, Morrone and Maffei, 1981; Hammond and McKay, 1981) by visual noise that only produces action potentials in complex cells (Hammond and McKay, 1977). In the case of the facilitated response, the input of the complex cells to simple cells is insufficient to bring the simple cell to threshold. The reason for this may-be due to a general principle of neuronal connectivity, operating not only in the cortex but also in the spinal cord. The number of synaptic connections we have found from one cell to any other single cell in the visual cortex is always less than 20 (Freund *et al.*, in preparation; Somogyi, Freund and Cowey, 1982; Somogyi *et al.*, 1983). A similar order of magnitude has been found for afferents in the spinal cord (Brown and Fyffe, 1978; Brown and Noble, 1982; Burke, Walmsley and Hodgson, 1979; Redman and Walmsley, 1983). Estimates of the potential change at the soma due to each bouton give a figure of about 100 μV per bouton per postsynaptic potential for the motoneuron (Jack, Redman and Wong, 1981; Redman and Walmsley, 1983) and for hippocampal pyramidal cells (McNaughton, Barnes and

Andersen, 1981). If the same figures apply to cells in the neocortex, then between 100 and 300 excitatory synapses activated simultaneously would be required to bring the cell to the threshold firing level. Thus the contribution of any single cell to the RF of any other cell may be small. Nevertheless, this sort of principle of connectivity would allow an anatomically minor input to exert the facilitatory influences of complex on simple cells seen with visual noise (Hammond and McKay, 1981).

Although it has been implicit from the first functional models of area 17 that new information is added at each synapse (Hubel and Wiesel, 1962, 1965), there are pathways whose role may be more concerned with feedback and feedforward mechanisms. The layer 4 cells receive a monosynaptic input from the LGN but do not themselves feed back to the LGN relay cells. By contrast, the layer 6 pyramidal cells that are monosynaptically driven do project to the LGN, where they directly excite the relay cells (Ahlsén, Grant and Lindström, 1982; Friedlander *et al.*, 1981). The layer 6 pyramidal cells have a rich projection almost exclusively to layer 4, where they form asymmetric (putative excitatory) synapses with different cell types (Wiesel and Gilbert, 1983). The conduction velocity of the axons projecting to the LGN is slow and thus the excitatory feedback will be delayed relative to the much shorter intracortical path to layer 4. The role of both paths may be similar: to facilitate transmission of the retinocortical impulses.

CONCLUSION

It is possible that all the circuits so far proposed (e.g. Ferster and Lindström, 1983; Gilbert and Wiesel, 1979, 1981; Hubel and Wiesel, 1962; Mitzdorf and Singer, 1977; Toyama *et al.*, 1974; Wiesel and Gilbert, 1983) do have an anatomical basis, although for the present these circuits remain speculative. What we also do not know is which circuits constitute the principal pathways and which constitute the minor pathways. The approach taken by Malpeli (1983) of examining the effect of removing afferent paths is one useful way of dissecting out the principal paths. Further combinations of techniques, e.g. intracellular HRP labelling and Golgi-staining, have been developed (Freund and Somogyi, 1983), and will extend the capabilities of the anatomical analyses. What is already clear is that input arrives and output leaves at many different levels within the laminae of area 17 and this must involve some parallel organization of circuits. Different degrees of processing may take place depending on the different destinations of the output. Much further work along the lines described is required to decide between the alternative models.

ACKNOWLEDGEMENT

Supported by the MRC.

REFERENCES

Ahlsén, C., Grant, K. and Lindström, S. (1982). Monosynaptic excitation of principal cells in the lateral geniculate nucleus by corticofugal fibres. *Brain Res.*, **234**, 454–458.

Benevento, L. A., Creutzfeldt, O. D., and Kuhnt, U. (1972). Significance of intracortical inhibition in the visual cortex. *Nature New Biol.*, **238**, 124–126.

Brown, A. G., and Fyffe, R. E. W. (1978). The morphology of group 1a afferent fibre collaterals in the spinal cord of the cat. *J. Physiol.*, **274**, 111–127.

Brown, A. G., and Noble, R. (1982). Connexions between hair follicle afferent fibres and spinocervical tract neurons in the cat; the synthesis of receptive fields. *J. Physiol.*, **323**, 77–91.

Bullier, J., and Henry, G. H. (1979a). Ordinal position of neurons in cat striate cortex. *J. Neurophysiol.*, **42**, 1251–1263.

Bullier, J., and Henry, G. H. (1979b). Neural path taken by afferent streams in striate cortex of the cat. *J. Neurophysiol.*, **42**, 1264–1270.

Bullier, J., and Henry, G. H. (1979c). Laminar distribution of first order neurons and afferent terminals in cat striate cortex. *J. Neurophysiol.*, **42**, 1271–1281.

Burke, R. E., Walmsley, B., and Hodgson, J. A. (1979). HRP anatomy of group 1a afferent contacts on alpha motoneurons. *Brain Res.*, **160**, 347–352.

Burr, D., Morrone, C., and Maffei, L. (1981). Intracortical inhibition prevents simple cells from responding to textured visual patterns. *Exp. Brain Res.*, **43**, 455–458.

Duysens, J., Orban, G. A., Van der Glas, H. W., and Maes, H. (1982). Receptive field structure of area 19 as compared to area 17 of the cat. *Brain Res.*, **231**, 293–308.

Ferster, D., and LeVay, S. (1978). The axonal arborizations of lateral geniculate neurons in the striate cortex of the cat. *J. comp. Neurol.*, **182**, 923–944.

Ferster, D., and Lindström, S. (1983). An intracellular analysis of geniculo-cortical connectivity in area 17 of the cat. *J. Physiol.*, **342**, 181–215.

Freund, T. F., and Somogyi, P. (1983). The section-Golgi impregnation procedure. 1. Description of the method and its combination with histochemistry after intracellular iontophoresis or retrograde transport of horseradish peroxidase. *Neurosci.*, **9**, 463–474.

Friedlander, M. J., Lin, C-S., Stanford, L. R., and Sherman, S. M. (1981). Morphology of functionally identified neurons in the lateral geniculate nucleus of the cat. *J. Neurophysiol.*, **46**, 80–129.

Gilbert, C. D. (1977). Laminar differences in receptive field properties of cells in cat primary visual cortex. *J. Physiol.*, **268**, 391–421.

Gilbert, C. D. (1983). Microcircuitry of the visual cortex. *Ann. Rev. Neurosci.*, **6**, 217–247.

Gilbert, C. D., and Wiesel, T. N. (1979). Morphology and intracortical projections of functionally characterized neurons in the cat visual cortex. *Nature*, **280**, 120–125.

Gilbert, C. D., and Wiesel, T. N. (1981). Laminar specialization and intracortical connections in cat primary visual cortex. In *The Organization of the Cerebral Cortex* (Eds F. O. Schmitt, F. G. Worden, G. Adelman and S. G. Dennis), MIT Press, Cambridge, Mass., pp. 163–191.

Gilbert, C. D., and Wiesel, T. N. (1983). Clustered intrinsic connections in cat visual cortex. *J. Neurosci.*, **3**, 1116–1133.

Hammond, P., and MacKay, D. M. (1977). Differential responsiveness of simple and complex cells in cat straite cortex to visual texture. *Exp. Brain Res.*, **30**, 275–296.

Hammond, P., and MacKay, D. M. (1981). Modulatory influences of moving textured backgrounds on responsiveness of simple cells in feline striate cortex. *J. Physiol.*, **319**, 431–442.

Henry, G. H., Harvey, A. R., and Lund, J. S. (1979). The afferent connections and laminar distribution of cells in the cat striate cortex *J. comp. Neurol.*, **187**, 725–744.

Hoffman, K. P., and Stone, J. (1971). Conduction velocity of afferents to cat visual cortex: a correlation with cortical receptive field properties. *Brain Res.*, **32**, 460–466.

Hornung, J. P., and Garey, L. J. (1981). The thalamic projection to cat visual cortex: ultrastructure of neurons identified by Golgi impregnation or retrograde horseradish peroxidase transport. *Neurosci.*, **6**, 1053–1068.

Hubel, D. H., and Wiesel, T. N. (1962). Receptive fields, binocular interaction and functional architecture in the cat's visual cortex. *J. Physiol.*, **160**, 106–154.

Hubel, D. H., and Wiesel, T. N. (1965). Receptive fields and functional architecture in two non-striate visual areas (18 and 19) of the cat. *J. Neurophysiol.*, **28**, 229–289.

Jack, J. J. B., Noble, D., and Tsien, R. W. (1975). *Electric Current Flow in Excitable Cells*, Oxford University Press, pp. 197–222.

Jack, J. J. B., Redman, S. J., and Wong, K. (1981). The components of synaptic potentials evoked in cat spinal motoneurones by impulses in single group 1a afferents. *J. Physiol.*, **321**, 65–96.

Kato, H., Bishop, P. O., and Orban, G. A. (1978). Hypercomplex, and simple/complex cell classifications in cat striate cortex. *J. Neurophysiol.*, **41**, 1071–1095.

Kelly, J. P., and Van Essen, D. C. (1974). Cell structure and function in the visual cortex of the cat. *J. Physiol.*, **238**, 515–547.

Lin, C-S., Friedlander, M. J., and Sherman, S. M. (1979). Morphology of physiologically identified neurons in the visual cortex of the cat. *Brain Res.*, **172**, 344–348.

Lorente de Nó, R. (1934). Studies on the structure of the cerebral cortex. II. Continuation of the study of the ammonic system. *J. Psychol. Neurol. (Leipzig).*, **46**, 113–177.

Lorente de Nó, R. (1949). Cerebral cortex: architecture, intracortical connections, motor projections. In *Physiology of the Nervous System* (Ed. J. F. Fulton), Oxford University Press, pp. 288–312.

Lund, J. S., Henry, G. H., MacQueen, C. L., and Harvey, A. R. (1979). Anatomical organisation of the primary visual cortex (area 17) of the cat. A comparison with area 17 of the macaque monkey. *J. comp. Neurol.*, **184**, 599–618.

McNaughton, B. L., Barnes, C. A., and Andersen, P. (1981). Synaptic efficacy and EPSP summation in granule cells of rat fascia dentata studied in vitro. *J. Neurophysiol.*, **46**, 952–966.

Malpeli, J. (1983). Activity of cells in area 17 of the cat in absence of input from layer A of lateral geniculate nucleus. *J. Neurophysiol.*, **49**, 595–610.

Martin, K. A. C. (1984). Neuronal circuits in cat striate cortex. In *Cerebral Cortex*. (Eds A. Peters and E. G. Jones), Vol. 2, Plenum Press, New York.

Martin, K. A. C., and Whitteridge, D. (1981). Morphological identification of cells in the cat's visual cortex, classified with regard to their afferent input and receptive field type. *J. Physiol.*, **320**, 14–15P.

Martin, K. A. C., and Whitteridge, D. (1982). The morphology, function and intracortical projections of neurons in area 17 of the cat which receive monosynaptic input from the lateral geniculate nucleus. *J. Physiol.*, **328**, 37–38P.

Martin, K. A. C., and Whitteridge, D. (1984). Form, function, and intracortical projections of spiny neurons in the striate visual cortex of the cat. *J. Physiol.*, **353**, 463–504.

Mitzdorf, U., and Singer, W. (1977). Prominent excitatory pathways in the cat visual cortex. (A 17 and A 18): a current source density analysis of electrically evoked potentials. *Exp. Brain Res.*, **33**, 371–394.

Peters, A., and Proskauer, C. C. (1980). Synaptic relationships between a multipolar stellate cell and a pyramidal neuron in the rat visual cortex. A combined Golgi-electron microscope studv. *J. Neurocytol.*, **9**, 163–184.

Ramón Y Cajal, S. (1911). *Histologie du Systeme Nerveux de l'Homme et des Vertebres*, Maloine, Paris.

Redman, S., and Walmsley, B. (1983). The time course of synaptic potentials evoked in cat spinal motoneurons at identified group 1a synapses. *J. Physiol.*, **343**, 117–133.

Singer, W., Tretter, F., and Cynader, M. (1975). Organization of cat striate cortex: a correlation of receptive-field properties with afferent and efferent connections. *J. Neurophysiol.*, **38**, 1080–1098.

Somogyi, P. (1977). A specific 'axo-axonal' interneuron in the visual cortex of the rat. *Brain Res.*, **136**, 345–350.

Somogyi, P. (1979). An interneuron making synapses specifically on the axon initial segment (AIS) of pyramidal cells in the cerebral cortex of the cat. *J. Physiol.*, **296**, 18–19P.

Somogyi, P., Freund, T. F., and Cowey, A. (1982). The axo-axonic interneuron in the cerebral cortex of the rat, cat and monkey. *Neurosci.*, **7**, 2577–2608.

Somogyi, P., Kisvárday, Z. F., Martin, K. A. C., and Whitteridge, D. (1983). Synaptic connections of morphologically identified and physiologically characterized large basket cells in the striate cortex of cat. *Neurosci.*, **10**, 261–294.

Toyama, K., Kimura, M., and Tanaka, K. (1981). Organization of cat visual cortex as investigated by cross-correlation technique. *J. Neurophysiol.*, **46**, 202–214.

Toyama, K., Maekawa, K., and Takeda, T. (1977). Convergence of retinal inputs onto visual cortical cells. 1. A study of the cells monosynaptically excited from the lateral geniculate body. *Brain Res.*, **137**, 207–220.

Toyama, K., Matsunami, K., Ohno, T., and Tokashiki, S. (1974). An intracellular study of neuronal organization in the visual cortex. *Exp. Brain Res.*, **21**, 45–66.

Wiesel, T. N., and Gilbert, C. D. (1983). Morphological basis of visual cortical function. *Quart.J. Exp. Physiol.*, **68**, 525–543.

Winfield, D., and Powell, T. P. S. (1983). Laminar cell counts and geniculo-cortical boutons in area 17 of cat and monkey. *Brain Res.*, **277**, 223–229.

Models of the Visual Cortex
Edited by D. Rose and V. G. Dobson

CHAPTER 54

Cortical circuitry underlying inhibitory processes in cat area 17

PETER SOMOGYI and KEVAN A. C. MARTIN
Department of Experimental Psychology, South Parks Road, Oxford, OX1 3UD, UK

INTRODUCTION

The early models of cortical organization did not emphasize inhibitory mechanisms, and suggested that specific excitatory connections were sufficient for properties like orientation tuning (Hubel and Wiesel, 1962). Following the discovery of end-inhibition on some receptive fields (RFs; see Hubel and Wiesel, 1965), inhibitory cells were introduced as neuronal components for increasing the selective properties of cortical cells. More recent work using intracellular recording (Benevento, Creutzfeldt and Kuhnt, 1972) or pharmacological agents (Rose and Blakemore, 1974; Sillito, 1975, 1977, 1979; Sillito *et al.*, 1980; Sillito and Versiani, 1977) have shown that most, if not all, specific properties are produced or strongly augmented by inhibitory processes. The inhibitory cells must therefore form the basis of the columnar organization of functional properties in the visual cortex (Hubel and Wiesel, 1963). GABA (γ-aminobutyric acid) is the most likely candidate for the inhibitory transmitter (Krnjević and Schwartz, 1967; Ribak, 1978; Sillito, 1975) and there is strong evidence that it is contained in the cells with smooth dendrites that have been proposed as the inhibitory interneurons (Freund *et al.*, 1983; Ribak, 1978; Somogyi *et al.*, 1983a; Somogyi *et al.*, 1984). The proportion of inhibitory cells in the cortex is small, perhaps only 20 per cent. yet there are a great variety of forms (Lorente de Nó, 1949; O'Leary, 1941; Peters and Regidor, 1981; Ramón y Cajal, 1911; Szentágothai, 1973), suggesting a diversity of function.

The principal reasons for using inhibitory mechanisms rather than excitatory mechanisms for producing RF specificity may be twofold. First, the system becomes more flexible. Different inhibitory systems superimposed on the same basic pattern of excitation could produce quite different patterns of output. Thus, as new demands are made on the processing system, further inhibitory systems could evolve. Ramón y Cajal (1911) has commented on the apparent increase through the phyla in the number of smooth cell types in the neocortex, with man having the greatest proportion. However, putative inhibitory cells form only about 10 to 20 per cent of the total cells in the cortex (Martin and Whitteridge, 1984; Tömböl, 1974; Winfield, Gatter and Powell, 1980), so the number of additional cells used for additional functions remains relatively small.

A second and related reason for using an inhibitory system is that fewer demands on the specificity of connections are required for each cell if inhibitory processes etch out specific RF properties from a general pool of excitation for each cell. In this manner the inhibitory tuning curves can be broad (as has been found by Benevento, Creutzfeldt and Kuhnt, 1972; Morrone, Burr and Maffei, 1982; Orban, Kato and Bishop, 1979; Sillito, 1979) and yet produce highly selective RF characteristics. To achieve the same degree of selectivity solely by excitatory connections would place great demands on the precision of the intercolumnar excitatory connections that form a large fraction of the intracortical connections. By requiring only a generalized excitation of the cell, only the minimum number of excitatory synapses would be necessary, whereas selectivity produced solely by excitatory connections would require many more connections to achieve specific patterns of excitatory input.

THEORETICAL CONSIDERATIONS

As has been emphasized by theoretical work (Blomfield, 1974; Jack, Noble and Tsien, 1975; Koch and Poggio, 1983), the position of the inhibitory input onto a single neuron is an important factor in determining the net response of the cell. If all the excitation arriving on the cell is to be inhibited, then the optimum position for locating the inhibitory synapses is on the cell soma and axon initial segment. However, if a specific input is to be inhibited, then the optimum location for the inhibitory synapses is just proximal to the excitatory input. The more distal the location of this excitatory/inhibitory combination on the dendrites, the more specific will be its effect (Jack, Noble and Tsien, 1975). Dendritic spines may also receive an inhibitory input that will have its major effect on the single excitatory synapse on the same spine. It has been emphasized that different logical operations can be carried out by particular local interactions of excitatory and inhibitory inputs to the dendritic tree (Diamond, Gray and Yasargil, 1970; Jack, Noble and Tsien, 1975; Koch and Poggio, 1983). Obviously, the output of any single inhibitory/excitatory pair will have little effect on the potential at the soma.

Only the net result of hundreds of such local interactions will determine whether the cell fires or not. In addition, the operations carried out on the distal portions of the dendrites may be further modulated by events occurring more proximally, especially around the soma and axon hillock region. Blomfield (1974) has emphasized one further aspect of the location of inhibitory synapses: if the inhibitory conductances are large, then inhibitory synapses located at the soma will produce a division-like change in the cell's response, whereas inhibitory synapses located on the dendrites will produce a subtractive change in response.

Thus in assessing the significance of a particular type of inhibitory neuron it is essential as a first step to identify the particular location of the synapses on its postsynaptic target. In attempting to apply these theories to cortical organization, one must appreciate that there are only a few instances where the synaptology of the putative inhibitory cells has been examined. The studies described below show that there are very marked differences in the postsynaptic targets of different cell types.

THE SYNAPTIC CONNECTIONS MADE BY THE AXONS OF PUTATIVE INHIBITORY CELLS IN THE CAT

The Axo-axonic cell

The axo-axonic cell, so named because it forms synapses on the initial segment of the axon of pyramidal cells (Somogyi, 1977, 1979), is the most selective of the putative inhibitory cells in its choice of postsynaptic target. Although the physiological properties of the axo-axonic cells are not known, there is now direct evidence that the terminal boutons of the axo-axonic cell contain the synthesizing enzyme for GABA, glutamate decarboxylase (GAD), indicating that this cell is probably inhibitory in function (Freund *et al.*, 1983). About five axo-axonic cells converge on a single pyramidal cell axon, and a single axo-axonic cell may contact several hundred pyramidal cells (Somogyi, Freund and Cowey, 1982). The axo-axonic cell seems to provide the bulk of its innervation to pyramidal cells in the superficial layers of the cortex (Freund *et al.*, 1983; Somogyi, Freund and Cowey, 1982). Since these pyramidal neurons provide most of the corticocortical connections it has been suggested that the axo-axonic cell controls the transfer of information between different cortical areas (Peters, Proskauer and Ribak, 1982; Somogyi, Hodgson and Smith, 1979). This may not be their only role since many of the pyramidal cells projecting to other cortical areas also provide a substantial collateral input to the deep layers of the same cortical area (Martin and Somogyi, Chapter 53 in this volume). Inhibition of the output of the pyramidal cells of the superficial layers could then markedly affect the activity of cells in the deep layers and hence the output to subcortical regions. Thus, given the appropriate input, the axo-axonic cells could provide a most effective control of the output of projection neurons. If this control were applied selectively to specific cortical regions, it could provide a basis for the mechanism of selective attention.

If the cortex uses positive feedback circuits (the pyramidal cells of layer 6 that project to layer 4 and the LGN may be one such example) then it may be necessary to insert inhibitory cells as governors to prevent epileptiform activity being initiated by positive feedback. If the axo-axonic cell had a relatively high threshold for activation then it would be a suitable candidate for such an inhibitory cell (Freund *et al.*, 1983). Physiological evidence has been obtained for the presence of an inhibitory mechanism that operates only at high levels of discharge of pyramidal cells (Sillito, 1979).

The Basket cell

The main input to the soma of pyramidal cells probably comes from the large basket cell (Kisvárday *et al.*, 1983; Martin, Somogyi and Whitteridge, 1983; Somogyi *et al.*, 1983b), although one other cell has now been found which provides somatic input to pyramidal cells (Kisvárday *et al.*, in preparation; see below). The evidence that the basket cells are GABAergic is indirect, but the finding that most if not all of the boutons providing somatic synapses on pyramidal cells contain GAD is strongly suggestive of their inhibitory role (Freund *et al.*, 1983). So far only the basket cells of the superficial layers have been investigated in detail at the electron microscopic (EM) level, but our observations of the basket cells of layer 5 suggest they have similar postsynaptic targets. The synapses of the basket cell are concentrated on the soma (30 to 40 per cent. of the basket cell's synapses) and proximal dendritic shafts (24 per cent.) of pyramidal cells. This location of the synapses suggests that their action may produce division-like changes in the cells response. Two RF properties, orientation tuning and directionality, have been examined for this purpose (Dean, Hess and Tolhurst, 1980; Morrone Burr and Maffei, 1982; Rose, 1977). In both instances increasing amounts of inhibition produce division-like changes in the cell's firing rate, suggesting that the relevant inhibitory synapses are positioned around the perisomatic region of the cell and may therefore have the basket cells as their source. We have suggested that different subsets of basket cells may be involved in producing many of the specific RF properties, like orientation tuning, directionality, binocular depth tuning and end-inhibition, that are known to be under inhibitory control (Martin, Somogyi and Whitteridge, 1983; Somogyi *et al.*, 1983b). This is because the circuitry required to generate these properties in a single cell becomes very complicated if the same set of basket cells are used as the final common path for all inhibitory mechanisms. It may also be useful for cells to use the same subset of inhibitory cells for those properties they have in common (e.g. end-inhibition), but would obviously have to use different inhibitory cells for generating dissimilar properties (e.g. different orientation selectivities).

The axon of the basket cell does not contact every pyramidal cell in the region occupied by its axon. Instead it appears to pick out columns of pyramidal cells located in small patches of cortex. Thus although the axon covers about 1.5 mm of cortex in the anteroposterior and mediolateral dimensions (the largest tangential spread of any smooth type so far encountered), only about 200 to 300 pyramidal cells are contacted by a single basket cell (Kisvárday *et al.*, 1983; Martin, Somogyi and Whitteridge, 1983; Somogyi *et al.*, 1983b). Each basket cell only provides about four or five of the synapses (about 5 to 10 per cent. of the total somatic input) on the soma of a pyramidal cell. If, as we have suggested (Martin and Somogyi, Chapter 53 this volume), 100 to 300 active excitatory synapses are required to bring the cell to threshold for firing, then it seems unlikely that the input of a single basket cell would be sufficient to produce a significant reduction in the response of the target cell, even if the inhibition is producing a shunt rather than a hyperpolarization of the membrane (Jack, Noble and Tsien, 1975). Thus controlling the pyramidal cell output by basket cell inhibition may require the coordinated action of several basket cells.

The basket cells also provide a substantial input to dendritic spines (20 per cent. of basket cell synapses; Somogyi *et al.*, 1983b). An inhibitory input to a spine probably has a very localized effect on the excitatory input (Jack, Noble and Tsien, 1975) and that effect will be subtractive. As yet a subtractive change in firing rate has not been related to a specific RF property, but the localized nature of spine inhibition suggests that a specific excitatory pathway is being inhibited by the basket cell input to the spines.

The Clutch cell

The other cell that gives a strong somatic input to its postsynaptic targets is located in layer 4. It is a multipolar cell (Martin, Somogyi and Whitteridge, 1983) which we have called a 'clutch' cell. Although the general distribution of its synapses between soma and dendrites is similar to that of the large basket cell, its axon is very much more restricted than that of the basket cell. Unlike the basket cell, the major portion of the axonal arborization of the clutch cell is in layer 4 and it probably makes synapses on many spiny stellate cells. Some collaterals of the clutch cell extend into lower layer 3 and into layer 5 and synapse on pyramidal cells (Kisvárday *et al.*, in preparation). Many of the spiny cells in layer 4A have widespread connections to other layers (Martin, 1984). Thus an inhibitory cell with a localized axon system within layer 4, such as the clutch cell, could have a widespread influence. The much more extensive collateral system of the basket cells in deep and superficial layers may in part be required to inhibit cells with a common excitatory source in layer 4.

Physiologically the clutch cells and the basket cells are heterogenous (Martin, Somogyi and Whitteridge, 1983). They can have either S- or C-type RFs and be activated monosynaptically or polysynaptically by X- or Y-like LGN afferents. In addition, two basket cells with RFs near the vertical meridian were driven by callosal afferents. This heterogeneity of RF type and serial position is not surprising, given that the cortex may be organized in a parallel fashion and that similar inhibitory processes act on different parallel paths at all levels.

The Double Bouquet cell

The fourth putative inhibitory cell whose synaptology has been investigated in the cat is the double bouquet cell (Somogyi and Cowey, 1981, 1984). While the three types described above concentrate their input at the perisomatic region of spiny cells, the double bouquet cell shows just the opposite trend. The vast majority of contacts it makes are onto small- or medium-sized dendritic shafts of non-pyramidal cells that may themselves be inhibitory. As has been pointed out (Somogyi and Cowey, 1981, 1984), this would provide a mechanism for the disinhibition seen in physiological studies. However, disinhibition may not be the primary role of the cell. It has been found (Martin, Somogyi and Whitteridge, 1983) that the tuning curves and RF properties of putative inhibitory cells are qualitatively no different from those of the pyramidal and spiny stellate cells. Since the RF properties of the inhibitory cells are presumably also produced by inhibition, it is inevitable that they should themselves receive an inhibitory input and that this would produce the disinhibition effect seen using electrical stimulation. The convergence of a number of inhibitory cells with different RF properties probably accounts for the broad inhibitory tuning curves that have been found (Burr, Morrone and Maffei, 1981).

Although both the basket and the clutch cell also contact putative inhibitory cells (Kisvárday *et al.*, in preparation; Martin, Somogyi and Whitteridge, 1983; Somogyi *et al.*, 1983a) these constitute 10 per cent. or less of the postsynaptic targets. The double bouquet cell, by contrast, contacts a far higher proportion of putative inhibitory cells, perhaps as high as 60 to 70 per cent. (Somogyi and Cowey, 1981). The reason for this may be that the activity of inhibitory cells needs to be coordinated because many converge on the same target cell. The double bouquet cell, with its localized vertically oriented axon passing through several layers, would be well-suited to such a coordinating role, especially for specific properties that are arranged in a columnar fashion.

CONCLUSION

It seems likely from the admittedly small amount of data available that any single cell receives inhibitory input from more than one inhibitory cell type, as in the case of the input to pyramidal cells from both the basket cell and the axo-axonic cell. Presumably the activities of these different convergent inhibitory paths must be coordinated in some way, and this would require rich interconnections between inhibitory cells. Also, those inhibitory cells that are involved in more than one inhibitory mechanism may receive their excitatory input from several independent local circuits. This organization would necessitate a much greater synaptic input to inhibitory cells than excitatory cells, and there is some evidence that cells with smooth dendrites do have a higher synaptic density on their dendrites than cells with spiny dendrites (Freund *et al.*, 1983; Ribak, 1978; Somogyi, Freund and Cowey, 1982).

All of the four putative inhibitory cells described above connect either to different cell types or to different positions on the same type of cell. There are many varieties of putative inhibitory cells whose synaptology has yet to be investigated. It is probable, on the basis of the survey to date, that they will also have particular cell types and postsynaptic sites of preference, reflecting yet further functional differentiation. We have suggested that some inhibitory cells, like the axo-axonic cell, may not be directly involved in producing RF selectivity. Other cells, like the double bouquet cells, because of the distal location of their particular input, may be involved only in the fine tuning of the cell's responses, with the coarse tuning being carried out by cells whose input is at a more strategic location on the cell, like the basket and clutch cells. Clearly we need to know a great deal more about the dendritic and somatic location of inputs from different inhibitory and excitatory sources and their postsynaptic targets before we can devise a realistic and integrated model of area 17. Nevertheless, the power of the methods described here is cause for optimism that the previously intractable problems in studying local circuitry are now potentially soluble.

ACKNOWLEDGEMENT

Supported by the MRC.

REFERENCES

Benevento, L. A., Creutzfeldt, O. D., and Kuhnt, U. (1972). Significance of intracortical inhibition in the visual cortex. *Nature, New Biol.*, **238**, 124–126.

Blomfield, S. (1974). Arithmetical operations performed by nerve cells. *Brain Res.*, **69**, 115–124.

Burr, D., Morrone, C., and Maffei, L. (1981). Intracortical inhibition prevents simple cells from responding to textured visual patters. *Exp. Brain Res.*, **43**, 455–458.

Dean, A. F., Hess, R. F., and Tolhurst, D. J. (1980). Divisive inhibition involved in directional selectivity. *J. Physiol.*, **308**, 84–85P.

Diamond, J., Gray, E. G., and Yasargil, G. M. (1970). The function of the dendritic spine: an hypothesis. In *Excitatory Synaptic Mechanisms* (Eds P. Anderson and J. K. S. Jansen), Universitets Forlaget, Oslo, pp. 213–222.

Freund, T. F., Martin, K. A. C., Smith, A. D., and Somogyi, P. (1983). Glutamate decarboxylase-immunoreactive terminals of Golgi-impregnated axo-axonic cells and of presumed basket cells in synaptic contact with pyramidal neurons of the cat's visual cortex. *J. comp. Neurol.*, **221**, 263–278.

Hubel, D. H., and Wiesel, T. N. (1962). Receptive fields, binocular interaction and functional architecture in the cat's visual cortex. *J. Physiol.*, **160**, 106–154.

Hubel, D. H., and Wiesel, T. N. (1963). Shape and arrangement of columns in cat's striate cortex. *J. Physiol.*, **165** 559–568.

Hubel, D. H., and Wiesel, T. N. (1965). Receptive fields and functional architecture in two non-striate visual areas (18 and 19) of the cat. *J. Neurophysiol.*, **28**, 229–289.

Jack, J. J. B., Noble, D., and Tsien, R. W. (1975). *Electric Current Flow in Excitable Cells*, Oxford University Press, pp. 197–222.

Kisvárday, Z. F., Martin, K. A. C., Somogyi, P., and Whitteridge, D. (1983). The physiology, morphology and synaptology of basket cells in the cat's visual cortex. *J. Physiol.*, **334**, 21–22P.

Koch, C., and Poggio, T. (1983). A theoretical analysis of electrical properties of spines. *Proc. Roy. Soc. Lond. B.*, **218**, 455–477.

Krnjević, K. and Schwartz, S. (1967). The action of γ-aminobutyric acid on cortical neurons. *Exp. Brain Res.*, **3**, 320–336.

Lorente de Nó, R. (1949). Cerebral cortex: architecture, intracortical connections, motor projections. In *Physiology of the Nervous System* (Ed. J. F. Fulton), Oxford University Pres, pp. 288–312.

Martin, K. A. C. (1984). Neuronal circuits in cat striate cortex. In *Cerebral Cortex* (Eds A. Peters and E. G. Jones), Vol. 2, Plenum Press, New York. pp. 241–284.

Martin, K. A. C., Somogyi, P., and Whitteridge, D. (1983). Physiological and morphological properties of identified basket cells in the cat's visual cortex. *Exp. Brain Res.*, **50**, 193–200.

Martin, K. A. C., and Whitteridge, D. (1984). Form, function, and intracortical projections of spiny neurons in the striate visual cortex of the cat. *J. Physiol.*, **353**, 463–504.

Morrone, M. C., Burr, D. C., and Maffei, L. (1982). Functional implications of crossorientation inhibition of cortical visual cells. I. Neurophysiological evidence. *Proc. Roy. Soc. (Lond.), B*, **216**, 335–354.

O'Leary, J. L. (1941). Structure of area striata of the cat. *J. comp. Neurol.*, **75**, 131–161.

Orban, G. A., Kato, H., and Bishop. P. O. (1979). Dimensions and properties of end-zone inhibitory areas in receptive fields of hypercomplex cells in cat striate cortex. *J. Neurophysiol.*, **42**, 833–849.

Peters, A., Proskauer, C. C., and Ribak, C. E. (1982). Chandelier cells in rat visual cortex. *J. comp. Neurol.*, **206**, 397–416.

Peters, A., and Regidor, J. (1981). A reassessment of the forms of nonpyramidal neurons in area 17 of cat visual cortex. *J. comp. Neurol.*, **203**, 685–716.

Ramón y Cajal, S. (1911). *Histologie du Systeme Nerveux de L'Homme et des Vertebres*, Maloine, Paris.

Ribak, C. E. (1978). Aspinous and sparsely-spinous stellate neurons in the visual cortex of rats contain glutamic acid decarboxylase. *J. Neurocytol.*, **7**, 461–478.
Rose, D. (1977). On the arithmetical operation performed by inhibitory synapses onto the neuronal soma. *Exp. Brain Res.*, **28**, 221–223.
Rose, D., and Blakemore, C. B. (1974). Effects of bicuculline on functions of inhibition in visual cortex. *Nature*, **249**, 375–377, 869.
Sillito, A. M. (1975). The contribution of inhibitory mechanisms to the receptive field properties of neurons in the striate cortex of the cat. *J. Physiol.*, **250**, 305–329.
Sillito, A. M. (1977). Inhibitory processes underlying the directional specificity of simple, complex and hypercomplex cells in the cat's visual cortex. *J. Physiol.*, **271**, 699–720.
Sillito, A. M. (1979). Inhibitory mechanisms influencing complex cell orientation selectivity and their modification at high resting discharge levels. *J. Physiol.*, **289**, 33–53.
Sillito, A. M., Kemp. J. A., Milson, J. A., and Berardi, N. (1980). A re-evaluation of the mechanisms underlying simple cell orientation selectivity. *Brain Res.*, **194**, 517–520.
Sillito, A. M., and Versiani, V. (1977). The contribution of excitatory and inhibitory inputs to the length preference of hypercomplex cells in layers II and III of the cat's striate cortex. *J. Physiol.*, **273**, 775–790.
Somogyi, P. (1977). A specific 'axo-axonal' interneuron in the visual cortex of the rat. *Brain Res.*, *136*, 345–350.
Somogyi, P. (1979). An interneuron making synapses specifically on the axon initial segment (AIS) of pyramidal cells in the cerebral cortex of the cat. *J. Physiol.*, **296**, 18–19P.
Somogyi, P., and Cowey, A. (1981). Combined Golgi and electron microscopic study on the synapses formed by double bouquet cells in the visual cortex of the cat and monkey. *J. comp. Neurol.*, **195**, 547–566.
Somogyi, P., and Cowey, A. (1984). Double bouquet cells. In *Cerebral Cortex* (Eds. E. G. Jones and A. Peters), Vol. I, Plenum Press, New York, pp. 337–360.
Somogyi, P., Freund, T. F., and Cowey, A. (1982). The axo-axonic interneuron in the cerebral cortex of the rat, cat and monkey. *Neurosci.*, **7**, 2577–2608.
Somogyi, P., Freund, T. F., Wu, J-Y., and Smith, A. D. (1983a). The section-Golgi impregnation procedure. 2. Immunocytochemical demonstration of glutamate decarboxylase in Golgi-impregnated neurons and in their afferent synaptic boutons in the visual cortex of the cat. *Neurosci.*, **9**, 475–490.
Somogyi, P., Hodgson, A. J., and Smith, A. D. (1979). An approach to tracing neuron networks in the cerebral cortex and basal ganglia. Combination of Golgi staining, retrograde transport of horseradish peroxidase and anterograde degeneration of synaptic boutons in the same material. *Neurosci.*, **4**, 1805–1852.
Somogyi, P., Kisvárday, Z. F., Freund, T. F., and Cowey, A. (1984). Characterization by Golgi impregnation of neurons that accumulate ^{3}H-GABA in the visual cortex of monkey. *Exp. Brain Res.*, **53**, 295–303.
Somogyi, P., Kisvárday, Z. F., Martin, K. A. C., and Whitteridge, D. (1983b). Synaptic connections of morphologically identified and physiologically characterized large basket cells in the striate cortex of cat. *Neurosci.*, **10**, 261–294.
Szentágothai, J. (1973). Synaptology of the visual cortex. In *Handbook of Sensory Physiology. Central Processing of Visual Information* (Ed. R. Jung), Vol. VII/3B, Springer, Berlin, pp. 269–324.
Tömböl, T. (1974). An electron microscope study of the neurons of the visual cortex. *J. Neurocytol.*, **3**, 525–531.

Winfield, D. A., Gatter, K. C., and Powell, T. P. S. (1980). An electron microscopic study of types and proportions of neurons in the cortex of the motor and visual areas of the cat and rat. *Brain*, **103**, 245–258.

Models of the Visual Cortex
Edited by D. Rose and V. G. Dobson

CHAPTER 55

Microcircuitry of layer IVab in cat cortical area 17

THOMAS L. DAVIS and PETER STERLING
Department of Anatomy, School of Medicine, University of Pennsylvania, Philadelphia, PA 19104, USA

Our goal in studying microcircuitry is to learn how the receptive fields of cortical neurons are generated by their extrinsic and intrinsic synaptic connections. To achieve this will obviously require a synthesis at the cellular level of anatomical, physiological and chemical information. The first synthetic idea started with the observation by Lund that spiny stellate neurons occur uniquely in layer IV, a principle terminal zone of the geniculate axons (Lund, 1973; Lund *et al.*, 1979). Hubel and Wiesel (1962) had reported layer IV to contain mainly simple cells, and Garey and Powell (1971) had reported that 80 per cent. of the lateral geniculate terminals in layer IV contact spines. It was natural to think, therefore, that simple cells are spiny stellates which receive most of the geniculate input and relay it to complex cells, possibly pyramidal, in accordance with the hierarchical model.

It was in the context of this hypothesis that our own studies of circuitry in layer IV began. We asked which cells actually *do* receive geniculate input. It was clear from the experience of Garey and Powell and previous work with the ventrolateral thalamus to motor cortex projection (Strick and Sterling, 1974) that in single thin sections it is rare to find a thalamic terminal on a spine or dendrite that can be traced to an identifiable soma. Therefore, we partially reconstructed neurons from serial sections of a patch of upper layer IVab from a cat which survived a lateral geniculate lesion by four days (Davis and Sterling, 1979). The patch contained 32 neurons.

We were startled to discover that many of the partially reconstructed neurons received geniculate contacts. This included stellate cells and also large pyramidal neurons with somas at the III–IVab border, small 'star pyramids' (Lorente de Nó, 1938) lying deeper and a fusiform cell with varicose dendrites. There was variation amongst the neurons in the density and pattern of geniculate contacts. Thus, geniculate contacts were distributed sparsely to pyramids and restricted to the basilar dendrites but were distributed generously to large, nonspiny stellates and were widespread on their somas and dendrites at least as far as the tertiary branches (Fig. 1). We have confirmed and extended these observations using anterograde labelling of the geniculate terminals, which is several times more sensitive than the method of degeneration (Einstein, Davis and Sterling, 1985). Our approach provided few observations on distal dendrites which bear most of the spines; therefore, we could not tell whether the geniculate massively contacts the spiny stellates. It was clear, however, that the lateral geniculate pathway diverges widely in layer IVab, apparently providing various specific patterns of contacts in this layer, possibly to neurons of every category.

This result forced a shift in our question from 'Which neurons in IVab receive geniculate input?' to 'How many categories of neuron are present, and how should they be defined?'. This now seemed crucial, since if the connections are as precise as hinted at by this study, it would be necessary, as in the retina, to focus the study of microcircuitry to particular neuron types (Sterling, 1983). We scrutinized the 32 partially reconstructed neurons and, based on associations between neuron size, morphology, cytology and synaptic pattern, distinguished seven categories (Davis and Sterling, 1979). These included the large pyramid, the star pyramid, the large nonspiny stellate and several categories of smaller stellate (Fig. 2). It was obvious that with more extensive reconstructions and with observations of the synaptic and axonal patterns of neurons obtainable with a Golgi-like filling, the seven categories might well be further subdivided. The observations seemed at least, however, to put a lower bound on the number of categories.

The question was probed next by partially reconstructing IVab neurons labelled by their selective accumulation of the inhibitory transmitter, GABA (Hamos, Davis and Sterling, 1983). Such neurons formed about 10 per cent of all neurons in the layer. Based on associations of size, soma morphology, cytology and synaptic pattern, they were subdivided into four categories: GABA 1 was large and dark with a dense distribution of synaptic contacts to the soma, a substantial percentage of which were from the geniculate; GABA 2 was small and pale, also with a dense distribution of somatic contacts but none from the geniculate; GABA 3 was radially fusiform with varicose dendrites and a sparse distribution of somatic contacts; GABA 4 was a medium stellate with a moderate distribution of somatic contacts and a heavy accumulation of GABA. The seven categories identified in the first study (Davis & Sterling, 1979) did not overlap the four categories of the GABA-accumulating neuron, so the number of categories of layer IVab neuron distinguished with this approach stands at 11.

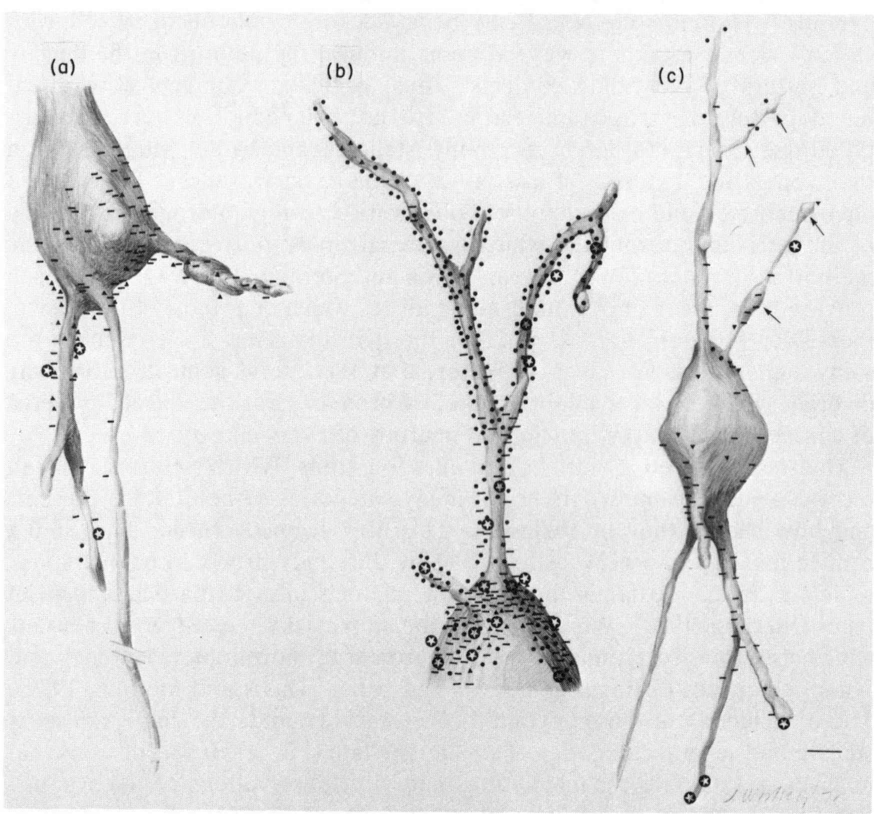

Figure 1 Three neurons reconstructed from serial sections through layer IVab (Davis and Sterling, 1979, reproduced by permission of Alan R. Liss, Inc.). (a) Pyramidal cell from the layer III–IV border, receiving a moderate distribution of flat vesicle contacts on soma (11 terminals/100 μm^2), round and flat vesicle contacts on basilar dendrites and geniculate contacts on basilar Dendrites. (b) Large stellate, receiving a heavy distribution of round and flat vesicle contacts on soma (48 terminals/100 μm^2), round and flat vesicle contacts on all dendrites and geniculate contacts on soma, and primary, secondary and tertiary Dendrites. (c) Varicose stellate receiving a light distribution of round and flat vesicle contacts on soma (7.7 terminals/100 μm^2), round and flat vesicle contacts on dendrites and geniculate contacts on dendrites. Bars, terminals with flat vesicles and symmetrical contacts; solid circles, terminals with round vesicles and asymmetrical contacts; triangles, unclassified terminals; circled stars, degenerating geniculate terminals; Bar μm = $5 \mu m$.

By this stage of our investigation of cortical layer IVab it had been established in the adult cat retina that, at a given eccentricity, neurons are of many distinct forms, each of which is associated with specific connectional, physiological and chemical features. These associations define in the retina

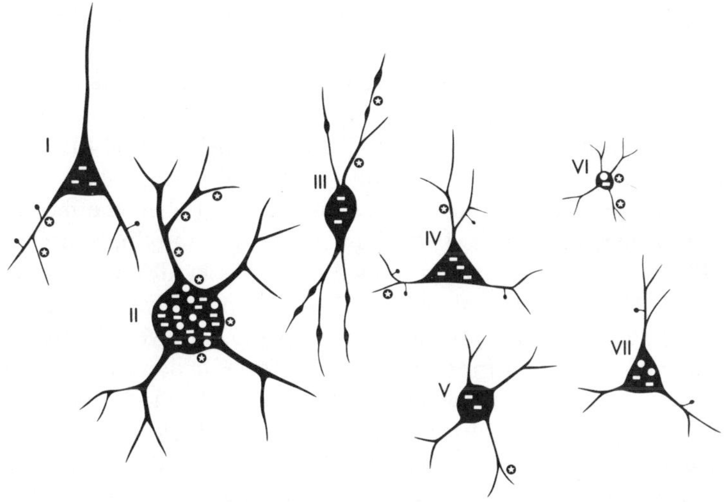

FIGURE 2 Diagram representing the major characteristics of seven categories of neuron found by reconstruction in layer IVab (Davis and Sterling, 1979, reproduced by permission of Alan R. Liss, Inc.). Symbols are explained in Fig. 1.

fundamental cell types (Sterling, 1983). Overall there are about 60 types, including 23 types of ganglion cell (Kolb, Nelson and Mariani, 1981). It occurred to us that the categories of IVab neuron we defined by associations of form, cytology, transmitter accumulation and synaptic pattern might also represent discrete types and that layer IVab, like the retina, might contain many more types than previously suspected. We sought a way to estimate how many types there might be.

We selected three categories of neuron believed from our previous work to represent discrete types and counted them in serial sections (Solnick, Davis and Sterling, 1984). These were the GABA 1, GABA 3 (Hamos, Davis and Sterling, 1983) and class II (Davis and Sterling, 1979). Each proved to represent but a small fraction of layer IVab neurons (0.6, 1.1 and 1.6 per cent.). We then asked whether the dendritic fields of the neurons in each category overlap and so 'tile' (Crick, personal communication, 1983) the layer in the tangential plane. Our estimates suggest that each category does indeed 'tile' (Solnick, Davis and Sterling, 1984) and, as in the retina (Sterling, 1983), only by a small factor. Obviously, other types of neuron not yet characterized might be more numerous and account for a much larger share of the total. If not, however, the number of neuron types in layer IVab might be as many as 50 or more. This would be consistent with our finding eleven categories of neuron in a few relatively small samples.

The hypothesis of discrete types in layer IVab was developed from partial reconstructions which provide rather little information regarding a cell's dendritic arborization, the number and morphology of its dendritic spines or the morphology of its axon. A comprehensive study of the neuronal forms in this layer at specified eccentricities in the adult such as have been accomplished for cat retina (e.g. Boycott and Wassle, 1974; Kolb, Nelson and Mariani, 1981) does not exist. Golgi studies of area 17 have employed mostly neonates and kittens up to three months old (when the morphology is rapidly evolving; see Cragg, 1975; LeVay, Stryker and Shatz, 1978; Lund *et al.*, 1979) and have been concerned with neurons in all layers rather than with a detailed comparison of the forms in a single layer. These studies demonstrate, nevertheless, that the neuronal forms in layer IV are diverse. More than half a dozen forms differing in size and dendritic and axonal pattern are evident in the drawings of O'Leary (1941) and seven forms of nonpyramidal neuron were observed by Peters and Regidor (1981).

We have studied the morphology of 30 layer IVab neurons in the adult cortex near the representation of the central area. Golgi-like filling was accomplished by uptake from small, local extracellular deposits of horseradish peroxidase (HRP) or by intracellular injection. Neurons were drawn and grouped according to their associations between soma size and shape, form and orientation of the dendritic arbor, and morphology and distribution of dendritic spines. For example, in Fig. 3a the cell is a medium-sized stellate with richly arborized dendrites and a heavy distribution of spines. Figure 3e and g show cells which are also medium stellates; however, Fig. 3e has varicose spine-free dendrites, while Fig. 3g has moderately arborizing dendrites with relatively few spines. In this manner, eleven morphological categories of neuron were distinguished, seven of which are shown in Fig. 3. Within certain of these categories we were able to find several examples that appeared strikingly similar (Kay *et al.*, 1982).

These categories were distinct in the sense that no one of them would be confused with any of the others. This far from establishes, however, that each category represents a discrete type. For one thing, the sample is small and there may well be additional forms that are not so easily distinguished from the established categories and which might be viewed according to a different hypothesis as part of a continuum of form (Gilbert, 1983). For another, the distinctions are so far entirely on anatomical criteria which prcvide no hint as to which are meaningful and which represent nonspecific biological variation. The results are certainly consistent, however, with the findings from reconstruction which include differences in synaptic pattern and transmitter accumulation.

It might be asked whether the receptive fields of layer IVab neurons show a diversity comparable to what has been observed anatomically. Hubel and Wiesel noted that simple cells, which predominate in layer IV, differ in the number of flanking on- and off-regions. It has recently been discovered that such cells form a series: S_1 with a single excitatory region, S_2 with two excitatory regions, and so on, up to S_4 (Palmer and Davis, 1981). Others have shown (Mullikin, Jones and Palmer, 1984b) subsequently that the series extends at least to S_6. The cells with odd numbers of excitatory regions can be either on or off, providing an additional basis for subdivision. These categories can be subdivided still further according to whether the response is 'X-like' or 'Y-like' (Mullikin, Jones and Palmer, 1984a). Altogether this suggests that the physiological and anatomical categories may be equally as diverse.

In conclusion, it is our hypothesis that the neurons in layer IVab are discrete in type, where type is defined as a unique association of morphological, cytological, chemical, synaptic and physiological features. Evidence so far suggests that there may be many types, perhaps 50 or more, each of which might 'tile' the layer in the tangential plane. One approach to testing this hypothesis would be to extend the observations on those categories of neuron in IVab that we have tentatively identified as types. The hypothesis predicts, for example, that each of the four categories of GABA-accumulating neuron in IVab will have a distinctive dendritic pattern and a distinctive distribution of synaptic contacts on the dendrites. This prediction could now be tested by the method recently introduced by Somogyi *et al.* (1983) for electron microscopy of Golgi-impregnated neurons following immunocytochemical staining for glutamic acid decarboxylase. Ultimately it will be critical to determine whether there is an exact correspondence between particular anatomical and physiological categories. We are now searching for such a correspondence by examining in the electron microscope neurons filled with HRP following quantitive characterization of their receptive fields. If this hypothesis is true, the strategy so useful in the retina for working out circuitry, i. e. to choose a few specific types and work them out in detail, might be adopted for layer IVab.

ACKNOWLEDGEMENTS

We thank John Dashe, Jill Einstein, James Hamos, Mathew Kay, Walter Mullikin and Bennett Solnick for their collaboration on the studies summarized here, and Harriet Abriss for typing the manuscript.

Supported by grants EY-00828 from the National Eye Institute, BNS 81–19839 from the National Science Foundation and a fellowship to T. L. Davis from the Alfred P. Sloan Foundation.

REFERENCES

Boycott, B. B., and Wassle, H. (1974). The morphological types of ganglion cells of the domestic cat retina. *J. Physiol.*, **240**, 397–419.

Cragg, B. G. (1975). The development of synapses in the visual system of the cat. *J. comp. Neurol.*, **160**, 147–166.

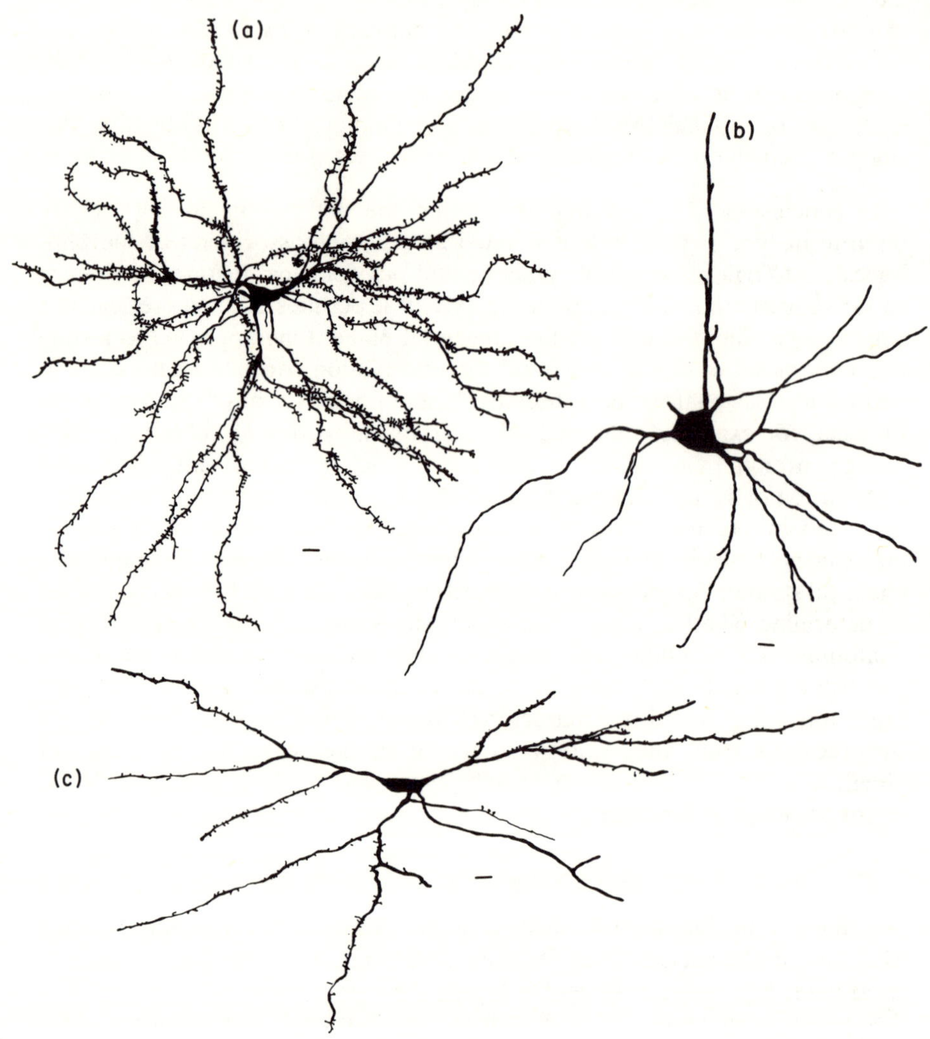

Figure 3 Light microscope drawing of seven different morphological forms of nonpyramidal cells found in layer IVab (Kay *et al.*, 1982): (a) medium stellate richly arborized, high spine density; (b) large stellate, spine-free, smooth dendrites; (c) medium fusiform cell, tangentially oriented dendrites, moderate spine density; (d) small, ovoid cell, richly arborized, moderate spine density; (e) medium stellate, spine-free, varicose dendrites; (f) medium stellate, moderately arborized, high spine density; (g) medium stellate, moderately arborized, low spine density. Scale 5 μm.

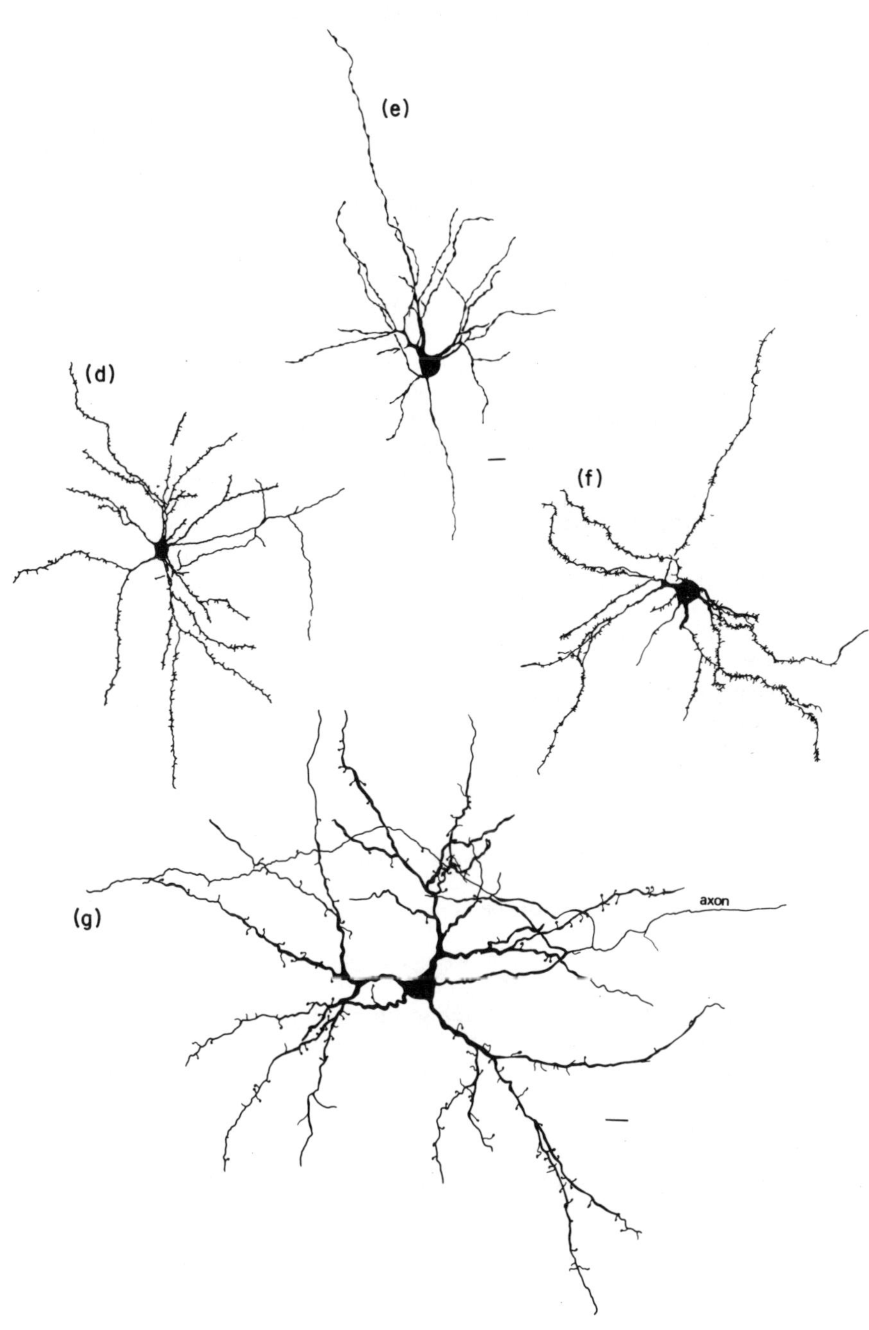
(e)
(d)
(f)
(g)
axon

Davis, T. L., and Sterling, P. (1979). Microcircuitry of cat visual cortex: classification of neurons in layer IV of area 17, and identification of the patterns of lateral geniculate input. *J. comp. Neurol.*, **188**, 599–627.

Einstein, G., Davis, T. L., and Sterling, P. (1985). The pattern of lateral geniculate synaptic contacts on neurons in layer IVab of the cat striate cortex (submitted).

Garey, L. J., and Powell, T. P. S., (1971). An experimental study of the termination of the lateral geniculo-cortical pathway in the cat and monkey. *Proc. Roy. Soc. (Lond.) B*, **179**, 41–54.

Gilbert, C. D. (1983). Microcircuitry of visual cortex. *Ann. Rev. Neurosci.*, **6**, 217–247.

Hamos, J. E., Davis, T. L., and Sterling, P. (1983). Four types of neuron in layer IVab of cat cortical area 17 accumulate ^{3}H-GABA. *J. comp. Neurol.*, **217**, 449–457.

Hubel, D. H., and Wiesel, T. N. (1962). Receptive fields, binocular interaction and functional architecture in the cat's visual cortex. *J. Physiol.*, **160**, 106–154.

Kay, M. D., Dashe, J. F., Mullikin, W. H., and Davis, T. L. (1982). Morphology of stellate cells in layer IV of cat area 17. *Neurosci. Abstr.*, **8**, 676.

Kolb, H., Nelson, R., and Mariani, A. (1981). Amacrine cells, bipolar cells, and ganglion cells of the cat retina: a Golgi study. *Vision Res.*, **21**, 1081–1114.

LeVay, S., Stryker, M. P., and Shatz, C. J. (1978). Ocular dominance columns and their development in layer IV of the cat's visual cortex: a quantitative study. *J. comp. Neurol.*, **179**, 223–244.

Lorente de Nó, R. (1938). The cerebral cortex: architecture, intracortical connections and motor projections. In *Physiology of the Nervous System* (Ed. J. F. Fulton), Oxford University Press, pp. 291–325.

Lund, J. S. (1973). Organization of neurons in the visual cortex, area 17, of the monkey (*Macaca mulatta*). *J. comp. Neurol.*, **147**, 455–496.

Lund, J., Henry, G. H., McQueen, C. L., and Harvey, A. R. (1979). Anatomical organization of the primary visual cortex (area 17) of the cat. A comparison with area 17 of the macaque monkey. *J. comp. Neurol.*, **184**, 599–619.

Mullikin, W. H., Jones, J. P., and Palmer, L. A. (1984a). Receptive field properties and laminar distribution of X-like and Y-like simple cells in cat area 17. *J. Neurophysiol.*, **52**, 350–371.

Mullikin, W. H., Jones, J. P., and Palmer, L. A. (1984b). Periodic simple cells in cat area 17. *J. Neurophysiol.*, **52**, 372–387.

O'Leary, J. L. (1941). Structure of the area striata of the cat. *J. comp. Neurol.*, **75**, 131–164.

Palmer, L. A., and Davis, T. L. (1981). Receptive-field structure in cat striate cortex. *J. Neurophysiol.*, **46**, 260–276.

Peters, A., and Regidor, J. (1981). A reassessment of the forms of non-pyramidal neurons in area 17 of cat visual cortex. *J. comp. Neurol.*, **203**, 685–716.

Solnick, B., Davis, T. L., and Sterling, P. (1984). Numbers of specific types of neuron in layer IVab of cat striate cortex. *Proc. Natl. Acad. Sci.*, **81**, 3898–3900.

Somogyi, P., Freund, T. F., Wu, J.-Y., and Smith, A. D. (1983). The section-Golgi impregnation procedure. 2. Immunocytochemical demonstration of glutamate decarboxylase in Golgi-impregnated neurons and their afferent synaptic boutons in the visual cortex of the cat. *Neurosci.*, **9**, 475–490.

Sterling, P. (1983). Microcircuitry of the cat retina. *Ann. Rev. Neurosci.*, **6**, 149–185.

Strick, P. L., and Sterling, P. (1974). Synaptic termination of afferents from the ventrolateral nucleus of the thalamus in the cat motor cortex. A light and electron microscope study. *J. comp. Neurol.*, **153**, 77–105.

Models of the Visual Cortex
Edited by D. Rose and V. G. Dobson

CHAPTER 56

Methodological solutions for neuroscience

David Rose
Department of Psychology, University of Surrey, Guildford, Surrey
GU2 5XH, UK
and
Vernon G. Dobson
Department of Psychology, University of Brunel, Uxbridge, Middlesex
UB8 3PH, UK

In the Introduction we pointed out that conceptual difficulties are rife in the field of research on the visual cortex, in part beause of poor communications. Maxwell (Chapter 2 in this volume) has argued that these problems follow inevitably in the brain sciences from the absence of any sufficiently explicit and systematic theoretical approach. In this chapter we discuss the sources of these problems and try to show how the treatment that Maxwell prescribes can be put into practice.

HUMPTY DUMPTY AND THE CHARGE OF THE LIGHT BRIGADE

Humpty Dumpty sat on a wall,
Humpty Dumpty had a great fall;
All the King's horses,
And all the King's men,
Couldn't put Humpty together again.
— children's nursery rhyme

'Forward the Light Brigade!'
Was there a man dismay'd?
Not tho' the soldier knew
Someone had blunder'd:
Their's not to make reply,
Their's not to reason why,
Their's but to do and die:
Into the valley of Death
Rode the six hundred.
—Alfred, Lord Tennyson

According to Maxwell, the use of inductivist or falsificationist scientific methods both imply that nothing should enter science without an empirical basis; all the rest is 'mere speculation'. However, the rigid application of empiricist principles, with the exclusion of all *a priori* assumptions, leads to two kinds of irrational behaviour.

First, without some kind of overview, or set of alternative overviews, there are no strategic guidelines as to how to dismantle or resolve systematically the complex and difficult problems with which we are faced into simpler, solvable subproblems (as required by Maxwell's rule 3 for rational science). As any physical system can be analysed in a potentially infinite number of different ways (Ashby, 1956), there is a risk that each field will fragment haphazardly into numerous camps or 'schools', each tackling different aspects of the overall problem with different empirical techniques, without reference to their relationships to each other or to the system as a whole.

Unfortunately, once these divisions occur, they tend to be self-sustaining, with rivalries over priority of access to scarce resources, and different groups becoming ego-involved with their own particular points of view. Janis (1982) has coined the term 'groupthink' to describe decision-making errors based on group conformity thinking under such circumstances. Typically, there is a delusion of invulnerability, a feeling of superior morality, and an illusion of unanimity and consensus due to self-censorship of criticism and the emergence of self-appointed 'mindguards' to pressurize deviants. Rivals are stereotyped as stupid, unreliable or irrational. Even at best, communication with other groups, and understanding of their points of view, is constrained by the social and cognitive surroundings within which each group exists (Knorr, Krohn and Whitley, 1980; Gilbert and Mulkay, 1984). Clearly, once started, the fragmentation process is difficult to reverse, and for this reason we refer to it as the *Humpty Dumpty Effect*. The almost universal occurrence of this effect has even led some philosophers of science to suggest that anarchy and irrationality must be the best procedure for science (Feyerabend, 1975).

Maxwell's second criticism of pure empiricism is that it implies that the actual process of building theories lies outside science and reason, so that theoretical skills cannot rationally be defined, studied or improved upon. Theorizing is thus seen as either a matter of luck or as an elementary, unskilled process which is best left to the empirical scientists who first discover the phenomena to be accounted for and who deal with them at first hand. Underlying this is the feeling that the world is a simple place, so that theorizing is largely a matter of pointing to obvious commonsense solutions. This is attested by the observation that research discoveries are almost always described in the literature in inductivist terms (Gilbert and Mulkay, 1984). Therefore, once a model is proposed for a new set of observations it tends to be accepted uncritically (particularly if it is actually proposed by someone in a position of authority). Much time and money is devoted to the model, and the reputations of its proponents and their supporters become staked upon it. This continuous build-up of momentum by a theory or paradigm is what we call the *Charge of the Light Brigade Effect.* It may continue to the point where rival or innovative ideas are crowded out by one 'dominant paradigm', a state which Kuhn (1970) describes as 'normal science', i.e. the situation when one consensus view is imposed by the dominant group or school. Unfortunately, the model chosen is sometimes the wrong model! The outcome of this can be that an erroneous consensus is formed which locks the field onto a false hypothesis for many years, at tremendous cost. In addition, since the solution to one scientific problem may provide a premise for several others, the acceptance of a false theory in one field may cause negative repercussions which are widespread.

Yet another sign of the inadequacy of pure empiricism is its failure to achieve perfect consistency. The irreproducibility and unreliability of experimental evidence is attested by the proliferation of conflicting data chronicled in the Introduction to this volume. This profusion results not only from the lack of a rational strategy for generating and evaluating models (as discussed by Maxwell, Chapter 2 in this volume), but also from the fact that laboratory skills, like any other skill, depend on practice, understanding, interpretations and equipment availability, which differ from one laboratory or individual scientist to another. Such skills cannot be readily and accurately articulated in the Methods sections of published articles (Gilbert and Mulkay, 1984) and may be 'as much an art as a science' (Robson, 1983, p. 81). Indeed, laboratories can be seen as 'factories' in which skills and practices have to be put together deliberately in order to 'make the right things happen'; for example, when you first set up a new experiment it never works until you have fiddled around with it for a while (Knorr-Cetina, 1981). The replicability of experiments, on which empiricism is based, is thus a fickle mistress.

The above consequences of the application of empiricist principles to science are complicated by the fact that, while paying lip-service to empiricism, many scientists actually use *a priori* criteria to evaluate their models, such as assuming that comprehensibility, coherence and simplicity will apply to the correct model (Maxwell, Chapter 2, and the Introduction to this volume). The problem here is that these criteria are derived from covert or intuitive metaphysical assumptions, and have different meanings from each metaphysical perspective (Rowe and Stone, 1979; Stone, 1983; Dobson and Rose, Chapter 3 in this volume). In consequence, a model which one scientist sees as obvious or simple can appear incomprehensible or complex to another (cf. the problem of incommensurability described in the Introduction to this volume). Thus there are further opportunities for misunderstanding and conflict.

AN EXPLICIT PROCEDURE FOR MODEL BUILDING

The Humpty Dumpty Effect and the Charge of the Light Brigade Effect can occur because the framework of empiricism does not call for the systematic and exhaustive exposition of all possible hypotheses about how a system may be functioning; nor does it show how conflicting hypotheses can be tested and selected in a rational and efficient manner (rule 2 of Maxwell, Chapter 2 in this volume). Even if an attempt is made to think rationally and logically, this does not guarantee that all possibilities will be considered comprehensively and tested efficiently: many people do not use such modes of thought spontaneously and correctly (Bruner, Goodnow and Austin, 1956; Johnson-Laird and Wason, 1977). In this section we therefore present in outline a procedure which we hope will act as a guide towards a rational methodology for science.

Moving up and down the hierarchy of problems

For this procedure to be effective, the initial approach should be top-down rather than bottom-up. The procedure should therefore start at the highest level of analysis with alternative metaphysical positions and then work down through the system. Thus the different metaphysical positions specify different views as to what the function of the brain is, since the metaphysical viewpoint describes the type of environment within which the brain has to operate and the tasks it must perform (Dobson and Rose, Chapter 3 in this volume). Similarly, the way in which the whole brain is seen as working will determine the function which a visual system is needed to perform. That function of the visual system will then require a visual cortex as one of the subsystems which is needed in order for that overall function to be carried out. The cortex in turn will require certain properties of the local circuits and single cells within it, so that the cortex can perform its own function successfully.

Another key feature of the procedure is that the various possible views should be articulated and defined at each level of the analysis. Thus at any particular level, a range of different ideas can be generated about what the function(s) might be at that level. The key to avoiding the Light Brigade Effect is to formulate at every level an exhaustive range of these functional hypotheses, so that the best is not missed out.

However, for each of these hypothesized functions, there may be several different mechanisms by which it can be achieved. Therefore models have to be built of these mechanisms, to see which (if any) will work, and which of these works best. Since it is not always intuitively obvious which models will work and which will not, *all* possible models should be built and tested.

This leads to the problem of excessive numbers. Several ways exist to prevent the number of these models being infinitely large. First, each functional hypothesis will indicate which properties of the problem system are essential to the functioning of that system and which properties are accidental epiphenomena (Gregory, 1959). In the initial stages of the analysis, all the essential properties must be included, but the accidental properties need not be modelled yet. Without a functional hypothesis, much time and effort might be lost in building models which, although capable of explaining phenomena (e.g. visual illusions) which are incidental or peripheral to the system's function, do not explain the important, everyday functions of the system (e.g. perceptual constancies). We are suggesting that initially only models that account for essential properties should be studied, and only later should the best of these be extended to see which of them can account also for the accidental properties. To give some further examples of the significance of functional hypotheses in vision research: the approach of Marr (1982) suggests that the detection of points of inflexion in the luminance profile of the image is essential while the exact breadth of orientation tuning of units in the visual cortex is not; MacKay and Barlow (Chapters 5 and 4 respectively in this volume), however, argue that the breadth of tuning of single units is moderately important; information processing hypotheses which assume that Fourier or Gabor mathematics are relevant to the function of the visual cortex will see spatial frequency tuning as one of the most essential properties of single unit behaviour; some recent evidence implicates phase encoding as more important than amplitude (e.g. Oppenheim and Lim, 1981); and so on.

A second way to reduce the number of models is to test the functional hypotheses at each level and to eliminate most of them before going on to the next lowest (more specific) level. Otherwise there will be an exponential growth in the number of hypotheses and models as we go down the system from problem to sub-problem to sub-sub-problem. Thus for each functional hypothesis there will be a range of models generated and tested, and a small number of these models will be found to be able to provide plausible explanations for that function. Only those functional hypotheses which do generate at least one plausible model, and all of these hypotheses, must be passed on to the next lower level. Each of these functional hypotheses will suggest that certain essential subfunctions should exist in order to carry out the overall function. It is these subfunctions, and only these functions, which are then the worthwhile subjects of the next stage of the model-building procedure, i. e. the 'preliminary, simpler, analogous, subordinate, specialized problems' of Maxwell's rule 3 (Chapter 2 in this volume). In the next chapter we will present some specific examples of the application of this procedure to the visual system. For example, if a Fourier representation is considered likely which requires the accurate encoding of phase as essential but not amplitude, then hypotheses should be formulated about how phase can be transmitted (e.g. by the relative spatial locations of the receptive fields of the inputs to a cortical cell, or via lateral inhibition in the cortex), while proposals about possible gain control mechanisms are unnecessary.

The clever twist is that by finding out which low-level, specialized models are successful and which are not, the merits of higher-level solutions and even alternative metaphysical perspectives can be assessed and appraised. Thus Maxwell's requirement (rule 4) for interconnecting problem, solving at fundamental and detailed levels of analysis leads us to feed information back up the hierarchy about which choices of functional hypotheses give rise to useful and successful solutions to lower-level problems. These higher-level solutions can then be fed back down the hierarchy to guide the solving of other, different low-level problems, on the assumption that in a logically coherent world similar principles will apply in solving similar problems. For example, if Heraclitean–Darwinian suppression of unwanted information models are found to provide the best explanations for orientation selectivity, then they are likely to do so also for stereopsis, spatial frequency selectivity, and so on.

A word should also be said here about 'linking hypotheses', which are ideas about the way in which, for example, single-unit firing is related to perception. These are similar to, but more general than, functional hypotheses, in that linking hypotheses attempt to specify the reciprocal translation of an event from one level of description to another, while functional hypotheses make suggestions only about translation from a higher to a lower level. Thus functional hypotheses might be about what single-unit behaviours are necessary to explain perception, in the top-down manner of reasoning.

Building and testing models

The detailed procedure to be carried out for building models at each level of the analysis and under each functional hypothesis at that level is as follows. Once the problem system has been identified and circumscribed, there is an initial process of formalization to enable models to be specified in easily manipulable symbols, and then a process of testing and assessment of each model.

Thus it is important *first* to attempt to identify and specify all the aspects or *parameters* of the system which are considered essential or in some degree relevant to its function. These parameters are variables or qualities which will be used to form the model of that level of the system. One way in which these parameters can be discovered in the first place is from empirical observations (turning over stones in a field to see what you find underneath: MacKay, 1983). However, just as was the case with the search for functional hypotheses (see above), it is necessary also to use a top-down approach here, because only with a functional hypothesis can a distinction be made between essential and accidental parameters. Moreover, further essential parameters which have not previously been discovered empirically may be identified in this way. Some of the parameters in the final list may be more important or critical than others, and this order of ranking should be made explicit too. This is to enable later testing to examine each parameter one at a time systematically starting with the most important. It may also prove helpful at the third step (see below) when the exclusion of the more trivial parameters may be necessary to simplify the preliminary running of the analysis.

The *second* step is to simplify the representation of each parameter, by reducing the number of possible values to a minimum number of discrete states, preferably only two. Thus the parameters should be expressed as dichotomies whenever possible, i.e. as extreme or opposite conditions, or as the presence or absence of a condition. One reason for this dichotomization is that it reduces the number of possible alternative models to a manageable number, and another is that it enables subsequent testing to discover the best model as rapidly as possible. This is achieved by designing each test to decide between the two alternative states of each of the dichotomized parameters in the model, i.e. by halving uncertainty with every test, in the same way that a number can be identified in a problem of information transmission by first identifying the most significant bit, then the next most significant bit, and so on.

The *third* step is to combine all the relevant parameter states (or at least the most important ones) so that all possible models are formed and are considered in the analysis. For example, if there are six relevant parameters, each expressed as being in one of two possible states, then there are $2^6 =$ 64 models which can be formed.

The *fourth* step is to test or assess these models according to prespecified criteria such as those listed below (1 to 8). First, whole classes of model may be eliminated by identifying those with relatively implausible parameter states, while other models can be eliminated individually on logical grounds as not capable of performing the overall function required by the functional hypothesis. The remainder can next be ranked according to their plausibility. Perhaps the best way to define relative plausibility is to say that the more plausible models are those that leave fewer anomalies between the predicted behaviour of the modelled system and optimal performance on the criteria. The most plausible models will later be accepted for further development.

The criteria to be used include those in the following list. Some of these criteria derive from the functional hypothesis and ultimately depend on the metaphysical point of view, while others involve empirical evidence and observations.

1. Efficiency.

Efficiency can be assessed along a number of dimensions, including the capacity of the model for information-processing, the biochemical energy required to work it, or the amount of gene space demanded to reproduce it. However, the meaning of the word 'efficient' can depend on the metaphysical perspective taken. As described by Dobson and Rose (Chapter 3 in this volume), Parmenidean–Cartesian metaphysics suggests that the most mathematically precise and orderly mechanisms will be used to accomplish the overall function of the system. In contrast, Heraclitean–Darwinian metaphysics suggests that the least precise or orderly mechanisms (which will still achieve the overall function) are the ones which will be found, because they require less energy to maintain them against the ravages of entropy. Thus accurate and orderly networks would require a lot of genetic, anatomical and physiological 'effort' to perform mathematically exact functions, whereas the 'as sloppy as possible' Heraclitean networks would be easier to maintain biologically but harder to describe mathematically.

2. Reliability.

This can be assessed by, for example, constructing computer simulations of the complete model and introducing random perturbations or errors in the connectivity, or random informational 'noise' into the inputs or into the circuitry, and then measuring the extent to which the model's ability to perform its overall function deteriorates.

3. Simplicity.

A relatively simple model can be defined as one which incorporates fewer parameters than another model, and may therefore be thought more likely to mirror Nature because each parameter of a natural system requires 'effort' to set it up, and complex systems are therefore less efficient and reliable. However, given that intelligent systems are more complex than the systems encountered in physics, it is a mistake which can lead to the Light Brigade Effect to maintain simple models when there are reasons for believing them to be inadequate; the rule is simply that models should be no more complicated than necessary.

4. Developmental coherence.

Can the modelled system develop from previous stages in ontogeny or phylogeny? This question is emphasized more in Heraclitean systems, because they are not creationist. The issue is discussed in more detail by Dobson and Rose (Chapter 3 in this volume).

5. Working coherence.

Does the model readily fit into and link up with other models at the same and different functional levels? Can the system act as a whole, cooperatively, to give overall emergent patterns of adaptive behaviour, or will the component subsystems pull in different directions or otherwise interact destructively?

6. Logical coherence.

Does the model function within the same metaphysical and other fundamental parameter states as other models of related systems in the brain and particularly in the visual cortex? In other words, are the processes and functional hypotheses consistent from one modelled subsystem to its neighbour? The existence of a coherent relationship with higher, more global models is demanded by Maxwell's rule 4 (Chapter 2 in this volume) and indeed there are instances within this book where such connections are used to test lower level concepts, for example Daugman's argument that Hubel and Wiesel's model of simple cells is incompatible with general models of information filtering, Nelson's and Singer's arguments about methods for figure–ground segregation, and Barlow's and Dobson and Rose's conceptions of environment-modelling systems. Lateral coherence at the same level can also be considered; for example it could be argued that orientation selectivity and direction selectivity are likely to be generated by mechanisms which work on similar principles.

7. Completeness.

What, if any, other subsystems must exist in addition to the modelled one, i.e. how many empirically observed phenomena or functionally required properties are accounted for by the model? For example, does a model of orientation selectivity also account for direction selectivity, spatial frequency selectivity, after-effects, and so on?

8. Empirical evidence.

The values or states of particular parameters can of course be investigated by direct empirical observation in many instances, particularly once the important parameters have been identified. Research on the visual cortex is nowadays at a relatively advanced stage, beyond the point at which the new procedure we describe should ideally have been applied. However, although not all the empirical evidence that has accrued over the past 25 years has been collected with the aim of distinguishing between different functional hypotheses or parameter states, much of this evidence can be used retrospectively in this way. For example, the anatomical evidence for the reciprocity of connections between area 17 and the other cortical areas (and with the LGN) suggests that models of visual cortex function which do not incorporate feedback cannot be complete models.

Thus the testing of models may involve the use of reasoning and argument, mathematics, computer simulation, and experiments. Note that no single one of these can provide an absolute definitive solution, in the way that pure empiricism calls on experimental evidence as the only and ultimate arbiter of truth and falsity (or that pure rationalism similarly relies on reasoning alone). The unreliability and skill dependence of empirical evidence has already been described in this article; but computer simulations, mathematics and complex logical reasoning are similarly dependent on the skills and judgements of individual scientists. For example, the ability of human beings to think logically is easily shown to be limited (Johnson-Laird and Wason, 1977). The final assessment of a model's plausibility therefore requires a balanced and judicious use of various arguments and lines of evidence (see also Toulmin, 1972, pp. 222–260).

Further development of models

Further development of the models under consideration may proceed by a process of elaboration, i.e. by incorporating extra factors. Those parameters which were originally considered of relatively low importance (being either accidental or trivial), and which were therefore not included (for simplicity) in the model-building process, can now be added to the limited group of models which have passed the previous formulation and testing procedure. The procedure is then re-run on these models to see if any significant improvements or deteriorations in performance occur as a result of including the extra factors. Consider the following hypothetical example. Suppose that 15 parameters were identified as relevant to the system under investigation, each of which could be expressed as in one of two states (step 2 in the above procedure). Then, a fully exhaustive model building (step 3) would generate 2^{15} models, which would be considered too many to handle. Therefore, only the six most important parameters might be used to generate models, i.e. those which are likely to have the biggest effects on the system's functioning. This would give 2^6 models, i.e. 64. After application of the criteria at step 4, five models might be left, each of which could provide a plausible explanation of the system. The next most important set of parameters, say four in number, could then be added to these five models, making $5 \times 2^4 = 80$ models for assessment against the criteria at step 4. This reiterative process of analysis, of reducing the numbers of possibilities and then adding more minor factors, can proceed as many times as necessary until a satisfactory state is achieved. Note that this incorporates the 'positive heuristic' of Lakatos (1970, pp. 135–136) for the *progressive* development of models.

Under purely empiricist methodologies of science it often transpires that many models exist simultaneously, and then workers sometimes use various techniques to try to circumvent their disagreements. Are any of these techniques relevant or useful to our new procedure? Two such techniques include either superimposing or subtracting the elements of different models (i.e. taking the union or the intersection of the sets of elements). For example, given one model in which LGN cells drive simple cells which drive complex cells, and another model in which LGN cells drive both simple and complex cells, it could be suggested either that LGN cells drive simple and complex cells and also that simple drive complex, or that LGN cells drive simple (the consensus of 'what everyone agrees on so far') while the drive to complex cells is uncertain and should be the subject of further empirical research. After following the explicit procedures described in this article then a similar line of reasoning might possibly be applied if different workers have used different sets of parameters to generate models of the same system. Compromises may be attempted by combining, or by taking the consensus of, the parameter choices of the different theorists. Such techniques are, however, illogical and are not permissible without clear and good outside arguments for any change in the choice of parameters. Consider the well-known fable of the judge presiding over an inquiry into a hit-and-run accident at a crossroads. One witness has said that the accident was caused by a red car heading south, but another witness said it was a black car heading west, a third said it was a blue car heading north and a fourth witness said it was a green car heading east. What should the judge conclude? He should not use the consensus technique and say: 'Well at least the witnesses are agreed on one thing: it was a car and not a lorry or a motorbike'. Nor should the judge use the all-including technique and order the police to search for a multicoloured car travelling in all directions at once. What he should conclude is that the evidence presented is thoroughly unreliable. Faced with the analogous situation in science, however, researchers are reluctant to reject all the evidence in this fashion; instead, they make subjective assessments of the reliability of each witness or laboratory, which is yet another way in which social factors enter into science.

As a final stage in the development of models, the division of each parameter into a limited number of discrete states may be relaxed. It was necessary to impose this division or dichotomization in order to simplify the procedure for model building and model choice, as we have already described. Once the general classes of model which work best have been identified then for each class we can entertain for each parameter the possibility that intermediate cases may exist along a continuum between the extremes that the dichotomy represented, or at least that the dichotomous categories may have fuzzy edges. Again, this should not be done gratuitously (the judge in the hit-and-run case blaming the accident on a stationary grey car), but only with explicit reasoning or evidence to guide each consideration, such as that given by Rowe and Stone (1980) and Tolhurst and Dean (Chapter 32 in this volume).

SUMMARY

We have described an explicit procedure for model building, which embodies Maxwell's rules for rational science and which should be applied at an early stage in research into any particular problem area. Although it may follow an initial blind groping for ideas in the new field (MacKay, 1983), *it must precede the main empirical thrust*, which is, along with logical and other forms of reasoning, only a part of the process of testing the possible underlying concepts and the solutions to the problems within the field which are generated by the procedure.

The procedure described in this paper would ideally enable all possible models of a system to be considered exhaustively, in contrast to the conventional methodology which permits (indeed encourages) workers to develop models in isolation under the assumption that the truth can be induced from the empirical evidence by an authority figure of sufficient genius. The approaches of Lakatos (1970) and Feyerabend (1975) are like ours in that they advocate a pluralistic consideration of theories, but theirs are 'shotgun' techniques which rely on the unsystematic genesis of hypotheses by irrational and unspecified means. Under our procedure, however, the inclusion of every possible model is something which obviously cannot be achieved in practice because the number of possibilities is too great. What the new procedure does ensure is that all *classes* of model are considered systematically, with a range of possible models studied in detail rather than just one model, and moreover that this range of possibilities at least includes all the plausible models. Thus we do not wish to claim that the procedure is an algorithm-like prescription for finding the Right Answer. The procedure consists of heuristic guidelines which should lead to a state in which a reasonably small set of clear and plausible alternatives has been identified. Creative debate will then still be possible over a number of issues, such as where to draw the boundary between plausible and non-plausible models, and what further tests would provide the most useful information.

In this way, the procedure will make the Humpty Dumpty and the Charge of the Light Brigade Effects less likely to occur, by facilitating agreement over general strategy and by systematically increasing the chances that a correct or useful model will be generated and will be amongst those given serious consideration right from the start.

ACKNOWLEDGEMENTS

VGD was supported by MRC.

REFERENCES

Ashby, W. R. (1956). *An Introduction to Cybernetics*, Methuen, London.
Bruner, J. S., Goodnow, J. J., and Austin, G. A. (1956). *A Study of Thinking*, Wiley, New York.
Feyerabend, P. (1975). *Against Method*, NLB, London.
Gilbert, G. N., and Mulkay, M. (1984). *Opening Pandora's Box. A Sociological Analysis of Scientists' Discourse*, Cambridge University Press, Cambridge.
Gregory, R. L. (1959). Models and the localisation of function in the central nervous system. In *Symposium on the Mechanization of Thought Processes*, HMSO, London, pp. 669–689.
Janis, I. L. (1982). *Groupthink*, 2nd ed., Houghton, Mifflin, Boston.
Johnson-Laird, P. N., and Wason, P. C. (Eds.) (1977). *Thinking*, Cambridge University Press, Cambridge.
Knorr, K. D., Krohn, R., and Whitley, R. (Eds.) (1980). *The Social Process of Scientific Investigation. Sociology of the Sciences, Vol. IV*, Reidel, Dordrecht.
Knorr-Cetina, K. D. (1981). *The Manufacture of Knowledge*, Pergamon, Oxford.
Kuhn, T. S. (1970). Reflections on my critics. In *Criticism and the Growth of Knowledge*. (Eds. I. Lakatos and A. Musgrave), Cambridge University Press, Cambridge, pp. 231–278.
Lakatos, I. (1970). Falsification and the methodology of scientific research programmes. In *Criticism and the Growth of Knowledge* (Eds. I. Lakatos and A. Musgrave), Cambridge University Press, Cambridge, pp. 91–196.
MacKay, D. (1983). In love with science. *Nature*, **302**, 357–358.
Marr, D. (1982). *Vision*. Freeman, San Francisco.
Oppenheim, A. V., and Lim, J. S. (1981). The importance of phase in signals. *Proc. IEEE*, **69**, 529–541.
Robson, J. G. (1983). Frequency domain visual processing. In *Physical and Biological Processing of Images* (Eds. O. J. Braddick and A. C. Sleigh), Springer, Berlin, pp. 73–87.
Rowe, M. H., and Stone, J. (1979). The importance of knowing our own presuppositions. *Brain Behav. Evol.*, **16**, 65–80.
Rowe, M. H., and Stone, J. (1980). The interpretation of variation in the classification of nerve cells. *Brain Behav. Evol.*, **17**, 123–151.
Stone, J. (1983). *Parallel Processing in the Visual System*, Plenum, New York.
Toulmin, S. E. (1972). *Human Understanding, Vol. 1*, Clarendon, Oxford.

Models of the Visual Cortex
Edited by D. Rose and V. G. Dobson

CHAPTER 57

Application of an explicit procedure for model building to the visual cortex

VERNON G. DOBSON
Department of Psychology, University of Brunel, Uxbridge, Middlesex UB8 3PH, UK
and
DAVID ROSE
Department of Psychology, University of Surrey, Guildford, Surrey, GU2 5XH, UK

There is a general feeling among scientists that philosophy has very little to do with the everyday practice of scientists in the laboratory. Even the most famous of living philosophers of science has had only slight and inconsistent effects at the workplace (Mulkay and Gilbert, 1981) and, as Chalmers admits (1982, p. 169), 'philosophy or methodology of science is of no help to scientists'. The prescriptions of those who wish to assert how science *should* proceed, quoting in support scattered examples of successful research in history, will not be accepted by scientists unless those philosophies are actually put into practice and shown to lead to successful and productive research: the proof of the pudding is in the eating. The proposals for an explicit procedure for scientific methodology that we made in the previous chapter must therefore be shown to be applicable to specific areas of research, to generate a comprehensive range of models which includes new and original models where necessary and which certainly includes models which provide plausible explanations or solutions, and to suggest critical tests and a programme or direction for further research.

As a first step towards fulfilling this necessity, we therefore apply the first three steps of the prescribed procedure to two related problems within the field of research on the visual cortex: the generation of orientation selectivity in cortical cells and the global pattern of orientation columns. In addition,

Dobson (Chapter 18 in this volume, and 1985) has applied the procedure to associative pattern-learning networks. Here, we identify the relevant parameters of the problem systems, and then we discuss how certain combinations of parameter states represent some of the commonly known models, while other combinations may represent as yet unconsidered possibilities. We hasten to say that the final step, the evaluation and testing of all these models against the various possible criteria, and thus the total run-through and final outcome of the whole procedure, remains to be completed by future research; at this moment in time we can only point to the direction in which this progressive research programme will head if it is put into practice.

PREAMBLE

A brief review of higher levels of analysis is a necessary prelude to this exercise. First, overall world-views were dichotomized by Dobson and Rose (Chapter 3 in this volume) as either Parmenidean–Cartesian or Heraclitean–Darwinian. The former assumes that the Universe is fundamentally orderly, predesigned, simple, perfect and benign. The view of overall brain function which stems from this world-view is that the brain exists to house object-centred models of reality for purposes defined by the Creator. The development of these models, when considered, is by connecting components up in the right way to cause the right things to happen. The view of the visual system which follows from this is that it is a passive pattern-recognition device which serves to generate a complete and faithful representation of the current environment. The Heraclitean–Darwinian point of view, in contrast, assumes the Universe to be fundamentally disorderly, unplanned, complex, chaotic and dangerous. The brain exists to construct the minimum model of the environment which will effectively enable the predictive suppression of unwanted events, through active interaction with the environment. The view of the visual system that follows from this is that it is an active system which seeks out the information that it needs about the environment for its own purposes, and guides action within that environment. Thus we have simplified the overall functional hypotheses about the visual system down to just two for the purposes of the present discussion: *pattern recognition* and *pattern seeking*.

By what mechanisms might the visual system perform these functions? One line of reasoning is that visual inputs may contain features or *points of high information content*, which have to be located and analysed by the visual system. For pattern *recognition*, these features have to be integrated in such a way that some of them form aggregates or sets which either match those of known patterns or which have intrinsic 'good form' (Pomerantz, 1981) and which are segregated from sets of features which form other patterns or no pattern (e.g. noise). For pattern *seeking*, either individual features or sets of features may form key cues to distinguishing between the presence or absence of a particular object in the environment. Thus our simplifying hypothesis is that the analysis of features will be one subfunction which exists within any visual system, whatever prior decisions or assumptions are made about metaphysics or whole-brain function. The details of how features are analysed and of how they are integrated will, of course, differ under different prior assumptions.

At any point in the visual scene there may be a particular luminance, rate of change of luminance, acceleration of luminance and higher derivatives of luminance, all with respect to three-dimensional space and time. The same is true for colour (rate of change of colour, and so on). The first derivative of luminance across the two-dimensional (tangent) plane (of the retina) yields information which is known as the feature *oriented edge*, because it has a direction or orientation in which the derivative is taken, as well as a magnitude.

We will discuss here the mechanism by which orientation selectivity arises, for a number of reasons. These are: (a) it enables more efficient encoding of the information input than a full description of the luminance alone (Barlow, 1959) and is less affected by changes in the level of ambient illumination, but it does not necessarily require the more elaborate circuitry that higher directional derivatives do (Marr, 1982, p. 57); (b) empirical evidence suggests that it is a common feature in the visual systems of many species; (c) it is a convenient feature for the visual system to begin working with, being useful to trace the outlines of stimulus objects, and it can be used as a basis for spatial frequency selectivity, direction of movement selectivity, stereopsis, etc. (although other mechanisms may exist for these prior to, simultaneously with or in parallel with orientation selectivity, e.g. Daugman, Chapter 10 in this volume, stresses that orientation and spatial frequency should be considered simultaneously).

The overall problem can be cast in the following form. Visual cells are divided into two arrays. A set of non-oriented cells in the 'presynaptic' array projects to a set of cells in the 'postsynaptic' array, some or all of which are orientation selective. For the present discussion, we will only consider models in which the projection between the two arrays is topographically ordered in two dimensions, with a small local scatter. The precise locus of the two arrays in the visual system is not in principle important (although it may be in practice); the arrays could be assumed to represent the geniculocortical mapping, or different retinal or cortical laminae, or different cell types within the same lamina. All models attempt to account for the generation of orientation preference in terms of the pattern of connectivity between the two arrays and local lateral interactions within arrays.

The problem can be further subdivided into questions about the local connectivity and about global patterns or arrangements of the connectivity across the arrays. Thus the local connections may not be everywhere the same, but may vary from one orientation-specific cell to another, if only to enable different orientations to be analysed at each point in the image. The arrangement of these differences may itself be an integral part of models of orientation selectivity, since it may be more efficient for the nervous system to set up the global patterns and allow the local connections to form according to a few local rules, rather than specifying the local circuitry for every cell individually. We will therefore present parameters of models of orientation selectivity at both local and global levels.

LOCAL PARAMETERS

First, we will consider in what ways the various inputs which converge on a given cell in the postsynaptic array can differ from one another in their functional properties.

Excitation versus inhibition

One difference is that the firing rate of the cell can either be increased or decreased by an input to it, i.e. each input is either excitatory or inhibitory. Thus for each input we can specify a dichotomous quality: excitation versus inhibition. This is our first parameter.

Locations of receptive fields: same or different

Next, consider that each input fibre will respond to visual stimuli falling within its own receptive field. However, the location of the receptive field may differ from one input to another. This forms a second dichotomy. We will not go into the various ways in which the locations can differ (e.g. along a straight line, separated by various distances, etc.) but will leave this to a more formal exposition (in preparation).

Phase

Another difference between the inputs may occur in the sign of the first derivative with respect to time, in other words whether their firing rates increase or decrease when the luminance within their receptive fields increases (and vice versa for decreases in luminance). This is commonly known as the distinction between on-responses and off-responses. Since the sign of the response to a grating stimulus varies with its phase relative to the receptive field, we can call our third parameter 'phase'. This can be specified in absolute terms for each input to a given cell. For intracortical connections, phase becomes a more complicated parameter (see, for example, Pollen, Foster and Gaska, Chapter 28 in this volume).

Orientation selectivity: present or absent

Finally in this section, remember that, by definition of the two arrays, some of the inputs to a cell in the second array may be non-oriented while others may be orientation selective, i.e. those from within the second array itself.

Next we turn to a factor which is of importance in many models, namely the shapes of the anatomical components. If these are prespecified in the brain, the shapes would affect the connections that can form. This factor first concerns the shapes of the axonal arborizations of the axons which run from the first array to the second. Second, it concerns the shapes of the dendritic trees, and also the shapes of the axonal arborizations within the local circuitry of each array. (For the present purposes it can be assumed that each array is a two-dimensional sheet of cells with topographic representation across it, so that the 'shapes' that we are referring to mean the tangential or laminar projections of the structures into the visual field. In the real, three-dimensional cortex the anatomical locations are not important as long as the topographic projection is maintained in principle.)

Thus for each class of functional component the overall shape must be described, if it is relevant. In each case, this has to be done by specifying a number of parameters:

Relevance of shape

First, it has to be made explicit whether the shape of each component actually is of relevance to orientation selectivity, or whether it has no relevance. (If shape is not relevant, then orientation selectivity may depend on circuitry that grows according to activity-dependent processes, or by recognition of cytospecific labels on individual nerve cells, as described later in the section on the further development of models.)

Isotropy

If shape is relevant, then the shape can be classified as either isotropic (circularly symmetrical) or non-isotropic. Of the latter possibilities, the ratio between the longest diameter and the diameter orthogonal to it is the relevant quantity for orientation selectivity.

Size

The range over which connections may form within an array can be specified relative to the scatter in the topography, in the first instance as greater than this value or not (in all the functionally relevant directions for anisotropic components).

We can put together these parameters in various combinations to show how complete models form. To illustrate this procedure (step 3 as described in the previous paper) we present some well-known examples.

SOME LOCAL CIRCUIT MODELS

The best-known model is of course Hubel and Wiesel's (1962) model for simple cells. This has all the inputs to a simple cell excitatory, with different receptive field locations offset along a straight line, all either on-centre or off-centre, all non-oriented, and with no specified anatomical shapes of the component cells and axons. However, the model of Benevento, Creutzfeldt and Kuhnt (1972) has converging non-oriented excitation and orientation-selective inhibition from inputs with coincident receptive fields but no specified phase, and again with no specified anatomical shapes. The more recent formulation of Heggelund (1981 and Chapter 37 in this volume) incorporates converging excitation and inhibition from offset receptive field locations, with both types of input non-oriented and either on-centre or off-centre, but no specified anatomical shapes. Braitenberg's model (Chapter 50 in this volume) also contains converging non-oriented excitation and inhibition from offset receptive field locations, but without specifying phase and definitely specifying shape to be isotropic on a scale larger than the scatter in the topography of the projection. Finally, the model of Schiller, Finlay and Volman (1976) has non-oriented excitatory inputs of one phase, and non-oriented inhibitory inputs of mixed phases with different receptive field locations from the excitatory inputs, converging onto dendrites which are anisotropic on a scale larger than the scatter of the topography.

These are some representative examples and their inclusion should not be taken to imply any evaluative decision about them. A full run-through of all possible combinations of parameter states and their evaluation is not possible in the space we have here.

GLOBAL PARAMETERS

In the preceding sections we suggested that orientation selectivity might be required for most visual system functions, and we described the local circuits which could generate orientation selectivity in terms of anisotropies in parameters such as receptive field location (i.e. topographic projection), phase, excitation-inhibition, and the shape and range of axons and dendrites. Next we consider the ways in which this local circuitry could be related to the retinotopic map, such that each visual field locus is analysed by a set of cells covering a comprehensive range of orientations.

Vertical versus horizontal organization

The comprehensive analysis of all orientations at each point in the image could be achieved first by systematically rotating local circuit anisotropies with depth through the postsynaptic array, or 'cortex', so that a full cycle of orientation preferences would be encountered during vertical penetrations through a cortical 'column'. The alternative is to dedicate the local circuits at each cortical locus to the production of a single orientation preference and to rotate the local circuit anisotropies responsible for orientation selectivity with distance across the cortical surface. If a complete cycle of orientation preferences is produced within the range of scatter in the retinocortical map, then each point in the visual field will be analysed by a complete set of orientation-selective cells.

While a vertical organization might appear to provide the simpler and more compact solution, it presents complicated developmental problems for local circuits. It would be simpler, developmentally, to organize the cortex horizontally. We will therefore continue to discuss here only models with horizontal organization across the array.

Mosaic versus continuously rotating organization

There are two ways of rotating the local circuit anisotropies: discontinuously or continuously. Thus either the retinocortical projection can be divided up into a mosaic of separate 'modules' within which local circuits rotate independently, or there may be no mosaic structure and the rotation is smooth and continuous over the whole map.

COMPREHENSIVE MODELS

In this section we aim to specify all the main classes of models which can account comprehensively for the production of columns of orientation-selective cells and the organization of these columns into cyclically repeating sequences, or 'hypercolumns', analysing corresponding visual field regions, or 'hyperfields'. Each of these models makes specific predictions about local circuit structure and the pattern produced on the cortical surface by amalgamating columns of cells with the same orientation preferences into iso-orientation bands or columns. We will deal with mosaic and continuously rotating models in turn.

Mosaic models

These can be divided according to which array(s) contains a mosaic pattern.

Presynaptic array mosaics

In these models the presynaptic array is assumed to consist of a mosaic of circular (Schwartz, Chapter 14 in this volume; Frisby, 1979) or hemi-circular (Dobson, Chapter 18 in this volume) hyperfields, with partially shifted overlap so that each point in the visual field is analysed by several hyperfields. Each hyperfield projects independently to a corresponding hypercolumn in the postsynaptic array, which is thus also divided in a mosaic pattern. This projection contains a rotation in the topography so that hyperfield radii map onto parallel orientation columns as shown in Fig. 1. This topological rotation will be produced if the density of cells in the postsynaptic array remains constant but the number of postsynaptic cells representing a given area of hyperfield decreases linearly with eccentricity from the hyperfield centre. This can either be achieved by having the receptive fields at the hyperfield centres smaller, or of constant size but overlapping more.

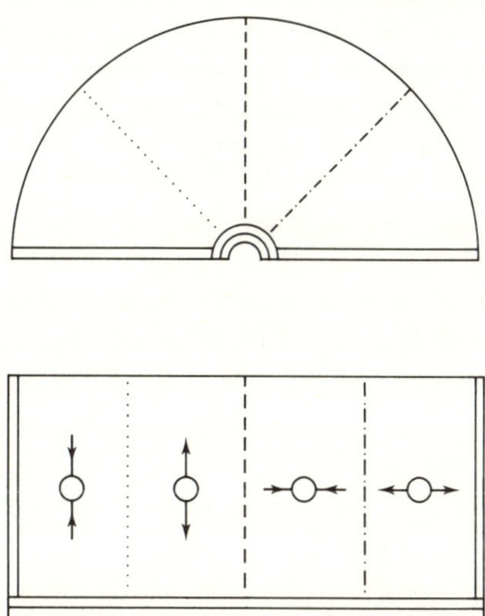

FIGURE 1 Local retinocortical mapping with rotating topography from a hemi-circular hyperfield (top) to a rectangular hypercolumn (bottom). Adjacent hyperfield radii project to adjacent parallel orientation columns within the corresponding hyper-column. Local circuit anisotropies are also aligned in parallel. There are four possible kinds of local circuit anisotropy: either axons or dendrites can be either orthogonal to, or parallel with, the orientation column system, as shown schematically in this figure in different parts of the hypercolumn. In this and subsequent figures the cell body is shown as a circle, outward-pointing arrows represent axonal processes and inward-pointing arrows represent dendrites.

In all of these models orientation selectivity can be produced by aniso-tropies in either excitatory or inhibitory postsynaptic circuitry, with elongated dendritic or axonal trees aligned either parallel with or orthogonal to the local orientation columns. If hyperfields and hypercolumns are arranged in a square mosaic then orientation columns in adjacent hypercolumns would line up 'in register' to form extended parallel iso-orientation bands, and the local circuit anisotropies would also line up in parallel over wide areas of cortex. These local anisotropic circuits determine whether the orientation preferences of cells in the same column are aligned with (Schwartz and Dobson, Chapters 14 and 18 respectively in this volume) or orthogonal to (Frisby, 1979) the corresponding hyperfield radius.

Postsynaptic array mosaics

These models require no topographic transformations in the retinocortical map. It is assumed that the postsynaptic array consists of a mosaic of circuit modules or hypercolumns. The circuitry within each hypercolumn has a radially symmetrical structure like those shown in Fig. 2, so that circuits may be aligned with hypercolumn radii, forming a star pattern, or they may be orthogonal to hyperfield radii, running round concentric circles, or they may be at the outer edge of the module projecting inwards or at the centre projecting outwards (Braitenberg, Chapter 50 in this volume; Dow and Bauer, 1984). These circuits may be implemented either by axons or dendrites and, depending on whether they are excitatory or inhibitory, generate orientation preferences either orthogonal to or aligned with the hypercolumn radii. The pattern of iso-orientation bands produced by these models tends to be interlacing and criss-crossing, regardless of whether the hypercolumns are arranged in hexagonal or square arrays.

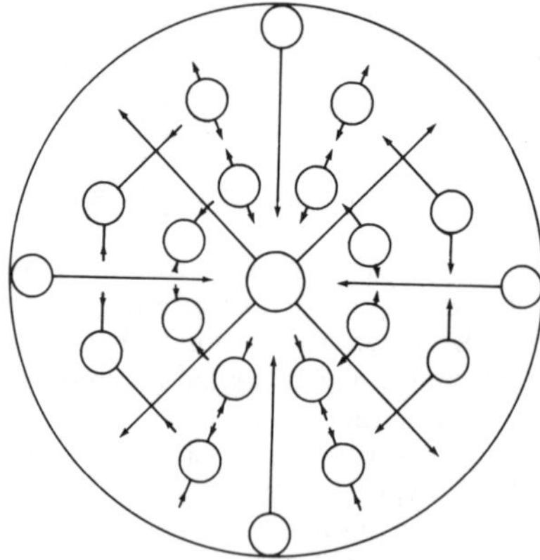

Figure 2 Cortical module in which local circuit anisotropies produce orientation columns radially organized around the module centre. These local circuits could be either axons or dendrites oriented either radially or circumferentially, as illustrated in different parts of the module.

Models with continuously rotating circuitries

These can be divided according to which anatomical components rotate across the cortex.

Dendritic mechanisms

Figure 3a represents columns of postsynaptic array cells sampling the retinotopic input with elongated dendrites. Cells in the same column have the same dendritic orientation, and this shifts progressively between columns. Assuming that the scatter in the mapping is small compared with the range of the dendrites the cells will have orientation preferences parallel to the dendrites receiving excitation and orthogonal to those receiving inhibition (Schiller, Finlay and Volman, 1976; Tieman and Hirsch, Chapter 45 in this volume).

Axonal mechanisms

In Fig. 3b cells in the presynaptic array project to elongated bands of cells in the postsynaptic array. Parallel rows of cells in the presynaptic array send projections which rotate progressively from column to column across the postsynaptic array. The receptive field structure produced by this mechanism for all the cells in the same postsynaptic array column is shown in Fig. 3b below each column, with crosses to indicate the position of the postsynaptic cell. More complex receptive fields are generated if axons are longer or the shift in orientation between columns is reduced. However, it will be evident from Fig. 3b that receptive field structures produced by rotating axons differ systematically between columns in aspects such as shape and size, in addition to orientation preference.

Figure 3c shows the pattern of axonal connectivity required to produce columns of cells with systematically shifting orientation preferences like those generated by the rotating dendrites in Fig. 3a. Note that a different pattern of circuitry would be required for cells in different columns.

FURTHER DEVELOPMENT OF MODELS

In the previous paper we discussed how models can be developed. The number of possible models which can be generated by the procedure can be extended by including more parameters in the formulation.

For example, all models have to have a developmental rationale, and so we can include a parameter which defines whether the model's growth depends on anatomical prespecification or physiological and environmental functional factors. This parameter is: *cytospecificity versus activity-dependent growth*. Thus the development of the circuits shown in Fig. 3 can be accounted for in terms of either of these processes. Cytospecificities are

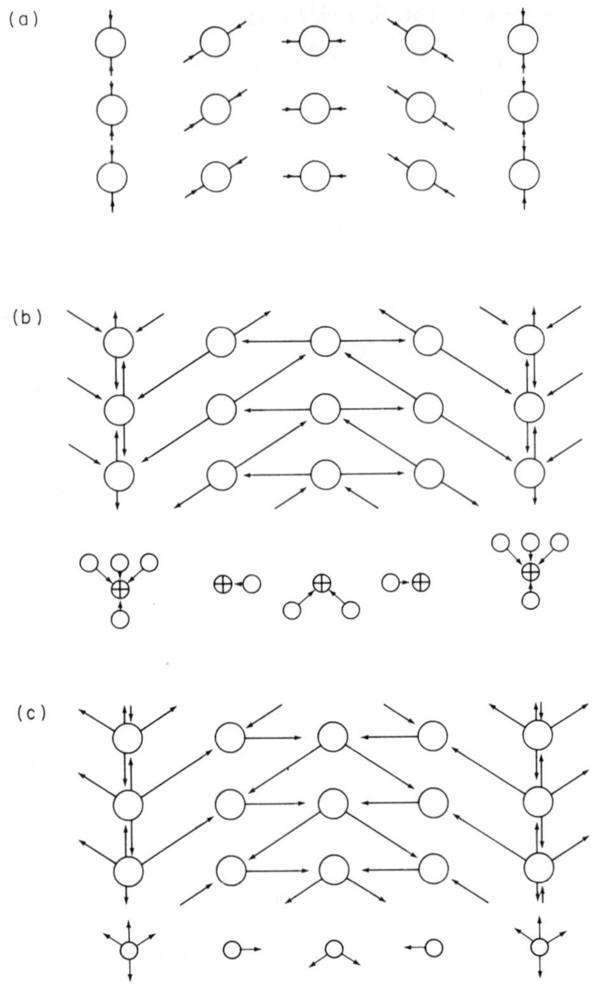

Figure 3 Continuously rotating local circuits. (a) Columns of cells with dendritic processes which rotate with distance to produce progressive variations in orientation selectivity alone. (b) Rotating axonal circuits; these produce receptive field structures which vary rather unsystematically in shape, size and orientation between columns, as shown for the cells marked with crosses below each column. (c) Illustrates how axonal circuits must vary between columns to produce parallel columns of cells.

chemical labels or signals which a cell can acquire from its visual field position within an array of nerve cells. These labels may set a cell to orient the growth of its axons or dendrites within chemical gradients or towards specific cell types. For example, Fig. 4 represents a presynaptic array within which the concentration of a substance 's' varies sinusoidally with horizontal distance.

Cells in the postsynaptic array are assumed to grow anisotropic dendrites elongated with the long axis vertical when projected into the visual field, unless they receive from presynaptic cells with high concentrations of 's', in which case they grow with horizontal projected dendrites. This would produce alternating parallel bands of cells with vertically and horizontally oriented dendrites in the postsynaptic array. With two spatially separate sources of substance 's', interference patterns would be produced, and iso-orientation bands would be interlacing rather than parallel. If two different cytochemical substances are involved their concentrations could signal components of a vector required to produce a full range of parallel or interlacing iso-orientation bands (Swindale, Chapter 47 in this volume).

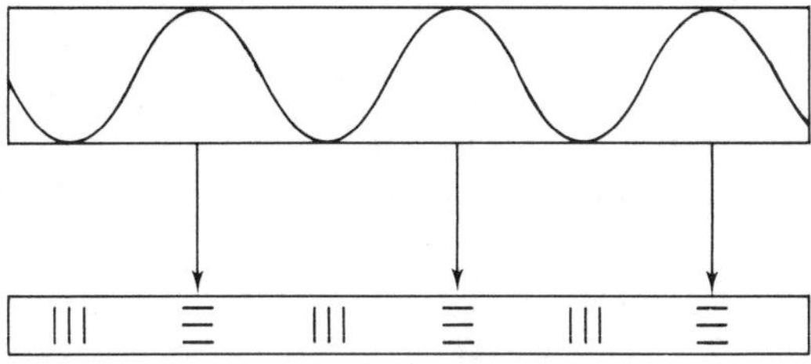

FIGURE 4 Cyclically varying cytospecificity model. The cyclic variation in the concentration of a substance 's' is represented by the sinusoidal curve in the presynaptic array (top). Cells in the postsynaptic array (bottom) normally grow dendrites oriented vertically to the visual field projection, but grow horizontally oriented dendrites when they receive inputs from presynaptic cells with relatively high 's' concentrations.

Alternatively, anisotropic local circuitry can be generated simply by the effects of experience with oriented stimuli on arrays linked by modifiable Hebb-type synapses, if there are local isotropic circuits mediating 'surround inhibition' in the postsynaptic array (Amari, Chapter 15 in this volume; von der Malsberg, 1973). The pattern of iso-orientation bands produced by neural activity models varies between parallel and swirling with blobs and bifurcations, depending on the exact circuitry assumed.

Several models suggest that different developmental rationales exist for different orientations. In these models, populations of cells with certain orientation preferences (e.g. horizontal and vertical, or radial) are produced by innate cytospecificities or circuit geometries. These are then used to 'seed' a neural activity-dependent process in the cortex, which first sets up a system of alternating columns of cells with those orientation preferences (e.g. vertical and horizontal) and then 'interpolates' columns with intermediate preferences between them (Cowan and von der Malsberg, Leventhal, and Vidyasagar, Chapters 48, 40 and 41 respectively in this volume).

At the local circuit level, some additional parameters of relevance which can be included concern further properties of the inputs to a given orientation-selective cell. We have already mentioned polarity (excitation—inhibition), receptive field location, phase and orientation selectivity; to these we can add *receptive field size* and *colour selectivity*. For example, Hubel and Wiesel's model has implicit that the lateral geniculate inputs to a given cortical cell have similar sizes, and in the monkey similar colour tuning too. The *linearity* of the inputs is also important in many of today's quantitative models. Finally, *ocular dominance*, both at the local circuit level and at the columnar level, can be incorporated.

Further anatomical parameters such as *synapse and spine structures and densities* and *axonal and dendritic branching patterns and densities* can also be considered, as can variations in all parameters with *eccentricity*.

Another technique for the development of models is the superimposition of mechanisms within the same cell. One example of this is given (with the necessary and appropriate reasoning) by Nelson (Chapter 11 in this volume), in whose model several mechanisms exist which enhance orientation selectivity, each added to the previous one to improve the fit against the criteria he uses.

DISCUSSION

In the above examples of the application of this new procedure we have tried to identify the essential parameters necessary for building models of orientation selectivity in single units and of orientation columns, and we have indicated how these parameters can be used to specify all possible classes of model. In the process we have generated some previously unpublished models of orientation selectivity and columnar organization. We have aimed to avoid the Humpty Dumpty and the Charge of the Light Brigade Effects, described in the previous article, by systematically generating and interrelating all possible models, and giving fair consideration to each.

If our simplifying assumptions and our choice of parameters are appropriate, then at least one model within the classes we have described should be producing orientation selectivity in the visual system. However, the interpretation of which models are expected to be valid and how that validity is to be assessed will depend, as we pointed out in the previous paper, on the functional hypotheses and fundamental assumptions which are made about the role of orientation selectivity and the purposes of the visual cortex and of the brain. For example, representation and encoding of the image would require the most efficient possible retention of the information in the image, with spatial frequency and phase encoded as well as orientation. On the other hand, selective attention to aspects of the environment which might signal unwanted events would require the minimalistic categorization of the presence or absence of features and of their distance, direction and motion. The model of orientation selectivity chosen by an individual worker may thus be covertly, and indeed should be overtly, influenced by these 'ulterior' considerations. This is because our overall aim and need is to have a functionally coherent view of the entire brain. The role of this view is both to guide and to be guided by our solving of specific research problems. Thus the tests which are made to decide between models of the visual cortex should most rationally and efficiently be aimed at examining the qualities of the possible ulterior meta-views, rather than simply consisting of the accumulation of larger and larger numbers of detailed observations which may or may not match the predictions of the various specific models.

We hope that in this brief article we have at least been able to convey some idea of the method and its power as a model-generating procedure. We have tried to make this procedure explicit and open to rational criticism and improvement. Ideally, we hope that this article will be considered as a worked example of how the procedure can begin to be put onto effect, because as Mulkay and Gilbert (1981) suggest, one way in which individual scientists can be guided in their day-to-day choices of action is to provide them with a set of examples of how similar, successful activities have been carried out by their colleagues in the recent past. If we have succeeded in this, we will have contributed to the two areas of scientific endeavour about which philosophers of science have least to say, yet which are of most importance to the working scientist: how to generate models and how to choose between them.

The completion of the process of formulating all the possible models that can be made by combining the parameters that we have presented, and the assessing of each parameter and model by reasoning, simulation and experiment, will occupy theorists for many years to come. Certainly, this task is beyond the scope of the present book, in which we and our authors have only begun the massive task of untangling, clarifying and making sense of this complicated field of research. We feel, however, that the future does look brighter than the scene that was depicted in the first chapter of this volume, and we are confident that with the increasing awareness of more efficient methodologies of science, such as that put forward by Maxwell in Chapter 2 in this volume, research will proceed surely (and perhaps even rapidly) towards a rational understanding of the visual cortex.

ACKNOWLEDGEMENTS

VGD was supported by the MRC.

REFERENCES

Barlow, H. B. (1959). Sensory mechanisms, the reduction of redundancy, and intelligence. In *Symposium on the Mechanization of Thought Processes*, HMSO, London, pp. 535–574.

Benevento, L. A., Creutzfeldt, O. D., and Kuhnt, U. (1972). Significance of intracortical inhibition in the visual cortex. *Nature New Biol.*, **238**, 124–126.
Chalmers, A. F. (1982). *What is this Thing called Science? 2nd ed.*, Open University Press, Milton Keynes.
Dobson, V. G. (1985). Superior accuracy of pattern retrieval in randomly connected associative nets using decrementing learning rules. *Neurosci. Lett.* **Suppl. 21**, 517.
Dow, B. M., and Bauer, R. (1984). Retinotopy and orientation columns in the monkey: a new model. *Biolog. Cybernetics*, **49**, 189–200.
Frisby, J. P. (1979). *Seeing*, Oxford University Press, Oxford.
Heggelund, P. (1981). Receptive field organization of simple cells in cat striate cortex. *Exp. Brain Res.*, **42**, 89–98.
Hubel, D. H., and Wiesel, T. N. (1962). Receptive fields, binocular interaction and functional architecture in the cat's visual cortex. *J. Physiol.*, **160**, 106–154.
Marr, D. (1982). *Vision*, Freeman, San Francisco.
Mulkay, M., and Gilbert, G. N. (1981). Putting philosophy to work: Karl Popper's influence on scientific practice. *Phil. Soc. Sci.*, **11**, 389–407.
Pomerantz, J. R. (1981). Perceptual organization in information processing. In *Perceptual Organization* (Eds. M. Kubovy and J. R. Pomerantz). Lawrence Erlbaum Associates, Hillsdale, New Jersey, pp. 141–180.
Schiller, P. H., Finlay, B. L., and Volman, S. F. (1976). Quantitative studies of single-cell properties in monkey striate cortex. V. Multivariate statistical analyses and models. *J. Neurophysiol*, **39**, 1362–1374.
von der Malsberg, C. (1973). Self-organization of orientation-sensitive cells in the striate cortex. *Kybernetik*, **14**, 85–100.

Models of the Visual Cortex
Edited by D. Rose and V. G. Dobson

GLOSSARY

Adaptation

The process of adjustment of the properties of a sensory system that occurs with changes in the prevailing levels of environmental stimulation. This adjustment is probably such as to optimize the system's performance of its function.

After-effects

These are phenomena which occur after there is a sudden decrease in the level of a particular stimulus. They may include: elevation of threshold for detecting a similar stimulus, decreased apparent strength of a similar (but less intense) stimulus, distortions of the apparent qualities of non-identical stimuli (they appear shifted along the relevant *domain* away from the initial adapting stimulus) and hallucinations of stimuli that are different from the inducing stimulus (see Georgeson, Chapter 22 in this volume). The reported durations of after-effects can vary from fractions of a second up to several months, for reasons which are not understood.

Algorithm

A procedure or set of rules which will always enable you to find the solution to a problem in a finite number of stages (cf. *heuristic*).

Amblyopia

A deficiency of vision which persists even when the image is in perfect focus on the retina. Common causes of this condition include visual problems during childhood, such as strabismus (squint; misalignment of one or both eyes), anisometropia (unequal focussing of the two eyes) or occlusion (blockage of visual input by a cataract, drooping eyelid or an eyepatch applied in an attempt to correct a strabismus). The pattern of deficiencies differs depending on which of these misfortunes is the cause of the amblyopia.

Analogue encoding

An analogue computer represents outside events by the corresponding physical states of its components. A brain which encodes information in analogue fashion would similarly have information stored and handled in the form of real and continuously variable (as opposed to discrete integer) states of its components (cf. *propositional encoding*).

Analytical processing

The type of computation performed by a device (e.g. a computer or a brain) in which events are represented in terms of symbolic accounts or arithmetical codes. These symbols do not interact directly with each other; they are operated on step by step by a formal mathematical or logical processor (cf. *synthetical processing*).

Anisotropic

Not the same in all directions.

Arcsin

Arcsin x is an angle whose sine is x.

Arctan

Arctan x is an angle whose tangent is x.

Avian pancreatic polypeptide

A putative neurotransmitter.

Bandpass filter

An information-carrying *channel* which will only convey information about one restricted region of a *domain*.

Bandwidth

The extent of the *domain* which a *bandpass filter* deals with. This may be quantified in a number of ways, e.g. the range (of frequencies, orientations, colours or whatever) over which the *channel's sensitivity* is at least half of its peak sensitivity, or l/*e* of its peak sensitivity.

Boolean logic

A two-valued logic wherein everything is either true or false, on or off, present or absent, etc. It is used as the basis of most modern computers. Combinations of inputs to a Boolean logic device determine the output of the device according to set rules. For example, a cortical cell which has two inputs A and B, and which is implementing the Boolean function NAND (short for 'not and') will give an all-or-nothing output if input A is firing and B is not, or if B is firing but A is not, or if neither of them is firing; it will not give an output if both inputs are firing simultaneously. Further, the Boolean function A AND NOT B will only give an output if A is active and B is not active (e.g. if A is an excitatory synapse and B is inhibitory); this is obviously not interchangeable with B AND NOT A.

Cartesian space coordinates

These are the normal, conventionally used coordinates of space, usually expressed for a plane surface as horizontal (x) and vertical (y) distances relative to some fixed origin.

Cell assembly

A set of neurones which fire together or in sequence because the activity of some of them is sufficient to fire the whole set.

Channel

An information-carrying structure, which may or may not be identified with a particular neural pathway. The word usually refers to a *bandpass* mechanism, i.e. one which only transmits information about a restricted part of some *domain* (see also *bandwidth*).

Cholecystokinin

A *peptide* and putative neurotransmitter, originally discovered in the gut, from where it is released to stimulate the secretion of bile into the intestine.

Choline acetyltransferase

The enzyme which synthesizes the neurotransmitter acetylcholine and which is found in the cytoplasm of presynaptic boutons.

Conformal mapping

See *mapping*.

Convolution integral

In a narrow sense, if an *operator* is applied sequentially at all points in an image and the outputs of the operator at each point are added up, then the sum is the convolution integral. More generally, the 'operator' may be any function, not necessarily limited to examining a small region of the image (e.g. a spatial frequency-selective *filter*).

Cooperative network

Strictly, a cooperative network consists of a large number of similar neurones, reciprocally connected within a local neighbourhood, and exchanging signals and changing states according to fixed rules. Sets of cells interact to reduce mismatch between the states of equivalent elements in each set of cells, until a stable, consistent, globally ordered state emerges. For example, a cell responding to a signal within a particular *channel* would facilitate responses in neighbouring cells in the same channel, and inhibit responses in different channels. More generally, the term is used to refer to a collection of nerve cells whose properties (both individual and collective) are determined at least in part by the interactions between those cells, rather than just by the properties of the afferent inputs to each cell from outside the network.

Cross-correlation

When the firing patterns of two nerve cells are recorded simultaneously, it is possible to see if the occurrence of an action potential in one cell is followed by an increase or a decrease in the probability that there will be an action potential in the other cell, within some time interval. If such an increase is seen it may be that the first cell drives the second; if a decrease, then an inhibitory connection may exist. Alternatively, both cells may be found to tend to fire simultaneously, in which case they may both receive excitation from a common source. The analysis can be repeated for connections from the second cell to the first. The computing procedures for correlating the firing patterns of the two cells in this way are well chronicled in the literature (see Toyama, Chapter 38 in this volume, for further description and references).

Cytochrome oxidase

An enzyme located in mitochondria which is important in several of the later steps in the metabolic breakdown of glucose. Brain areas with a high concentration of cytochrome oxidase are likely to contain many dendrites which receive excitatory depolarizing synapses, and hence neurons with high spontaneous rates of firing and wide *bandwidths* in their stimulus selectivities, according to M. Wong-Riley and E. Carroll (*J. comp. Neurol.*, **222**, 1–17 and 18–37, 1984). In the steady state these areas match with those found to have high activity by the *2-deoxyglucose* technique, but changes in the latter can occur more quickly in response to experimental manipulations of the input to the cortex (A.L. Humphrey and A. E. Hendrickson, *J. Neurosci.*, **3**, 345–358, 1983). See, for example, Mitchison's and Swindale's articles, Chapters 46 and 47 respectively in this volume, for further discussion.

Cytospecificities

Innate or acquired chemical markers on cell membranes which enable cells to recognize and distinguish one another.

2-Deoxyglucose

This is a variation on the normal glucose molecule in which there is no oxygen atom attached to one of the carbon atoms (that in the position conventionally labelled '2') in the glucose molecule. Nerve cells derive their energy entirely from the breakdown of glucose (as opposed to the breakdown of fats or proteins which can be used as well in the rest of the body). The enzyme which catalyses the first step in this breakdown functions even if it encounters a molecule of 2-deoxyglucose. However, the second enzyme does not, so a build-up of partially-metabolized 2-deoxyglucose occurs in the cell. Giving radioactively labelled 2-deoxyglucose therefore enables the identification of cells with a high metabolic rate, which is presumed to indicate a high rate of neural activity. See also the entry for *cytochrome oxidase*.

Difference of Gaussians

In the context of vision research, this applies to the subtraction of one *Gaussian* distribution from another, the latter having a lower standard deviation and a greater peak amplitude than the first. The total area under the latter distribution is usually set equal to or greater than that under the first, so that the resultant profile has zero or positive total area beneath it. This profile approximates the receptive field profile of many visual cells, wherein the influence of the 'surround' is assumed to be subtracted from that of the 'centre' to determine the cell's firing rate, and both regions in isolation are assumed to have Gaussian profiles. The difference of Gaussian profiles is also known as the 'Mexican hat'.

Differential operator

See *operator*.

Directional derivative

See *operator*.

Domain

A synonym for dimension, parameter, or feature space. Basically, anything which can form one axis of a graph can be called a domain. (Sometimes, two-dimensional planes are also given this label, e.g. the Fourier domain; see *Fourier analysis*.) The usefulness of considering cortical cells to interact specifically with other cells which differ in their location along a particular domain is described in this volume by Nelson (Chapter 11); see also Singer (Chapter 12) and Georgeson (Chapter 22) and D. H. Ballard *et al.* (*Nature*, **306**, 21–26, 1983).

Empiricism

The belief that all knowledge, scientific or personal, stems from experience, observation and experiment.

Entropy

A measure of the disorder of a system and of the unavailability of energy. Within a closed system the total entropy can only increase. The entropy S of a system in a given state is related to the probability W of finding it in that state by the equation $S = k.\ln(W)$, where k is a constant and ln stands for logarithm to the base e.

Epigenetic

Relating to the development of the embryo in gradual stages of differentiation from a relatively unstructured egg.

Epiphenomenon

A phenomenon which is manifested spuriously by a system and which is without causal effect on (or functional role for) the system; a useless side-effect.

Filter

A mechanism which alters the distribution of activity along a *domain*. A common use of the term is to describe a device which suppresses almost entirely all of the information entering a *channel* except that in a narrow band (see *bandpass filter*).

Fourier analysis

A method whereby any waveform can be described as the sum of a set of sinewaves. For example, most people will be familiar with the idea that different musical instruments playing the same note (e.g. middle C) produce sounds which have the same fundamental frequency (e.g. 256 Hz) but a piano and a violin sound different because they include different sets of *harmonics* (e.g. 512 Hz, 768 Hz, 1,024 Hz, etc.) in the sound waveforms they generate. Just as any such sound waveform (fluctuations in air pressure as a function of time) can be analysed into its component frequencies, so any luminance waveform (fluctuations in light intensity as a function of spatial location) can similarly be analysed. This yields a description of a (static) visual scene as the sum of a set of sinewaves, each having a spatial frequency (cycles per degree of angle subtended at the eye), amplitude (intensity or contrast), phase (either expressed relative to the other spatial frequencies in the image or relative to some absolute coordinate frame such as the fixation point or the fovea) and orientation. The amplitude at or near the spatial frequency of zero cycles per degree can be taken to represent the mean luminance of the display.

Fourier space or the Fourier plane or domain is most easily conceived as a polar graph with spatial frequency on the radial axis (with zero cycles per degree at the origin) and the orientation axis representing the orientation of the sinewave (zero degrees may be for horizontal or vertical lines in the image, depending on convention). Amplitude is then represented as the intensity of marking on the graph, or the height above the Fourier plane in a three-dimensional representation. Phase is not normally represented directly in this graph, but requires a separate encoding.

Functional hypothesis

A hypothesis about the purpose or function which a system exists in order to carry out.

Fusiform

Spindle-shaped, like a fat cigar.

GABA-transaminase (GABA-T)

The enzyme which breaks down the inhibitory neurotransmitter GABA (*gamma-aminobutyric acid*). It is found in association with mitochondria.

Gabor's theory of communication

Heisenberg pointed out that the position and momentum of an electron could not both be known with certainty, and he developed a mathematical principle which specified that when the uncertainties about the two variables were quantified and these quantities were multiplied together, there was a minimum value which the product of this multiplication could never be less than. Gabor applied this same principle to the joint uncertainties about the time of occurrence of a signal travelling along a communication line and the frequency components of its waveform (see *Fourier analysis*). Thus a very brief signal might only include a portion of one cycle of a sinewave, making it difficult to measure exactly the frequency of the sinewave. To do the latter accurately requires that several cycles be received, but this takes a longer time, so the estimated time of occurrence of the signal becomes less precise. Marcelja pointed out that the same principle could be applied in the spatial *domain* (to uncertainty about spatial frequency and spatial location) as in the time domain. In this volume, Daugman (Chapter 10), Kulikowski and Murray (Chapter 25), Palmer, Jones and Mulliken (Chapter 27), Pollen, Foster and Gaska (Chapter 28) and Nielsen (Chapter 39) all provide excellent descriptions of this application. The effect of the uncertainty introduced by the measuring instrument (J. Maddox, *Nature*, **308**, 601, 1984) remains to be worked out for the case of spatial vision (i.e. not just uncertainty about the positions of the eyes but also the uncertainty of the rest of the brain about exactly what spatial location and spatial frequency a given cell in the visual cortex is responsive to; e.g. how accurately labelled a *labelled line* is).

Gamma-aminobutyric acid

Our current best guess as to what is the principal inhibitory neurotransmitter in the visual cortex. It is synthesized by the enzyme *glutamic acid decarboxylase* and broken down by *GABA-transaminase*.

Gaussian

The basic shape of a Gaussian (or normal) distribution is given by exp $[-x^2]$, but it is more usually expressed with constants that set its amplitude and standard deviation σ, e.g. $(1/\sqrt{2\pi}\ \sigma)\ \exp[-x^2/\sigma^2]$ (which has unit area beneath it), or in two dimensions $(1/2\pi\ \sigma^2)\ \exp[-(x^2+y^2)/\sigma^2]$.

Glutamic acid decarboxylase

The enzyme which synthesizes the inhibitory neurotransmitter *gamma-aminobutyric acid*. It occurs in high concentrations in the cytoplasm of presynaptic boutons.

Harmonic

A wave at a frequency which is an integer multiple of some fundamental, baseline or reference frequency. Thus the second harmonic is at twice the frequency of the fundamental, the third harmonic at three times, and so on.

Hebb synapses

These are synapses postulated by Hebb which increment their gains when relatively high rates of presynaptic cell firing have been accompanied by high levels of synergistic postsynaptic cell activity, i.e. postsynaptic inhibition at an inhibitory synapse, or postsynaptic excitation at an excitatory synapse.

Heuristic

A guideline or pointer towards the general area in which the solution to a problem may be found (cf. *algorithm*).

Horseradish peroxidase

An enzyme which when injected into the brain can run both up and down inside axons to fill the terminal boutons and the cell bodies and dendrites of the appropriate neurones. The presence of the enzyme can later be detected by appropriate staining of histologically prepared sections of the tissue.

Incommensurability

The inability to decide which is the better of two qualitatively different things (as in the saying 'as different as chalk and cheese').

Isotropic

The same in all directions.

Labelled line

The theory that different nerve cells have 'specific nerve energies' was articulated by Müller in the 1830s to explain why stimulation of, for example, the optic nerve would only evoke visual sensations and not auditory or somatosensory sensations. Young and Helmholtz developed a similar idea for colour vision, in which, for example, the activation of 'red' neurones elicited sensations of redness, and so on. This general principle is commonly invoked to explain the encoding of location on a receptor surface (such as the retina or the skin), and has recently been extended to suggest that labelled lines may exist for such properties as the preferred spatial frequency of a cortical cell, so that other cells can tell whether a particular spatial frequency exists in the image by seeing if the cell(s) labelled for that spatial frequency is firing.

Laplacian

See *operator*.

Level of description

A structure or an event can be described briefly by using words which apply to the whole system or to large parts of it (i.e. at a high level of description) or by an exhaustive enumeration of the minutiae (i.e. at a low level of description). Translation between levels must be possible in principle, even if exhausting in practice. Thus an 'action potential', at the physiological level of description, can be translated into a description of the same event at the chemical or molecular level, as a list of the movements of ions and the changing conformations (shapes) of certain molecules in the cell membrane, with the pathways and timings of these events fully specified for each ion or molecule. Similarly, a central concern for brain research is to know how to translate, in principle, between the firing patterns of millions of nerve cells (physiological–anatomical level of description) and psychological descriptions of events (e.g. 'the recognition of grandmother').

Linking hypothesis

A hypothesis about how to translate from one *level of description* to another, for example between descriptions of the physiological activities of nerve cells and a psychological event such as object perception and recognition.

Logarithmic conformal mapping

See *mapping*.

Logos

The word used by the ancient Greek philosophers to mean reason, logic or wisdom: the principle which governs both existence and thought.

Mapping

A system or function for converting a set of real numbers into a different set of real numbers. The term is particularly used when these numbers are coordinates on a graph or image. Conformal mapping is a type of mapping in which the coordinates of one plane (conventionally referred to as the z plane) are mapped onto those of another (the w plane) in such a way that any two lines drawn in the z plane which intersect each other at an angle Θ also intersect at an angle Θ when they are mapped onto the w plane. In mapping between the visual image (the z plane) and the visual cortex (the w plane) the *Cartesian coordinates* x and y in the image become transformed into two coordinates u and v on the surface of the cortex. In logarithmic conformal mapping the mapping function can be symbolized as $w = \log_b(z + a)$ where a is a constant and b is the base to which the logarithms are taken. In this case, $u = \frac{1}{2}\log_b(x^2+y^2+2ax+a^2)$ and $v=\log_b(e).\arctan[y/(x+a)]$. If the value of a is small, then u approximates to the radial eccentricity of the point (x,y) in the image from the origin, and v is approximately proportional to its polar angle from the horizontal meridian.

Masking

A destructive form of interaction between two stimuli, wherein they make each other less visible. In a typical masking experiment, one stimulus is strong (e.g. of high contrast) and its effects on the other, weaker stimulus are measured, usually by finding the detection threshold for the latter and noting any change relative to the threshold measured in the absence of the strong (masking) stimulus. If the two stimuli are identical (apart from in their strengths) then the threshold measured becomes the increment threshold (or discrimination threshold). The masking effect is then explained by one of the theories about Weber's law (which states that the increment threshold is directly proportional to the strength of the masking stimulus) such as the existence of a non-linear gain in the relevant *channel*. However, if the two stimuli activate different channels, then cross-channel inhibition may be postulated to explain any masking effects.

Materialism

The belief that all phenomena can best be explained ultimately in terms of the movement of matter according to deterministic laws.

Matricial equation

A shorthand notation for a set of equations. The arrangement is such that, for example, the equations

$$A = Bx + Cy$$
$$D = Ex + Fy$$

can be written in matrix form as

$$\begin{pmatrix} A \\ D \end{pmatrix} = \begin{pmatrix} B & C \\ E & F \end{pmatrix} \cdot \begin{pmatrix} x \\ y \end{pmatrix}.$$

Metaphysics

These are general *a priori* assumptions and systems of belief about knowledge and existence.

Mexican hat

The shape of the *difference of Gaussians* function.

Model

A system of symbols or processes which is in some way analogous to systems in the real world and which is such that manipulation of the model enables predictions to be made about events in the real world.

NAND

See *Boolean logic*.

Nativism

The belief that knowledge is inborn (cf. *empiricism*).

Nonius lines

Two lines presented one to each eye in such a way that when the eyes are fixating correctly the two lines appear to the subject to be colinear.

Ontogeny

The development of the individual.

Operator

An abbreviation for a mathematical process or operation which is applied, for example to the luminance at some point (or to the luminance distribution within a restricted area) in an image. Thus a differential operator has as its output the rate of change (slope) of the luminance distribution at a point in the image (or in quantized images, the difference between the luminances at two adjacent points or pixels). Because the slope has a direction (in two-dimensional *Cartesian space*) this output is also known as a directional derivative. This operator can be implemented by lateral inhibition in the nervous system. The second derivative (rate of change of slope or acceleration of luminance) can be calculated from the luminances at three points in a row, i.e. from the difference between the luminance at the central point and the average of that at the two end-points. This can be implemented independently of the direction in which it is measured across the image, by adding together the accelerations along two orthogonal directions (say, horizontal and vertical) or by comparing the luminance at a point with that in an annular surrounding region; this then is the Laplacian operator, which is similar in shape to a concentric receptive field with a *difference of Gaussians* cross-sectional profile.

Opponent colour processing

Opponent processes or cells are those that respond to light of a certain wavelength or wavelengths with an excitatory response, but give inhibitory responses to light of some other wavelength(s). (Furthermore, removal of the stimulus may induce the opposite response to that given at light 'on'.) Double-opponent cells have spatially separate areas in their receptive fields, with each area having one of two different opponent processes driving it (e.g. red light giving excitatory responses when shone into the field centre, but green light giving excitation in the surround).

Orthogonal

At right angles; uncorrelated.

Peptides

Short chains of amino-acids.

Phylogeny

The evolution of a species.

Pleomorphic

Of multiple shapes.

Poisson distribution

This is a description of the probabilities of observing particular numbers of events within a given sample. Thus if the mean number of events that occur within each sample is m (found by averaging across a large number of samples), then the probability of observing x events within any given sample is $m^x e^{-m}/x!$, where $x!$ is factorial x, i.e. $x(x-1)(x-2)\cdots 1$.

Positron emission tomography

A method of brain scanning in which radioactively labelled *2-deoxyglucose* is administered to a conscious subject and accrues in the brain in proportion to the local levels of metabolic activity. The positrons emitted from the 2-deoxyglucose collide with electrons in the immediately surrounding tissues, causing high-energy electromagnetic rays to be emitted which can be sensed by an array of detectors outside the head. The location of the source of the rays is then calculated by a computer, which thus produces a picture of the pattern of metabolic activity within the brain.

Propositional encoding

A proposition or assertion is a statement that, in formal logic, is either true or false (cf. *Boolean logic*). The theory of propositional encoding suggests that the brain reduces its knowledge of the world to a series of such statements (e.g. 'some swans are white') and also its hypotheses about the world (e.g. 'all swans are white', 'I am looking at a swan', 'I am looking at a high spatial frequency grating' or 'I detect an edge'). At the neural level, a single unit or a *cell assembly* which represents such a statement would for functional purposes either be active or not active (cf. *analogue encoding*).

Rectification

A sinewave that oscillates about the value zero is rectified if all its negative values are converted into their equivalent positive values, i.e. into the modulus of each value in the original sinewave. Simply eliminating all negative values (i.e. replacing them by zeros) is called half-wave rectification. (Rectifiers in electronics are used to convert alternating currents to direct currents.) Thus a cell which responds to a stimulus such as a drifting sinewave grating with a firing rate which oscillates at double the frequency of stimulation is sometimes described as giving a rectified response (although this is not mathematically accurate), while the term half-wave rectification is often applied to simple cells with no spontaneous activity which give a single burst of firing once during every cycle of the stimulus.

Scalar

This is a quantity which has magnitude but no particular direction. Examples include mass, temperature and real numbers (cf. *vector*).

Sensitivity

The reciprocal of threshold. Since contrast is usually defined as the difference between the maximum and minimum luminances in the image divided by the sum of the maximum and minimum luminances, a value which can vary between 0 and 1, sensitivity to contrast cannot be measured if it is below unity, and can rise towards plus infinity.

Somatostatin

A *peptide* and putative neurotransmitter, originally discovered in the pituitary gland where it suppresses the release of growth hormone.

Spiritualism

The belief that all phenomena are ultimately attributable to the operation of supernatural agents or spirits.

Stochastic

A stochastic process is one in which for each variable involved there is a definite probability of occurrence of each possible value of the variable.

Synthetical processing

The type of computation performed by a device which represents events in terms of the direct physical nature of its components. These representations interact with one another continuously and in parallel to represent interactions between events in the real world (cf. *analytical processing*).

Threshold neurone

A logical device that roughly models the input–output behaviour of a single neurone. Inputs to the synapses are multiplied by their synaptic weights and if the sum across all the synapses exceeds the cell's threshold then its output will be active; if they do not exceed threshold, the cell will be inactive.

V1, V2, etc.

A system of numbering of the visual areas of the cerebral cortex. V1 is the primary visual cortex (Brodmann area 17), V2 is the neighbouring area, and so on. The higher numbered areas have not necessarily been found to correspond with the traditional demarcations of the cortex based on histological appearance.

Varicose

With regions of dilation; e.g. a varicose axon has a varying diameter along its length.

Vasoactive intestinal polypeptide

A putative neurotransmitter, originally discovered in the intestines where it increases blood flow.

Vector

This is a quantity which has both magnitude and direction in space, e.g. force or velocity.

Weighting function

The weighting attached to something (i.e. the number by which it is to be multiplied) will often vary as a function of some other variable. For example, the output of a linear cell with a particular sensitivity profile to its receptive field might be calculated by multiplying the luminance at each point in the image by the sensitivity at that point in the receptive field and then integrating across all points. The sensitivity profile can thus be called a weighting function, in which weighting varies across space.

W-, X- and Y-cells

Three classes into which relay cells in the retina and lateral geniculate nucleus are divided. The exact definitions and delimitations of the properties of each class are still not universally agreed. (Broadly, W-cells have slowly conducting axons, sluggish responses, often with complex or 'primitive' non-concentric receptive fields, while X-cells have small receptive fields and give sustained and linear responses to the colour and fine detail in images, whereas Y-cells are few in number but have fast-conducting axons and large receptive fields and respond well to transients or movements in the image.) In this volume, the most detailed description is given by Sherman (Chapter 8).

Index

Associative Index

This index is designed to facilitate the discovery of cross-correlations between the themes of the book in a way which the conventional index cannot. Thus it contains less specific items and headings than the detailed index, and it addresses the reader to the chapters in which those topics are a *major* theme rather than to individual page numbers. It also includes recommendations as to which chapters should be consulted next after reading each chapter, so that the interested reader can quickly find associated material and ideas which are linked in with those that he has just read about. The associations presented are not comprehensive, and are not always two-way, but are given as a suggested aid to guiding oneself along the thematic pathways in the book.

Author(s)	Perception	Action	Psychophysics	Simulation	Physiology	Anatomy	Development	Orientation	Binocularity	Receptive fields	Colour	Direction	Topography	Columns	Microcircuitry	Parallel and WXY	Outputs	Fourier and Gabor	Page numbers	Associated chapters (page numbers)
Dobson and Rose	○	○					○												22	37, 546
Barlow	○						○												37	22, 47, 123, 164, 462, 473
MacKay	○	○																○	47	37, 62, 96, 326
Creutzfeldt		○														○	○		54	47, 62, 71
Schiller		○															○		62	47, 54
Sherman	○															○			71	54, 242, 351, 380
Cavanagh	○																○	○	85	96, 146, 253, 265, 281
Daugman			○	○			○											○	96	253, 265, 273, 281
Nelson	○				○			○	○	○				○	○				108	123, 200, 223, 292, 326, 396, 443

Singer	○			○		○	○	○										123	37, 108, 172, 443, 473
Hartmann			○						○								○	137	273, 334, 479
Schwartz	○		○		○							○	○					146	85, 182, 310, 546
Amari			○			○						○	○					157	37, 164, 473
Cooper			○			○	○	○										164	123, 157, 172, 473
Frégnac			○	○		○	○											172	37, 47, 123, 164, 432, 462
Dobson			○			○	○		○			○	○	○				182	37, 123, 146, 292
Wolfe and Blake		○						○										192	200, 211
Levi		○				○		○										200	108, 192, 211
Sloane		○						○										211	192, 200, 301, 334, 514
Georgeson		○					○	○			○		○				○	223	108, 146, 292, 396
Wright and Johnston		○										○				○		233	146, 310
Gouras		○		○					○	○					○			242	253, 301
Kulikowski and Murray	○			○					○	○					○	○	○	253	185, 96, 265, 273, 281, 301, 320, 326
Glezer		○		○					○								○	265	85, 96, 253
Palmer *et al.*				○					○					○	○		○	273	85, 96, 253, 265, 524
Pollen *et al.*				○													○	28	85, 96, 265, 273, 334, 341, 374
Bonds and de Bruyn				○			○						○					292	108, 334, 390, 396, 432
Michael				○					○	○			○	○	○			301	242, 253, 452
Orban				○								○						310	146, 233
Tolhurst and Dean				○					○									320	253, 273, 358
Hammond				○							○				○			326	47, 108, 334
Maffei				○										○	○			334	211, 292, 326, 396
Hochstein and Spitzer				○					○						○			341	71, 253, 281, 351
Henry				○											○			351	71, 341, 358
Heggelund			○	○			○		○					○	○			358	351, 396
Toyama				○										○	○			366	334, 396, 504, 514
Nielsen			○				○		○								○	374	281, 408

Author(s)	Perception	Action	Psychophysics	Simulation	Physiology	Anatomy	Development	Orientation	Binocularity	Receptive fields	Colour	Direction	Topography	Columns	Microcircuitry	Parallel and WXY	Outputs	Fourier and Gabor	Page numbers	Associated chapters (page numbers)
Leventhal					○		○	○					○	○	○	○			380	172, 292, 390, 432, 462, 504, 514
Vidyasagar					○			○								○			390	47, 380, 396, 462
Sillito					○	○		○				○		○	○	○			396	292, 358, 380, 390, 408, 504, 514, 524
Koch and Poggio				○				○				○							408	374, 396, 420
Emerson *et al.*					○	○						○			○				420	408, 504, 514, 524
Tieman and Hirsch					○	○	○	○		○					○	○			432	123, 292, 380, 390, 396
Mitchison				○										○					443	108, 123, 479, 485, 504
Swindale				○										○					452	146, 301, 479, 485
Cowan and v. d. Malsberg				○			○	○						○		○			462	172, 292, 380, 390, 443, 473, 479
Legéndy				○			○							○					473	37, 123, 157, 164, 462
Braitenberg				○				○		○		○		○				○	479	137, 432, 443, 452, 462, 485, 504
Albus					○	○								○					485	443, 452, 462, 479
Peters						○									○	○			492	396, 408, 432, 504, 524
Martin and Somogyi						○									○	○			504	273, 358, 366, 380, 492, 514, 524
Somogyi and Martin						○		○				○		○	○				514	182, 380, 408, 492, 524
Davis and Sterling					○	○									○	○			524	273, 396, 420, 432, 492, 504, 514
Rose and Dobson				○	○	○		○		○				○	○				546	22, 37, 47, 182, 432, 443, 462, 479